Foliations II

Foliations II

Alberto Candel
Lawrence Conlon

Graduate Studies
in Mathematics
Volume 60

American Mathematical Society
Providence, Rhode Island

2000 *Mathematics Subject Classification.* Primary 57R30.

The first author was supported by NSF Grants DMS-9973086 and DMS 0049077

For additional information and updates on this book, visit
www.ams.org/bookpages/gsm-60

Library of Congress Cataloging-in-Publication Data

Candel, Alberto, 1961–
Foliations I / Alberto Candel, Lawrence Conlon.
p. cm. — (Graduate studies in mathematics, ISSN 1065-7339 ; v. 23)
Includes bibliographical references and index.
ISBN 0-8218-0809-5 (alk. paper)
1. Foliations (Mathematics) I. Title. II. Series. III. Conlon, Lawrence, 1933–

QA613.62.C37 1999
514′.72 21—dc21 99-045694

∞ The paper used in this book is acid-free and falls within the guidelines
established to ensure permanence and durability.
Visit the AMS home page at http://www.ams.org/

10 9 8 7 6 5 4 3 2 1 08 07 06 05 04 03

To our wives,
Juana and Jackie

Contents

Preface

For this second volume of *Foliations*, we have selected three special topics: analysis on foliated spaces, characteristic classes of foliations, and foliated 3-manifolds. Each of these is an example of deep interaction between foliation theory and some other highly developed area of mathematics. In all cases, our aim is to give useful, in-depth introductions.

In Part 1 we treat C^*-algebras of foliated spaces and generalize heat flow and Brownian motion in Riemannian manifolds to such spaces. The first of these topics is essential for the "noncommutative geometry" of these spaces, a deep theory originated and pursued by A. Connes. The second is due to L. Garnett. While the heat equation varies continuously from leaf to leaf, its solutions have an essentially global character, making them hard to compare on different leaves. We will show, however, that leafwise heat diffusion defines a continuous, 1-parameter semigroup of operators on the Banach space $C(M)$ and, following Garnett [**77**], we will construct probability measures on M that are invariant under this semiflow. These are called *harmonic measures*, and they lead to a powerful ergodic theory for foliated spaces. This theory has profound topological applications (*cf.* Theorem 3.1.4), but its analytic and probabilistic foundations have made access difficult for many topologists. For this reason, we have added two survey appendices, one on heat diffusion in Riemannian manifolds and one on the associated Brownian flow. For similar reasons, we have added an appendix on the basics of C^*-algebras. We hope that these will serve as helpful guides through the analytic foundations of Part 1.

Part 2 is devoted to characteristic classes and foliations. Following R. Bott [**9**], we give a Chern-Weil type construction of the exotic classes based on the Bott vanishing theorem (Theorem 6.1.1). The resulting theory can be viewed either as a topic in algebraic topology, motivated by foliation theory, or as a deep application of algebraic topology to the study of foliations. We take the latter viewpoint, emphasizing qualitative aspects such as G. Duminy's celebrated vanishing theorem for the Godbillon-Vey class (unpublished) and S. Hurder's analogous theorems for the generalized Godbillon-Vey classes in higher codimension [**102**]. We begin Part 2 with a chapter on the "grandfather" of all characteristic classes, the Euler class of oriented circle bundles, giving complete proofs of the applications, due to J. Milnor [**129**] and J. Wood [**189**], concerning obstructions to the existence of foliations transverse to the fibers of circle bundles over surfaces.

In Part 3, we study compact 3-manifolds foliated by surfaces, a topic that has been popular since the advent of the Reeb foliation of S^3. The special methods of 3-manifold topology yield existence theorems and qualitative properties unique to dimension three. The theorem of S. P. Novikov [**141**] on the existence of Reeb components has the consequence that "Reebless foliations" carry important topological information about the ambient 3-manifold. Together with a theorem of W. Thurston [**175**] on compact leaves of Reebless foliations, this led to D. Gabai's groundbreaking work in which taut foliations are used as powerful tools for studying 3-manifold topology. We develop this theory up to Gabai's constructions of taut, finite depth foliations on certain sutured 3-manifolds, giving details only in the disk decomposable case (depth one). This will bring the reader to the threshold of the "modern age" of essential laminations. These laminations are generalizations simultaneously of taut foliations and incompressible surfaces, and are the object of much current research. Essential laminations, however, need a book of their own and we hope that one or more of the specialists will provide such.

Appendix D pertains to Part 3, being a detailed account of Palmeira's theorem that the only simply connected n-manifold foliated by leaves diffeomorphic to $\mathbb{R}^{n-1}$ is $\mathbb{R}^n$. In fact, if $n \geq 3$, the foliated manifold is diffeomorphic to $\mathbb{R}^2 \times \mathbb{R}^{n-2}$ in such a way that the foliation is the product of a foliation of $\mathbb{R}^2$ by the space $\mathbb{R}^{n-2}$. Although valid in all dimensions $n \geq 3$, this result has important applications to Reebless-foliated 3-manifolds.

The bibliography is not intended to be a comprehensive list of all publications on these areas of foliation theory. Only references explicitly cited in the text are included, with the result that many important papers and books are omitted (with apologies to the authors).

The three parts of this book can be read independently. One minor exception to this is that certain standard properties of the Euler class, proven in Part 2, are needed in Part 3. Of course, all parts depend in various ways on material in Volume I. All references to that volume will be of the form [I,...].

Finally, the first named author expresses his sincere thanks and appreciation to the second for his invitation to join in this journey through the theory of foliations, and for seeing that it got to an end.

Part 1

Analysis and Geometry on Foliated Spaces

Foreword to Part 1

This part of the book emphasizes the use of analytic tools for extracting geometric information out of foliated spaces.

The first chapter describes the construction, due to A. Connes, of the C^*-algebra of a foliated space. Also introduced here is the fundamentally important concept of the graph of a foliated space. Our treatment of these topics is only introductory, being intended as an aid to readers who would like to begin studying some of the vast literature on this subject.

The second chapter is devoted to the harmonic measures of L. Garnett. These measures provide a universally applicable tool for studying ergodic-theoretic problems on foliated spaces.

In the third chapter, we treat topological and geometric applications, featuring a beautiful theorem of E. Ghys (Theorem 3.1.4) concerning the ends of almost all leaves.

The first two chapters require significant analytic background, namely C^*-algebras, diffusion on manifolds and Brownian motion. While there are numerous texts treating these topics, the authors feel that the reader will be well served by a concise survey of each of the theories in the form of three appendices. These appendices are intended to provide a guided tour of these theories and are no substitute for the extensive literature to which they refer.

Chapter 1

The C^*-Algebra of a Foliated Space

The topic of this chapter is analysis on foliated spaces, and how it relates to their dynamics. Specifically, the chapter describes in detail the construction of the C^*-algebra of a foliated space, introduced by A. Connes. For basics about foliated spaces, together with many examples, see [I, Chapter 11].

For locally compact Hausdorff spaces there is a well established dictionary that gives a translation of topological properties of the space into analytical properties of the commutative C^*-algebra of continuous functions on the space, and *vice versa*. Such a dictionary is inadequate for translating into analysis features of spaces with poor topological properties, as is the case of the space of leaves of a foliation. (For example, if the foliated space has a dense leaf, then the space of leaves has no interesting continuous function.)

To transfer geometric properties into analysis the dictionary must be extended to include the language of noncommutative C^*-algebras. Thus the foliation C^*-algebra is meant to be a replacement for the continuous functions on the space of leaves. It has a rich structure (so that there are sufficiently many "functions") but, unlike the algebra of continuous functions on a locally compact Hausdorff space, it is generally noncommutative, a fact that reflects the typically complicated structure of the leaf space.

Besides describing the construction of this C^*-algebra, its relation to dynamical systems will be exemplified via a theorem of T. Fack and G. Skandalis. The proof of this theorem requires a discussion of the basic facts pertaining to the representation theory of the foliation C^*-algebra, most of which will be developed *ad hoc*. To make this chapter essentially self-contained, an appendix has been added providing those basic facts of C^*-algebras that the discussion requires.

One of the most important results of the theory of Connes is undoubtedly the foliation index theorem, generalizing the Atiyah-Singer index theorem. Unfortunately, lack of space prevents us from treating this topic, although most of the tools required to prove the index theorem for the de Rham operator are developed in this and other chapters. Fortunately, besides Connes' original papers, the topic is covered in a textbook fashion by C. Moore and C. Schochet [**134**]. The theory of foliation groupoids and C^*-algebras has nowadays become a piece of the so-called *noncommutative geometry*, heralded by Connes; his book [**38**] is a masterful and broad exposition of the theme.

The material presented in this chapter is primarily based on the papers by Connes [**34, 36**] and by Fack and Skandalis [**63**]. An overview of Connes' (earlier) work in foliations is given in his Bourbaki Seminar lecture [**35**].

1.1. Twisted Forms and Densities

Let $(M, \mathcal{F})$ be a foliated space with leaves of dimension $\dim \mathcal{F}$. The orientation bundle of M is the line bundle $\mathcal{O}$ whose transition functions are given by the sign of the determinant of the Jacobian of the transition functions of the tangent bundle. More precisely, if $\{U_i, \varphi_i\}$ is a covering by foliated charts $\varphi_i : U_i \xrightarrow{\sim} D_i \times Z_i$, where D_i is a disc in $\mathbb{R}^{\dim \mathcal{F}}$, then, on the intersection $U_i \cap U_j$, the coordinate changes have the form

$$\varphi_{ij} = \varphi_i \circ \varphi_j^{-1}(x, z) = (\varphi_{ij}(x, z), h_{ij}(z)).$$

The cocycle given by

$$\psi_{ij} = \operatorname{sign} \det J(\varphi_{ij}),$$

where J denotes the Jacobian, is locally constant and defines the orientation line bundle $\mathcal{O}$ over M.

Definition 1.1.1. A *twisted form* of degree k on M is a section of the bundle $\Lambda^k TM^* \otimes \mathcal{O}$.

It should be clear that the exterior product of a p-form and a twisted q-form is a twisted $(p+q)$-form.

Since the transition functions describing the line bundle $\mathcal{O}$ are locally constant, the exterior differential on forms extends to a differential operator

$$d : \Lambda^k TM^* \otimes \mathcal{O} \to \Lambda^{k+1} TM^* \otimes \mathcal{O}$$

satisfying $d \circ d = 0$, as in the case of the usual de Rham operator.

Let E be a (real or complex) vector bundle of rank n over the foliated space M with fiber F. There is an associated line bundle $\Lambda^n E$ over M whose typical fiber is the exterior product $\Lambda^n F$. For $\alpha > 0$, there is an associated bundle induced by the representation $x \in \mathbb{R}^* \mapsto |x|^\alpha \in \mathbb{R}^*$. This bundle is called the bundle of densities of order α (or α-densities) on E. This is a bundle whose structure group is the multiplicative group of positive real numbers. Thus it makes sense to talk about positive densities. Note that these bundles of densities are all trivial because the base space M admits smooth partitions of unity (*cf.* [I, Theorem 11.4.1]).

When E is the complexification of the tangent bundle of M, the associated bundle of α-densities is denoted by $\mathcal{D}^\alpha M$. An *α-density* on M is a section of the bundle of α-densities, and the space of compactly supported sections of this bundle is denoted by $\Gamma_c(M, \mathcal{D}^\alpha)$ (see Section 1.2 for additional information). Thus a density of order α on M is a map ϕ that assigns to each $x \in M$ a map

$$\phi_x : \Lambda^{\dim \mathcal{F}} T_x M \to \mathbb{C}$$

such that

$$\phi_x(\lambda v) = |\lambda|^\alpha \phi_x(v)$$

for all $\lambda \in \mathbb{R}$ and $v \in \Lambda^{\dim \mathcal{F}} T_x M$.

In terms of the transition functions associated to an atlas for M, the bundle of densities of order α is given by the cocycle $g_{ij} = |J(\phi_{ij})|^{-\alpha}$. Thus the line bundle of densities of order 1 is the complexification of the line bundle $\Lambda^{\dim \mathcal{F}} T^* M \otimes \mathcal{O}$ described at the beginning of this section.

A positive density (of order one) on a manifold M defines a positive measure that is equivalent to Lebesgue measure on coordinate charts. Unlike a differential form of top degree $\dim M$, a density transforms with a multiplicative factor equal to the absolute value of the determinant of a coordinate change. Thus a compactly supported density on a manifold can be integrated, and the usual Stokes' formula continues to hold. Details are given in R. Bott and L. Tu [**11**] and de Rham [**44**].

An examination of the coordinate changes of the bundles $\mathcal{D}^\alpha M$ immediately shows that the product of a density of order α and a density of order β is a well defined density of order $\alpha + \beta$. Also, if σ is a density of order α, its absolute value $|\sigma|$ is a density of order α.

A half-density (density of order 1/2) σ on a manifold M is square integrable if

$$\int_M |\sigma|^2 < \infty.$$

Since the product of two half-densities is a 1-density, there is a naturally defined inner product on $\Gamma_c(M, \mathcal{D}^{1/2})$. The completion of $\Gamma_c(M, \mathcal{D}^{1/2})$ with respect to the norm defined by this inner product is the Hilbert space of half-densities, and is denoted by $L^2(M)$.

It is of course possible to construct $L^2(M)$ from functions. Since the bundle $\mathcal{D}^{1/2}M$ is trivial (because M admits partitions of unity), a choice of an everywhere positive density σ_0 allows us to identify $\Gamma_c(M, \mathcal{D}^{1/2})$ with $C_c^\infty(M)$ (the compactly supported functions) and to define integrals of compactly supported functions, thus producing a Hilbert space $L^2(M, \sigma_0)$. (Observe that, in Riemannian geometry, the metric tensor defines a half-density.) A different choice of trivializing density σ_1 will of course produce another Hilbert space $L^2(M, \sigma_1)$. These two spaces are canonically isomorphic. Indeed, the quotient σ_1/σ_0 is an everywhere positive function on M, and multiplication by its positive square root induces a unitary operator $L^2(M, \sigma_0) \to L^2(M, \sigma_1)$.

Proposition 1.1.2. *If f is a diffeomorphism of X, then f induces a unitary operator of $L^2(X)$.*

Proof. This is the change of variable formula. □

Exercise 1.1.3. Let $f : X \to Y$ be a quasi-isometry between Riemannian manifolds. Show that f induces a bounded operator, with bounded inverse, $f^* : L^2(Y, \sigma_Y) \to L^2(X, \sigma_X)$, where σ_X and σ_Y are the half-densities induced on X and Y by the respective metric tensors.

1.2. Functions on Non-Hausdorff Spaces

The definition of foliated space makes sense even when the underlying topological space fails to satisfy the Hausdorff separation axiom. Non-Hausdorff spaces appear naturally in the theory of foliations. In particular, the main topic of this chapter requires use of the graph of a foliated space. This graph, also called the holonomy groupoid, is a foliated space naturally attached to a given foliated space, but it fails to be Hausdorff in many examples.

It will be necessary to use functions with compact support on such spaces. However, a non-Hausdorff space may not have sufficiently many such functions, the basic reason being that compact subsets of a Hausdorff space are not necessarily closed. The non-Hausdorff spaces that will appear here have a particularly simple local structure, and even when it is possible to construct appropriate functions using this local structure, the standard operation of "extension by 0" of local objects to the full space does not produce continuous functions. M. Crainic and I. Moerdijk [**39**] proposed a very natural way of dealing with this problem, and this preliminary section describes it. (That paper develops an extended sheaf theory for non-Hausdorff manifolds. It was originally an appendix to [**40**], where they apply it to the study of the cyclic type homologies of non-Hausdorff smooth groupoids.)

In this section, X will denote a separable topological space having the structure of a foliated space, but it is not required that the topology be Hausdorff. It is only required that X can be covered by countably many open sets homeomorphic to a product $D \times Z$, where D is an open ball in Euclidean n-space and Z is a separable locally compact Hausdorff space, and such that the coordinate changes are as described on page 6 (essentially, matching the D factors on overlapping charts).

Let $\mathcal{C}^\infty$ denote the structure sheaf of the foliated space X, that is, the sheaf of smooth functions on X. Let $\mathcal{A}$ be a sheaf of $\mathcal{C}^\infty$-modules over X, for instance, the sheaf of differential forms or other sheaves of smooth sections of foliated vector bundles.

For such a sheaf $\mathcal{A}$ over X, let $\mathcal{A}'$ denote its Godement resolution: $\mathcal{A}'(U)$ is the set of all sections (continuous or not) of $\mathcal{A}$ over $U \subseteq X$.

For a Hausdorff open subset W of X, let $\Gamma_c(W, \mathcal{A})$ denote the set of continuous compactly supported sections of $\mathcal{A}$ over W. If $W \subset U$, then "extension by 0" induces a well defined homomorphism

$$\Gamma_c(W, \mathcal{A}) \to \mathcal{A}'(U).$$

For an open subset U of X, let $\Gamma_c(U, \mathcal{A})$ denote the image of the homomorphism

$$\bigoplus_W \Gamma_c(W, \mathcal{A}) \to \mathcal{A}'(U),$$

where W runs through the family of open Hausdorff subsets $W \subset U$.

The advantage of this definition of compactly supported smooth functions is that Γ_c is functorial on $\mathcal{A}$, and that constructions performed in coordinate charts patch globally. Thus, for instance, if $\mathcal{A}^k$ is the sheaf of differential k-forms along the leaves of X, there is a natural differential operator $d : \Gamma_c(X, \mathcal{A}^k) \to \Gamma_c(X, \mathcal{A}^{k+1})$ extending $d : \mathcal{A}^k \to \mathcal{A}^{k+1}$.

The computing machine is the Mayer-Vietoris sequence, as is described next.

Lemma 1.2.1. *Let $U \subset X$ be an open set and let $\{W_i\}$ be a cover of U by Hausdorff open sets. Then the sequence*

$$\bigoplus_i \Gamma_c(W_i, \mathcal{A}) \to \Gamma_c(U, \mathcal{A}) \to 0$$

is exact.

Proposition 1.2.2. *Let $X = \bigcup_i U_i$ be an open cover of X indexed by an ordered set I. Then there is a long exact sequence*

$$\cdots \to \bigoplus_{i_0<i_1} \Gamma_c(U_{i_0 i_1}, \mathcal{A}) \to \bigoplus_{i_0} \Gamma_c(U_{i_0}, \mathcal{A}) \to \Gamma_c(X, \mathcal{A}) \to 0$$

where $U_{i_0 i_1 \cdots i_n} = U_{i_0} \cap U_{i_1} \cdots \cap U_{i_n}$.

The index set $I = \{i_0, i_1, \cdots\}$ is assumed to be ordered. Indices in any order are allowed, even with repetitions, subject to the convention that when two indices are interchanged, the function becomes its negative:

$$f_{ij} = -f_{ji}.$$

In particular, a function with repeated indices is zero.

Proof. The map $\Phi : \bigoplus_{i_0<i_1} \Gamma_c(U_{i_0 i_1}, \mathcal{A}) \to \bigoplus_{i_0} \Gamma_c(U_{i_0}, \mathcal{A})$ is defined as follows. If $f = \sum_{i_0 i_1} f_{i_0 i_1} \in \bigoplus_{i_0<i_1} \Gamma_c(U_{i_0 i_1}, \mathcal{A})$, then $\Phi(f) = \sum_{i_0} \Phi(f)_{i_0}$ and

$$\Phi(f)_{i_0} = \sum_i f_{i i_0}$$

where $f_{i i_0}$ is the extension by 0 of $f_{i i_0}$ to U_{i_0}.

The result is standard when X is Hausdorff (*cf.* [**11**]). The general case proceeds by reducing to the case where each U_i is Hausdorff. The details are given in [**39**]. □

Corollary 1.2.3. *Let $Y \subset X$ be a closed subfoliated space. Then there is an exact sequence*

$$0 \to \Gamma_c(X \setminus Y, \mathcal{A}) \xrightarrow{i} \Gamma_c(X, \mathcal{A}) \xrightarrow{r} \Gamma_c(Y, \mathcal{A}) \to 0,$$

where i is extension by 0 and r is the restriction.

Connes introduced in [**36**] a concept of compactly supported smooth function on a non-Hausdorff manifold X. These are the elements of the linear span of the family of functions on X that have compact support in a coordinate chart and are extended by zero outside the chart. These functions are not necessarily continuous on X, and this family is not closed under products. If this space is denoted by $C^\infty(X)$, then there is a homomorphism

$$\Gamma_c(X, \mathcal{C}^\infty) \to C^\infty(X)$$

given by pointwise evaluation: to an element $f = \sum_i f_i \in \Gamma_c(X, \mathcal{C}^\infty)$ it associates the function $x \in X \mapsto \sum_i f_i(x)$. This homomorphism is always surjective, but it fails to be injective in general unless, of course, X is Hausdorff. Its kernel consists of those elements f that evaluate to 0 at each $x \in X$.

The following example, mentioned in [**39**], illustrates this point.

Example 1.2.4. Let X be the line with two origins, the quotient space of $\mathbb{R} \times \{0,1\}$ by the equivalence relation $(x,0) \sim (x,1)$ if $x \neq 0$. If f is a function in $C_c^\infty(\mathbb{R})$ such that $f(0) = 0$ and $f'(0) = 1$, then $F = \{f, -f\}$ is a nontrivial element in $\Gamma_c(X, \mathcal{C}^\infty)$, but it is trivial in $C^\infty(X)$.

1.3. The Graph of a Foliated Space

This section describes the construction and basic properties of the graph, or holonomy groupoid, of a foliated space. This is an object canonically associated to a foliated space, and it represents a desingularization of the graph of the equivalence relation on the foliated space given by "lying in the same leaf." The graph of a foliated space already appears in the early papers on foliation theory (C. Ehresmann [**59**]). More recently its geometric properties were studied by E. Winkelnkemper [**186, 187**], specifically in the context of Riemannian foliations. The study of its analytic significance started with Connes [**34**].

Let $(M, \mathfrak{F})$ be a foliated space. If L is a leaf of M, x a point in L and Z_x a transversal through x containing x in its interior, there is a well defined homomorphism of the fundamental group $\pi_1(L, x)$ into the group of germs at x of local homeomorphisms of Z_x that fix x. This homomorphism is called the *holonomy representation*, and was discussed in [I, Section 11.2]. It depends of course on the choice of the transversal Z_x, but a different choice produces an equivalent representation. Thus the concept of trivial germinal holonomy for a loop is well defined. The image of the holonomy representation of L is called the *holonomy group* of L, and the corresponding covering space is called the *holonomy covering* of L.

Let $\Pi(M, \mathfrak{F})$ be the space of paths on leaves, that is, maps $\alpha : [0,1] \to M$ that are continuous with respect to the leaf topology on M. For such a path α let $s(\alpha) = \alpha(0)$ be its source or initial point and let $r(\alpha) = \alpha(1)$ be its range or terminal point. The space $\Pi(M, \mathfrak{F})$ has a partially defined multiplication: the product $\alpha \cdot \beta$ of two elements α and β is defined if the terminal point of β is the initial point of α, and the result is the path β followed by the path α. (Note that this is the opposite to the usual composition of paths $\alpha \# \beta = \beta \cdot \alpha$ used in defining the fundamental group of a space.)

Definition 1.3.1. The *graph*, or holonomy groupoid, of the foliated space $(M, \mathfrak{F})$ is the quotient space of $\Pi(M, \mathfrak{F})$ by the equivalence relation that identifies two paths α and β if they have the same initial and terminal points, and the loop $\alpha \cdot \beta^{-1}$ has trivial germinal holonomy.

The graph of $(M, \mathfrak{F})$ will be denoted by $G(M, \mathfrak{F})$, or simply by $G(M)$ or by G when all other variables are understood.

The source and range maps defined above on $\Pi(M, \mathcal{F})$ induce source and range maps $s, r : G(M, \mathcal{F}) \to M$. The partially defined product on $\Pi(M, \mathcal{F})$ induces one on $G(M, \mathcal{F})$: two elements γ_1 and γ_2 are composable if $s(\gamma_1) = r(\gamma_2)$, and the product $\gamma = \gamma_1 \cdot \gamma_2$ can be represented by a path obtained by first following a path representing γ_2 starting at its source, and then following a path representing γ_1 until reaching its range. Thus $s(\gamma_1 \cdot \gamma_2) = s(\gamma_2)$ and $r(\gamma_1 \cdot \gamma_2) = r(\gamma_1)$.

For subsets A, B of M let

$$G_A = \{\gamma \mid s(\gamma) \in A\},$$
$$G^B = \{\gamma \mid r(\gamma) \in B\},$$

and $G_A^B = G_A \cap G^B$. Thus, if $x \in M$, then $G_x = \{\gamma \in G \mid s(\gamma) = x\}$ is the covering space of the leaf L_x through x corresponding to the holonomy representation of L at x, and G_x^x is isomorphic to the holonomy group of L.

A topology for the graph $G(M, \mathcal{F})$ will be introduced presently which, together with the product just defined, will make it into a topological groupoid. The following example illustrates what this topology should be like.

Example 1.3.2. Let $M = D \times Z$ be a trivial foliated space with leaves $D \times \{z\}$, where D is an open ball in Euclidean space. There is no holonomy; hence an element in the graph of M is completely determined by specifying its initial and final points. The graph of M is thus identified with the space $D \times D \times Z$, and the source and range maps are given by

$$s(x, y, z) = (y, z) \qquad r(x, y, z) = (x, z).$$

The product $(x, y, z) \cdot (x', y', z')$ is defined when $z = z'$ and $x' = y$, in which case

$$(x, y, z) \cdot (y, y', z) = (x, y', z).$$

The graph itself is a trivial foliated space with leaves $D \times D \times \{z\}$.

Motivated by this example, we construct a topology for $G(M)$ as follows. Let $\gamma \in G$ be an element with $s(\gamma) = x$ and $r(\gamma) = y$ in the leaf L. Represent γ by an immersion $\alpha : [0, 1] \to L$. Embed $[0, 1]$ in $\mathbb{R}^{\dim L}$ as a closed straight line segment I from the point p_0 to the point p_1 in $\mathbb{R}^{\dim L}$. Then, for some $\varepsilon > 0$, α can be extended to an immersion f of the open set $B = \{w \in \mathbb{R}^{\dim L} \mid \|w - I\| < \varepsilon\}$ so that $f|B_\varepsilon(p_0)$, respectively $f|B_\varepsilon(p_1)$, is a coordinate chart around x, respectively y, in L. By applying the Reeb stability argument (*cf.* [I, Proposition 11.4.8]), there is a space Z homeomorphic to a transversal through x in M so that f can be extended to an immersion F of the trivial foliated space $B \times Z$ into M, and so that $F|B_\varepsilon(p_0) \times Z$, respectively $F|B_\varepsilon(p_1) \times Z$, is a foliation chart around x, respectively y, in M. Let $\alpha_{x,w}$ denote the straight path from 0 to w in $B_\varepsilon(p_0)$, and let $\alpha_{y,w}$ be the straight path from p_1 to w in $B_\varepsilon(p_1)$. The collection of elements $\gamma \in G$ that have a representative of the form

$$t \in I \mapsto F((\alpha_{x,u}\#\alpha\#\alpha_{y,v}^{-1})(t), z) \in M$$

for some $u \in B_\varepsilon(p_0)$, $v \in B_\varepsilon(p_1)$ and $z \in Z$, is a neighborhood $U(\gamma)$ of γ in G. (As above, $\alpha\#\beta$ denotes the path α followed by the path β.) Note that this makes sense because an element of G can be represented by at most one such path. This neighborhood $U(\gamma)$ is diffeomorphic to $B_\varepsilon(p_1) \times B_\varepsilon(p_0) \times Z$ (as a foliated space with leaves $B_\varepsilon(p_1) \times B_\varepsilon(p_0) \times \{z\}$).

Exercise 1.3.3. The graph of a foliated space M without holonomy is, as a set, the graph of the equivalence relation on M given by the foliation, namely, $G = \{(x, y) \in M \times M \mid x \text{ and } y \text{ are in the same leaf}\}$, but the topology of G is in general different from the topology induced from $M \times M$.

Proposition 1.3.4. *Let $(M, \mathcal{F})$ be a foliated space of leaf dimension $\dim \mathcal{F}$. The neighborhoods described above give the graph G of M a topology and a structure of a foliated space of dimension $2 \dim \mathcal{F}$. Its holonomy pseudogroup is equivalent to the holonomy pseudogroup of M.*

Proof. Two things remain to be verified. First, the intersection of two neighborhoods $U(\gamma)$ and $U(\gamma')$ of γ and γ' in G of the type constructed above must be shown to contain a neighborhood of like form around each point. Second, the coordinate changes between $U(\gamma)$ and $U(\gamma')$ respect their trivial foliated structure.

Let $\sigma \in U(\gamma) \cap U(\gamma')$ be represented by a path from x to y. The construction of $U(\gamma)$ and $U(\gamma')$ shows that $s(U(\gamma))$ and $s(U(\gamma'))$ are foliated charts whose intersection contains x, and that $r(U(\gamma))$ and $r(U(\gamma'))$ are foliated charts whose intersection contains y. The previous construction applied to σ and a sufficiently small ε (together with the Reeb stability argument) shows that there is a neighborhood $U(\sigma)$ of the same type contained in the intersection $U(\gamma) \cap U(\gamma')$. Writing the details is straightforward but cumbersome, so they are left to the reader.

Furthermore, if the coordinate changes around x and y between the foliations charts $s(U(\gamma))$ and $s(U(\gamma'))$, and between $r(U(\gamma))$ and $r(U(\gamma'))$, have the form

$$(a', z') = (\phi_1(a, z), h(z)) \qquad \text{and} \qquad (b', z') = (\phi_2(b, z), h(z)),$$

respectively, then the coordinate change around σ is of the form

$$(a', b', z) = (\phi_1(a, z), \phi_2(b, z), h(z)),$$

as desired. □

The topology on $G(M)$ is in general not Hausdorff, a phenomenon caused by the transverse structure of M. The simplest foliated space with non-Hausdorff graph is constructed as follows.

Exercise 1.3.5. Let Z be the subset of $\mathbb{R}$ given by $Z = \{0, \pm 2^n\}$, where n denotes the integers, and let $f : Z \to Z$ be the homeomorphism of Z defined by

$$f(z) = \begin{cases} z & \text{if } z \leq 0, \\ 2z & \text{if } z > 0. \end{cases}$$

Let $(M, \mathcal{F})$ be the foliated space obtained by suspension of f, that is, M is the quotient of $\mathbb{R} \times Z$ under the action of $\mathbb{Z}$ given by $(x, z) \cdot n = (x + n, f^n(z))$.

Show that M is a foliated space of dimension one that has exactly one noncompact leaf and a countable number of compact leaves, exactly one of them with nontrivial holonomy. The leaves of the graph G are countably many tori, a cylinder, and a plane. The graph G is the quotient of $\mathbb{R}^2 \times Z$ by the action of the group $\mathbb{Z} \times \mathbb{Z}$ given by

$$(y, x, z) \cdot (m, n) = \begin{cases} (y + m, x + n, z) & \text{if } z < 0, \\ (y, x + n, f^n(z)) & \text{if } z \geq 0. \end{cases}$$

Here is a condition equivalent to the graph being Hausdorff.

Definition 1.3.6. The holonomy pseudogroup of a foliated space is *pseudo-analytic* if the following holds. If $h : Z \to Z'$ is a holonomy transformation between two transversals Z and Z', with $Z \subset Z'$, and $W \subset Z$ is an open subset such that $h|W = \mathrm{id}$, and if $x \in \overline{W}$ is such that $h(x) = x$, then $h = \mathrm{id}$ on a neighborhood of x.

Proposition 1.3.7. *Let M be a foliated space. Then the graph of M is Hausdorff if and only if the holonomy pseudogroup of M is pseudo-analytic.*

Proof. Since M is Hausdorff, the problem is to find disjoint neighborhoods of two distinct elements γ and γ' of G having the same source x and the same range y. Let α and α' be paths respectively representing γ and γ'.

Suppose that the pseudo-analyticity condition holds. Choose transversals Z_x and Z_y through x and y respectively so that the holonomies $h(\alpha)$ and $h(\alpha')$ are homeomorphisms of Z_x onto Z_y. Then $h(\alpha)^{-1} \circ h(\alpha')$ is not germinally the identity at x. The pseudo-analyticity property then implies that there is a neighborhood of x in Z_x such that the germ of $h(\alpha)^{-1} \circ h(\alpha')$ is not the identity at any point of this neighborhood. From this it is easy to construct disjoint foliated charts for γ and γ'.

Conversely, suppose that G is Hausdorff. Let $h : Z \to Z'$ be a holonomy transformation having an open set V in its domain such that $h|V = \mathrm{id}$ and a point $x \in \overline{V}$ such that germ of h at x is not the germ of the identity.

By reducing the domain and range of h around x, it may be assumed to be induced by a loop α based at x. Then α determines an element γ in G that cannot be separated from the element represented by the constant path at x. □

Exercise 1.3.8. Prove or decide the following.

(1) The graph of a foliated space all leaves of which are simply connected is Hausdorff.

(2) The graph of a foliated space all leaves of which have trivial holonomy is Hausdorff.

(3) The graph of the Reeb foliation of the three-sphere is not Hausdorff.

(4) True or false: The graph of a codimension one foliated three-manifold containing a Reeb component is not Hausdorff.

Exercise 1.3.9. Let f be a homeomorphism of the circle that is real analytic. Let Z act on the product $\mathbb{R} \times S^1$ via $n(x, \theta) = (x + n, f^n(\theta))$, and let M be the quotient space under this action. This M has a foliated space structure whose leaves are the projection of the subsets $\mathbb{R} \times \{\theta\}$. Show that the graph of the foliated space can be identified to the quotient of $\mathbb{R} \times \mathbb{R} \times S^1$ under the action of $\mathbb{Z}$ given by $n(x, y, \theta) = (x + n, y, f^n(\theta))$.

The leaves of the graph G of M are as follows. Let L be a leaf of M and let $G(L)$ be the corresponding leaf of G, i.e., $G(L)$ consists of those elements of G represented by a path in L. If $x \in L$, then there is a covering map

$$p : G_x \times G_x \to G(L)$$

given by $p : (\gamma_1, \gamma_2) \mapsto \gamma_1 \cdot \gamma_2^{-1}$. The holonomy group G_x^x acts as group of covering transformations by $(\gamma_1, \gamma_2) \cdot \gamma = (\gamma_1 \cdot \gamma, \gamma_2 \cdot \gamma)$. In particular, if L has no holonomy, then p is a diffeomorphism (where $G(L)$ has its natural manifold structure).

Exercise 1.3.10. Show that the holonomy group of $G(L)$, as leaf of the foliated space G, is isomorphic to the holonomy group of L.

Besides the foliated structure described above, the graph $G(M)$ has two other foliated structures (which are in fact subfoliations of the given one). These foliations are given by the fibers of the range and source maps, as follows from the next proposition.

Proposition 1.3.11. *The source and range maps $s, r : G \to M$ are topological submersions. In particular, the fibers of each map are proper Hausdorff submanifolds of G.*

Proof. A map $f : X \to Y$ between topological spaces is called a topological submersion (with fibers of dimension d) if for every $x \in X$ there exist an open neighborhood U such that $V = f(U)$ is open in Y and a homeomorphism $h : V \times \mathbb{R}^d \to U$ such that $f(h(v, z)) = v$ for every $(v, z) \in V \times \mathbb{R}^d$.

This property is immediate from the construction of the topology on the holonomy groupoid. Indeed, an element $\gamma \in G$ has a foliation chart neighborhood U diffeomorphic to $D \times D' \times Z$ where the range and source maps $r|U$ and $s|U$ take the form $r(x, x', z) = (x, z)$ and $s(x, x', z) = x'$, respectively. □

Thus each leaf $G(L)$ of G is foliated by the fibers $\{r^{-1}(x) \mid x \in L\}$ and by $\{s^{-1}(x) \mid x \in L\}$. The foliations of $G(L)$ induced by the source and range maps are as follows. There is a covering map

$$(\gamma_1, \gamma_2) \in G_x \times G^x \to \gamma_1 \cdot \gamma_2 \in G(L)$$

with covering transformation group G_x^x acting by $(\gamma_1, \gamma_2) \cdot \gamma = (\gamma_1 \cdot \gamma, \gamma^{-1} \cdot \gamma_2)$ (compare the discussion after Exercise 1.3.8). This action preserves the vertical and horizontal foliations, and they induce the foliations of $G(L)$ corresponding to the range and source maps, respectively.

Exercise 1.3.12. Show that the leaves of the foliation of G given by the fibers of the range map (and similarly for the source map) have trivial holonomy.

The following theorem summarizes the information collected so far on the structure of the graph of a foliated space.

Let $G \times_M G$ be the fibered product induced by the source and range maps, that is,

$$G \times_M G = \{(\gamma_1, \gamma_2) \mid s(\gamma_1) = r(\gamma_2)\},$$

which is the space that makes the following diagram commutative:

$$\begin{array}{ccc} G \times_M G & \xrightarrow{p_2} & G \\ {\scriptstyle p_1}\downarrow & & \downarrow{\scriptstyle r} \\ G & \xrightarrow[s]{} & M \end{array}$$

Theorem 1.3.13. *The graph G of $(M, \mathcal{F})$ is a groupoid with unit space $G^0 = M$, and this algebraic structure is compatible with a foliated structure on G and M. Furthermore, the following properties hold.*

(1) *The range and source maps $r, s : G \to M$ are topological submersions.*

(2) *The inclusion of the unit space $M \to G$ is a smooth map.*

(3) *The product map $G \times_M G \to G$, given by $(\gamma_1, \gamma_2) \mapsto \gamma_1 \cdot \gamma_2$, is smooth.*

(4) *There is an involution $i : G \to G$, given by $i(\gamma) = \gamma^{-1}$, which is a diffeomorphism of G, sends each leaf to itself, and exchanges the foliations given by the range and source maps.*

The following proposition provides a canonical way of passing from a foliated atlas of $(M, \mathcal{F})$ to one of G. It will became relevant when constructing the convolution algebra of G in the next section.

Proposition 1.3.14. *Let $\mathcal{U} = \{U_i\}$ be a regular foliated atlas of M. For each finite sequence of indices $\{i_0, \dots, i_k\}$, the product*

$$G(U_{i_0}) \cdots G(U_{i_n})$$

is either empty or a foliated chart for the graph G. The collection of all such finite products is a covering of G by foliated charts.

Proof. The case of a single $G(U_i)$ was already considered. The case of a product $G(U_i) \cdot G(U_j)$ is as follows. Assume that $U_i = D_i \times Z_i$ and $U_j = D_j \times Z_j$ are two foliated charts for M that intersect. Then there are an open subset Z_{ji} of the transversal Z_i and an open subset Z_{ij} of the transversal Z_j so that the plaques of U_i that meet U_j are in bijective correspondence with the points of Z_{ji} and with the points of Z_{ij}. That is, the holonomy transformation h_{ji} determined by the intersection of the foliated charts U_i and U_j has maximal domain $Z_{ji} \subset Z_i$ and $h_{ji}(Z_{ji}) = Z_{ij}$. Define a map $D_i \times D_j \times Z_{ji} \to G$ by sending the point (x_i, x_j, z) to the product

$$(x_i, y, z) \cdot (y, x_j, h_{ji}(z)),$$

where y is an arbitrary point in the intersection of the plaque through z in U_i and the plaque through $h_{ji}(z)$ in U_j. The choice of y is irrelevant because it can be assumed that the intersection of such plaques is connected and simply-connected (Exercise 1.3.15). Hence the choice of another point y' will produce another representation of the same element of $G(U_i) \cdot G(U_j)$. It is evidently injective, because the image determines the range and source points; thus it defines a diffeomorphism of foliated spaces

$$D_i \times D_j \times Z_{ji} \to G(U_i) \cdot G(U_j).$$

The case of products of arbitrary length is handled similarly. □

Exercise 1.3.15. Definitions of *regular* foliated atlases differ. In [I, Definition 11.2.8] it was not required that the intersection of two plaques be connected and simply connected. Show that these requirements can be added to the definition and that it remains true that every foliated atlas has a coherent refinement that is regular [I, Lemma 11.2.9]. (Hint. Choose a Riemannian metric and show that the atlas can be chosen so that the plaques are geodesically convex.)

Exercise 1.3.16. Let G be the graph of a foliated space and let $\mathcal{T}_G$ be the topology of G, that is, the collection of open subsets of G. Show that $\mathcal{T}_G$, endowed with this product, is an associative semigroup.

Let H be a connected and simply-connected Lie group acting locally freely (on the right) on a locally compact Hausdorff space X. It was shown in [I, 11.3.B] that X admits a foliated space structure with each leaf diffeomorphic to a homogeneous space of H.

Exercise 1.3.17. Show that if the action of H on X is free, then the graph of this foliated space is diffeomorphic and isomorphic to the groupoid $X \times H$ with source and range maps $s(x, h) = xh$ and $r(x, h) = x$, respectively, and product $(x, h_1) \cdot (xh_1, h_2) = (x, h_1 h_2)$.

If the action is not free, then there is holonomy and the graph of X may not be of this form. However, it is not necessary that the action be free for the graph to be of this form.

Exercise 1.3.18. Suppose that the action of H on X is locally free and satisfies the following property: for every $h \in H$ other than the identity, either h or h^{-1} is a strict contraction at each of its fixed points, if there are any. Then show that the graph of the foliated space given by the action is diffeomorphic and isomorphic to the groupoid $X \times H$ described above.

1.4. The C^*-algebra of a Foliated Space

This section describes the construction of a C^*-algebra associated to a foliated space. The construction utilizes the groupoid structure of the graph, and it may be seen as a generalization of the construction of the C^*-algebra of a group (which is briefly mentioned in Section A.6). For a general treatment of C^*-algebras of topological groupoids, see Renault [**155**].

Let $(M, \mathcal{F})$ be a foliated space and let $G = G(M, \mathcal{F})$ be its graph. Let $\mathcal{D}^\alpha G$ be the bundle of densities of order α over G. The source and range maps $s, r : G \to M$ induce a canonical splitting of this bundle as

$$\mathcal{D}^\alpha G = s^* \mathcal{D}^\alpha M \otimes r^* \mathcal{D}^\alpha M,$$

where $\mathcal{D}^\alpha M$ is the bundle of densities of order α on M. Thus the fiber of $\mathcal{D}^\alpha G$ over a point $\gamma \in G$ is the tensor product $\mathcal{D}^\alpha M_{s(\gamma)} \otimes \mathcal{D}^\alpha M_{r(\gamma)}$.

Besides the canonical identification $(\mathcal{D}^\alpha G)_\gamma = (\mathcal{D}^\alpha M)_{s(\gamma)} \otimes (\mathcal{D}^\alpha M)_{r(\gamma)}$, there are also canonical identifications induced by the range and source maps. If $x \in M$, then $G_x = s^{-1}(x)$ is the holonomy cover of the leaf L_x through x, the range map $r : G_x \to L_x$ being the covering projection. Hence, if $\gamma \in G_x$, there is a canonical identification $\mathcal{D}^\alpha G_x = r^* \mathcal{D}^\alpha M$, so that the fiber at each point $\gamma \in G_x$ is $(\mathcal{D}^\alpha G_x)_\gamma = (\mathcal{D}^\alpha M)_{r(\gamma)}$.

Let $\Gamma_c(G, \mathcal{D}^{1/2})$ be the space of compactly supported smooth sections of $\mathcal{D}^{1/2}G$ over G; its elements are the compactly supported half-densities on G. When G is not Hausdorff, the meaning of $\Gamma_c(G, \mathcal{D}^{1/2})$ is that which was described in Section 1.2. This space will now be given the structure of an algebra with involution. This structure is first described when G is Hausdorff, and the details will then be given when G is arbitrary.

If f_1, f_2 are two elements of $\Gamma_c(G, \mathcal{D}^{1/2})$, define their convolution $f_1 * f_2$ by

$$(f_1 * f_2)(\gamma) = \int_{\gamma_1 \cdot \gamma_2 = \gamma} f_1(\gamma_1) f_2(\gamma_2).$$

The integral is well defined because, for fixed $\gamma \in G$ with $s(\gamma) = x$ and $r(\gamma) = y$, the integrand is the density

$$\gamma_1 \mapsto f_1(\gamma_1) f_2(\gamma_1^{-1} \cdot \gamma)$$

on $G^y = \{\gamma_1 \mid r(\gamma_1) = y\}$, which is compactly supported as it vanishes if γ_1 is not in the support of f_1 or $\gamma_1^{-1} \cdot \gamma$ is not in the support of f_2.

Define an involution $f \mapsto f^*$ on $\Gamma_c(G, \mathcal{D}^{1/2})$ by

$$f^*(\gamma) = \overline{f(\gamma^{-1})}.$$

It is elementary to verify that this convolution operation is associative: $(f_1 * f_2) * f_3 = f_1 * (f_2 * f_3)$, and that the involution satisfies $(f_1 * f_2)^* = f_2^* * f_1^*$. The details are left for the reader.

The case in which the graph G is not Hausdorff will now be discussed. Convolution of two half-densities on G cannot be defined as above because, when G is not Hausdorff, elements of the space $\Gamma_c(G, \mathcal{D}^{1/2})$ are not really functions. As was described in Section 1.2, this space of half-densities is given by an exact sequence

$$\bigoplus_{i_0 i_1} \Gamma_c(V_{i_0 i_1}, \mathcal{D}^{1/2}) \to \bigoplus_{i_0} \Gamma_c(V_{i_0}, \mathcal{D}^{1/2}) \to \Gamma_c(G, \mathcal{D}^{1/2}) \to 0,$$

where the V_{i_0} are the elements of an open cover $\mathcal{V}$ of G.

The first step is to construct an open cover $\mathcal{V} = \{V_i\}$ of G by foliated charts and to define a convolution on $\bigoplus_i \Gamma_c(V_i, \mathcal{D}^{1/2})$. The construction of this open cover $\mathcal{V}$ is given by Proposition 1.3.14. Let $\mathcal{U} = \{U_n\}_{n \in N}$ be a regular cover of M by foliated charts, indexed by an ordered set N. Associated to $\mathcal{U}$ there is an open cover $\mathcal{V} = \{V_i\}_{i \in I}$ of G, indexed by an ordered set I, constructed as follows. The ordered set I is the set of finite sequences of elements of N, the order being the lexicographical order. For each finite sequence $i = \{n_1, n_2, \ldots, n_k\}$ of elements of N, let $V_i \subset G$ be the product $V_i = G(U_{n_1}) \cdot G(U_{n_2}) \cdots G(U_{n_k})$. As shown in Proposition 1.3.14, each V_i is a foliated chart for the graph.

Furthermore, note that the ordered set I is itself an associative semigroup with the product given by concatenation of finite sequences, and written $i \cdot j$. The collection of open subsets of G is itself a semigroup, and it is also associative because the multiplication of G is associative. Thus this construction shows that a regular cover $\mathcal{U} = \{U_n\}$ of M induces a semigroup homomorphism $i = \{n_1, \ldots, n_k\} \mapsto V_i = G(U_{n_1}) \cdot G(U_{n_2}) \cdot \ldots \cdot G(U_{n_k})$ from the finite sequences of elements of N into the open subsets of the graph.

The space of compactly supported half-densities on G is then taken as given by the exact sequence

$$\bigoplus_{i_0 i_1} \Gamma_c(V_{i_0 i_1}, \mathcal{D}^{1/2}) \to \bigoplus_{i_0} \Gamma_c(V_{i_0}, \mathcal{D}^{1/2}) \to \Gamma_c(G, \mathcal{D}^{1/2}) \to 0$$

associated to a regular cover for $(M, \mathcal{F})$ as above. The first step for defining a convolution is to do it at the level of $\bigoplus_i \Gamma_c(V_i, \mathcal{D}^{1/2})$, as the following lemma indicates.

Lemma 1.4.1. *If $f_1 \in \Gamma_c(V_{i_1}, \mathcal{D}^{1/2})$ and $f_2 \in \Gamma_c(V_{i_2}, \mathcal{D}^{1/2})$, then their convolution is a well-defined element $f_1 * f_2 \in \Gamma_c(V_{i_1} \cdot V_{i_2}, \mathcal{D}^{1/2})$.*

Proof. If $f_1 \in \Gamma_c(V_{i_1}, \mathcal{D}^{1/2})$ and $f_2 \in \Gamma_c(V_{i_2}, \mathcal{D}^{1/2})$, then the half-density F defined by $F(\gamma_1, \gamma_2) = f_1(\gamma_1) f_2(\gamma_2)$ belongs to $\Gamma_c(V_{i_1} \times_M V_{i_2}, \mathcal{D}^{1/2})$. The convolution $f_1 * f_2$ is

$$f_1 * f_2(\gamma) = \int_{\gamma_1 \cdot \gamma_2 = \gamma} F(\gamma_1, \gamma_2)$$

and it defines an element $\Gamma_c(V_{i_1} \cdot V_{i_2}, \mathcal{D}^{1/2})$. Indeed, in local coordinates, if $V_{i_1} = D_1 \times D_1' \times Z_1$ and $V_{i_2} = D_2 \times D_2' \times Z_2$, then the convolution takes the form

$$f_1 * f_2(x_1, x_2', z) = \int_y f_1(x_1, y, z) f_2(y, x_2', z),$$

which is a smooth half-density with support in $V_{i_1} \cdot V_{i_2} = D_1 \times D_2' \times Z_{12}$. □

This lemma suffices to justify the fact that convolution is a well defined operation in the Hausdorff case, that is, that the convolution of two compactly supported smooth half-densities is itself a compactly supported smooth half-density. In general, it shows that convolution of two elements $f_1 = \sum_i f_{1i}$ and $f_2 = \sum_i f_{2i}$ of $\bigoplus_i \Gamma_c(V_i, \mathcal{D}^{1/2})$ can be defined by

$$f_1 * f_2 = \sum_{ij} f_{1i} * f_{2j},$$

and that the result is an element of $\bigoplus_i \Gamma_c(V_i, \mathcal{D}^{1/2})$. The associative property of the index set for the cover $\mathcal{V}$ shows that convolution is an associative operation. The next proposition shows that this operation induces a convolution on $\Gamma_c(G, \mathcal{D}^{1/2})$.

Proposition 1.4.2. *The convolution defined on* $\bigoplus_i \Gamma_c(V_i, \mathcal{D}^{1/2})$ *induces a well defined convolution on* $\Gamma_c(G, \mathcal{D}^{1/2})$.

Proof. What remains to be shown is that if $g \in \bigoplus_{i_0} \Gamma_c(V_{i_0}, \mathcal{D}^{1/2})$ represents the zero element in $\Gamma_c(G, \mathcal{D}^{1/2})$, and if $f \in \Gamma_c(V_j, \mathcal{D}^{1/2})$, then $f * g$ also represents the zero element. That g represents the zero element means that g is in the image of

$$\Phi : \bigoplus_{i_0 i_1} \Gamma_c(V_{i_0 i_1}, \mathcal{D}^{1/2}) \to \bigoplus_{i_0} \Gamma_c(V_{i_0}, \mathcal{D}^{1/2}),$$

hence that g can be written in the form

$$\sum_{i_1} \sum_{i_0} g_{i_0 i_1},$$

where the half-density $g_{i_0 i_1} \in \Gamma_c(V_{i_1}, \mathcal{D}^{1/2})$ is the extension by zero of an element of $\Gamma_c(V_{i_1 i_0}, \mathcal{D}^{1/2})$. By the semigroup properties of the cover $\mathcal{V}$, the open set $V_j \cdot V_{i_r} = V_{j \cdot i_r} = V_{j_r}$. Then the convolution $f * g$ has the form

$$f * g = \sum_{i_1} \sum_{i_0} f * g_{i_0 i_1},$$

and $f * g_{i_0 i_1}$ is an element of $\Gamma_c(V_{j \cdot i_0}, \mathcal{D}^{1/2})$ that, in fact, belongs to the set $\Gamma_c(V_j \cdot (V_{i_0 i_1}), \mathcal{D}^{1/2})$. But

$$V_j \cdot V_{i_0 i_1} = V_j \cdot (V_{i_0} \cap V_{i_1}) \subset (V_j \cdot V_{i_0}) \cap (V_j \cdot V_{i_1}) = V_{j_0 j_1},$$

where $V_{j_r} = V \cdot V_{i_r} \in \mathcal{V}$, $r = 0, 1$, because $\mathcal{V}$ is closed under products. Therefore, the densities $f * g_{i_0 i_1}$ define densities $h_{j_0 j_1} \in \Gamma_c(V_{j_0 j_1}, \mathcal{D}^{1/2})$, and $h = \sum_{j_0 j_1} h_{j_0 j_1}$ has the property that $\Phi(h) = f * g$, as required. □

The involution $f \mapsto f^*$ on $\Gamma_c(G, \mathcal{D}^{1/2})$, given by $f^*(\gamma) = \overline{f(\gamma^{-1})}$ as in the Hausdorff case, is defined at the level of $\bigoplus_i \Gamma_c(V_i, \mathcal{D}^{1/2})$. It is apparent that it induces a well-defined involution on $\Gamma_c(G, \mathcal{D}^{1/2})$.

Proposition 1.4.3. *The convolution product* $(f_1, f_2) \mapsto f_1 * f_2$ *and the involution* $f \mapsto f^*$ *make* $\Gamma_c(G, \mathcal{D}^{1/2})$ *into an involutive (associative) algebra.*

It is a straightforward matter to show that this structure on $\Gamma_c(G, \mathcal{D}^{1/2})$ is independent of the choice of regular cover by foliated charts of the foliated space $(M, \mathcal{F})$.

Remark. The use of half-densities explicitly shows how canonical the construction of the algebra is. However, working with densities is not always as transparent as working with functions. The construction can be defined in terms of functions by choosing once and for all a positive half-density on M (*e.g.*, the canonical density of a metric tensor along the leaves) that in turn induces a trivialization of the bundle of half-densities over the graph.

Thus, if σ is an everywhere positive half-density on M, then $s^*\sigma \otimes r^*\sigma$ trivializes the bundle $\mathcal{D}^{1/2}G$. Convolution is now defined for two functions $f_1, f_2 \in \Gamma_c(G, \mathcal{C}^\infty)$ by

$$f_1 * f_2(\gamma) = \int_{G_{s(\gamma)}} f_1(\gamma \cdot \gamma'^{-1}) f_2(\gamma') \cdot r^*\sigma(\gamma').$$

This device of passing from densities to functions will be utilized when convenient.

Exercise 1.4.4. Let $W = G(U_1) \cdots G(U_n)$. Then the collection of linear combinations of elements of the form $f = f_1 * \cdots * f_n$, $f_i \in G(U_i)$, is dense in the space of compactly supported continuous half densities on W, in the inductive limit topology.

For each point x in M there is a natural representation R_x of the involutive algebra $\Gamma_c(G, \mathcal{D}^{1/2})$ on the Hilbert space of half-densities on the holonomy covering of the leaf L through x. Indeed, the holonomy cover of L can be identified with $G_x = \{\gamma \in G \mid s(\gamma) = x\}$ with covering projection $r : G_x \to L$, and then the representation R_x is given by

$$R_x(f)\xi(\gamma) = \int_{\gamma_1 \cdot \gamma_2 = \gamma} f(\gamma_1)\xi(\gamma_2).$$

This is well defined because, for fixed x and $\gamma \in G_x$, there are canonical identifications $(\mathcal{D}^{1/2}G_x)_\gamma = \mathcal{D}^{1/2}M_{r(\gamma)}$, so that for $\gamma_1 \cdot \gamma_2 = \gamma$, $f(\gamma_1)\xi(\gamma_2)$ is a density of order one with coefficients in $\mathcal{D}^{1/2}M_{r(\gamma_1)} = (\mathcal{D}^{1/2}G_x)_{s(\gamma)}$, and thus the integral over $\{\gamma_1 \cdot \gamma_2 = \gamma\}$ is a half density on G_x. This is also well defined in the non-Hausdorff case because, for each γ, the fiber $G^{r(\gamma)}$ is a proper Hausdorff submanifold of G, hence the restriction $f|G^{r(\gamma)}$ is smooth with compact support in the usual sense.

Proposition 1.4.5. *If $V \subset G$ is a foliated chart for the graph of $(M, \mathcal{F})$ and $f \in \Gamma_c(V, \mathcal{D}^{1/2})$, then $R_x(f)$ is a bounded integral operator on $L^2(G_x)$.*

Proof. Assume that V is of the form $V = D \times D' \times Z$, where $r(V) = D \times Z$ and $s(V) = D' \times Z$ are foliated charts in M. The expression

$$R_x(f)\xi(\gamma_1) = \int_{G_x} f(\gamma_1 \cdot \gamma_2^{-1})\xi(\gamma_2)$$

shows that $R_x(f)$ is given by the smooth kernel $F(\gamma_1, \gamma_2) = f(\gamma_1 \cdot \gamma_2^{-1})$. Thus $(\gamma_1, \gamma_2) \in \operatorname{supp}(F)$ if and only if $\gamma_1 \cdot \gamma_2^{-1} \in \operatorname{supp}(f)$, which in turn can happen only when there is $z \in Z$ such that $\gamma_1 \cdot \gamma_2^{-1} = (r(\gamma_1), r(\gamma_2), z)$. Therefore there are two sequences $\{A_n\}$ and $\{B_n\}$ of mutually disjoint compact domains in G_x such that the kernel F can be written as a sum $F = \sum_n F_n$, where F_n is a smooth kernel with compact support in $A_n \times B_n$. Indeed, the kernel F decomposes as a sum $F = \sum_n F_n$, where the F_n are compactly supported kernels on $G_x \times G_x$ with support on one of the countably many components of the preimage of the intersection of $G(L)$ with the foliated chart V via the covering projection $p : G_x \times G_x \to G(L)$. Each of these components is diffeomorphic to a product $D \times D'$ via the projection p, and the relative position of two such components $A_1 \times B_1$ and $A_2 \times B_2$ can be of two different forms: either they project to two different plaques of V (in which case there is nothing else to add), or they project to the same, in which case there is an element γ in G_x^x such that $(A_1 \times B_1) \cdot \gamma = A_2 \times B_2$, from which it follows that $A_1 \cap A_2 = B_1 \cap B_2 = \emptyset$.

Therefore, if $\xi \in L^2(G_x)$, then ξ can be written as an orthogonal sum $\xi = \xi_0 + \sum \xi_n$, where $F_n\xi_0 = 0$ for every n, and $F_n\xi_m = 0$ if $n \neq m$. Moreover, if $n \neq m$, then the inner product $\langle F_n\xi | F_m\xi\rangle = \langle \xi | F_n^* F_m \xi\rangle = 0$, because the domains A_n and A_m are disjoint. (To simplify notation, F_n denotes both the kernel and the operator that it defines.)

Since each F_n is a compactly supported smooth kernel, its norm is finite and given by

$$\|F_n\|^2 = \int_{G_x \times G_x} |F_n(\gamma_1, \gamma_2)|^2.$$

Let $\lambda = \sup_n \|F_n\|$. It is evident from the definition of the norm of an operator that $\|F\| \geq \lambda$. If $\xi \in L^2(G_x)$, then, taking into account the observations made in the previous paragraph,

$$\begin{aligned}
\|R_x(f)\xi\|^2 &= \|R_x(f)(\xi_0 + \sum_n \xi_n)\|^2 \\
&= \sum_n \|F_n\xi_n\|^2 \\
&\leq \lambda^2 \sum_n \|\xi_n\|^2 \\
&\leq \lambda^2 \|\xi\|^2,
\end{aligned}$$

so that $\|R_x(f)\| \leq \lambda$ also; hence $\|R_x(f)\| = \lambda$. Finally, $\lambda < \infty$. Indeed,

$$\lambda^2 \leq \sup_{z \in Z} \int_{D \times D' \times \{z\}} |f(x, x', z)|^2$$

and this is finite because, as f is continuous with compact support in the space $D \times D' \times Z$, the integral is a compactly supported continuous function of the transverse parameter $z \in Z$. □

Corollary 1.4.6. *Each R_x is an involutive representation of $\Gamma_c(G, \mathcal{D}^{1/2})$ in $L^2(G_x)$.*

Proof. The fact that R_x is an involutive representation means that it satisfies $R_x(f_1 * f_2) = R(f_1)R(f_2)$ and $R_x(f^*) = R_x(f)^*$, and this is a formal calculation. The previous lemma, linearity of R_x, and the definition of $\Gamma_c(G, \mathcal{D}^{1/2})$ complete the proof. □

Summarizing, associated to a foliated space $(M, \mathcal{F})$ there is the graph $G = G(M, \mathcal{F})$. Associated to this graph there is the space of compactly supported smooth densities $\Gamma_c(G, \mathcal{D}^{1/2})$, which has the structure of an associative algebra with involution. For each $x \in M$, there is a representation R_x of this algebra on the Hilbert space of square integrable half-densities on the holonomy cover of the leaf through x. This brings about the following definition.

Definition 1.4.7. The *reduced* C^*-algebra of the foliated space $(M, \mathcal{F})$ is the completion of $\Gamma_c(G, \mathcal{D}^{1/2})$ with respect to the pseudonorm

$$\|f\| = \sup_{x \in M} \|R_x(f)\|.$$

This C^*-algebra is denoted by $C_r^*(M, \mathcal{F})$.

An obvious consequence of the construction of $C_r^*(M, \mathcal{F})$ is the following.

Corollary 1.4.8. *Let M be a foliated space and let $\mathfrak{U}$ be a regular cover by foliated charts. Then the algebra generated by the convolution algebras $\Gamma_c(G(U), \mathcal{D}^{1/2})$, $U \in \mathfrak{U}$, is dense in $C_r^*(M, \mathcal{F})$.*

It will be shown in the next section that the structure of the C^*-algebra $C_r^*(U, \mathcal{F}|U)$, where U is a foliated chart, is rather elementary, being essentially the continuous functions on the leaf space of U vanishing at infinity. On the other hand, the structure of the C^*-algebra $C_r^*(M, \mathcal{F})$ can be rather complicated, reflecting the complicated way in which the foliated charts are assembled.

The following lemma shows that the norm $\|R_x(f)\|$ depends only on the leaf through x, which may be seen as a naive justification for thinking of the C^*-algebra $C_r^*(M, \mathcal{F})$ as playing the role of the algebra of continuous functions on the leaf space of M.

Lemma 1.4.9. *Each element $\gamma \in G$ induces a unitary operator*

$$R_\gamma : L^2(G_{s(\gamma)}) \to L^2(G_{r(\gamma)})$$

that conjugates the operators $R_{s(\gamma)}(f)$ and $R_{r(\gamma)}(f)$. In particular, the norm of $R_x(f)$ is independent of the point in the leaf through x.

Proof. An element $\gamma \in G$ induces a diffeomorphism $G_{r(\gamma)} \to G_{s(\gamma)}$ by right multiplication: $\gamma_1 \mapsto \gamma_1 \cdot \gamma$. This in turn induces a unitary operator $R_\gamma : L^2(G_{s(\gamma)}) \to L^2(G_{r(\gamma)})$ by $R_\gamma \xi(\gamma_1) = \xi(\gamma_1 \cdot \gamma)$. It is straightforward to verify that the identity $R_{s(\gamma)}(f) R_\gamma = R_\gamma R_{r(\gamma)}(f)$ holds for every density $f \in \Gamma_c(G, \mathcal{D}^{1/2})$. □

Note also that the proof of Proposition 1.4.5 shows that the norm $\|f\|$ is finite for each $f \in \Gamma_c(G, \mathcal{D}^{1/2})$. In principle, there could be elements f other than the trivial one with $\|f\| = 0$. The next lemma explains when this happens.

Lemma 1.4.10. *If $f \in \Gamma_c(G, \mathcal{D}^{1/2})$ does not evaluate to zero at each $\gamma \in G$, then there exists a point x in M such that $R_x(f) \neq 0$.*

Proof. If f evaluates to 0 at each $\gamma \in G$, then clearly $R_x(f) = 0$ for all $x \in M$, because $R_x(f)$ only depends on the evaluation of f.

Suppose that $f(\gamma) \neq 0$ for a particular γ. It may be assumed that f is a positive half-density on a neighborhood of γ. Let $x = s(\gamma)$. Then f restricts to a function on the space G_x that is positive on a neighborhood of γ in G_x. This neighborhood can be taken to be the intersection of a foliation chart around γ with G_x. Let ξ be a positive function with compact support on the chart around $s(x)$. Then the expression for $R_x(f)(\xi)$ immediately gives the nontriviality of $R_x(f)$, and hence that $\|f\| \neq 0$. □

Remark. Lemma 1.4.10 and Example 1.2.4 show that the C^*-algebra thus obtained is the same as the one constructed via the original concept of smooth function of Connes [**36**].

By construction, for each $x \in M$, the involutive representation R_x of $\Gamma_c(G, \mathcal{D}^{1/2})$ extends to a representation of $C_r^*(M, \mathcal{F})$. Whether these representations are irreducible or not is important and will be examined next.

Theorem 1.4.11. *Let $(M, \mathcal{F})$ be a foliated space and let $x \in M$. Then the representation R_x is irreducible if and only if the leaf through x has no holonomy.*

Proof. The irreducibility will be studied by analyzing the operators on $L^2(G_x)$ that commute with R_x (*cf.* Theorem A.10.7).

If the leaf L through x has no holonomy, the map $\Pi : G_x \times G_x \to G(L)$ given by $\Pi(\gamma_1, \gamma_2) = \gamma_1 \cdot \gamma_2^{-1}$ is a diffeomorphism. Let σ_1, σ_2 be compactly supported, smooth half-densities on G_x, and let $\langle\sigma_1| \otimes |\sigma_2\rangle$ be the operator on $L^2(G_x)$ given by $\xi \mapsto \langle\xi|\sigma_2\rangle\sigma_1$. If D_1, D_2 are relatively compact domains in G_x containing the supports of σ_1 and σ_2, respectively, then, using the fact that the leaf $G(L)$ has no holonomy as well as the locally triviality of the foliation of G, it is possible to construct a foliation chart V in G of the form $D_1 \times D_2 \times Z$, for an appropriate transversal Z.

Let z_0 be a point of Z so that the plaque $D_1 \times D_2 \times \{z_0\}$ is the image of $D_1 \times D_2 \subset G_x \times G_x$ under the map Π. If h_n is a sequence of compactly supported continuous functions on Z, with $0 \leq h_n \leq 1$ and $h_n(z_0) = 1$, and with supports shrinking to $\{z_0\}$ as $n \to \infty$, then

$f_n(x_1, x_2, z) = \sigma_1(x_1)\overline{\sigma_2(x_2)}h_n(z)$ is an element of $\Gamma_c(G, \mathcal{D}^{1/2})$, and the sequence of operators $R_x(f_n)$ converges in the weak operator topology to the operator $\langle\sigma_1| \otimes |\sigma_2\rangle$ on $L^2(G_x)$. Thus the weak closure of the operators $R_x(f)$ contains all the finite rank operators $\sum_{ij}\langle\sigma_i| \otimes |\sigma_j\rangle$, where σ_i, σ_j belong to $\Gamma_c(G_x, \mathcal{D}^{1/2})$, hence all finite rank operators on the space $\mathcal{B}(L^2(G_x))$ of bounded operators.

Therefore, an operator on $L^2(G_x)$ that commutes with all the operators $R_x(f)$, $f \in \Gamma_c(G, \mathcal{D}^{1/2})$, must commute also with all the finite rank operators. Hence it must be a scalar multiple of the identity operator because the finite rank operators are weakly dense in the bounded operators. The irreducibility of R_x then follows from Theorem A.10.7.

If the leaf L through $x \in M$ has holonomy, then the representation R_x is not irreducible. Indeed, because of Lemma 1.4.9, the action of G_x^x on G_x induces a unitary representation of G_x^x on $L^2(G_x)$ by right translation that commutes with the representation R_x.

This unitary representation of G_x^x is nondegenerate; in fact, it is equivalent to a multiple of the regular representation of G_x^x. Indeed, let

$$y \in L \mapsto \gamma_y \in G_x$$

be a Borel section of the covering projection $r : G_x \to L$. (That is, pick a fundamental domain $D \subset G_x$ for the action of G_x^x on G_x. Then r is invertible on D, and so $\gamma_y = r^{-1}(y) \in D$.) The map $G_x \to G_x^x \times L$ given by $\Phi(\gamma) = (\gamma_{r(\gamma)}^{-1} \cdot \gamma, r(\gamma))$ conjugates the action of G_x^x on G_x with the action of G_x^x on $G_x^x \times L$ given by right multiplication on the first factor and identity on the second one. This map Φ therefore conjugates the representation of G_x^x on $L^2(G_x)$ with the representation of G_x on $\ell^2(G_x^x) \oplus L^2(L)$ given by right transition on the first factor and identity on the second, namely, $\gamma(\phi, f) = (\phi \circ \gamma, f)$; hence it is a multiple of the right regular representation of G_x^x. □

The index r in the notation $C_r^*(M, \mathcal{F})$ is meant as a reminder of the similarity of this algebra to the reduced C^*-algebra of a locally compact group, as described in Appendix A. It is possible to define a C^*-algebra for a foliated space which corresponds to the full C^*-algebra of a group. This is achieved by considering all involutive representations that are continuous in a certain sense. Since $\Gamma_c(G, \mathcal{D}^{1/2})$ is generated (as a convolution algebra) by $\Gamma_c(G(U), \mathcal{D}^{1/2})$, where the U are foliated charts for M, this continuity requirement can be introduced at a local level. The natural topology on $\Gamma_c(G(U), \mathcal{D}^{1/2})$ is the inductive limit topology.

Definition 1.4.12. The *full* C^*-algebra of the foliated space $(M, \mathcal{F})$, denoted by $C^*(M, \mathcal{F})$, is the completion of $\Gamma_c(G, \mathcal{D}^{1/2})$ with respect to the pseudonorm

$$\|f\| = \sup_{\pi} \|\pi(f)\|,$$

where π runs through all the involutive representations of $\Gamma_c(G, \mathcal{D}^{1/2})$ on a separable Hilbert space whose restrictions to the graph $G(U)$ of each foliated chart U for $(M, \mathcal{F})$ are weakly continuous for the inductive limit topology on $\Gamma_c(G(U), \mathcal{D}^{1/2})$.

Because each R_x, $x \in M$, is an involutive representation of the convolution algebra of G, and is continuous as the definition requires, there is an obvious surjection $C^*(M, \mathcal{F}) \to C_r^*(M, \mathcal{F})$; in general, as with groups, it is not an isomorphism. The concept of amenability that arises in this situation is studied in Renault [**155**], in the general setting of topological groupoids. The algebra $C^*(M, \mathcal{F})$ is relevant for the representation theory of the holonomy groupoid, as will be discussed in Section 1.7.

1.5. The Basic Examples

As its construction shows, the C^*-algebra of a foliated space is built up from the C^*-algebras of very simple foliated spaces, namely foliated charts. This section examines this particular example in increasing levels of generality.

1.5.A. One leaf. Let $(M, \mathcal{F})$ be a foliated space consisting of exactly one leaf, so that M is a connected manifold. Since there is no holonomy, a path γ is completely determined by its source and range. Thus the graph $G(M) = M \times M$, with $s(x, y) = y$ and $r(x, y) = x$, and product

$$(x, z) \cdot (z, y) = (x, y).$$

If $x \in M$, then the range map $r : G_x \to M$ induces a unitary operator $L^2(G_x) \to L^2(M)$, and the representation R_x is equivalent to the representation of $\Gamma_c(G, \mathcal{D}^{1/2})$ on $L^2(M)$ given by

$$R(f)(\xi)(y) = \int_M f(y, z)\xi(z).$$

The norm of $R(f)$ is the norm of the kernel $f(y, x)$ as an integral operator acting on $L^2(M)$. Therefore, the C^*-algebra $C_r^*(M, \mathcal{F})$ is the algebra of compact operators on the separable Hilbert space $L^2(M)$.

Proposition 1.5.1. *The reduced C^*-algebra of a foliated space M consisting of exactly one leaf is the algebra $\mathcal{K}(L^2(M))$ of compact operators on $L^2(M)$.*

Of course, if M is just one point, then $L^2(M) = \mathbb{C}$ and $\mathcal{K}(\mathbb{C}) = \mathbb{C}$.

The following proposition will determine the structure of the full C^*-algebra of M.

Proposition 1.5.2. *Let $\mathcal{D} = \Gamma_c(X \times X, \mathcal{D}^{1/2})$ be the convolution algebra of a connected manifold X. Let π be an involutive representation of $\mathcal{D}$ on a Hilbert space H. Then π is equivalent to a direct sum of the zero representation and copies of the canonical representation of $\mathcal{D}$ on $L^2(X)$.*

Proof. Let $\{\alpha, \beta, \gamma, \cdots\}$ be the family of compactly supported half-densities on X, of norm 1 in $L^2(X)$. Let P_α and $U_{\alpha\beta}$ be the elements of $\mathcal{D}$ defined by $P_\alpha = \langle\alpha| \otimes |\alpha\rangle$ and $U_{\alpha\beta} = \langle\alpha| \otimes |\beta\rangle$. These operators possess the following properties:

(1) $P_\alpha = U_{\alpha\alpha}$ is a projection operator,

(2) $U^*_{\alpha\beta} = U_{\beta\alpha}$,

(3) $U_{\alpha\beta}U_{\gamma\delta} = 0$ if $\langle\beta|\gamma\rangle = 0$, and

(4) the cocycle relations holds:

$$U_{\alpha\beta}U_{\beta\gamma} = U_{\alpha\gamma}.$$

If T is a bounded operator on $L^2(X)$, then

$$P_\alpha T P_\beta(\xi) = \langle T\phi_\beta|\phi_\alpha\rangle U_{\alpha\beta}(\xi).$$

This relation is preserved by the representation π, namely

$$\pi(P_\alpha)\pi(T)\pi(P_\beta) = \langle T\phi_\beta|\phi_\alpha\rangle \pi(U_{\alpha\beta}).$$

Since P_α is a projection, so is $\pi(P_\alpha)$. If $H_\alpha = \pi(P_\alpha)H$ is the range of this projection, then $\pi(U_{\alpha\beta})$ is an isometry of H_β onto H_α. Therefore all subspaces H_α have the same dimension. If this dimension equals zero, then the operator $\pi(P_\alpha) = 0$. Thus also $\pi(U_{\alpha\beta}) = 0$ because the representation π is involutive and because of properties (1) and (2).

If $K \in \mathcal{D}$, then K is the limit, in the inductive limit topology, of a net of sums of finite linear combinations of operators $c_{\alpha\beta}U_{\alpha\beta}$. Since the representation π is continuous with respect to the inductive limit topology on $\mathcal{D}$, the image $\pi(K) = 0$, and so the representation π is the zero representation.

Assume then that the subspaces H_α have positive dimension. Choose a unit vector $\xi_{\alpha_0} \in H_{\alpha_0}$ and let $\xi_\alpha = \pi(U_{\alpha\alpha_0})\xi_{\alpha_0}$, which is also a unit vector. Let V be the closed subspace of H generated by the vectors ξ_α. Then $\pi(U_{\alpha\beta})\xi_\gamma = \langle\beta|\gamma\rangle\xi_\alpha$, so that V is invariant under all the operators $\pi(U_{\alpha\beta})$. On this subspace, P_α is the projection onto the one-dimensional subspace generated by ξ_α. If $T = \sum c_{\alpha\beta}U_{\alpha\beta}$, then the relations (1)–(4) above also imply that $\pi(T)$ leaves V invariant. Therefore, by continuity, V is invariant under $\pi(\mathcal{D})$.

Extend α_0 to a complete orthonormal system $\alpha_0, \alpha_1, \ldots$ for $L^2(X)$. Define an operator $T : L^2(X) \to V$ by setting $T(\alpha_k) = \xi_{\alpha_k}$. Then T is injective by property (3) above. The vectors ξ_{α_k} form an orthonormal system (by property (3) above) that also spans V. Thus T is an isometry of $L^2(X)$ onto V. Moreover, T conjugates the representation π on V with the canonical representation of $\mathcal{D}$ as compact operators on $L^2(X)$.

Consider the representation π on the orthogonal complement of V. If this representation is different from zero, there is a subspace V_1 on which the representation is equivalent to the canonical representation of $\mathcal{D}$. A standard maximality argument completes the proof of the statement. □

Corollary 1.5.3. *The full C^*-algebra of a foliated space consisting of exactly one leaf is the same as the reduced C^*-algebra.*

1.5.B. Foliated chart. The second basic example is when the foliated space is a foliated chart, or, slightly more generally, a trivial foliated space $M = N \times Z$, where N is a connected manifold.

The graph G of $N \times Z$ can be described at once as the space $N \times N \times Z$ with product

$$((z,y),t) \cdot ((y,x),t) = ((z,x),t)$$

and range and source maps given by

$$r((x,y),t) = (x,t) \qquad \text{and} \qquad s((x,y),t) = (y,t).$$

Proposition 1.5.4. *The reduced C^*-algebra $C_r^*(N \times Z)$ of the trivial foliated space $N \times Z$ is the tensor product $\mathcal{K} \otimes C_0(Z)$, where $\mathcal{K}$ is the algebra of compact operators on $L^2(N)$ and $C_0(Z)$ is the space of continuous functions on Z that vanish at infinity.*

Proof. An element f of $\Gamma_c(G, \mathcal{D}^{1/2})$ is a compactly supported function on Z with values in $\Gamma_c(N \times N, \mathcal{D}^{1/2})$ that is continuous for the inductive limit topology.

The representation $R_{(x,z)}$ assigns to the smooth density $f \in \Gamma_c(G, \mathcal{D}^{1/2})$ the compactly supported smooth kernel $R_{(x,z)}(f)$ acting on the space $L^2(N)$ of square integrable half-densities on N in the standard way. The norm $\|R_{(x,z)}(f)\|$ is the norm of the kernel operator $R_{(x,z)}(f)$ acting on $L^2(N)$. The compactly supported smooth kernel operators are dense in the compact operators in the operator norm topology; hence the completion of $\Gamma_c(G, \mathcal{D}^{1/2})$ with respect to the representations $R_{(x,z)}$, $(x,z) \in D \times Z$, is the algebra $C_0(Z, \mathcal{K})$ of continuous functions on Z, vanishing at infinity, with values in the algebra $\mathcal{K}$ of compact operators on the Hilbert space $L^2(N)$. By Lemma A.9.2, this algebra $C_0(Z, \mathcal{K})$ is the tensor product $\mathcal{K} \otimes C_0(Z)$. □

Let $(M, \mathcal{F})$ be an arbitrary foliated space and let $U \subset M$ be an open subset. Then $(U, \mathcal{F}|U)$ is a foliated space and the inclusion $U \hookrightarrow M$ induces a homomorphism of groupoids $G(U) \hookrightarrow G$, hence a mapping

$$i : \Gamma_c(G(U), \mathcal{D}^{1/2}) \to \Gamma_c(G, \mathcal{D}^{1/2})$$

that is a homomorphism of involutive algebras.

Proposition 1.5.5. *Let U be an open subset of the foliated space M. Then the inclusion $U \hookrightarrow M$ induces an isometry of $C_r^*(U, \mathcal{F}|U)$ into $C_r^*(M, \mathcal{F})$.*

Proof. The proof will only be given for the case in which U is a foliated chart for M of the form $U = D \times Z$. The general case is left for the reader.

It will be shown that, for each $x \in M$, the representation $R_x \circ i$ of $\Gamma_c(G(U), \mathcal{D}^{1/2})$ is equivalent to a direct sum of representations R_{x_α}, $x_\alpha \in U$, and the trivial representation. The statement follows from this; the basic idea is implicit in the proof of Lemma 1.4.5.

The space $L^2(G_x)$ can be decomposed into a direct sum by means of a partition of G_x as follows. Since $G(U)$ is an open subgroupoid of G, then the relation "$\gamma_1 \sim \gamma_2$ if and only if $\gamma_1 \cdot \gamma_2^{-1} \in G(U)$" is an equivalence relation on $r^{-1}(U) \cap G_x$. Each equivalence class is open and connected (because $G(U)$ is open and each plaque of U is connected). Each component is in fact canonically identified with a component of the intersection of the leaf through x with U. This equivalence relation decomposes G_x into a countable collection of sets, namely

$$(*) \qquad G_x = (B \cap G_x) \cup \bigcup_{\alpha \in \pi_0(L_x \cap U)} L_\alpha,$$

where $B = G \smallsetminus r^{-1}(U)$ is the complement of $r^{-1}(U)$ in G, and where $\{L_\alpha\}$ are the countably many components of $G_x \cap r^{-1}(U)$. The range map r identifies each of these components to each of the plaques determined by the intersection of the leaf through x with U.

Let $f \in \Gamma_c(G(U), \mathcal{D}^{1/2})$ and $\xi \in L^2(G_x)$ be such that

$$R_x(i(f))(\xi)(\gamma) = \int_{G_x} f(\gamma \cdot {\gamma'}^{-1})\xi(\gamma').$$

If $f \in \Gamma_c(G(U), \mathcal{D}^{1/2})$, then $f(\gamma \cdot {\gamma'}^{-1})$ can be different form 0 only when $\gamma \cdot {\gamma'}^{-1} \in G(U)$. Thus $R_x(i(f))$ preserves the decomposition of $L^2(G_x)$ induced by the partition G_x given in $(*)$. The restriction of $R_x(i(f))$ to $L^2(B \cap G_x)$ is zero. Each of the other pieces L_α corresponds to a plaque of U; hence the restriction of the representation $R_x \circ i$ to $L^2(L_\alpha)$ is equivalent to a representation of the form R_{x_α} for some $x_\alpha \in U$.

It follows (*cf.* Lemma 1.4.5) that the norm $\|f\|$ in $C_r^*(U, \mathcal{F}|U)$ equals the norm $\|i(f)\|$ in $C_r^*(M, \mathcal{F})$. □

The full C^*-algebra of $M = N \times Z$ can also be explicitly described. Let $\mathcal{D} = \Gamma_c(N \times N, \mathcal{D}^{1/2})$ denote the convolution algebra of the manifold N. Then the space of compactly supported half-densities on the graph G of M is the space of maps $C_c(Z, \mathcal{D})$ with the inductive limit topology. There is a natural map from the algebraic tensor product $\mathcal{D} \otimes_{\mathbb{C}} C_c(Z)$ into $C_c(Z, \mathcal{D})$ that assigns to an element $k \otimes f$ the map $z \in Z \mapsto f(z)k \in \mathcal{D}$, and is extended by linearity. This map is in fact an injective homomorphism of involutive algebras, the involution on $\mathcal{D} \otimes_{\mathbb{C}} C_c(Z)$ being given by $(k \otimes a)^* = k^* \otimes a^*$, and the product being $(k \otimes a) * (h \otimes b) = (k * h) \otimes (ab)$. The image of this map is dense in $C_c(Z, \mathcal{D})$, with respect to the inductive limit topologies, and, more generally:

Lemma 1.5.6. *The algebraic tensor product $\mathcal{D} \otimes_{\mathbb{C}} C_c(Z)$ is dense in the full C^*-algebra $C^*(M, \mathcal{F})$.*

Proof. By definition, $C^*(M, \mathcal{F})$ is the completion of $C_c(Z, \mathcal{D})$ with respect to the norm given by

$$\|f\| = \sup_{\pi} \|\pi(f)\|,$$

where π ranges through all the involutive representations that are continuous with respect to the inductive limit topology. The statement follows because $\mathcal{D} \otimes_{\mathbb{C}} C_c(Z)$ is dense in $C_c(Z, \mathcal{D})$ for this topology. □

Therefore, to understand the structure of $C^*(M, \mathcal{F})$ it suffices to analyze the structure of the involutive representations of $\mathcal{D} \otimes C_c(Z)$.

Lemma 1.5.7. *Let π be an involutive representation of the algebraic tensor product $\mathcal{D} \otimes_{\mathbb{C}} C_c(Z)$ on a Hilbert space H. Then there are representations π_1 of $\mathcal{D}$ and π_2 of $C_c(Z)$ on H such that*

$$\pi(f \otimes a) = \pi_1(f)\pi_2(a) = \pi_2(a)\pi_1(f),$$

for all $f \in \mathcal{D}$ and all $a \in C_c(Z)$.

Proof. Let A denote the algebraic tensor product $\mathcal{D} \otimes_{\mathbb{C}} C_c(Z)$. It may be assumed, without loss of generality, that the representation π is such that the subspace $\pi(A)H$ is dense in H.

Let $\xi \in \pi(A)H$. Then ξ admits a presentation as a finite sum of the form $\xi = \sum_{ij} \pi(f_{ij} \otimes a_{ij})\xi_i$. Let $Z^\bullet$ be the one-point compactification of Z. For $a \in C(Z^\bullet)$, set

$$\varphi(a) = \langle \sum_{ij} \pi(f_{ij} \otimes a a_{ij})\xi_i | \xi \rangle$$

Then φ is a linear function of $a \in C(Z^\bullet)$, which is well defined because $C_c(Z)$ is an ideal of $C(Z^\bullet)$. Moreover,

$$\varphi(a^* a) = \|\sum_{ij} \pi(f_{ij} \otimes a a_{ij})\xi_i\|^2 \geq 0.$$

Thus φ is a positive linear functional on $C(Z^\bullet)$ (*cf.* Definition A.5.9). It is then bounded, because $C(Z^\bullet)$ is a C^*-algebra with unit, and

$$|\varphi(a)| \leq \|a\|\varphi(1) = \|a\|\|\xi\|^2.$$

Thus $\varphi(a)$ is independent of the presentation of ξ, and there exists a bounded operator $\pi_2(a)$ on H such that $\pi(a)\xi = \sum_{ij} \pi(f_{ij} \otimes aa_{ij})\xi_i$ for $\xi \in \pi(A)H$ as above.

It is easy to verify that π_2 is a representation of $C_0(Z)$ on H and that it satisfies

$$\pi(f \otimes ab) = \pi_2(a)\pi(f \otimes b).$$

Of course, π_2 restricts to a representation of $C_c(Z)$.

Let u_n be an approximate unit for $C_0(Z)$ constructed as in Example A.5.8. If $f \in \mathcal{D}$ and if $\xi \in \pi(A)H$ is as above, let

$$\pi_1(f)\xi = \sum_{ij} \pi(f * f_{ij} \otimes a_{ij})\xi_i.$$

This is independent of the presentation of ξ, for if

$$\xi = \sum_{\alpha\beta} \pi(f_{\alpha\beta} \otimes b_{\alpha\beta})\zeta_\alpha$$

is another presentation, then $\{a_{ij}, b_{\alpha\beta}\}$ is a finite set of compactly supported functions on Z; hence there is n_0 such that $u_n a_i j = a_{ij}$ and $u_n b_{ij} = b_{ij}$ for all $n \geq n_0$ and all $a_{ij}, b_{\alpha\beta}$, and therefore

$$\sum_{ij} \pi(f f_{ij} \otimes a_{ij})\xi_i = \sum_{ij} \pi(f f_{ij} \otimes u_n a_{ij})\xi_i = \pi(f \otimes u_n)\xi.$$

This shows that, in fact, $\pi_1(f)\xi = \pi(f \otimes u_n)\xi$. Furthermore, the exercise below shows that the operator norm $\|\pi(f \otimes u_n)\|$ is bounded. Therefore $\pi_1(f) = \lim_n \pi(f \otimes u_n)$ extends to a bounded operator on H (*cf.* Theorem A.1.4).

The other properties listed in the statement are easy to verify. □

Exercise 1.5.8. This exercise completes the proof of the lemma above. Let $v_n = u_n^2$, and let $w_n = v_n^2$, which are also approximate units for $C_0(Z)$. Carry out the following.

(1) Let $\xi \in H$ and define $\rho(f) = \langle \pi(f \otimes w_n)\xi | \xi \rangle$, a positive linear functional on $\mathcal{D}$.

(2) Let $\mathcal{N} = \{f \in \mathcal{D} \mid \rho(f^* f) = 0\}$. Then $\mathcal{N}$ is a closed subspace of $\mathcal{D}$, and ρ induces a nondegenerate inner product on the quotient $\mathcal{D}/\mathcal{N}$. Let H_ρ be the Hilbert space completion.

(3) Using the scheme of the Gelfand-Naimark-Segal representation, construct a representation π_ρ of $\mathcal{D}$ on H_ρ that is continuous for the inductive limit topology on $\mathcal{D}$.

(4) Using Proposition 1.5.2, show that

$$\|\pi(fg \otimes v_n)\xi\| \leq \|f\|_2 \|\pi(g \otimes v_n)\xi\|,$$

where $\|f\|_2$ is the norm of $f \in \mathcal{D}$ as a compact operator on $L^2(N)$.

(5) It follows that

$$\|\pi(f^* f \otimes v_n)\| \leq \|f\|_2 \|\pi(f \otimes v_n)\|.$$

(6) Show that $\lim_n \pi_2(v_n) = I$. In particular, $\|\pi_2(v_n)\|$ is bounded.

(7) From the previous fact, and from $\pi(f \otimes ab) = \pi_2(a)\pi(f \otimes b)$ proved in the lemma, conclude that

$$\|\pi(f \otimes u_n)\| \leq \|f\|_2 \|\pi_2(u_n)\|.$$

Corollary 1.5.9. *Let π be a nondegenerate representation of $C_c(Z, \mathcal{D})$ on H. Then π extends to a representation of the C^*-algebra $C_0(Z, \mathcal{K})$, where $\mathcal{K}$ is the algebra of compact operators on $L^2(N)$.*

Proof. The restriction of π to the subspace $\mathcal{D} \otimes_{\mathbb{C}} C_c(Z)$ can be expressed as the product of representations π_1 of $\mathcal{D}$ and π_2 of $C_c(Z)$ on H. By Proposition 1.5.2 and Corollary A.8.3, each extends to representations of $\mathcal{K}(L^2(N))$ and $C_0(Z)$ on H that still commute, and thus define a representation of $\mathcal{K} \otimes C_0(Z)$ on H, as Theorem A.9.2 guarantees. □

Corollary 1.5.10. *The full C^*-algebra of a trivial foliated space $N \times Z$ is the tensor product $\mathcal{K}(L^2(N)) \otimes C_0(Z)$.*

1.5.C. Fibrations. In this example $(M, \mathcal{F})$ is a foliated space whose leaves are the fibers of a locally trivial fibration $p : M \to B$. Thus B has a covering by open sets $\{B_i\}$ so that $p^{-1}(B_i) \cong L \times B_i$. The C^*-algebra of M is built by assembling the C^*-algebras of the trivial foliated spaces $L \times B_i$.

Lemma 1.5.11. *Let X be a manifold and let* Diff(X) *be the group of diffeomorphisms, with the topology of uniform convergence on compact subsets. Then the homomorphism of* Diff(X) *into the group of unitary operators on $L^2(X)$, given by Proposition* 1.1.2, *is continuous for the strong operator topology.*

Proof. A net $\{U_i\}$ of unitary operators on $L^2(X)$ converges to U in the strong operator topology if the operator norm $\|U_i(\xi) - U(\xi)\| \to 0$, for every $\xi \in L^2(X)$.

Let $f_n \to f$ uniformly on compact subsets of X, let $\xi \in L^2(X)$, and let $\varepsilon > 0$. There is a compactly supported half-density φ such that $\|\xi - \varphi\| < \varepsilon$

in $L^2(X)$. Since φ is compactly supported, $f_n^*\varphi \to f^*\varphi$ uniformly; thus $\|f_n^*\varphi - f^*\varphi\| < \varepsilon$ in $L^2(X)$, if $n > n(\varepsilon)$. Therefore

$$\begin{aligned}\|f_n^*\xi - f^*\xi\| &\leq \|f_n^*(\xi - \varphi)\| + \|f_n^*\varphi - f^*\varphi\| + \|f^*(\xi - \varphi)\| \\ &\leq 3\varepsilon,\end{aligned}$$

if $n > n(\varepsilon)$. □

Theorem 1.5.12. *Assume that $(M, \mathcal{F})$ is given by the fibers of a fibration $p : M \to B$ with fiber F. Then $C^*(M, \mathcal{F})$ is isomorphic to $C_0(B) \otimes \mathcal{K}(L^2(F))$.*

Proof. It should be clear from the discussion carried out so far in this section that the C^*-algebra $C^*(M, \mathcal{F})$ is as follows. There is a locally trivial bundle E over B with fiber $\mathcal{K} = \mathcal{K}(L^2(F))$ so that $C_r^*(M, \mathcal{F})$ is isomorphic to the algebra of continuous sections $s : B \to E$ vanishing at infinity on B with the norm $\|s\| = \sup_{x \in B} \|s(x)\|$. The structure of this bundle E will now be examined.

The bundle E is associated to a locally trivial Hilbert bundle $H \to B$ over B with fiber $L^2(F)$ and transition functions given by the cocycle that defines $M \to B$. Indeed, if $\{B_i\}$ is a cover of B that trivializes M, then there is a cocycle

$$g_{ij} : B_i \cap B_j \to \mathrm{Diff}(F)$$

that is continuous for the topology on $\mathrm{Diff}(F)$ of convergence on compact subsets. By the continuity of the representation $\mathrm{Diff}(F) \to U(L^2(F))$, $\{g_{ij}\}$ induces a cocycle for the Hilbert bundle H.

Since the group of unitary operators on an infinite dimensional Hilbert space is contractible (*cf.* [**47**, Lemma 10.8.2]), the Hilbert bundle H is, in fact, trivial, which means that, perhaps after passing to a refinement of the covering, there are continuous maps

$$h_i : B_i \to U(L^2(F))$$

such that $g_{ij} = h_i^* h_j$. This in turn implies that the operator bundle $E \to B$ is, in fact, trivial; that is, it is isomorphic to $\mathcal{K}(L^2(F)) \times B$, and thus the C^*-algebra $C_r^*(M, \mathcal{F})$ is isomorphic to the tensor product $C_0(B) \otimes \mathcal{K}(L^2(F))$. □

Remark. The locally trivial C^*-algebra bundles over a topological space B with fiber $\mathcal{K}$ are classified by a cohomology class in $H^3(B, \mathbb{Z})$, which is called the Dixmier-Douady invariant (*cf.* [**49**], [**47**, Ch. 10, §8]). What was shown above is that the C^*-algebra of a foliated space given by the fibers of a locally trivial fibration over B defines a C^*-algebra bundle over B with trivial Dixmier-Douady invariant.

1.5.D. Group actions. If a connected Lie group H acts locally freely (on the right) on a separable compact space X, then X inherits a foliated space structure whose leaves are homogeneous spaces of H by a discrete subgroup. If moreover the action is free, then the graph of this foliated space is diffeomorphic to the product $X \times G$, with source and range maps $s(x, g) = xg$ and $r(x, g) = x$, and product $(x, g) \cdot (xg, h) = (x, gh)$.

Even if the action is not locally free, the groupoid $X \times G$ still makes sense and it is possible to construct a C^*-algebra out of it [**82**]. More generally, the groupoid $X \times G$ still makes sense when X is a Borel space and the action is Borel. This particular example of groupoid has played a significant role in ergodic theory and representation theory, notably in the work of G. W. Mackey [**124**] and his disciples.

1.5.E. Foliated bundles. Unlike the first three constructions in this section, foliated bundles give a variety of examples of foliated spaces.

A foliated bundle $(M, \mathcal{F})$ over a manifold B with fiber F is determined by the global holonomy representation $H : \pi_1(B) \to \text{Homeo}(F)$ and can be identified with the quotient of $\widetilde{B} \times F$ by the action of $\pi_1(B)$ given by $(b, z) \cdot \gamma = (\gamma z, H(\gamma)(z))$. However, as with group actions, the structure of the graph of a foliated bundle is not readily ascertained from the structure of M given by the global holonomy representation.

Assume that the elements of the representation H are such that if $H(\gamma)$ is the identity on an open set $V \subset F$ and $x \in \overline{V}$, then $H(\gamma)$ is also the identity on a neighborhood of x. This condition guarantees that the holonomy pesudogroup is pesudo-analytic.

Exercise 1.5.13. If H is pseudo-analytic as above, show that the graph G of the foliated bundle M is diffeomorphic to $(\widetilde{B} \times \widetilde{B} \times F)/\Gamma$.

Let X be a locally compact, separable Hausdorff space, and let Γ be a group of homeomorphisms of X. For simplicity it will be assumed that Γ is countable. The bi-invariant Haar measure on the group Γ is the counting measure.

Let $C_c(X)$ denote the algebra of compactly supported, continuous functions on X. If Γ acts on the right on X, the action denoted by $x \mapsto x\gamma$, then there is an induced left action of Γ on $C_c(X)$ given by

$$(\gamma f)(x) = f(x\gamma)$$

The space $X \times \Gamma$ becomes a topological groupoid by defining the source $s(x, \gamma) = x\gamma$ and range $r(x, \gamma) = x$, and the product

$$(x, \gamma_1) \cdot (x\gamma_1, \gamma_2) = (x, \gamma_1\gamma_2).$$

Let $C_c(X \times \Gamma)$ be the space of compactly supported, continuous functions on the product $X \times \Gamma$. This space is endowed with the structure of an involutive algebra by defining a convolution by

$$f_1 * f_2(x, \gamma) = \sum_{\gamma_1\gamma_2=\gamma} f_1(x, \gamma_1) f_2(x\gamma_1, \gamma_2)$$

and an involution by

$$f^*(x, \gamma) = \overline{f(x\gamma, \gamma^{-1})}.$$

These operations are evidently well defined and continuous because they only involve finite sums of compactly supported, continuous functions.

For each $x \in X$ there is a representation of the involutive algebra $C_c(X, \Gamma)$ in the Hilbert space $\ell^2(\Gamma)$ given by

$$R_x(f)\xi(\gamma) = \sum_{\gamma_1\gamma_2=\gamma} f(x\gamma^{-1}, \gamma_1)\xi(\gamma_2).$$

Then the reduced C^*-algebra of the dynamical system (X, Γ) is the completion of $C_c(X, \Gamma)$ under the norm

$$\|f\| = \sup_{x\in X} \|R_x(f)\|.$$

This C^*-algebra is denoted by $C_0(X) \ltimes_r \Gamma$.

The full C^*-algebra is defined by taking the supremum over all the involutive representations of the convolution algebra $C_c(X, \Gamma)$, and is denoted $C_0(X) \ltimes \Gamma$.

If the pair (X, Γ) arises from the global holonomy representation of a foliated bundle $(M, \mathcal{F})$, then it is natural to expect that this C^*-algebra should be related to the foliation C^*-algebra. The precise relation is as follows (although this is not the most general statement that can be obtained).

Theorem 1.5.14. *Let M be a foliated bundle over B with fiber F and global holonomy representation $\Pi = \pi_1 B \to \mathrm{Homeo}(F)$. If the holonomy representation is pesudo-analytic and faithful, then the C^*-algebra $C_r^*(M, \mathcal{F})$ is Morita equivalent (see the following remark) to the crossed product $C_0(F) \ltimes_r \Pi$. Similarly, the full C^*-algebra $C^*(M, \mathcal{F})$ is Morita equivalent to $C_0(F) \ltimes \Pi$.*

Remark. Since the definition and basic properties of Morita equivalence require a somewhat lengthy discussion, we omit this discussion. However, the importance of this concept cannot be overestimated. Morita equivalence for C^*-algebras was introduced by M. A. Rieffel [**156**], and is an equivalence relation between C^*-algebras that says essentially that equivalent algebras have equivalent representation theory. The concept of Morita equivalence is very important because it guarantees that equivalent C^*-algebras have isomorphic K-theory. In fact, two separable C^*-algebras A, B are Morita equivalent if and only if they are stably isomorphic, namely $A \otimes \mathcal{K} \cong B \otimes \mathcal{K}$, where

$\mathcal{K}$ is the algebra of compact operators on a separable Hilbert space [**14**]. From the foliation point of view, it has the immediate advantage of making a simplification of the graph, because integrals can now be replaced by sums and the analysis becomes easier. Perhaps its topological analogue is best exemplified by the relation between the structure of a foliated space and its leaves, and that of one of the associated holonomy pseudogroups.

Example 1.5.15. An example of fundamental importance in foliation C^*-algebras is the noncommutative torus. This is the C^*-algebra generated by two unitary operator S and T subject to the relation $ST = e^{2\pi\sqrt{-1}\theta}TS$, where $\theta \in (0,1)$ is an irrational number. It is Morita equivalent to the C^*-algebra of the foliation of the two-dimensional torus $\mathbb{R}^2/\mathbb{Z}^2$ induced by the foliation of the plane by lines of slope θ.

The construction of the C^*-algebra associated to a discrete group acting on a space extends, with minimal modifications, to the case of a pseudogroup of homeomorphisms of a topological space, as is the case of the holonomy pseudogroup of a foliated space. The following result, due to M. Hilsum and G. Skandalis [**97**], insures that the algebra thus obtained is Morita equivalent to the algebra of the foliated space.

Theorem 1.5.16. *Let $(M, \mathcal{F})$ be a foliated space and let $(Z, \mathcal{H})$ be the holonomy pseudogroup associated to a complete transversal Z. Then the foliation C^*-algebra is Morita equivalent to the pseudogroup C^*-algebra.*

We note, to conclude this section, that there are several papers dealing with the particular structure of the C^*-algebra of certain foliated spaces. Sometimes the C^*-algebra is built up from elementary pieces reflecting similar geometric features of the foliated space. For instance, X. Wang [**181**] studied foliations of the plane, A. M. Torpe [**176**], and T. Fack and X. Wang [**64**] studied Reeb foliations, and one of the authors [**17**] studied related codimension one foliations. At other times the foliation has a structure related to that of a transformation group, and this is reflected in the foliation algebra. For instance, T. Natsume [**140**] studied foliations without holonomy, and H. Takai [**166**] studied Anosov foliations.

1.6. Quasi-invariant Currents

This section discusses some aspects of the general concept of quasi-invariant measures for a foliated space. This will be used to state some results on the representation theory of the C^*-algebra of a foliated space in a later section. A particular kind of quasi-invariant measures, harmonic measures, will receive a detailed treatment in Chapter 2.

Let $(M, \mathcal{F})$ be a foliated space and let G be its graph. Let $\mathcal{D}^1 G$ be the bundle of densities of order one on G. This bundle splits as

$$\mathcal{D}^1 G = s^* \mathcal{D}^1 M \otimes r^* \mathcal{D}^1 M,$$

where $\mathcal{D}^1 M$ is the bundle of densities of order one on M. Thus the fiber at $\gamma \in G$ is the tensor product $\mathcal{D}^1 M_{s(\gamma)} \otimes \mathcal{D}^1 M_{r(\gamma)}$. Compactly supported densities are elements of $\Gamma_c(G, \mathcal{D}^1)$.

Definition 1.6.1. A *current* on the foliated space $(M, \mathcal{F})$ is a positive linear functional m on the space $\Gamma_c(M, \mathcal{D}^1)$ of compactly supported densities on M.

A current induces a (positive) Radon measure on M, and conversely. Indeed, let σ be an everywhere positive density on M and define a positive linear functional I on $C_c^\infty(M)$ by

$$I(f) = \int f(x)\sigma(x) \cdot m(x).$$

Let U be a foliated chart that has compact closure in a larger foliated chart V. Then there exists a smooth function ϕ with support in V such that $0 \leq \phi \leq 1$ and $\phi = 1$ on U. If f is a smooth real valued function with compact support in U, then

$$\pm f(x) \leq \|f\| \phi(x),$$

and so

$$\pm I(f) \leq \|f\| I(\phi).$$

Since every compact subset of M can be covered by finitely many foliated charts like U, and since the compactly supported smooth functions on M are dense in the compactly supported continuous functions, this shows that I extends to a positive linear functional on $C_c(M)$. The Riesz representation theorem then implies that I is represented by a Radon measure on M.

The measure induced by a current depends on the trivializing density chosen, but the equivalence class of the measure does not.

Definition 1.6.2. Two currents on $(M, \mathcal{F})$ are equivalent if the measures that they induce are equivalent (*i.e.*, mutually absolutely continuous).

Two things follow immediately from this discussion. First, a current is defined on all the Borel densities on M, *i.e.*, Borel sections of the density bundle. Second, if two currents m and m' are equivalent, then there is a positive Borel function h on M such that

$$\int_M f \cdot m' = \int_M fh \cdot m$$

for every density f on M. The function h is obtained by applying the Radon-Nikodym theorem to the measures associated to the currents, and a straightforward calculation shows that it is independent (in the almost everywhere sense) of the trivializing density. These comments justify the following definition.

Definition 1.6.3. Let m be a current on $(M, \mathcal{F})$. A Borel subset N of M has m-measure zero if the measure of N is zero with respect to one (hence, any) of the measures associated to m.

The discussion below can be carried out in the context of measures but, not having a natural choice of nowhere vanishing density, it is more canonical to use the language of densities and currents. In a later chapter a class of measures, harmonic measures, will be discussed. In that situation there will be a canonical density, which is the one provided by a metric tensor, and so the language of functions and measures will be adopted.

A current m on M induces two currents $m \circ s_*$ and $m \circ r_*$ on G, that is, two positive linear functionals on $\Gamma_c(G, \mathcal{D}^1)$, as follows. There are natural projections $s_*, r_* : \Gamma_c(G, \mathcal{D}^1) \to \Gamma_c(M, \mathcal{D}^1)$ induced by integration along the fibers of the source and range map. Thus if $f \in \Gamma_c(G, \mathcal{D}^1)$, then

$$s_* f : x \mapsto \int_{G_x} f(\gamma)$$

is a density on M, that can now be integrated with respect to m, the result being $\int_G f \cdot m \circ s_*$, and similarly for $m \circ r_*$.

These currents $m \circ s_*$ and $m \circ r_*$ are related because the involution $i(\gamma) = \gamma^{-1}$ of G exchanges them, namely

$$\int_G f(\gamma^{-1}) \cdot m \circ s_*(\gamma) = \int_G f(\gamma) \cdot m \circ r_*(\gamma).$$

Remark. Two observations need to be made at this point. One of the standard notations for the action of a current m on a density f is $\langle m, f \rangle$. Here both this notation and the measure notation $\int f \cdot m$ will be used.

The second observation concerns the meaning of measure on a non-Hausdorff space such as (generally) the graph. Typically measure theory is developed under the assumption that the underlying space is locally compact and Hausdorff (but see [**69**]). There are no problems in carrying out the measure theory needed to deal with the locally compact Hausdorff spaces appearing here, and unless something requires special attention, no further mention of this point will be made.

There are many positive linear functionals on $\Gamma_c(M, \mathcal{D}^1)$, but not all of them are of interest. For instance, pick $x \in M$ and a $(\dim \mathcal{F})$-vector $v \in \Lambda^{\dim \mathcal{F}} T_x M$. Then the evaluation map $\sigma \mapsto \sigma(v)$ is a current on $(M, \mathcal{F})$.

(In the language of measures, it is equivalent to the point mass measure at x.)

Exercise 1.6.4. Determine the precise situation in which the currents $m \circ s_*$ and $m \circ r_*$ above are equivalent.

Definition 1.6.5. A current m on the foliated space $(M, \mathcal{F})$ is called a *quasi-invariant current* when the currents $m \circ s_*$ and $m \circ r_*$ on G are equivalent. It these currents are equal, then m is called *invariant.*

Hence, if a current m is quasi-invariant, there is a Borel function

$$\delta : G \longrightarrow \mathbb{R}_+,$$

unique in the almost everywhere sense, such that

$$\delta \cdot m \circ s_* = m \circ r_*.$$

Given that G is a groupoid, this Radon-Nikodym derivative is going to satisfy an extra property.

Definition 1.6.6. A Borel function $\delta : G \longrightarrow \mathbb{R}_+$ such that

$$\delta(\gamma_1 \cdot \gamma_2) = \delta(\gamma_1)\delta(\gamma_2)$$

is called a *modular function.*

If m is a current on M and δ satisfies $\delta \cdot m \circ s_* = m \circ r_*$, then δ is called a modular function for m. Thus, if m has a modular function, then it is quasi-invariant. The converse will be the content of Proposition 1.6.13.

Example 1.6.7. A modular function is a Borel homomorphism of G into the multiplicative group of positive real numbers. A geometric way of obtaining such a homomorphism starts with a one-form α on M that is of class C^1 on the leaves but only measurable on M, that is closed and such that, for each leaf L, the homomorphism

$$\int \alpha : \pi_1 L \longrightarrow \mathbb{R}$$

given by integration of α vanishes on loops in L with trivial germinal holonomy. Then the function

$$\gamma \in G \mapsto \exp \int_\gamma \alpha$$

is well defined and has the properties required of a modular function.

Example 1.6.8. Since compactly supported densities can be integrated, a manifold always has a canonical current, called the Lebesgue current and denoted by ℓ.

Let M be a connected manifold, viewed as a foliated space with one leaf. Its graph G is the product $M \times M$ with source $s(x, y) = y$ and range $r(x, y) = x$. If m is a current on M, the currents on G induced by m are

$$m \circ s_* = \ell \otimes m \qquad \text{and} \qquad m \circ r_* = m \otimes \ell.$$

It follows that m is quasi-invariant if and only if the current m and the Lebesgue current ℓ are mutually absolutely continuous; therefore, if and only if there exists an everywhere positive function ϕ on M such that m is of the form

$$\int_M f(x) \cdot m(x) = \int_M f(x)\phi(x) \cdot \ell(x),$$

for every compactly supported density f. A modular function for m is given by

$$\delta(x, y) = \frac{\phi(x)}{\phi(y)}.$$

Exercise 1.6.9. Show that a foliated manifold $(M, \mathcal{F})$ (of class C^1, say) admits a quasi-invariant current whose measure is equivalent to the Lebesgue measure on the manifold M.

Exercise 1.6.10. Let m and m' be two equivalent currents on $(M, \mathcal{F})$. Prove the following.

(1) The current m is quasi-invariant if and only if m' is quasi-invariant.

(2) If δ is a modular function for m, then

$$\delta'(\gamma) = \delta(\gamma)\frac{h(r(\gamma))}{h(s(\gamma))},$$

where h is a Radon-Nikodim derivative of m' with respect to m, is a modular function for m'.

Lemma 1.6.11. *Let $M = L \times Z$ be a trivial foliated space, where Z is a locally compact separable metrizable space and L is a connected manifold. Let m be a current on $L \times Z$. Then the following are equivalent:*

(1) *The current m is quasi-invariant.*

(2) *The current m admits a disintegration of the form*

$$m = \int_Z \lambda_z \, \mu(z),$$

where μ is a measure on Z and λ_z is equivalent to the Lebesgue current on L, for almost all $z \in Z$ with respect to μ.

(3) *There exists a positive Borel function f on $D \times T$ such that the current $\mu' = f \cdot \mu$ is invariant.*

Proof. Because $M = L \times Z$ is locally compact and separable, a current m on $L \times Z$ has an associated measure that is finite. Thus Theorem A.2.9 applies and the current m can be disintegrated: there exist a measure μ on Z and a measurable family of currents β_z on L such that

$$\int f(x,z) \cdot m(x,z) = \int_Z \int_L f(x,z) \cdot \beta_z(x)\mu(x),$$

for every compactly supported half-density f.

If m is a quasi-invariant current, then there exists a function $\phi(x,y,z)$ on the graph $G = L \times L \times Z$ that is positive for almost all z (with respect to μ) and such that

$$\phi \cdot m \circ s_* = m \circ r_*.$$

Writing out the expression for this identity, and utilizing Exercise 1.6.8, it follows that for almost every $z \in Z$, with respect to μ, the current β_z is equivalent to the Lebesgue current on L.

Given (2) and a disintegration of m as $\int \beta_z \mu(z)$, there exists a positive function $f(x,t)$ such that $\beta_z(x) = f(x,z) \cdot \ell(x)$, where ℓ is the Lebesgue current, and it is immediate from the expressions above that $\ell \otimes \mu$ is an invariant current. Thus (3) holds.

The rest of the statement is self-evident. □

Exercise 1.6.12. Let m be a quasi-invariant current on the foliated space $(M, \mathcal{F})$.

(1) Let U be a foliated chart and let $B \subset U$ be a Borel saturated subset of U. Show that the saturation of B is a Borel set.

(2) Let $B \subset U$ be as in (1). Show that if B has measure zero, then the saturation of B also has measure zero.

(3) Show that, in general, the saturation of a Borel subset of M is not necessarily a Borel set.

(4) A *Borel transversal* is a Borel subset of M that meets every leaf in an at most countable set. Show that the saturation of a Borel transversal is a Borel set.

Proposition 1.6.13. *Let m be a current on the foliated space M. The following conditions are equivalent:*

(1) *there exists a modular function $\delta : G \to \mathbb{R}_+$ such that m is of modulus δ;*

(2) *the current m is quasi-invariant.*

Proof. It is evident that (1) implies (2). Let $\mathcal{U} = \{U_i\}$ be a locally finite regular cover of M by foliated charts U_i. The restricted currents $m|U_i$ are quasi-invariant, and, by Lemma 1.6.11, each $m|U_i$ is equivalent to an invariant current m_i for the foliated space $(U_i, \mathcal{F}|U_i)$. Let $\{\phi_i\}$ be a partition of unity subordinated to the cover $\mathcal{U}$ and let m' be the current $m' = \sum_i \phi_i m_i$. This current is equivalent to m; thus by Exercise 1.6.10 it suffices to consider m'.

By Lemma 1.6.11, the invariant current m_i is of the form $\ell \otimes \mu_i$, where ℓ is the Lebesgue current and μ_i is a measure on the transversal Z_i of the foliated chart U_i. Let f_{ij} be a Borel function on $U_i \cap U_j$ that is constant along the plaques and that realizes the Radon-Nikodym derivative $d\mu_i/d\mu_j$.

For each triple of indices i, j, k, the set of points $x \in U_i \cap U_j \cap U_k$ such that $f_{ij}(x) f_{jk}(x) \neq f_{ik}(x)$ is a Borel saturated subset of the intersection (because the functions involved are constant along the leaves) that has m'-measure zero (because Radon-Nikodym derivatives are determined up to measure zero). By Exercise 1.6.12, its saturation X_{ijk} also has measure zero; thus $X = \bigcup_{ijk} X_{ijk}$ is a saturated set of measure zero.

Define $\delta_i : G(U_i) \to \mathbb{R}_+$ by

$$\delta(\gamma) = \begin{cases} \dfrac{\sum_j \phi_j f_{ji}(r(\gamma))}{\sum_j \phi_j f_{ji}(s(\gamma))}, & \text{if } s(\gamma) \in M \smallsetminus X, \\ 1, & \text{if } s(\gamma) \in X. \end{cases}$$

By the cocycle property of the Radon-Nikodym derivatives, $\delta_i = \delta_j$ on $G(U_i) \cap G(U_j)$. Furthermore,

$$\delta_i(\gamma_1 \cdot \gamma_2) = \delta_i(\gamma_1)\delta_i(\gamma_2)$$

for every $\gamma_1, \gamma_2 \in G(U_i)$.

The idea now is to use the fact that every $\gamma \in G$ can be written as a finite product $\gamma = \gamma_1 \cdot \ldots \cdot \gamma_k$, with $\gamma_j \in G(U_{i_j})$ and define $\delta(\gamma) = \delta_{i_1}(\gamma_1) \cdots \delta_{i_k}(\gamma_k)$. For this to be well defined on G, a further subset of measure zero must be removed.

Claim. *There is a Borel saturated set Y of measure zero such that if $x \in M \smallsetminus Y$ and $\gamma_1, \ldots, \gamma_k$ is a finite sequence with $\gamma_j \in G(U_{i_j})$ and $\gamma_1 \cdots \gamma_k$ is the constant path at x, then $\delta_{i_1}(\gamma_1) \cdots \delta_{i_k}(\gamma_k) = 1$.*

Proof of the Claim. For a finite set of indices $I = \{i_1, \ldots, i_k\}$ with $i_1 = i_k$, define δ_I on $G(U_{i_1}) \cdots G(U_{i_k})$ by $\delta_I(\gamma) = \delta_{i_1}(\gamma_1) \cdots \delta_{i_k}(\gamma_k)$ if $\gamma = \gamma_1 \cdots \gamma_k$ with $\gamma_j \in G(U_{i_j})$. The definition of $\delta_I(\gamma)$ is independent of the decomposition of γ because δ_i is a homomorphism on $G(U_i)$.

The set $G(U_{i_1}) \cdots G(U_{i_k})$ is open and invariant under right and left multiplication by $G(U_{i_1}) = G(U_{i_k})$. Therefore, there exists an open saturated subset $V \subset U_{i_1}$ such that $G(U_{i_1}) \cdots G(U_{i_k}) \cap G(U_{i_1}) = G(V)$.

Let $A_I = \{\gamma \in G(V) \mid \delta_I(\gamma) \neq \delta_{i_1}(\gamma)\}$. Then A_I is a Borel set invariant under right and left multiplication by $G(V)$. Thus there is a Borel saturated subset B_I of the foliated space V such that $A_I = G(V) \cap s^{-1}(B_I)$.

If $f \in \Gamma_c(G(V), \mathcal{D}^{1/2})$ is a half-density of the form $f = f_i * \cdots * f_k$ with $f_j \in \Gamma_c(G(U_{i_j}), \mathcal{D}^{1/2})$, then, on the one hand,

$$\int f(\gamma) \cdot m' \circ s_*(\gamma) = \int f(\gamma^{-1})\delta_{i_1}(\gamma^{-1}) \cdot m' \circ s_*(\gamma)$$

and, on the other hand,

$$\int (f_1 * \cdots * f_k)(\gamma) \cdot m' \circ s_*(\gamma) = \int (f_1 * \cdots * f_k)(\gamma^{-1})\delta_I(\gamma^{-1}) \cdot m' \circ s_*(\gamma).$$

It follows that B_I has m'-measure zero, because half-densities of the form considered are dense in the space of continuous half-densities with compact support in $G(V)$ (*cf.* Exercise 1.4.4).

Thus the saturation Y_I of B_I is a Borel set of measure zero, and so is the countable union $Y = \bigcup_I Y_I$. □

Setting

$$\delta(\gamma) = \begin{cases} \delta_{i_1}(\gamma_1) \cdots \delta_{i_k}(\gamma_k), & \text{if } \gamma = \gamma_1 \cdots \gamma_k \notin s^{-1}(Y), \\ 1, & \text{otherwise,} \end{cases}$$

one sees that δ is a modular function for the current m'. Since m is equivalent to m', Exercise 1.6.10 completes the proof. □

If δ is a modular function for G, then δ induces a homomorphism of groups $\delta : G_x^x \to \mathbb{R}_+^*$. If δ is the modular function of a quasi-invariant current, then δ is a Radon-Nikodym derivative and, by manipulations with sets of measure zero, it should then be possible to make δ the trivial homomorphism on each G_x^x, for all x but a measure zero set. This indeed is the case, and the following exercise outlines a proof, following Fack and Skandalis [**63**]. The result is in fact a sort of measure-theoretic counterpart to [I, Theorem 2.3.12].

Exercise 1.6.14. The statement to be proved is the following. Let m be a quasi-invariant current on M with modular function δ. Then there is a modular function δ' such that $\delta'(\gamma) = 1$ for every $\gamma \in G$ with $r(\gamma) = s(\gamma)$, and m is of modulus δ'. Proceed as follows:

Pick a nowhere vanishing density σ on M and interpret m as a measure on M as described above. Then the induced current on G gives rise to measures on G, say ν_s and ν_r, and the Radon-Nikodym derivative $d\nu_r/d\nu_s(\gamma) = \delta(\gamma)$.

(1) Let $N = \{x \in M \mid \delta|G_x^x = 1\}$. Show that N is a Borel saturated subset of M.

(2) Let U be a foliated chart for M and let B be the set of $\gamma \in G$ for which there exists an element $\tau(\gamma) \in G(U)$ with $r(\tau(\gamma)) = r(\gamma)$ and $s(\tau(\gamma)) = s(\gamma)$. Show that B is a Borel subset of G.

(3) If f is a Borel function on G, with compact support, then show that

$$\int_B f(\gamma) \cdot \nu_r(\gamma) = \int_{G(U)} \sum_{g \in G_{r(\gamma)}^{r(\gamma)}} f(g\gamma) \cdot \nu_r(\gamma),$$

and hence that

$$\int_B f(\gamma^{-1})\delta(\gamma^{-1}) \cdot \nu_r(\gamma) = \int_B f(\gamma^{-1})\delta(\tau(\gamma^{-1})) \cdot \nu_r(\gamma).$$

(4) Conclude from (3) that $\delta(\gamma) = \delta(\tau(\gamma))$ for almost every $\gamma \in B$.

(5) Let $B' = \{\gamma \in B \mid \delta(\gamma) \neq \delta(\tau(\gamma))\}$. Show that $\nu_r(B) = 0$ and that B' is both left and right invariant under $G(U)$.

(6) Show that $r(B')$ has m-measure zero. Since $r(B') = N \cap U$, deduce that N also has measure 0.

Modifying δ by setting $\delta' = 1$ on G^N concludes the proof.

Exercise 1.6.15. It is assumed that the reader knows how to construct a closed, nowhere dense subset E of the unit circle having positive Lebesgue measure.

(1) Show that there exists a diffeomorphism f of the circle whose set of fixed points is exactly E.

(2) Let g be an irrational rotation of the circle and let G be the group of diffeomorphisms of S^1 generated by f and g. Show that G preserves the Lebesgue measure class on S^1 and, in fact, that the action is ergodic (that is, the Lebesgue measure $\lambda(B)$ of every G-invariant Borel subset B of S^1 is either $\lambda(B) = 0$ or $\lambda(B) = \lambda(S^1)$).

(3) Use the group G and the method of suspension to construct a foliated three-manifold M with a quasi-invariant current inducing the Lebesgue measure class on M and such that this measure class is quasi-invariant and ergodic, and the union of leaves without holonomy has measure zero.

(4) Conciliate this example with the result outlined in the previous exercise.

On a foliated space $(M, \mathcal{F})$ with leaves of dimension $\dim \mathcal{F}$, an invariant current can be characterized by the property that $\int d\beta \cdot m = 0$, for every twisted $(\dim \mathcal{F}-1)$-form β. The following proposition is an extension of this cohomological fact.

Proposition 1.6.16. *Let m be a quasi-invariant current whose modular function is of the form $\log \delta(\gamma) = \int_\gamma \alpha$, where α is a closed one-form on M. Then*

$$\int d\beta \cdot m = -\int_M \alpha \wedge \beta \cdot m,$$

for every twisted $(\dim \mathcal{F} - 1)$-form $\beta \in \Gamma_c(M, A^{\dim \mathcal{F}-1} \otimes \mathcal{O})$.

Proof. Suppose that β has support in a foliated chart $U = D \times Z$. The current m restricted to U is of the form $\exp(h)\ell \otimes \mu$, where μ is a measure on Z and h is a measurable function on U such that

$$h(x,z) - h(y,z) = \int_y^x \alpha,$$

on almost every plaque of U. Thus h is differentiable along the plaques of U and $dh = \alpha$.

Then, for μ-almost every plaque $D \times \{z\}$,

$$\int_{D\times\{z\}} e^h d\beta = \int_{D\times\{z\}} e^h \alpha \wedge \beta,$$

because $d(e^h\beta) = e^h\alpha\wedge\beta + e^h d\beta$ and β has compact support in $D\times\{z\}$. The identity stated follows by integrating this one with respect to μ on Z. $\square$

Exercise 1.6.17. Let $(M, \mathcal{F})$ be a foliated space with quasi-invariant current m. Suppose that the modular function δ of m is given by integration of a continuous one-form α on M, as in Exercise 1.6.7.

(1) There is a regular cover $\mathcal{U} = \{U_i\}$ of M by foliated charts so that the restriction $\alpha|U_i = d\log h_i$, where h_i is a positive function on U_i, and $h_{ij} = h_i/h_j$ is constant along the plaques of $U_i \cap U_j$.

(2) Let $\widetilde{M} = M \times \mathbb{R}_+$. Then $\widetilde{M}$ has a foliated structure given by foliated charts $V_i = U_i \times \mathbb{R}_+$ and coordinate changes given by

$$(x_j, t, z_j) = (x_i, t\,h_{ij}, z_i).$$

(3) Show that $\widetilde{M}$ has a quasi-invariant current of the form $m\otimes\ell$, where ℓ is integration along the fibers of the projection $\widetilde{M} \to M$.

(4) Show that the graph $G(\widetilde{M})$ can be identified with $G(M)\times\mathbb{R}_+$, with source and range maps given by

$$s(\gamma,t) = (s(\gamma, t\delta(\gamma)) \qquad \text{and} \qquad r(\gamma,t) = (r(\gamma,t))$$

and product

$$(\gamma_1, t_1)\cdot(\gamma_2, t_2) = (\gamma_1\cdot\gamma_2, t_1).$$

In the following proposition, the word "transversal" refers to a transversal Z coming from a foliated chart $U \cong D \times Z$.

Proposition 1.6.18. *A quasi-invariant current on M induces a Radon measure on each transversal. This measure is not unique, but any two are in the same measure class. Moreover, the action of the holonomy pseudogroup preserves this measure class.*

Conversely, a Radon measure on the transversals, with the property that its measure class is invariant under the action of the holonomy pseudogroup, induces a quasi-invariant current on M.

Proof. Let Z be a transversal with $U \cong D \times Z$ a corresponding foliated chart. The current restricted to U decomposes as a product of a multiple of the Lebesgue current and a measure on Z, which is easily seen to be a Radon measure. The equivalence of these measures under the action of the holonomy pseudogroup follows from the quasi-invariance of m.

Conversely, suppose there is given a Radon measure on the transversals, quasi-invariant under the holonomy pseudogroup. Let $\mathfrak{U} = \{U_i\}$ be a locally finite regular cover of M by relatively compact foliated charts. Each U_i is diffeomorphic to the trivial foliated space $D \times Z_i$, for some locally compact metrizable space Z_i. Because the cover is regular, each Z_i is a relatively compact subset of a compact transversal. Let μ_i be the finite Radon measure on Z_i provided by the hypothesis. Let ϕ_i be a partition of unity subordinated to $\mathfrak{U}$. If σ is a compactly supported density on M, then $\phi_i \sigma$ is a compactly supported density on the foliated chart U_i. Define

$$\int \sigma \cdot m = \sum_i \int_{Z_i} \int_D \phi_i \sigma \cdot \mu_i.$$

Then m is a positive linear functional on compactly supported densities on M, that is, a current.

It remains to show that this current is quasi-invariant, but this is exactly like Proposition 1.6.13, using the fact that transverse measures μ_i are equivalent under the holonomy pseudogroup acting on the transversals Z_i. □

Exercise 1.6.19. Let $\mathfrak{B}$ denote the collection of Borel transversals in the foliated space $(M, \mathcal{F})$ (see Exercise 1.6.12 for the definition). Let m be a quasi-invariant current on M.

(1) Show that $\mathfrak{B}$ is a σ-ring of subsets of M.

(2) Show that m induces a measure on each Borel transversal.

(3) Show that if m is invariant, then it induces a measure on the σ-ring $\mathfrak{B}$.

The final result of this section interprets the concept of quasi-invariant current for a foliated space at the level of its C^*-algebra.

Theorem 1.6.20. *Let M be a foliated space with quasi-invariant current m. Then m induces a semi-continuous weight φ on the C^*-algebra $C_r^*(M, \mathcal{F})$, by setting*

$$\varphi(f) = \int_M \operatorname{tr} \delta^{-1} R_x(f) \cdot m(x),$$

for $f \in \Gamma_c(G, \mathcal{D}^{1/2})$. If the transverse current m is invariant, then φ is a trace.

This result is one of the ingredients of the foliations index theorem of Connes [**34**]. A weight on a C^*-algebra A is a function $\varphi : A^+ \to [0, \infty]$ such that $\varphi(a + b) = \varphi(a) + \varphi(b)$ and $\varphi(\lambda a) = \lambda\varphi(a)$ for all $a, b \in A^+$ and all $\lambda \in \mathbb{R}_+$, where $A^+ = \{a \in A \mid a = bb^* \text{ for some } b \in A\}$ is the cone of positive elements of A, and a trace is a weight such that $\varphi(a^*a) = \varphi(aa^*)$ (see [**145**]). The other unexplained symbol is $\operatorname{tr} \delta^{-1} R_x(f)$, which is the local trace of the operator $\delta^{-1} R_x(f)$. The proof is not difficult at this stage, but as it will not be required it is omitted.

1.7. Representations of the Foliation C^*-algebra

Let $(M, \mathcal{F})$ be a foliated space and let G be its graph. A representation of G consists of a family of Hilbert spaces $\{H_x\}_{x \in M}$ and a collection of maps $\{U(\gamma)\}_{\gamma \in G}$ such that

(1) $U(\gamma) : H_{s(\gamma)} \to H_{r(\gamma)}$ is a unitary operator,
(2) $U(\gamma_1 \cdot \gamma_2) = U(\gamma_1)U(\gamma_2)$,
(3) $U(\gamma^{-1}) = U(\gamma)^*$.

Basic examples of representations of the graph are the following.

Example 1.7.1. The *trivial representation* of G. The family of Hilbert spaces is $H_x = \mathbb{C}$ for every $x \in M$, and $U(\gamma) : \mathbb{C} \to \mathbb{C}$ is the identity for every $\gamma \in G$.

Example 1.7.2. The *regular representation* of G. The Hilbert space over $x \in M$ is $H_x = L^2(G^x)$, and G acts on this family by right translation:

$$[U(\gamma)(\xi)](\gamma') = \xi(\gamma^{-1} \cdot \gamma').$$

For a group S, there is a correspondence between unitary representations of S and nondegenerate representations of the full C^*-algebra $C^*(S)$. If ds denotes left-invariant Haar measure on S, then a unitary representation

$$\pi : S \to \mathcal{B}(H)$$

induces a representation of the C^*-algebra $C^*(S)$ by integration: if $f \in C_c(S)$ and $\xi \in H$, then

$$\pi(f)(\xi) = \int_S f(s)\pi(s)\xi \cdot ds.$$

Based on the analogy of the graph of a foliated space with a group, in order to promote a representation of the graph to a representation of the C^*-algebra $C^*(M, \mathcal{F})$, an extra piece of machinery is needed. This piece is going to be a quasi-invariant current m on M (which, in the group case, is the point mass measure at the identity).

Let $(M, \mathcal{F})$ have quasi-invariant current m with modular function δ, let $\{H_x\}$ be a Borel field of Hilbert spaces on M, and construct the field $\mathcal{D}_x^{1/2} \otimes H_x$ over M, where $\mathcal{D}_x^{1/2} = (\mathcal{D}^{1/2}M)_x$. A section of this field assigns to a point $x \in M$ a half-density with coefficients in H_x. Thus, given sections ξ and ζ of $\{\mathcal{D}_x^{1/2} \otimes H_x\}$, the pointwise inner product $\langle \xi_x | \zeta_x \rangle$ is a density on M, which can be integrated with respect to the current m. Let H be the Hilbert integral of $\{\mathcal{D}_x^{1/2} \otimes H_x\}$ with respect to the quasi-invariant current m,

$$H = \int_M^{\oplus} \mathcal{D}_x^{1/2} \otimes H_x \cdot m(x).$$

Let U be a representation of the graph G on $\{H_x\}$, as defined at the beginning of this section.

This data induces a representation of $C^*(M, \mathcal{F})$ on H, which is constructed as follows. If $f \in \Gamma_c(G, \mathcal{D}^{1/2})$ and $\xi \in \{\mathcal{D}_x^{1/2} \otimes H_x\}$, then

$$(\pi(f)\xi)_x = \int_{G^x} f(\gamma)\delta(\gamma)^{-1/2}U(\gamma)(\xi_{s(\gamma)}).$$

This is well defined, because for fixed $x \in M$ the assignment

$$\gamma \mapsto f(\gamma)\delta(\gamma)^{-1/2}U(\gamma)(\xi_{s(\gamma)})$$

is a density on G^x with coefficients on $\mathcal{D}_x^{1/2} \otimes H_x$, and so the integral over G^x can be taken and results in an element of H evaluated at x.

It is fairly straightforward to verify that π satisfies the following identities:

(1) $\pi(f_1 * f_2) = \pi(f_1)\pi(f_2)$, and
(2) $\pi(f^*) = \pi(f)^*$,

so that π is a representation of $\Gamma_c(G, \mathcal{D}^{1/2})$ on H. This representation is continuous for the inductive limit topology on the graph of each foliated chart, and thus it induces a representation of the full C^*-algebra $C^*(M, \mathcal{F})$.

The following theorem shows that the general representation of $C^*(M, \mathcal{F})$ is of this form.

Theorem 1.7.3. *Let π be a nondegenerate representation of $C^*(M, \mathcal{F})$ on a Hilbert space H. Then there exist a quasi-invariant current m with modular function δ and a representation U of G on a field of Hilbert spaces $\{H_x\}_{x \in M}$ such that:*

(1) *H is the Hilbert integral*

$$H = \int_M^{\oplus} \mathcal{D}_x^{1/2} \otimes H_x \cdot m(x),$$

and

(2) *for every $f \in C_c(G, \mathcal{D}^{1/2})$ and $\xi \in H$,*

$$\pi(f)(\xi)_x = \int_{G^x} f(\gamma)\delta(\gamma)^{-1/2} U(\gamma)\xi_{s(\gamma)}.$$

The proof of this result becomes rather technical and will not be given in full. It is proved in Fack and Skandalis [**63**]. The result is, in fact, valid for more general groupoids (*cf.* Hahn [**88**] and Renault [**155**]), but the particular local structure of the graph of a foliated space makes the proof more elementary.

Here are the two basic examples illustrating this theorem. Let $(M, \mathcal{F})$ have a quasi-invariant current m with modular function δ.

Example 1.7.4. *The trivial representation.* Let $\{H_x\}$ be the trivial field of Hilbert spaces $H_x = \mathbb{C}$, and let the action of the graph be the identity $U(\gamma) = \mathrm{id} : \mathbb{C} \to \mathbb{C}$. Then H is identified with $L^2(M, m)$, the space of half-densities on M that are square-integrable with respect to m. That is, $L^2(M, m)$ is the completion of $\Gamma_c(M, \mathcal{D}^{1/2})$ with respect to the inner product

$$\langle \xi | \zeta \rangle = \int_M \xi(x)\zeta(x) \cdot m(x).$$

The action of $f \in \Gamma_c(G, \mathcal{D}^{1/2})$ on $\xi \in L^2(M, m)$ is given by

$$\pi(f)\xi_x = \int_{G_x} f(\gamma)\delta^{-1/2}(\gamma)\xi_{s(\gamma)}.$$

Example 1.7.5. *The regular representation.* In this example the field $H_x = L^2(G^x)$ and the action of $\gamma \in G$ is given by $U(\gamma)(\xi)(\gamma') = \xi(\gamma^{-1} \cdot \gamma')$ for $\xi \in L^2(G^x)$. The representation is

$$[\pi(f)\xi]_x(\gamma') = \int_{G^x} f(\gamma)\delta(\gamma)^{-1/2}\xi_{s(\gamma)}(\gamma^{-1} \cdot \gamma')$$

for $f \in \Gamma_c(G, \mathcal{D}^{1/2})$.

If $m \circ r_*$ is the current induced on G by m, then the direct integral $H = \int^{\oplus} \mathcal{D}_x^{1/2} \otimes L^2(G^x) \cdot m(x)$ can be identified with $L^2(G, m \circ r_*)$. To a vector field $x \in M \mapsto \xi_x \in L^2(G^x)$ it assigns the element $\xi(\gamma) = \xi_{r(\gamma)}(\gamma)$, and conversely.

Exercise 1.7.6. This exercise gives a nicer expression for the regular representation of Example 1.7.5.

(1) If δ is the modular function of the quasi-invariant current m, show that the mapping $\xi \mapsto \delta^{1/2}\xi$ is an isometry from $L^2(G, m \circ r_*)$ onto $L^2(G, m \circ s_*)$.

(2) Show that the isometry obtained in (1) conjugates the regular representation constructed in Example 1.7.5 to the representation of $C^*(M, \mathcal{F})$ on $L^2(G, m \circ s_*)$ given by left convolution: $\xi \mapsto f * \xi$, for $f \in \Gamma_c(G, \mathcal{D}^{1/2})$.

Exercise 1.7.7. If $x \in M$, then R_x is a representation of $C^*(M, \mathcal{F})$ (*cf.* Section 1.4). Determine the quasi-invariant current and field of Hilbert spaces that corresponds to it via the above theorem.

The beginning of the proof of Theorem 1.7.3 will be given below, and it consists in studying the local situation, namely the case of a trivial foliated space (this is the case that will be needed in the next section). The general situation proceeds by appropriately gluing these local pieces.

Proposition 1.7.8. *Let $M = N \times Z$ be a trivial foliated space, where N is a connected manifold and Z is a locally compact, second countable space. Let π be a nondegenerate representation of $C^*(M, \mathcal{F})$. Then there exist an invariant current m on M and a field of Hilbert spaces over M so that π is equivalent to a representation of the form given in the conclusion of Theorem* 1.7.3.

Proof. The C^*-algebra $C^*(M, \mathcal{F})$ is isomorphic to $\mathcal{K} \otimes C_0(Z)$, where $\mathcal{K}$ is the algebra of compact operators on the Hilbert space $L^2(N)$. If π is a nondegenerate representation of $C^*(M, \mathcal{F})$ on a Hilbert space, then, by Theorem A.9.2, π induces nondegenerate, commuting representations π_1 of $\mathcal{K}$ and π_2 of $C_0(Z)$ on the same Hilbert space such that $\pi(f \otimes a) = \pi_1(f)\pi_2(a)$, for $f \otimes a \in \mathcal{K} \otimes C_0(Z)$.

By Corollary A.8.8, the representation π_2 is equivalent to the multiplication representation on a field of Hilbert spaces over Z. Hence there are a measure μ on Z and a field of Hilbert spaces $\{K_z \mid z \in Z\}$ such that π_2 takes the form $\pi_2(a)\xi = a\xi$, for $a \in C_0(Z)$ and $\xi \in \int^{\oplus} K_z \cdot \mu$.

Since the representation π_1 commutes with π_2, it is then equivalent to a representation of $\mathcal{K}$ on $\int^{\oplus} K_z \cdot \mu$ which commutes with all the multiplication operators. Therefore, by Theorem A.3.16, each operator $\pi_1(f)$ is decomposable, that is, of the form $\{\pi_{1,z}(f)\}$, with $\pi_{1,z}$ a representation of $\mathcal{K}$ on the Hilbert space K_z.

Because of the structure of the representations of the algebra $\mathcal{K}$ described in Proposition 1.5.2, there exists a measurable function

$$\phi : Z \to \{1, 2, \cdots, \infty\}$$

so that, for $z \in \{\phi = n\}$, $\pi_{1,z}$ is equivalent to the direct sum of n copies of the canonical representation of $\mathcal{K}$ on $L^2(N)$. It follows that there is a field of Hilbert spaces over Z so that the representation π_1 is equivalent to the diagonal representation on the Hilbert integral of the field $L^2(N) \otimes J_z$ with respect to the measure μ, where J_z is a Hilbert space of dimension $n(z)$.

The Hilbert space $L^2(N)$ is canonically isomorphic to the Hilbert integral $\int_N^{\oplus} \mathcal{D}_w^{1/2} \otimes \mathbb{C} \cdot \ell(w)$, where ℓ is the Lebesgue current on N. Thus, letting m be the invariant current $m = \ell \otimes \mu$, and $\{H_x\}$ be the field $H_{(w,z)} = \mathbb{C} \otimes J_z$, it follows that the representation π is equivalent to the representation of $C^*(M, \mathcal{F})$ on the Hilbert integral $\int_M^{\oplus} \mathcal{D}_x^{1/2} \otimes H_x \cdot m(x)$ as described by Theorem 1.7.3. □

The following observation is immediate from the structure of the representation described in the proof of the above proposition; it will be used explicitly in the next section.

Corollary 1.7.9. *Let π be a representation of the C^*-algebra $C^*(M, \mathcal{F})$ of a trivial foliated space $M = N \times Z$. Then there is an open subset W of Z such that $\pi(f) = 0$ for every $f \in \Gamma_c(M, \mathcal{D}^{1/2})$ with support in $N \times W$.*

There is another way in which operator algebras appear in dynamical systems and ergodic theory. These operator algebras are the von Neumann algebras, or rings of operators (*cf.* Appendix A for the definition), and in fact their presence in these areas dates to the first papers of F. J. Murray and J. von Neumann. The survey article by C. Moore [**133**] and the book by A. Connes [**38**] have a detailed discussion of these matters, as well as references to the main researches in the topic. The original definition of the von Neumann algebra of a foliation appears in Connes [**33**].

As was described in Example 1.7.5, a quasi-invariant current m on the foliated space $(M, \mathcal{F})$ induces a representation of the full C^*-algebra $C^*(M, \mathcal{F})$. This is the regular representation, and it is induced by the action of $f \in \Gamma_c(G, \mathcal{D}^{1/2})$ on $\xi \in L^2(G, m \circ s_*)$ given by left convolution.

Definition 1.7.10. The von Neumann algebra $W(M, \mathcal{F}, m)$ is the algebra generated by the regular representation of $C^*(M, \mathcal{F})$ associated to the quasi-invariant current m.

That is, $W(M, \mathcal{F}, m)$ is the weak closure of the operators $\xi \mapsto f * \xi$ on $L^2(G, m \circ s_*)$ given by left convolution by elements $f \in \Gamma_c(G, \mathcal{D}^{1/2})$.

Certain von Neumann algebras are broadly classified into three classes, types I, II and III, according to certain structural properties (which would take too long to describe here). A general von Neumann algebra W decomposes as a direct sum $W = W_{\mathrm{I}} \oplus W_{\mathrm{II}} \oplus W_{\mathrm{III}}$ of von Neumann algebras of the respective types. A von Neumann algebra is a factor if its center is trivial. A fundamental theorem is that a von Neumann algebra can be decomposed as a direct integral of factors. Factors are divided into three main groups, types I, II, and III (and each in turn into several subclasses). Examples of factors arising from foliations are as follows: the von Neumann algebra of a foliation consisting of exactly one compact leaf is a factor of type I, the von Neumann algebra of the irrational flow on the two-dimensional torus is a factor of type II, and that of the stable foliation of the geodesic flow on the unit tangent bundle of a compact surface of constant negative curvature is a factor of type III.

There is a dynamical partition of a foliated space with quasi-invariant current (or, more generally, of a measured equivalence relation) that mimics this qualification of factors. A quasi-invariant current decomposes M into ergodic components, and each ergodic component gives rise to a von Neumann algebra via the regular representation. An ergodic component is of type I if there is a Borel transversal that intersects almost every leaf in a single point. It is called of type II if the quasi-invariant current is equivalent to an invariant one, finite or infinite, but does not have an essential saturated set of type I, and otherwise it is called of type III.

Remark. The von Neumann algebra of a foliated space with a quasi-invariant ergodic current is a factor at least when the set of leaves with holonomy has measure 0; a proof is given in Hahn [**88**] in the general context of measure groupoids, *cf.* also Takesaki [**169**]. Chapter V.7 of [**168**] discusses this result for the von Neumann algebra associated to the action of a discrete group on a measure space.

The type of an ergodic quasi-invariant current is determined by a certain cohomology class. Let m have modular function δ, let $m \circ s_*$ be the induced current on G, and let $Z(G)$ be the space of Borel cocycles $\sigma : G \to \mathbb{R}_+^*$. A cocycle σ is a coboundary if there is a Borel map $f : G \to \mathbb{R}_+^*$ such that

$$\delta(\gamma) = \frac{f(r(\gamma))}{f(s(\gamma))}$$

almost everywhere with respect to $m \circ s_*$. The cohomology $H^1(G, m)$ is the quotient of cocycles modulo coboundaries.

Thus, an ergodic m is of type I if it is supported in a single leaf; if it is not of type I, then it is either of type II or type III according to whether the cohomology class of the modular function δ is trivial or not in $H^1(G, m)$.

The interplay between the geometry of a foliation and the type of its von Neumann algebra was analyzed in a series of papers by J. Heistch, S. Hurder and A. Katok, culminating in the paper by Hurder and Katok [**103**]. An illustration of this work is the following remarkable theorem of Hurder, showing the geometric significance of this ergodic concept for foliations.

Theorem 1.7.11. *If $(M, \mathcal{F})$ is a foliated manifold, of class C^2, endowed with the Lebesgue measure class, and the Godbillon-Vey class is not zero, then the von Neumann algebra contains a nontrivial component of type* III.

This theorem is proven by analyzing the Radon-Nikodym derivative cocycle given by an equivalent invariant current, and a related result will be described in Section 7.1 (Theorem 7.1.20). At the basis of this result is the work of W. Krieger [**115**] on ergodic theory of nonmeasurable equivalence relations, and the work of G. Duminy [**52**] on the Godbillon-Vey class. Connes [**37**] further elaborates on the matter, in relation to cyclic cohomology.

Another result in the same topic is the following, obtained by one of the authors [**18**].

Theorem 1.7.12. *If M has an ergodic harmonic measure that is not totally invariant, then $(M, \mathcal{F}, m)$ is of type* III.

The proof of this theorem also involves an analysis of cocycles, although this time they take place at a different level, namely, over the semiflow of Brownian motion, and the analysis is somewhat more "tangential." As we mention several times, the geometric output of the nonvanishing of the Godbillon-Vey class is the existence of a resilient leaf. It is also possible, by analyzing this Brownian flow, to show that the nonvanishing of certain "characteristic classes" implies the existence of a resilient leaf for quasi-conformal foliations.

1.8. Minimal Foliations and their C^*-algebras

The main result of this section is a theorem of Fack and Skandalis [**63**], stated immediately below, describing the structure of the C^*-algebra of a minimal foliated space (that is, a foliated space all of whose leaves are dense). Throughout this section there is one standing assumption: *the graph of the foliated space $(M, \mathcal{F})$ under consideration is Hausdorff.* The theorem to be proven is the following.

Theorem 1.8.1. *Let $(M, \mathcal{F})$ be a foliated space (with Hausdorff graph G). Then $(M, \mathcal{F})$ is minimal if and only if the C^*-algebra $C_r^*(M, \mathcal{F})$ is simple.*

The proof requires several preliminary facts. Minimality of the foliated space is not required for them, but the Hausdorff property of the graph is still required (although not for all of them).

Lemma 1.8.2. *Let $Y \subset M$ and $x \in M$. Then R_x is weakly contained in $\{R_y \mid y \in Y\}$ if and only if x belongs to the closure of the saturation of Y.*

Proof. Let W be the closure of the union of the leaves through the points of Y. If $x \notin W$, then there is an element $f \in \Gamma_c(G, \mathcal{D}^{1/2})$ such that f is not zero at the constant path at x, but $f(\gamma) = 0$ for every $\gamma \in W$. Then $R_x(f) \neq 0$ but $R_y(f) = 0$ for every $y \in Y$, and so, by Definition A.6.14, R_x is not weakly contained in $\{R_y\}_{y \in Y}$.

Conversely, suppose that $x \in W$. Then there is a sequence of points $z_n \to x$ with $z_n \in L_{y_n}$, $y_n \in Y$. For $f \in \Gamma_c(G, \mathcal{D}^{1/2})$ and $\xi \in \Gamma_c(G, \mathcal{D}^{1/2})$, $R_{s(\gamma)}(f)\xi(\gamma) = f * \xi(\gamma)$ is a smooth density on G. Since G is Hausdorff, the density is continuous in the usual sense, and so $\langle R_w(f)\xi|\xi\rangle$ is a continuous function of $w \in M$. Thus

$$\lim_n \langle R_{z_n}(f)\xi|\xi\rangle = \langle R_x(f)\xi|\xi\rangle,$$

and the result follows because the collection $\{\xi|G_x \mid \xi \in \Gamma_c(G, \mathcal{D}^{1/2})\}$ is dense in $L^2(G_x)$. □

Remark. If the graph G is not Hausdorff, then it is still true that if R_x is weakly contained in $\{R_y \mid y \in Y\}$ then x belongs to the closure of the saturation of Y, but the converse is not necessarily true.

Corollary 1.8.3. *Let π be a representation of $C_r^*(M, \mathcal{F})$. If $\|\pi(f)\| \geq \|R_x(f)\|$ for every x in a leaf without holonomy, then $\|\pi(f)\| \geq \|f\|$ for every $f \in C_r^*(M, \mathcal{F})$.*

Proof. The union of leaves without holonomy is a dense saturated subset of M (*cf.* [I, Theorem 2.3.12]). Hence the result is immediate from the previous lemma. □

Lemma 1.8.4. *If π is a representation of $C_r^*(M, \mathcal{F})$ such that, for each foliated chart U, the restriction of π to $C_r^*(U, \mathcal{F}|U)$ is faithful, then $\|f\| \geq \|R_x(f)\|$ for every point x in a leaf without holonomy.*

The proof of this lemma requires two preliminary facts, which are given as Claims 1 and 2 below.

Let x be a point in a leaf without holonomy and let $U = D \times Z$ be a foliated chart around x.

Claim 1. *Let $K \subset G$ be a compact set. Then there exists an open saturated subset V of U such that $x \in V$ and $K \cap G_V^V \subset G(U)$.*

Proof. Suppose that there is a sequence $\gamma_n \in K \cap G_U^U$ such that $r(\gamma_n) \to x_r$ and $s(\gamma_n) \to x_s$, for some points x_r and x_s in the plaque through x in U, and that $\gamma_n \notin G(U)$. Composing on the right and on the left with elements of $G(U)$, we obtain a sequence $\gamma_n' \in G(U) \cdot K \cdot G(U)$ such that the sequences $r(\gamma_n')$ and $s(\gamma_n')$ both converge to x, while still satisfying $\gamma_n' \notin G(U)$. Since $G(U) \cdot K \cdot G(U)$ is relatively compact, it may be assumed that this sequence γ_n' converges to certain element $\gamma \in G$. This γ belongs to G_x^x and is represented by the constant path at x because the leaf through x has no holonomy. In particular, $\gamma \in G(U)$. On the other hand, by construction, γ is a limit point of the complement of $G(U)$. This is a contradiction, because $G(U)$ is a neighborhood of the constant path at x. □

Let D_x be the plaque of U through x, and let $\xi \in L^2(G_x)$ be a smooth half-density with compact support in D_x and such that $\int_{D_x} |\xi|^2 = 1$. For $f \in \Gamma_c(G(U), \mathcal{D}^{1/2})$, set

$$\varphi(f) = \langle R_x(f)\xi | \xi \rangle.$$

Then φ is the state (Definition A.5.9) associated to the unit vector ξ with respect to the regular representation R_x of $\Gamma_c(G(U), \mathcal{D}^{1/2})$ on $L^2(D_x)$.

Claim 2. *The state φ has a unique extension to a state $\widetilde{\varphi}$ on $C_r^*(M, \mathcal{F})$. Furthermore,*

$$\widetilde{\varphi}(f) = \langle R_x(f)\xi | \xi \rangle,$$

for every $f \in C_r^*(M, \mathcal{F})$.

Proof. Let ψ be any extension of φ to $C^*(M, \mathcal{F})$ (such an extension always exists [**145**, Proposition 3.1.6]) and let $(\pi_\psi, H_\psi, \xi_\psi)$ be the Gelfand-Naimark-Segal representation associated to ψ (*cf.* Theorem A.6.21).

Let $f \in \Gamma_c(G, \mathcal{D}^{1/2})$ and let V be the open saturated set obtained by applying Claim 1 to the compact set $\operatorname{supp} f$. Then $V = D \times Z'$ for some open subset $Z' \subset Z$. Let h be a continuous function with compact support on Z', with $0 \le h \le 1$, and such that $h(z_x) = 1$, where z_x is the point of Z' representing the plaque through x in V. Then $g(y_1, y_2, z) = \xi(y_1)\overline{\xi(y_2)}h(z)$ is an element of $\Gamma_c(G(U), \mathcal{D}^{1/2})$. By construction, g is of unit norm in $C_r^*(U, \mathcal{F}|U)$ and, because of Proposition 1.5.5, it is also of unit norm in

$C_r^*(M, \mathcal{F})$. Then $\psi(g) = \varphi(g) = 1$, and thus $\pi_\psi(g)\xi_\psi = \xi_\psi$. Therefore

$$\begin{aligned}\psi(f) &= \langle \pi_\psi(f)\xi_\psi | \xi_\psi\rangle \\ &= \langle \pi_\psi(g * f * g)\xi_\psi | \xi_\psi\rangle \\ &= \varphi(g * f * g) \\ &= \langle R_x(g * f * g)\xi | \xi\rangle,\end{aligned}$$

as desired. This string of identities is straightforward, except perhaps for the third one. However, an examination of the construction given in the proof of Claim 1 shows that $g * f * g$ belongs to $\Gamma_c(G(U), \mathcal{D}^{1/2})$, where $\psi = \varphi$. □

Proof of Lemma 1.8.4. Let π be a representation of $C_r^*(M, \mathcal{F})$ such that its restriction to the foliation algebra $C_r^*(U, \mathcal{F}|U)$ of a foliated chart U is faithful. Let x be a point in U in a leaf without holonomy (such a point always exists because the union of leaves with holonomy is nowhere dense), and let φ be the state on $C_r^*(U, \mathcal{F}|U)$ constructed in Claim 2 above. Since π is injective on $C_r^*(U, \mathcal{F}|U)$, the composition $\varphi \circ \pi^{-1}$ is defined and is a state on the C^*-algebra $\pi(C_r^*(U, \mathcal{F}|U))$ (by Theorem 1.5.5, π is an isometry). It extends to a state ψ on $\pi(C_r^*(M, \mathcal{F}))$. Since π is norm-decreasing and injective on $C_r^*(U, \mathcal{F}|U)$, $\psi \circ \pi$ is a state on $C_r^*(M, \mathcal{F})$ that extends φ. In view of Claim 2, the associated Gelfand-Naimark-Segal representation $\pi_\psi \circ \pi^{-1}$ of $C_r^*(M, \mathcal{F})$ is contained in the representation R_x, and since R_x is irreducible (Lemma 1.4.11), they are equivalent. □

Proof of Theorem 1.8.1. By definition, a C^*-algebra is simple if it has no closed, two-sided, proper ideals. Because of Exercise A.6.13, this is equivalent to saying that every nonzero representation is faithful.

Suppose that there is a leaf L that is not dense in M. Then there are a point $x \in M \setminus \overline{L}$, a foliated chart U around x that does not meet $\overline{L}$, and so an element $f \in \Gamma_c(G, \mathcal{D}^{1/2})$ that is positive and has support in $G(U)$. The corresponding representation R_x has the property that $R_x(f) \neq 0$, but R_x has nontrivial kernel, and thus $C_r^*(M, \mathcal{F})$ is not simple.

Conversely, assuming that M is minimal, it will be shown that, if π is a representation of $C_r^*(M, \mathcal{F})$, then, for each foliated chart U in M, the restriction $\pi | C_r^*(U, \mathcal{F}|U)$ is injective.

Let X denote the set of points $x \in M$ for which there is a neighborhood V_x of x in G so that $\pi(f) = 0$ for every f with support in $G(V_x)$. By construction, this set X is open in M. Furthermore, given a foliated chart $U = D \times Z$ for M, the description of the representations of the C^*-algebra $\mathcal{K} \otimes C^*(Z)$ given in Corollary 1.7.9 implies that there exists an open saturated set $V \subset U$ such that the kernel of the representation π is contained in

$C^*_r(V, \mathcal{F}|V)$. Therefore, $X \cap U = V$, which means that X is saturated in each foliated chart, and hence in M.

Since $\Gamma_c(G, \mathcal{D}^{1/2})$ is generated by $\Gamma_c(G(V), \mathcal{D}^{1/2})$, V a foliated chart, there exists a foliated chart V such that $\pi|C^*(V, \mathcal{F}|V) \neq 0$. Thus $X \neq M$, and so, by minimality, $X = \emptyset$ and $\pi|C^*(U, \mathcal{F}|U)$ is injective.

By Lemma 1.8.4, the representation π satisfies $\|\pi(f)\| \geq \|R_x(f)\|$ for every $f \in \Gamma_c(G, \mathcal{D}^{1/2})$ and every point x in a leaf without holonomy. Consequently, Corollary 1.8.3 implies that $\|f\| \leq \|\pi(f)\|$ for every $f \in C^*_r(M, \mathcal{F})$, from which it follows that π is injective. □

Exercise 1.8.5. Let M be a foliated space, $N \subset M$ a proper closed saturated subset, and let $U = M \smallsetminus N$ and $G(N) = G(M) \smallsetminus G(U)$.

(1) Show that $G(N)$ is a closed subgroupoid of the graph of M, but that it is not necessarily diffeomorphic to the graph of N as a foliated space in its own right. When are they diffeomorphic?

(2) Show that the collection of elements $f \in \Gamma_c(G, \mathcal{D}^{1/2})$ that restrict to zero on $G(N)$ generates a closed two-sided ideal of $C^*_r(M, \mathcal{F})$.

Remark. As the authors of [**63**] note, the full C^*-algebra of a minimal foliated space is not necessarily simple. Their pertinent example is as follows. Let Π be a discrete subgroup of $\mathrm{PSl}(2, \mathbb{R})$ that acts on the upper half-plane $\mathbf{H}$ with cocompact orbit space. Let $\Pi \to \mathrm{SU}(2)$ be an injective homomorphism with dense image, and form the space $M = (\mathbf{H} \times \mathrm{SU}(2))/\Pi$ that is foliated by upper half-planes $\mathbf{H}$. The transverse structure corresponds to the action of Π on $\mathrm{SU}(2)$; hence M is a minimal foliated space, and so the C^*-algebra is simple. On the other hand, Subsection 1.5.E implies that the C^*-algebra $C^*(M, \mathcal{F})$ is Morita equivalent to that obtained from the action on $\mathrm{SU}(2)$, and since Γ is not amenable, this is not simple.

Fack and Skandalis [**63**] also described the ideal structure of the C^*-algebra of a foliated space with a dense leaf. Their result and its proof are outlined in the next exercise.

Exercise 1.8.6. A C^*-algebra is *primitive* if it admits a faithful irreducible representation.

The following steps outline a proof of the fact that the C^*-algebra of a foliated space is primitive if and only if the foliated space is *transitive* (*i.e.*, it has a dense leaf).

(1) Show that, if the foliated space $(M, \mathcal{F})$ has a dense leaf, then it also has a dense leaf without holonomy.

(2) Show that, if the leaf through $x \in M$ is dense in M and has no holonomy, then the regular representation R_x is faithful (and irreducible).

(3) If $(M, \mathcal{F})$ is not transitive, then show that there exist two disjoint, nonempty, open saturated subsets U_1, U_2 of M.

(4) For U_1, U_2 as in (3), show that, if π is an irreducible representation of $C_r^*(M, \mathcal{F})$, then π is trivial either on $\Gamma_c(G(U_2), \mathcal{D}^{1/2})$ or on $\Gamma_c(G(U_2), \mathcal{D}^{1/2})$.

Further results on how the dynamical structure of a foliated space is reflected in the ideal structure of its C^*-algebra were obtained by Fack [**62**]. We also note that C^*-algebras had appeared in dynamical systems prior to the works described here, specifically as operator algebras associated to tranformation groups [**82**]. There were moreover numerous papers dealing with the ideal structure of transformation group C^*-algebras. A sample of references (without any pretense of being exhaustive and offering apologies for not being so) includes E. G. Effros and F. Hahn [**58**], G. Zeller-Meier [**192**], J.-L. Sauvageot [**162**], E. C. Gootman and J. Rosenberg [**83**], and D. P. Williams [**184**]. The presence of holonomy, however, gives a distinctive flavor to the foliated space theory.

Chapter 2

Harmonic Measures for Foliated Spaces

In comparison to the ergodic theory for flows, the ergodic theory for foliations is at a rather underdeveloped stage. One reason for this is that foliations that have invariant measures are rather scarce. Garnett [**77**] introduced another type of measure for a foliation, proved that such measures always exist, and exhibited the basic facts of ergodic theory with respect to them. These measures, called harmonic measures, will be examined in this chapter.

A justification for this new concept and its relevance to the ergodic theory of foliations is as follows. When comparing a foliation with a classical dynamical system, leaves are the analogues of orbits. As seen when studying ends, it is possible to develop a concept for going to infinity along a leaf that is suitable for geometric applications. On the other hand, ergodic theorems are finer in the sense that they provide not only asymptotic information, but also metric information. For this, some kind of parameterization of the orbits of the dynamical system is required.

A way of dealing with this initial problem is to notice that a Riemannian manifold has a canonically associated semi-flow, namely the Brownian flow on the space of paths, that gives a well defined sense for going to infinity (although the flow lines may not actually go to infinity). These flows are then tied together via the ambient space, making it possible to determine important asymptotic properties of the leaves from the behavior of the Brownian flow. The way to tie together these flows, which occur independently along all leaves, is to use a measure on the ambient space that bears a relation to these independent processes. Such measures are precisely the harmonic measures. Since these measures generalize holonomy-invariant measures, but always exist, they allow for a natural setting in which to develop the ergodic theory of foliated spaces.

Locally, a harmonic measure m decomposes into a measure on a transversal and a measure along the plaques. The plaque measure is not only smooth, but it is the product of a positive harmonic function times the Riemannian volume of the plaque. While generally the associated transverse measure is not holonomy-invariant, the harmonic measure m on M is invariant under the leafwise heat diffusion associated to a Riemannian metric.

All of this relies heavily on the theory of heat diffusion and Brownian motion on complete Riemannian manifolds of bounded geometry. For the reader's convenience, a detailed review of this theory is offered in Appendices B and C.

In this chapter, measurability of a mapping on a topological space or of a subset of such a space refers to Borel measurability. The measures considered on locally compact Hausdorff spaces, barring some obvious or stated indication to the contrary, are Radon measures. The basic definitions are collected in Section A.2.

2.1. Existence of Harmonic Measures

Let $(M, \mathcal{F})$ be a foliated space. It is always possible to endow M with a smooth leafwise metric tensor. In a foliated chart $U = D \times Z$ for $(M, \mathcal{F})$, with coordinates $(x, z) = (x^1, \ldots, x^n, z)$ ($n = \dim \mathcal{F}$ being the leaf dimension), such a metric tensor g has the local expression

$$g = \sum_{i,j=1}^{n} g_{ij}(x^1, \ldots, x^n, z)\, dx^i \otimes dx^j,$$

where the matrix of smooth functions (g_{ij}) is symmetric and positive definite. If (g^{ij}) denotes the inverse of this matrix and $|g|$ its determinant, then the (leafwise) Laplacian $\triangle = \operatorname{div} \circ \operatorname{grad}$ has the local expression

$$\triangle f = \frac{1}{\sqrt{|g|}} \sum_{j=1}^{n} \frac{\partial}{\partial x^j} \left(\sum_{i=1}^{d} g^{ij} \sqrt{|g|} \frac{\partial}{\partial x^i} f \right)$$

(*cf.* formula (B.1)). Thus,

$$\triangle = \sum_{i,j=1}^{n} g^{ij} \frac{\partial^2}{\partial x^j \partial x^i} + \text{ first order terms},$$

and $\triangle$ is an elliptic operator that annihilates constants.

The metric tensor g on $(M, \mathcal{F})$ induces a metric tensor $g|_L$ on each leaf L of M, and thus a corresponding Laplacian $\triangle_L = \triangle|_L$. If f is a function on

M that is of class C^2 on each leaf, then $\triangle f$ is the aggregate of the leafwise Laplacians $\triangle_L f$. Thus $\triangle$ is defined (at least) on the continuous functions f on M that are of class C^2 on each leaf and are such that $\triangle f$ is a continuous function on M.

Definition 2.1.1. Let $(M, \mathcal{F})$ be a foliated space with metric tensor g and corresponding Laplacian $\triangle$. A measure m on M is said to be *harmonic* if

$$\int_M \triangle f(x) \cdot m(x) = 0,$$

for every $f \in C(M)$ that is of class C^2 along each leaf and is such that $\triangle f$ is continuous on M.

Thus harmonic measures are related to the differential operator $\triangle$ in a manner analogous to the way in which transverse, holonomy-invariant measures are related to the exterior derivative operator d.

The metric tensor on M induces a measure on each leaf, called the leafwise Riemannian measure vol. This is the collection of Borel measures vol_L on the leaves L of $\mathcal{F}$ which are uniquely determined by the requirement that, on each bounded domain $D \subset L$, $\mathrm{vol}_L(D)$ is the Riemannian volume of D, computed as the integral over D of the volume density of $g|L$. In integrals it will be denoted simply by dx.

Example 2.1.2. Let $M = L \times Z$, where L is a complete Riemannian manifold and Z is a locally compact metric space. The foliation $\mathcal{F}$ will be the product foliation with leaves $L \times \{z\}$, $z \in Z$. Let ν be a Radon measure on Z and let $h : L \to \mathbb{R}$ be a positive harmonic function. Let m be the measure on M given by

$$\int f(x,z) \cdot m(x,z) = \int f(x,z)h(x) \cdot dx \otimes \nu(z),$$

for all continuous functions $f : M \to \mathbb{R}$ with compact support. If f is of class C^2 on each leaf and $\triangle f$ is continuous on M, then

$$\begin{aligned}\int_M \triangle f(x) \cdot m(x) &= \int_Z \left(\int_L \triangle f(x,z)h(x) \cdot dx \right) \cdot \nu(z) \\ &= \int_Z \left(\int_L f(x,z)\triangle h(x) \cdot dx \right) \cdot \nu(z) = 0,\end{aligned}$$

by the Green-Stokes formula (Exercise B.1.4). That is, m is a harmonic measure.

Example 2.1.3. Let A be a 2×2 matrix of positive integers with $\det A = 1$ and $\operatorname{tr} A > 2$. The eigenvalues of this matrix are positive real numbers

λ and λ^{-1}. Choose this data so that $\lambda > 1$, and let v and w denote nontrivial eigenvectors of λ and λ^{-1}, respectively. Consider the group G of transformations of $\mathbb{R}^3 = \mathbb{R}^2 \times \mathbb{R}$ generated by

$$\begin{aligned} \varphi_v &: (u,t) \mapsto (u+v,t), \\ \varphi_w &: (u,t) \mapsto (u+w,t), \\ \varphi_A &: (u,t) \mapsto (Au,t+1). \end{aligned}$$

The quotient $M = \mathbb{R}^3/G$ is the suspension of $f_A : T^2 \to T^2$, the diffeomorphism induced by the unimodular matrix A. The flow Φ_t on M produced by this suspension is one of the standard examples of an "Anosov flow".

Introduce new coordinates on $\mathbb{R}^3$ by

$$(sv + rw, t) \leftrightarrow (s, y = \lambda^t, r).$$

The level sets $r = r_0$ are canonically identified with the upper half-plane $\mathbf{H}$ of complex numbers $z = s + iy$, $y > 0$. This identifies $\mathbb{R}^3$ with $\widetilde{M} = \mathbf{H} \times \mathbb{R}$, defining there a foliation $\widetilde{\mathfrak{F}}$ by leaves $\mathbf{H} \times \{r\}$, and these leaves are given the standard hyperbolic metric. The function $h(s,y,r) = y$ is harmonic along each leaf relative to the hyperbolic metric. If Ω denotes the 2-form on $\widetilde{M}$ that defines $\widetilde{\mathfrak{F}}$ and restricts to the hyperbolic volume form on each leaf, the 3-form $\omega = h\Omega \wedge dr$ is invariant by the action of G. Indeed,

$$\begin{aligned} \varphi_v(s,y,r) &= (s+1,y,r), \\ \varphi_w(s,y,r) &= (s,y,r+1), \\ \varphi_A(s,y,r) &= (\lambda s, \lambda y, \lambda^{-1} r) \end{aligned}$$

and the assertion follows easily. A similar calculation shows that the leafwise hyperbolic metric itself is G-invariant. Thus, the volume form ω induces a G-invariant measure $\widetilde{m}$ on $\widetilde{M}$. Let $D \subset \mathbf{H}$ be an open hyperbolic disk of finite radius, let $U = D \times (a,b)$, and let f be any $\widetilde{m}$-integrable function supported in U. Then

$$\int f \cdot m = \int_a^b \left(\int_{D\times\{r\}} f(z,r)h(z,r) \cdot dz \right) \cdot dr,$$

where dz denotes the measure of the hyperbolic metric on $\mathbf{H}$, and dr denotes Lebesgue measure on (a,b).

The foliation $\widetilde{\mathfrak{F}}$ is invariant under G, and so passes to a foliation $\mathfrak{F}$ of M. (The reader may recognize this as the unstable foliation $\mathfrak{F}^u$ on M of the Anosov flow Φ_t.) A foliated atlas $\mathfrak{U}$ for $\mathfrak{F}$ is obtained by choosing foliated charts for $\widetilde{\mathfrak{F}}$ of the form $U = D \times (a,b)$ such that the projection of $\widetilde{M}$ onto M carries U diffeomorphically to a foliated chart for $\mathfrak{F}$. The G-invariant measure $\widetilde{m}$ induces a measure m on M that, relative to the foliated atlas $\mathfrak{U}$, has the local form in Example 2.1.2. Thus, the condition in Definition 2.1.1

is satisfied for functions f with compact support in one or another of these foliated charts. This extends to arbitrary functions by means of a partition of unity, proving that m is a harmonic measure for the foliation $\mathcal{F}$. The roles of the eigenvectors v and w can be interchanged and, when this is done, a harmonic measure for the stable foliation $\mathcal{F}^s$ of Φ_t is obtained.

Exercise 2.1.4. Find a solvable Lie group structure on $\widetilde{M} = \mathbb{R}^3$ relative to which the measure $\widetilde{m}$ in Example 2.1.3 is Haar measure.

Exercise 2.1.5. Let M be the foliated space of [I, Example 11.3.20].

(1) Let Π be the group generated by two elements α, β, subject to the relation $\alpha\beta\alpha^{-1} = \beta^2$. Let $\mathbf{H}$ be the hyperbolic plane (viewed as the upper half-plane) and let $\mathbb{Q}_2$ be the field of dyadic numbers. Show that $\alpha(z, a) = (2z, 2a)$ and $\beta(z, a) = (z+1, a+1)$ define an action of Π on $\mathbf{H} \times \mathbb{Q}_2$ and that the quotient space is diffeomorphic to M.

(2) Let ℓ denote the measure induced by the hyperbolic metric of constant curvature -1 on $\mathbf{H}$, and let da denote Haar measure on the topological group $\mathbb{Q}_2$, normalized so that the measure of the unit ball $\mathbb{Z}_2$ is 1. Also let $\Im(z)$ denote the imaginary part of z. Show that the measure $\Im(z)\ell(z) \otimes da$ induces a harmonic measure m on M.

(3) Show that the total measure $m(M) = \log 2$.

The existence of harmonic measures for a compact foliated space is a consequence of ellipticity of the Laplacian and of the Hahn-Banach theorem. A different method, using a fixed point theorem, will be given later in an exercise. However, this presupposes a fairly deep analysis of the process of diffusion along the leaves of a foliated space.

Lemma 2.1.6. *Let X be a manifold with metric tensor h, and let $\triangle$ be the associated Laplacian. If f is a function on X that is of class C^2 and $y_0 \in X$ is a local maximum of f, then $\triangle f(y_0) \le 0$. Similarly, if y_0 is a local minimum of f, then $\triangle f(y_0) \ge 0$.*

Proof. Suppose that y_0 is a local maximum of f. In local coordinates about y_0, the Laplacian $\triangle f$ can be written as

$$\triangle f(x) = \sum_{i,j} h^{ij}(x) \frac{\partial^2 f}{\partial x^j \partial x^i}(x) + \text{ first order terms,}$$

where the matrix (h^{ij}) is symmetric and positive definite. A suitable choice of the coordinates guarantees that $(h^{ij}(y_0))$ is the identity matrix. Furthermore, since $\triangle$ annihilates constants and all first order derivatives of f vanish at y_0, we see that

$$\triangle f(y_0) = \sum_i \frac{\partial^2 f}{\partial x^i \partial x^i}(y_0) \leq 0,$$

the inequality following from the fact that the restriction of f to each coordinate axis through y_0 has a local maximum at y_0. The second assertion is proven similarly. □

When M is compact, $C(M)$, the space of (real valued) continuous functions on M, is a (real) Banach space with the uniform norm. Let $\mathbf{1}$ denote the constant function $\mathbf{1}(x) = 1$ on M.

Corollary 2.1.7. *Let M be a compact foliated space with metric tensor and corresponding Laplacian $\triangle$. Then the closure of the range of $\triangle$ in $C(M)$ does not contain the constant function $\mathbf{1}$.*

Proof. The range of $\triangle$ is, by definition, the collection of continuous functions on M that are of the form $\triangle f$, for f continuous on M and of class C^2 on each leaf.

If there is a sequence $\{f_n\}_{n=1}^{\infty}$ of such functions such that $\triangle f_n$ converges in $C(M)$ to the constant function $\mathbf{1}$, then there is an index n_0 such that $\triangle f_n \geq 1/2$ on M, for all $n \geq n_0$. On the other hand, since M is compact, Lemma 2.1.6 implies that $\triangle f_n \leq 0$ somewhere, a contradiction. □

The version of the Hahn-Banach theorem that is needed here is the following.

Theorem 2.1.8. *Let E be a linear subspace of the Banach space $C(M)$, and let $\phi : E \to \mathbb{R}$ be a linear functional of norm λ. Then ϕ extends to a continuous linear functional $C(M) \to \mathbb{R}$ with the same norm as ϕ.*

Proof. The fact that ϕ has norm λ on E implies that $|\phi(f)| \leq \lambda \|f\|$ for all $f \in E$. Thus ϕ is a uniformly continuous function on E, and it then admits a unique extension to the closure of E, which is obviously linear and of the same norm as ϕ.

By the above, it may be assumed that E is a closed subspace of $C(M)$. If $f \notin E$, then ϕ extends to the subspace $E + \mathbb{R}f$ by setting $\phi'(h + tf) = \phi(h)$. It is obvious that ϕ' is linear and of the same norm as ϕ. Let $Q = \{f_n\}$ be a countable dense subset of $C(M)$ and inductively apply this argument to the elements of Q that do not belong to E. The result is a linear extension of ϕ to a linear functional defined on a dense subspace of $C(M)$. The argument of the first paragraph gives the conclusion. □

Lemma 2.1.9. *A continuous linear functional $\phi : C(M) \to \mathbb{R}$ is given by the integral with respect to a probability measure m on M if and only if $\|\phi\| = 1$ and $\phi(\mathbf{1}) = 1$.*

Proof. Assume that $\|\phi\| = 1$ and $\phi(\mathbf{1}) = 1$. By the Riesz representation theorem (Theorem A.2.6), it is only necessary to show that $\phi(f) \geq 0$ whenever f is nonnegative. No generality is lost if it is assumed that $f(x) \leq 1$, for every $x \in M$. Then $h = 2f - \mathbf{1}$ satisfies $-\mathbf{1} \leq h \leq \mathbf{1}$ everywhere on M, and $\|\phi\| = 1$ implies that $|\phi(h)| \leq 1$. Thus $-1 \leq \phi(h)$, implying that $0 \leq 2\phi(f)$. The converse is immediate. □

Theorem 2.1.10. *A compact foliated metric space, endowed with a metric tensor, always admits a harmonic probability measure.*

Proof. Let $H \subset C(M)$ denote the closure of the range of $\triangle$ in the uniform norm, and let $a = \inf_{f \in H} \|\mathbf{1} - f\|$. Corollary 2.1.7 implies that $a > 0$. But in fact $a = 1$; indeed, $a \leq 1$ because H is a subspace, and $a \geq 1$ because Lemma 2.1.6 implies that, for every function g which is continuous and C^2 along the leaves, $\mathbf{1} - \triangle g \leq 1$ somewhere on M.

Let $\phi : H + \mathbb{R}\mathbf{1} \to \mathbb{R}$ be the linear functional defined by $\phi(h + t\mathbf{1}) = t$. If $v = h + t\mathbf{1} \in H + \mathbb{R}\mathbf{1}$ and $t = 0$, then $|\phi(v)| = 0 \leq \|v\|$. If $t \neq 0$, then

$$|\phi(v)| = |\phi(h + t\mathbf{1})| = |t| \leq |t|(\|(1/t)h + \mathbf{1}\|) = \|v\|.$$

Thus, the Hahn-Banach theorem gives a linear extension

$$\Phi : C(M) \to \mathbb{R}$$

of ϕ such that $|\Phi(g)| \leq \|g\|$, for all $g \in C(M)$. Also, $\Phi(\mathbf{1}) = 1$, and so $\|\Phi\| = 1$. By Lemma 2.1.9, Φ is the integral associated to a probability measure on M. Since $\Phi|H = \phi|H \equiv 0$, this measure is harmonic. □

Exercise 2.1.11. In the 30's the paper [**116**] by N. Kryloff and N. Bogoliouboff, showing the existence of invariant measures for flows on compact metric spaces, was a landmark. Their proof proceeded by showing that the sequence of Cesaro averages of a point evaluation measure converges to an invariant measure (mimicking the ergodic theorem). A different proof, following the outline of Theorem 2.1.10, is as follows.

(1) Let M be a compact metrizable space that has a one-dimensional foliation $\mathfrak{L}$. Show that, if the tangent bundle $T\mathfrak{L}$ has a nowhere vanishing section, then the foliation can be described by a topological action $\mathbb{R} \times M \to M$, and conversely.

(2) As you should know, not every one-dimensional foliation admits such a vector field. In any case, show that there is a finite cover N of M such that the lifted foliation does.

(3) Let $\phi : \mathbb{R} \times N \to N$ be the flow (without fixed points) describing the foliation of N. Show that $A : C^\infty(N) \to C^\infty(N)$, defined by

$$Af(x) = \lim_{t \to 0} \frac{f(\phi(x,t)) - f(\phi(x,0))}{t},$$

is well defined and vanishes on the functions that are constant along leaves.

(4) Prove that there is a probability measure m on N such that, for all $f \in C^\infty(M)$, $\int Af \cdot m = 0$. Show that

$$\int_N f(\phi(x,t)) \cdot m(x) = \int_N f(x) \cdot m(x),$$

for all $t \in \mathbb{R}$, which means that m is invariant under the flow. Use m to construct a transverse invariant measure for the foliated space $(M, \mathfrak{L})$.

Shortly after Kryloff and Bogoliuboff, A. Weil [**182**] (second part of the talk) gave a proof of the existence of an invariant measure for a map that is somewhat related to the proof given here for the existence of harmonic measures. The details are in the following exercise.

Exercise 2.1.12. Let T be a continuous map of a compact Hausdorff space X. A probability measure μ on X is invariant under T if

$$\int_X f(T(x)) \cdot \mu(x) = \int_X f(x) \cdot \mu(x),$$

for every continuous function on X.

(1) Show that the set of continuous functions of the form $f \circ T - f$ is a linear subspace of $C(X)$.

(2) Because every continuous function on X has a maximum and a minimum, show that the function $\mathbf{1}$ is at distance ≥ 1 from the subspace of (1).

(3) Use the Hahn-Banach theorem to construct an invariant measure μ.

2.2. The Diffusion Semigroup

While the existence of harmonic measures required only elementary calculus and soft analysis, the analysis of their structure requires the introduction of the semigroup of operators associated to the Laplacian, hence some hard(er) analysis.

If M is a compact foliated space with metric tensor g, then each leaf L with the metric $g|L$ is a complete Riemannian manifold of bounded geometry. That is, there is a lower bound for both the injectivity radius and the sectional curvature. (This lower bound is, in fact, independent of the leaf.)

On a manifold L of bounded geometry and with Laplacian Δ, the heat diffusion is introduced as follows. If f is a bounded continuous function on L, the heat equation on L with initial conditions f asks for a bounded solution $u \in C^{2,1}(L \times (0,\infty))$ to the parabolic differential equation

$$\frac{\partial}{\partial t}u(x,t) = \triangle u(x,t)$$

such that, uniformly on compact subsets of L, $\lim_{t\to 0} u(x,t) = f(x)$ (which is abbreviated by writing $u(x,0) = f(x)$). It is a fundamental theorem (Theorem B.6.8) that such a solution exists and is unique.

The heat equation on $(L, g|L)$ admits a fundamental solution, called the heat kernel. This is a function $p(x,y;t)$ that, for each $y \in L$, satisfies

$$\frac{\partial}{\partial t}p(x,y;t) = \triangle_x p(x,y;t)$$

and has the property that, if f is a bounded function on L, then

$$D_{L,t}f(x) = \int_L f(y)p(x,y;t) \cdot dy$$

is the bounded solution to the heat equation on L with initial conditions f. These operators $D_{L,t}$ form what is called the semigroup of diffusion operators of the manifold (L,g).

The aggregate of the various semigroups $D_{L,t}$, L a leaf of M, thus defines a semigroup D_t of operators on functions on M.

Definition 2.2.1. The semigroup D_t is called the diffusion semigroup of (M,g).

Thus, if f is a suitable function on the foliated space M, then $D_t f$ is defined to be the function that, at the point $x \in M$, has the value prescribed by the diffusion of f on the leaf L_x through x, *i.e.*,

$$D_t f(x) = \int_{L_x} f(y)p(x,y;t) \cdot dy,$$

where $p(x,y;t)$ is the heat kernel of the Riemannian manifold $(L_x, g|L_x)$.

Since the metric and Laplacian depend continuously on the transverse coordinate, it is reasonable to expect that $D_t f$ would have all the continuity properties that f has on M. Some concepts from the theory of semigroups of operators will be helpful in order to show that this is the case. K. Yosida [**191**] and E. Hille and R. S. Phillips [**96**] are thorough references for this material.

Definition 2.2.2. Let $(M,\mathcal{F})$ be a compact foliated space, and let $C(M)$ denote the Banach space of continuous functions with the supremum norm. A semigroup of operators on $C(M)$ is a family $T_r : C(M) \to C(M)$, $r \geq 0$, of positive linear operators on $C(M)$ (*i.e.*, $T_r f \geq 0$ if $f \geq 0$) with the following properties.

(1) For all $r \geq 0$, $\|T_r\| \leq 1$ and $T_0 =$ identity.
(2) For all $r, s \geq 0$, $T_{r+s} = T_r \circ T_s$.
(3) For every function $f \in C(M)$, the limit $\lim_{t\to 0}\|T_r f - f\| = 0$.

Remark. It makes perfect sense, of course, to consider the more general case of a noncompact foliated space. In this case, $C(M)$ is replaced by $C_0(M)$. The discussion and results below also apply, with some modifications, to this situation.

Exercise 2.2.3. Show that a semigroup of operators on $C(M)$ is a continuous homomorphism of the semigroup of nonnegative real numbers into the semigroup of bounded operators $C(M) \to C(M)$ of norm ≤ 1, the space of operators being endowed with the strong topology.

Associated to a semigroup of operators $\{T_r\}_{r\geq 0}$ as in the definition, there is the *infinitesimal generator*, which is the operator A defined by

$$Af = \lim_{t\to 0} \frac{1}{t}(T_t f - f),$$

when the function f is such that the above limit exists uniformly on M. The collection of all such functions f forms a linear subspace of $C(M)$, called the domain of A. In general, the domain of A is not the full Banach space $C(M)$, and it may be difficult to describe explicitly. The following exercise illustrates this concept.

Exercise 2.2.4. Let $\phi : \mathbb{R} \times M \to M$ be a continuous flow on a compact metrizable space M. Define a semigroup T_t on $C(M)$ by setting

$$T_t f(x) = f(\phi(t, x)).$$

(1) Verify that T_t has all the properties required in order to be a semigroup of operators on $C(M)$.
(2) Identify the infinitesimal generator, A, of this semigroup.
(3) Show that the domain of A is a proper, dense, linear subspace of $C(M)$.

The definition of semigroup implies that the function

$$r \in [0, \infty) \to T_r f \in C(M)$$

is norm continuous.

It will be helpful for the discussion that follows to recall two operations on vector valued functions $[a, b] \to C(M)$.

The first one is differentiation: a map $\phi : [a, b] \to C(M)$ is differentiable at $s \in (a, b)$ if the limit

$$\lim_{t\to 0} \frac{\phi(t+s) - \phi(s)}{t}$$

exists in $C(M)$. That is to say, there is a function $f \in C(M)$ such that the family of functions $(1/t)(\phi(t+s) - \phi(t))$ converges uniformly to f on M as $t \to 0$. The function $\phi : (a,b) \to C(M)$ is differentiable on (a,b) if it is differentiable at every point of (a,b). Its derivative is denoted by

$$\frac{d}{dt}\phi.$$

Right (respectively, left) derivatives at a (respectively, b) are also defined in the obvious fashion, giving the notion of differentiability on $[a,b)$, $(a,b]$ or $[a,b]$.

The second operation is integration: if $\phi : [a,b] \to C(M)$ is a continuous function, then there is an element of $C(M)$, denoted by

$$\int_a^b \phi(t) \cdot dt.$$

The definition is evident if ϕ is constant, and in general the integral is defined by approximating ϕ uniformly by piecewise constant functions (much the same as the definition of the Riemann integral of a continuous real valued function).

Standard theorems of calculus of real valued functions continue to hold true for vector valued functions. Two basic facts are as follows.

Proposition 2.2.5. *Let $\phi : [a,b] \to C(M)$ be a function.*

(1) *If ϕ is continuous, then*

$$\|\int_a^b \phi(t) \cdot dt\| \leq \int_a^b \|\phi(t)\| \cdot dt.$$

(2) *If ϕ is differentiable with continuous derivative $\frac{d}{dt}\phi$ on $[a,b]$, then $\frac{d}{dt}\phi(t)$ is integrable and, for $s \in [a,b]$,*

$$\phi(s) - \phi(a) = \int_a^s \frac{d}{dt}\phi(t) \cdot dt.$$

There are several facts regarding the infinitesimal generator A of a semigroup of operators T_r on $C(M)$ that will greatly simplify some calculations below. The proofs are part of the basic general theory of semigroups of operators, and they can be found in Appendix B (in a slighly different context).

Proposition 2.2.6. *Let T_r be a semigroup of operators on $C(M)$ with infinitesimal generator A and domain $\mathcal{D}_A$. Then $\mathcal{D}_A$ is invariant under T_r and*

$$AT_r = T_r A$$

on $\mathcal{D}_A$.

Proof. The proof given for Proposition B.10.10 also applies here. □

Proposition 2.2.7. *If $f \in \mathcal{D}_A$, then the function*

$$r \in [0, \infty) \to T_r f \in C(M)$$

has continuous derivative on $[0, \infty)$ and

$$\frac{d}{dr} T_r f = A T_r f.$$

Furthermore,

$$T_r f - f = \int_0^r \frac{d}{ds} T_s f \, ds.$$

Theorem 2.2.8. *The domain of the infinitesimal generator of a semigroup of operators is dense in $C(M)$.*

Proof. See Proposition B.10.10. □

Corollary 2.2.9. *If T_r is a semigroup acting on $C(M)$ with infinitesimal generator A, then $\bigcup_{r>0} T_r(\mathcal{D}_A)$ is dense in $\mathcal{D}_A$.*

Proof. By Proposition 2.2.6, $T_r(\mathcal{D}_A) \subset \mathcal{D}_A$ and, if $f \in C(M)$, then $f = \lim_{t \to 0} D_t f$; hence the result. □

A family of operators $T_t : C(M) \to C(M)$ is strongly continuous if $\|T_{t+h} f - T_t f\| \to 0$ as $h \to 0$, for each $f \in C(M)$, and it is strongly differentiable if the limit

$$\lim_{h \to 0} \frac{T_{t+h} f - T_t f}{h}$$

exists for every t (one-sided derivatives are defined accordingly).

If T_t is strongly continuous, then the integral $g = \int_a^b T_t f \cdot dt$ can be defined for each $f \in C(M)$. Then

$$\|g\| \leq \int_a^b \|T_t f\| \cdot dt \leq \|f\| \int_a^b \|T_t\| \cdot dt.$$

If T_t is a strongly continuous family of operators, then the function $t \mapsto \|T_t f\|$ is continuous. In general, the map $t \mapsto \|T_t\|$ is not necessarely continuous, but it is always locally bounded and lower semicontinuous, hence integrable. Therefore $f \mapsto g = \int_a^b T_t f \cdot dt$ defines a bounded linear operator $\int_a^b T_t \cdot dt$ on $C(M)$ with norm $\leq \int_a^b \|T_t\| \cdot dt$. This operator has the following properties.

(1) $(\int_a^b T_t \cdot dt) x = \int_a^b T_t x \cdot dt$

(2) The strong derivative $(d/dt) \int_a^t T_s \cdot ds = T_t$.

Definition 2.2.10. The convolution of two strongly continuous families of operators S_t and T_t on $C(M)$, with $t \geq 0$, is the family of operators $(S \times T)_t$, $t \geq 0$, defined by

$$(S \times T)_t f = \int_0^t S_{t-s}(T_s f) \cdot ds.$$

Compare Example A.4.7, where a general construction of involutive algebras is described.

Lemma 2.2.11. *The convolution of two strongly continuous families of operators is a strongly continuous family of operators.*

Proof. It is evident that the convolution of two strongly continuous families of operators takes continuous functions to continuous functions. Indeed, for $f \in C(M)$, the function

$$s \in [0, t] \mapsto A_{t-s} B_s f$$

with values in the Banach space $C(M)$ is continuous. Thus it is integrable, and the result is a continuous function of t. □

Lemma 2.2.12. *Convolution is an associative operation. That is, if R_t, S_t and T_t are three strongly continuous families of operators, then*

$$((R \times S) \times T)_t f = (R \times (S \times T))_t f,$$

for all $f \in C(M)$.

Proof. Note that by a change of variable we get

$$(A \times B)_t = \int_0^t A_s B_{t-s} \cdot ds,$$

and that the operators A_t commute with the integrals. The rest of the details are left as an exercise. □

With these basic concepts of functional analysis out of the way, the main theorem of this section can be stated and proved.

Theorem 2.2.13 (Continuity of Diffusion). *Let M be a compact foliated space with metric tensor g and associated Laplacian Δ. Then the leafwise diffusion operators $\{D_{t,L}\}$ coalesce to define a semigroup of operators D_t on $C(M)$ whose infinitesimal generator agrees with Δ on a dense subset of $C(M)$.*

Remark. Most of the results below continue to hold, with appropriate modifications, if M is not compact but the metric tensor is assumed to be such that the geometry of all the leaves is uniformly bounded.

The argument for continuity of diffusion of a function given here is an adaptation of the construction of the heat kernel on a compact Riemannian manifold; see [**7**] or [**27**] and the review given in Section B.6. The heat kernel on a foliated space does not exist as such (although it follows from the construction given here that it exists as an element of the C^*-algebra of the foliated space), but an adaptation of the method of construction permits us to define $D_t f$ for each continuous function f on M as a solution to the heat equation with initial condition given by f. Another method, based on the Hille-Yosida theorem for semigroups of operators, is given in [**18**]. These methods are different from that proposed in L. Garnett [**77**]. Her argument seems to apply if no leaf has holonomy, and to hinge on exponential decay of the heat kernel. In a manifold of bounded geometry the heat kernel $p(x, y; t)$ decays exponentially with the distance $d(x, y)$ only if the manifold is noncompact.

The proof of the theorem requires a geometric setup at the level of the foliated space. Let TM be the tangent bundle to the leaves. This is a foliated space whose leaves are the tangent spaces to the leaves of L. The exponential map $\exp : TM \to M$ is defined by combining the exponential maps of the leaves: if $x \in M$ belongs to the leaf L and $v \in T_xL$, then $\exp_x(v) \in L \subset M$. This is a smooth map because the metric tensor is smooth. Due to the smoothness of the metric and the compactness of M, it is evident that there exists $r > 0$ so that, for each $x \in M$, the exponential map $\exp_x$ is a diffeomorphism of $\{v \in T_xL_x \mid \|v\| \leq r\}$ onto the metric ball of radius r about x in the leaf L_x.

Let $M \times_{\mathcal{F}} M$ be the graph of the equivalence relation "to be in the same leaf." Then, for sufficiently small r, the exponential map defines a bijection of the disc bundle $T_rM = \{v \in TM \mid \|v\| \leq r\}$ onto the set $\{(x, y) \in M \times_{\mathcal{F}} M \mid d(x, y) \leq r\}$. Indeed, if $\pi : T_rM \to M$ is the bundle projection, this bijection sends the point $v \in T_rM$ to $(\pi(v), \exp_{\pi(v)}(v))$.

There is still one more geometric view of this convex neighborhood. If $G(M)$ is the graph of the foliated space, the map that sends the point $v \in TM$ to the element of $G(M)$ represented by the path $\gamma(t) = \exp_{\pi(v)} tv$, $t \in [0, 1]$, is smooth and is a diffeomorphism of the disc bundle T_rM onto a neighborhood of the diagonal embedding $M \hookrightarrow G(M)$.

The proof that D_t is a diffusion semigroup will be carried out by finding a suitable family of operators that converge to D_t in an appropriate way. Such operators will now be analyzed.

If K is a smooth function on $M \times_{\mathcal{F}} M$ with support in $\{d(x, y) < r\}$, then K defines an operator, denoted by the same symbol,

$$K : C(M) \to C(M)$$

by setting

$$Kf(x) = \int f(y)K(y,x) \cdot dy.$$

This type of operator has appeared earlier in Chapter 1. Indeed, given the identification of $T_r M$ with an open subset of the graph $G(M)$, the action Kf can be viewed as the convolution $f * K$ of the functions $f \circ s$ and K on the graph. Therefore K is a well-defined operator. More generally, the following fact is equally easy to verify.

Lemma 2.2.14. *Let F be a smooth function on $M \times_{\mathcal{F}} M \times (a,b)$ such that F_t has compact support in a fixed neighborhood of the diagonal. Then the one-parameter family of operators $F_t : f \mapsto F_t f$ is strongly continous.*

These abstract concepts will now be applied to two particular examples. The first one arises from the parametrix H_t for the heat kernel whose construction was described in Section B.6. This parametrix H_t was constructed on a single manifold as a function on the tangent bundle. A review of the construction (specifically, Lemma B.6.2) shows that it is constructed locally on the tangent bundle and that its smoothness properties only depend on those of the metric tensors. Specifically, its smoothness depends on the smoothness of the distance function within the cut-locus and on the smoothness of solutions to differential equations. It is clear that such properties continue to hold when considering a foliated space endowed with a smooth metric tensor.

Lemma 2.2.15. *In the coordinates given by the convex neighborhood of the diagonal in $M \times_{\mathcal{F}} M$, this parametrix H_t has the local expression*

$$H_t(x,y) = \frac{1}{(2\pi t)^{n/2}} e^{-d(x,y)^2/4t} h(x,y,t),$$

where $n = \dim \mathcal{F}$, $d(x,y)$ is the distance between the points x and y in the leaf containing them, and $h(x,y,t)$ is a continuous, compactly supported function that behaves as a polynomial in t, the coefficients of which are continuous, compactly supported functions on an $r_1 = r/3$-neighborhood of the diagonal.

Then H_t, $t > 0$, defines a strongly continuous family of operators on $C(M)$, also denoted by H_t, by

$$H_t f(x) = \int f(y) H_t(y,x) \cdot dy.$$

Because of the properties in the the construction of H_t (*cf.* Lemma B.6.1), the limit $\lim_{t \to 0} H_t f = f$ uniformly on M. Thus the family of operators that the function H_t defines, together with the identity operator H_0, is a strongly continuous family.

Let K_t be the function defined by

$$(\partial/\partial t - \triangle)(H_t f)(x) = K_t f(x).$$

This is a smooth function that has compact support in a convex $r_1 = r/3$-neighborhood of the diagonal. The properties of K_t listed below readily follow from the construction described in Section B.6, in particular, Lemma B.6.2.

Lemma 2.2.16. *In the coordinates given by the convex neighborhood, this function K_t is given by the formula*

$$K_t(x,y) = t^{j-n/2} e^{-d(x,y)^2/4t} k(x,y,t),$$

where $j \geq n/2 + 2$ ($n = \dim \mathcal{F}$) and the function $k(x,y;t)$ is also compactly supported in an $r_1 = r/3$-neighborhood of the diagonal. Furthermore, one has $\|K_t\| \leq Ct^{j-n/2}$.

This function is actually defined for $t = 0$, so it is immediate that K_t, $t \geq 0$, is a strongly continuous family of operators. In fact, $K_0 = 0$.

Lemma 2.2.17. *Let F_t, $t \geq 0$, be a strongly continuous family of operators on $C(M)$. Then*

$$\left(\frac{\partial}{\partial t} - \triangle\right)((H \times F)_t f) = F_t f + (K \times F)_t f.$$

Proof. Note that, if R is a smooth function on a convex neighborhood of the diagonal in $M \times_{\mathcal{F}} M$, then so is $\triangle R$ (the Laplacian acting on the second variable). Furthermore, the restriction of R to each leaf $L \times L$ is a proper kernel (as in Lemma B.4.8). Thus, $(\triangle R)f = \triangle(Rf)$, by that lemma.

By definition,

$$(H \times F)_t f(x) = \int_0^t H_{t-s}(F_s f)(x) \cdot ds.$$

Upon differentiation under the integral sign with respect to t, one sees that

$$\begin{aligned}\frac{\partial}{\partial t}(H \times F)_t f &= F_t f + \int_0^t \frac{\partial}{\partial t} H_{t-s}(F_s f) \cdot ds \\ &= F_t f + \int_0^t \triangle H_{t-s}(F_s f) \cdot ds + \int_0^t K_{t-s}(F_s f) \cdot ds,\end{aligned}$$

and the stated identity follows because the Laplacian also commutes with the integral. (The fact that differentiation can be performed under the integral sign follows by a standard application of the dominated convergence theorem, *e.g.*, [**67**, Theorem 2.27]. Details are left to the reader.) □

The idea now is to construct the diffusion $D_t f$ of a function $f \in C(M)$ in the form

$$D_t f = H_t f + (H \times F)_t f.$$

In order to have $(\partial/\partial t - \triangle)D_t f(x) = 0$, the operator F_t must satisfy

$$\left(\triangle - \frac{\partial}{\partial t}\right) H_t f = F_t f + (K \times F)_t f.$$

A formal solution to this equation is given by the operator F_t defined by

$$F_t f = \sum_{i=1}^{\infty} (-1)^i K_t^{\times i} f,$$

where $K_t^{\times i}$ denotes the i-fold convolution $(K\times \overset{i}{\cdots} \times K)_t$. It will be shown next, by estimating its terms, that the series $\sum_{i=1}^{\infty}(-1)^i K_t^{\times i} f$ converges uniformly on M.

Lemma 2.2.18. *There is a constant $B > 0$ such that*

$$\|K_t f\| \leq B t^{j-n/2} \|f\|,$$

for every $f \in C(M)$.

Proof. By compactness of M it is possible to choose a finite regular cover $\{U_1, \dots, U_N\}$ by foliated charts with the property that the diameter of the plaques of any of these flow boxes is $\leq r/12$. If f is a continuous function with support in such a flow box, the description of the kernel K_t in Lemma 2.2.16 guarantees that

$$|K_t f(x)| \leq B' \|f\| t^{j-n/2},$$

for any $x \in M$, where B' is a constant that depends only on the Riemannian metric. (It is a bound for the areas of the geodesic balls of radius $\leq r$ on the leaves, hence finite because of the compactness of M and continuity of the metric tensor.) The value of $Kf(x)$ depends only on the restriction of f to one of the plaques of the intersection of U with the leaf through x. Since the operator defined by K is linear and each continuous function f on M can be written as a finite sum $f = \sum_{i=1}^{N} f_i$, where N is the number of flow boxes in the cover, each f_i has compact support in the corresponding U_i, and $\|f_i\| \leq \|f\|$, the stated claim follows (with $B = NB'$). □

Consider now the convolution $(K \times K)_t f$, which was defined by

$$(K \times K)_t f(x) = \int_0^t K_{t-s}(K_s f) \cdot ds.$$

Then

$$\|K_{t-s}(K_s f)\| \leq B(t-s)^{j-n/2} \|K_s f\| \leq B^2 \|f\| ((t-s)s)^{j-n/2},$$

from which it follows that

$$\begin{aligned}\|K_t^{\times 2} f\| &\leq B^2\|f\| \int_0^t ((t-s)s)^{j-n/2} \cdot ds \\ &= B^2\|f\| \frac{\Gamma(1+j-n/2)^2}{\Gamma(2(1+j-n/2))} t^{2j-n+1}\end{aligned}$$

(where $\Gamma(\bullet)$ is the Euler Γ-function). Feeding this inequality back into the convolution and iterating, we see that

$$\|K_t^{\times i} f\| \leq B^i \|f\| \frac{\Gamma(j-n/2+1)^i}{\Gamma(i(j-n/2+1))} t^{i(j+1-n/2)-1}.$$

Since

$$\lim_{i\to\infty} \frac{\Gamma(i(j-n/2+1))}{\Gamma((i+1)(j-n/2+1))} = 0,$$

the ratio test implies that the series $\sum_{i=1}^{\infty}(-1)^i K_t^{\times i} f$ converges uniformly on $M \times [0,t]$, for all $t \geq 0$, *i.e.*, converges uniformly on compact subsets of $M \times [0,\infty)$.

Moreover, these estimates show in fact that, for a given $T \geq 0$, there is a constant $C = C(T)$ such that

$$\|\sum_{i=1}^{\infty}(-1)^i K_t^{\times i} f\| \leq C t^{k-n/2},$$

for $t \leq T$.

By the estimates for the derivatives of the kernels H and K, it follows that the function $D_t f$ is of class C^2 along the leaves, C^1 in the time variable t, and that $\triangle D_t f$ is a continuous function on M, for all $t \geq 0$. This is seen exactly as in the case of Riemannian manifolds, and thus will not be given here (*cf.* [**7**] or [**27**]).

Proposition 2.2.19. *Let f be a continuous function on M. Then there is a unique continuous function $D_t f$ on $M \times (0,\infty)$, of class C^2 on each leaf, such that*

$$\triangle D_t f(x) = \frac{\partial}{\partial t} D_t f(x)$$

for each $x \in M$ and $t > 0$.

Indeed, if $f \in C(M)$ and L is a leaf of M, then the restriction $D_t|L$ is a bounded solution to the heat equation on L with f as initial condition. Since M is compact, L has bounded geometry and $f|L$ and $D_t f|L$ are bounded. Thus, Theorem B.6.8 implies that such $D_t f$ is unique and given by

$$D_t f(x) = \int_L p(x,y;t) f(y) \cdot dy,$$

for every $x \in L$. This same theorem also gives the corollary that follows.

Corollary 2.2.20. *The operator $D_t : C(M) \to C(M)$ constructed in the proof satisfies $\|D_t\| \le 1$.*

Fixing t and applying the result to the initial value $f_t = D_t f$, it follows that $(f_t)_s = f_{s+t}$. Thus the operation $D_t : C(M) \to C(M)$, which to a function f assigns the solution f_t, satisfies $D_t D_s = D_{s+t}$. That is, D_t is a semigroup acting on $C(M)$.

Next it has to be shown that $\lim_{t\to 0} D_t f = f$ in $C(M)$, that is, that the convergence is uniform on M for each $f \in C(M)$.

Proposition 2.2.21. *The intrinsic domain of the semigroup D_t is $C(M)$. That is to say, if $f \in C(M)$, then $\lim_{t\to 0} D_t f = f$ in $C(M)$.*

Proof. From the construction of $D_t f$ it follows that

$$D_t f = H_t f + R_t f,$$

where $R_t f$ satisfies $\|R_t f\| \le Ct$ for some constant C (which depends on f), and where $\lim_{t\to 0} H_t f = f$, as was already said. □

The results obtained so far are summarized in the next proposition. (The last part of the statement follows from Theorem B.10.13 because a continuous function on M is bounded on each leaf.)

Proposition 2.2.22. *Let $(M, \mathcal{F})$ be a compact foliated space, endowed with a metric tensor and associated Laplace operator $\triangle$. Then there exists a semigroup of operators $D_t : C(M) \to C(M)$ with the properties listed in Definition 2.2.2. Furthermore, if f is a continuous function on M that is of class C^2 on each leaf, then*

$$\lim_{t\to 0} \frac{D_t f(x) - f(x)}{t} = \triangle f(x),$$

for every $x \in M$, and the convergence is uniform on each compact subset of each leaf of M.

The next result is important because it identifies a subset of the domain of the infinitesimal generator, thus completing the proof of Theorem 2.2.13.

Proposition 2.2.23. *Let $\triangle_{\mathrm{i}}$ be the infinitesimal generator of the semigroup D_t and let $\mathcal{D}$ be its domain. Then there is a dense subset of $\mathcal{D}$ consisting of functions that are of class C^2 along each leaf. Furthermore, for every such function f, $\triangle_{\mathrm{i}} f = \triangle f$.*

Proof. The infinitesimal generator $\triangle_{\mathrm{i}}$ of the semigroup D_t is defined by

$$\triangle_{\mathrm{i}} f = \lim_{t\to 0} \frac{D_t f - f}{t}.$$

By Corollary 2.2.9 and the properties of the operators D_t, there is a dense subset $\mathcal{E} \subset \mathcal{D}$ consisting of functions that are of class C^2 on each leaf. Since M is compact, these functions are bounded on each leaf. Thus, Theorem B.10.13 applies and proves that, for such functions,

$$\lim_{t \to 0} \frac{D_t f(x) - f(x)}{t} = \triangle f(x),$$

for every $x \in M$, uniformly on compact subsets on each leaf. Therefore, for functions $f \in \mathcal{E}$, $\triangle_i f(x) = \triangle f(x)$ for every $x \in M$. □

Remark. The infinitesimal generator and its domain can actually be identified, see [**18**]. In fact, the domain contains, but generally is not equal to, the collection of all continuous functions f on M that are C^2 along each leaf, and such that $\triangle f$ is continuous on M. Since this will not be needed, it will not be proven here.

2.3. The Markov Process

This section describes the construction of the Markov process and Brownian motion associated to the Laplace operator on a foliated space. Let $(M, \mathcal{F})$ be a foliated space with Laplace operator $\triangle$ and associated diffusion semigroup D_t. Associated to these objects, there is a space $\Omega(M)$ with a continuous semiflow of transformations $\theta = \{\theta_t \mid t \geq 0\}$, and this section is devoted to their description. The space $\Omega(M)$ is fibered over M, and the operator $\triangle$ (more precisely, the heat kernels on the leaves) will be used to define a family of probability measures P_x on the fibers. A harmonic measure m on M will be shown to induce an invariant measure for this semiflow by integration of the measures P_x over m.

Let D_t be the diffusion semigroup acting on $C(M)$, as constructed in the previous section. The action of the diffusion operators D_t on continuous functions extends to an action on Borel subsets of M. Indeed, to each $x \in M$ there is assigned the positive linear functional

$$f \in C(M) \mapsto D_t f(x),$$

which, by the Riesz representation theorem, is represented by a Radon probability measure on M. Let this measure be denoted by $P_t(x, \bullet)$. If L is the leaf through x, then the inclusion map $j : L \to M$ is continuous; hence the measure $j_* p(x, \bullet; t)dy$ is a Radon measure on M (*cf.* Exercise A.2.4). Since, for continuous functions on M, $D_t f(x) = \int_L f(y) \cdot p(x, y; t)dy$, it must be that $P_t(x, \bullet) = j_* p(x, \bullet; t)dy$. This is by the uniqueness of the Radon measure representing a positive linear functional on $C(M)$ (Theorem A.2.6).

Lemma 2.3.1. *Let B be a Borel subset of M. Then $x \mapsto P_t(x, B)$, for t fixed, is a Borel function of $x \in M$.*

Proof. If $U \subset M$ is an open subset, then the characteristic function χ_U is the limit of a decreasing sequence of continuous functions f_n with $0 \le f_n \le 1$. By the dominated convergence theorem, for each $x \in M$,

$$\lim_{n\to\infty} D_t f_n(x) = P_t(x, U).$$

Because each $D_t f_n$ is continuous and the sequence $D_t f_n$ is decreasing, the limit is a Borel function on M (in fact, lower semicontinuous).

Let $\mathfrak{A}$ be the family of Borel subsets B of M for which $P_t(x, B)$ is a measurable function of x. Then, by the previous paragraph, $\mathfrak{A}$ contains the open subsets of M. If $A, B \in \mathfrak{A}$ and $A \subset B$, then

$$P_t(x, B \smallsetminus A) = P_t(x, B) - P_t(x, A),$$

for each $x \in M$; hence $B \smallsetminus A \in \mathfrak{A}$. If $B_n \in \mathfrak{A}$ is an increasing sequence of elements of $\mathfrak{A}$, then $P_t(x, B_n)$ is an increasing sequence of functions that converge pointwise to $P_t(x, \bigcup_n B_n)$. The Monotone Class Theorem C.1.9 then implies that every Borel subset of M is in $\mathfrak{A}$. □

Corollary 2.3.2. *If f is a bounded measurable function on M, then $D_t f(x)$, defined as the integral of f with respect to the measure $P_t(x, \bullet)$, is a bounded measurable function of $x \in M$.*

Let $\mathfrak{B}$ be the σ-field of Borel subsets of M. Then the function

$$P_t : M \times \mathfrak{B} \to [0, 1]$$

just constructed possesses the following properties.

(1) For each t and $x \in M$, $P_t(x, \bullet) : \mathfrak{B} \to [0, 1]$ is a probability measure on $\mathfrak{B}$.

(2) For each t and $B \in \mathfrak{B}$, $P_t(\bullet, B)$ is a Borel function on M.

(3) The semigroup equation

$$P_{s+t}(x, B) = \int P_s(\bullet, B) \cdot P_t(x, \bullet)$$

holds.

Let $\Omega(M)$ denote the space of leaf paths on M. This is the space of maps $\omega : [0, \infty) \to M$ that are continuous for the leaf topology on M. That is, each $\omega \in \Omega(M)$ is a continuous, half-infinite path on a leaf. Because M is metrizable, the space of continuous maps from the infinite half-line $[0, \infty)$ into M, endowed with the compact-open topology, is also metrizable. Then $\Omega(M)$ is a closed subspace, hence metrizable, complete and separable. That is, $\Omega(M)$ is a polish space. Let the map $\pi_t : \Omega(M) \to M$ denote the position at time t, $\pi_t(\omega) = \omega(t)$.

Remark. When L is a leaf or even an abstract manifold, the notation $\Omega(L)$ will denote the space of continuous, half-infinite paths $\omega : [0,\infty) \to L$.

Exercise 2.3.3. If d is a distance on M, then show that the expression

$$d(\omega,\omega') = \sum_{n=1}^{\infty} \frac{1}{2^n} \sup_{0\le t\le n} d\big(\omega(t),\omega'(t)\big)$$

defines a bounded metric on $\Omega(M)$ such that the evaluations maps $\omega \mapsto \omega(t)$ are Lipschitz. Verify the other metrical properties of $\Omega(M)$ claimed in the last paragraph.

The construction of the measures on $\Omega(M)$ mentioned above needs to be done first in the space of all maps from the half-line $[0,\infty)$ into M, which is denoted by $M^{[0,\infty)}$. The natural topology of this space is the product topology; however, its associated Borel σ-field is too large for most purposes, and instead the σ-field $\mathfrak{C}$ generated by the *cylinder sets* is considered. The cylinder sets are those of the form

$$C = \{\omega \in M^{[0,\infty)} \mid \omega(t_1) \in B_1, \dots, \omega(t_n) \in B_n\},$$

where $B_1, \dots, B_n$ are Borel subsets of M, and $0 \le t_1 < \cdots < t_n$ is a finite set of times. That is, C consists of all elements of $M^{[0,\infty)}$ that can be found within B_i at time t_i.

The structure of the measure space $(M^{[0,\infty)}, \mathfrak{C})$ is best understood by viewing it as an inverse limit. To do so, let the collection of finite subsets of $[0,\infty)$ be partially ordered by inclusion. Associated to each finite subset F of $[0,\infty)$ is the measure space $(M^F, \mathfrak{B}^F)$, where $\mathfrak{B}^F$ is the Borel field of the product topology on M^F. Each inclusion of finite sets $E \subset F$ canonically defines a projection $\pi_{EF} : M^F \to M^E$ that drops the finitely many coordinates in $F \smallsetminus E$. These projections are continuous, hence measurable, and consistent, for if $E \subset F \subset G$, then $\pi_{EF} \circ \pi_{FG} = \pi_{EG}$. The family $\{M^F, \pi_{EF} \mid E, F \subset [0,\infty) \text{ finite}\}$ is an inverse system of spaces, and its inverse limit is $M^{[0,\infty)}$ with canonical projections $\pi_F : M^{[0,\infty)} \to M^F$. The σ-field generated by the cylinders sets is the smallest one making all the projections π_F measurable.

For each $x \in M$, define a probability measure P_x on the measure space $(M^{[0,\infty)}, \mathfrak{C})$ as follows. If $F = \{0 \le t_1 < \cdots < t_n\}$ is a finite subset of $[0,\infty)$ and $C^F = B_1 \times \cdots \times B_n$ is a cylinder set of $(M^F, \mathfrak{B}^F)$, define $P_x^F(C^F)$ by the formula

$$D_{t_1}(\chi_{B_1} D_{t_2-t_1}(\cdots(\chi_{b_{n-2}} D_{t_{n-1}-t_{n-2}}(\chi_{B_{n-1}} D_{t_n-t_{n-1}}(\chi_{B_n})))\cdots))(x),$$

where χ_{B_i} is the characteristic function of B_i and D_t is the diffusion operator associated to the Laplace operator $\triangle$ on M. Corollary 2.3.2 legitimizes this procedure.

It is an obvious consequence of the semigroup property of D_t that if $E \subset F$ are finite subsets of $[0,\infty)$ and C^E is a cylinder subset of M^E, then

$$P_x^E(C^E) = P_x^F(\pi_{EF}^{-1}(C^E)).$$

Thus it follows that a probability measure P_x on $(M^{[0,\infty)}, \mathfrak{C})$ can be defined so that it is consistent with the inverse limit structure (Kolmogoroff's theorem, see [**51**, Chapter 12]). This measure P_x gives probability one to the set of paths ω such that $\omega(0) = x$.

Cylinder sets can be used to define a σ-field of subsets of $\Omega(M)$ via the inclusion into $M^{[0,\infty)}$. Even when the topologies are unrelated, it happens that the Borel field $\mathfrak{F}$ of $\Omega(M)$ as a polish space is the one generated by the sets $C \cap \Omega(M)$, where C runs over all cylinder sets in $M^{[0,\infty)}$. Therefore, every probability measure on $(\Omega(M), \mathfrak{F})$ is uniquely determined by its values on the cylinder sets. Going from a probability measure on $M^{[0,\infty)}$ to one on $\Omega(M)$ is a more delicate task, because it usually happens (and, indeed, it does in the process of constructing the Wiener measure) that the outer and inner measures of $\Omega(M)$ have different values. In general, passing from the bigger space to the smaller one is achieved by invoking the following proposition:

Proposition 2.3.4. *Let $(\Omega, \mathfrak{F}, P)$ be a probability space and let Ω_0 be a subset of Ω with outer measure one. Let $\mathfrak{F}_0$ be the σ-field on Ω_0 consisting of the sets of the form $B \cap \Omega_0$, $B \in \mathfrak{F}$. Then $P_0(B \cap \Omega_0) = P(B)$ is a probability measure on $(\Omega_0, \mathfrak{F}_0)$.*

To show that these probability measures P_x give full measure to the smaller space of continuous paths $\Omega(M)$ requires further analytical work. This work consists of two steps: a first one for the passage from $M^{[0,\infty)}$ to the space of continuous paths in M, and a second one for going to the smaller class of continuous leaf paths $\Omega(M)$. The first one can be taken under a condition on the measures P_x of the following form:

$$\lim_{t\to 0} \frac{1}{t} \sup_{x\in M} P_x\left[\omega(t) \notin B_M(x,\varepsilon)\right] = 0,$$

for each $\varepsilon > 0$. A condition of the form

$$(*) \qquad \lim_{t\to 0} \frac{1}{t} \sup_{x\in M} P_x[\omega(t) \notin B_{L_x}(x,\varepsilon)] = 0,$$

for each $\varepsilon > 0$, guarantees the second restriction on paths. Here, $B_X(x,\varepsilon)$ denotes the metric ε-ball in the metric space X. (These types of conditions guaranteeing continuity of paths of a Markov process appear to originate with J. R. Kinney [**113**], and are further elaborated by E. Dynkin [**56**]).

If $\triangle$ is a Laplace operator on a compact foliated space, then the Gaussian estimate of Cheng, Li and Yau [**29**] (recalled in Theorem B.7.1) applies.

Indeed, as M is compact, Proposition B.2.1 applies and gives global bounds for the curvature and injectivity radius of the leaves, and so the Gaussian estimate holds uniformly for all the leaves. Moreover, it is always possible to construct a distance function d_M on M with respect to which all the leaf inclusions are Lipschitz maps of Lipschitz constant ≤ 1 (see [**16**]); thus the second condition implies the first. Once the Gaussian estimate for the heat kernel is available, a direct calculation shows that the second condition $(*)$ holds. This calculation is given in Proposition B.7.3. The proof of continuity is given, for the case of a single manifold, in Section C.4, and the general case is done in exactly the same way.

The construction of the Markov process associated to the Laplace operator on a compact foliated space requires the introduction of a family of σ-fields. Let $\mathfrak{F}_t$, $t \geq 0$, be the σ-field that is generated by all cylinder sets in $\Omega(M)$ with associated sequence $t_1, \ldots, t_n$ bounded above by t. Hence $\mathfrak{F}_t$ keeps track of happenings up to time t. It is clear that $\mathfrak{F}_s \subset \mathfrak{F}_t$ if $s \leq t$, and that the projection π_t is measurable with respect to $\mathfrak{F}_t$ for each $t \geq 0$.

The collection $(\Omega(M), \mathfrak{F}_t, \pi_t)$ is called a stochastic process with values on $(M, \mathfrak{B})$. Let $\{P_x \mid x \in M\}$ be a family of measures such that:

(1) P_x is a probability measure on $\Omega(M)$ concentrated on the fiber $\pi_0^{-1}(x)$.

(2) For each $B \in \mathfrak{F}$, the map $x \mapsto P_x(B)$ is measurable.

(3) For every $x \in M$, $t \geq s \geq 0$, Borel sets $A \in \mathfrak{F}_s$ and $B \subset M$,

$$P_x\left[A \cap \{\omega(t) \in B\}\right] = \int_A P_{v(s)}\left[\omega(t-s) \in B\right] \cdot P_x(v).$$

All this constitutes the Markov process associated to the Laplace operator Δ on M. Item (3) above is called the Markov property, and is a reflection of the semigroup property of the diffusion semigroup operators D_t. More details can be found in Appendix C (for the case of a single manifold, but the general theory works the same way with very minor obvious modifications).

Example 2.3.5. Let $(M, \mathcal{F})$ be as above and let B be a Borel subset of M. Then $\{\omega \mid \omega(t) \in B\}$ is the set of continuous leaf paths that visit the set B at time t, and

$$P_x\left[\{\omega(t) \in B\}\right] = P_t(x, B) = \int_{L_x \cap B} p(x, y; t) \cdot dy$$

is the probability that continuous leaf paths, issuing from $x \in M$ at time 0, will visit the set B at time t.

The space $\Omega(M)$ supports a dynamical system that is closely connected to the foliation dynamical system of M. This is the semigroup

$$\theta = \{\theta_t \mid t \geq 0\}$$

of shift transformations

$$\theta_t : \Omega(M) \to \Omega(M)$$

defined by

$$\theta_t(\omega)(s) = \omega(s+t).$$

This semigroup of transformations allows us to study recurrence properties of the foliation dynamical system on M as if it were given by a one-dimensional flow.

The space $\Omega(M)$ resembles a fiber bundle over M where the fiber over each point x has a probability measure and varies in a measurable way with x. Thus, if a measure m on M is given, one on $\Omega(M)$ is obtained by first integrating a function over the fibers and then integrating the resulting function over the base M. The plan, then, is to construct a measure on $\Omega(M)$ that is invariant under the shift transformations θ, and whose fiber marginals are related to the Wiener measures on the fibers. The following fact justifies the legitimacy of this plan.

Lemma 2.3.6. *The shift transformations $\theta_t : \Omega(M) \to \Omega(M)$ are measurable.*

Proof. In fact, as is easily verified, the shift transformations are continuous for the topology given to $\Omega(M)$. □

As shown next, the scheme proposed for constructing a measure on $\Omega(M)$ from one on M works perfectly when starting with a harmonic measure on M. As a matter of fact, this statement is just a rephrasing of the diffusion invariance of harmonic measures.

Theorem 2.3.7. *Let M be a compact foliated space with metric tensor, and let m be a harmonic measure for M. Then the measure μ on $\Omega(M)$ defined by*

$$\int_{\Omega(M)} F(w) \cdot \mu(\omega) = \int_M \left(\int_{\Omega(L_x)} F(\omega) \cdot P_x(\omega) \right) \cdot m(x),$$

for all bounded measurable functions $F : \Omega(M) \to \mathbb{R}$, is invariant under the shift θ.

Proof. First, it is an exercise for the reader to show that the right-hand side of the displayed formula above defines a measure μ on $\Omega(M)$.

Second, because of the measurable structure given to $\Omega(M)$ and the fact that $\pi_t = \pi_0 \circ \theta_t$, and the monotone class theorem C.1.10, it is enough to show the invariance of μ for functions F of the form $f \circ \pi_0$, where f is a bounded measurable function on M. Thus the claim is that, for such a function f,

$$\int_{\Omega(M)} f(\pi_0(\theta_t(\omega))) \cdot \mu(\omega) = \int_{\Omega(M)} f(\pi_t(\omega)) \cdot \mu(\omega).$$

From Appendix C, the diffusion $D_t f$ can be computed as an "expected value"

$$E_x [f \circ \pi_t] = \int_{\Omega(L_x)} f(\pi_t(\omega)) \cdot P_x(\omega) = D_t f(x).$$

Thus,

$$\int_M \left(\int_{\Omega(L_x)} f(\pi_t(\omega)) \cdot P_x(\omega) \right) \cdot m(x) = \int_M D_t f(x) \cdot m(x),$$

and the assertion follows from the invariance of m under diffusion. □

This structure will suffice for the results to be described in the rest of this chapter. It is possible to develop the ergodic theory for the foliated space with harmonic measure by studying the induced shift invariant measure (see [**18**]). It is to be remarked, and, indeed, is rather obvious, that not every measure on $\Omega(M)$ that is shift invariant is induced by a harmonic measure on M. However, these can be characterized as the shift-invariant measures that satisfy a Gibbs condition, as in statistical mechanics.

2.4. Characterizations of Harmonic Measures

Let $(M, \mathcal{F})$ be a foliated space with metric tensor and associated Laplacian Δ. By definition, a measure m on the foliated space $(M, \mathcal{F})$ is harmonic if $\int_M \Delta f \cdot m = 0$, for a suitable collection of continuous functions f on M. Two other characterizations of harmonic measures will be described in this section.

By Theorem 2.2.13, the diffusion operators obtained leaf by leaf coalesce to give well defined operators D_t taking continuous functions on M to continuous functions on M. By duality between continuous functions and measures, the operators D_t act on a measure m by the adjoint construction. That is, the measure $D_t m$ is defined by

$$\int_M f \cdot D_t m = \int_M D_t f \cdot m,$$

for all $f \in C(M)$. The fact that $D_t m$ is nonnegative follows from the definition of diffusion.

The first characterization of harmonic measures is that they are the fixed points of D_t acting on measures. The proof requires the following lemma.

Lemma 2.4.1. *Let f be a function of class C^2 on M. Then*

$$\lim_{t\to 0} \frac{D_t f(x) - f(x)}{t} = \triangle f(x),$$

for each $x \in M$, boundedly on M.

Proof. It is shown in Proposition B.10.13 that $(D_t f - f)/t \to \triangle f$ uniformly on compact subsets of each leaf. The proof of this lemma is an extension of those arguments.

It suffices to consider f with compact support in a flow box U of the form $U = \{x \in M \mid d_{L_x}(x, Z) < \delta\}$, where Z is a transversal in M, and $\delta > 0$ is a suitable small number.

If x is outside U, then

$$\frac{D_t f(x) - f(x)}{t} - \triangle f(x) = \frac{D_t f(x)}{t}$$

and

$$|D_t f(x)| \leq \int_{L_x \setminus B(x,\delta/2)} |f(y)| p(x, y; t) \cdot dy \leq \frac{A\|f\|}{t^{n/2+2}} \exp(-\delta/Bt),$$

and this bound only depends on bounds for the geometry of the leaf. By the compactness of M, the bound only depends on M and f and is finite because of the continuity of f.

If $x \in U$, then write $f = f_P + f_O$ on L_x, where f_P agrees with f on the plaque P of U containing x (and extends by 0 to L_x outside a neighborhood of P), and $f_O = f - f_P$ vanishes on a suitable neighborhood of P in L_x.

Then, for $x \in P$,

$$\frac{D_t f(x) - f(x)}{t} = \frac{D_t f_P(x) - f_P(x)}{t} + \frac{D_t f_O(x) - f_O(x)}{t}.$$

The second term on the right-hand side is treated as above. To analyze the first term, note that the proof of Proposition B.10.5 shows that

$$\left| \frac{D_t f_P(x) - f_P(x)}{t} - \triangle_P f_P(x) \right| \leq \sup_{0 \leq s \leq t} \|D_s \triangle f_P - \triangle f_P\|_{L_x}.$$

Moreover,

$$\begin{aligned}
\|D_s \triangle f_P - \triangle f_P\|_{L_x} &\leq \|D_s \triangle f_P\|_{L_x} + \|\triangle f_P\|_{L_x} \\
&\leq \|\triangle f_P\|_{L_x} + \|\triangle f_P\|_{L_x} \\
&= 2\|\triangle f_P\|_{L_x} \\
&\leq 2\|\triangle f\|.
\end{aligned}$$

The first inequality and the first equality are obvious. The second inequality uses the the fact the the norm of the diffusion operator D_s satisfies $\|D_s\| \leq 1$, and the last inequality holds because $\triangle f = \triangle f_P$ on P, by the local nature of the Laplacian operator, and $\triangle f_P \equiv 0$ outside P. □

Proposition 2.4.2. *The measure m on the compact foliated space M is harmonic if and only if $D_t m = m$, for all $t \geq 0$.*

Proof. Assume that m is harmonic. By Corollary 2.2.9, there is a dense subspace $\mathcal{E}$ of $C(M)$, contained in the domain of the infinitesimal generator of the semigroup D_t and consisting of functions f that are of class C^2 along the leaves and such that $\triangle f$ is continuous on M. Since $\mathcal{E}$ is dense in $C(M)$, it suffices to show that $\int f \cdot m = \int D_t f \cdot m$ for every $f \in \mathcal{E}$ and $t \geq 0$. For these functions the derivative $\frac{d}{dt} D_t f = \triangle D_t f$. The function

$$t \in [0, \infty) \mapsto \int D_t f \cdot m$$

is continuous and has continuous right derivative, hence it is differentiable. The derivative with respect to t can be taken inside the integral sign, yielding

$$\frac{d}{dt}\left(t \mapsto \int D_t f \cdot m\right) = \int \triangle D_f f \cdot m = 0.$$

Thus the function is constant; hence m is diffusion-invariant.

Suppose that m is diffusion-invariant, $D_t m = m$, for every $t \geq 0$. If f is of class C^2 on M, then Lemma 2.4.1 above shows that

$$\lim_{t \to 0} \frac{D_t f - f}{t} = \triangle f$$

boundedly on M. The result follows immediately by integrating this identity, because the bounded convergence theorem permits us to take the integral before taking the limit. □

Exercise 2.4.3. Show that a measure m on M is harmonic if and only if $\int \triangle_{\mathrm{i}} f \cdot m = 0$ for every f in the domain of the infinitesimal generator $\triangle_{\mathrm{i}}$ of the diffusion semigroup D_t.

Definition 2.4.4. A measure m on M is *absolutely continuous* with respect to a measure m', written $m \ll m'$, if, for Borel sets $B \subset M$, $m'(B) = 0$ implies that $m(B) = 0$. If $m \ll m' \ll m$, the measures are said to be equivalent or to belong to the same measure class. The measure class of m is denoted by $[m]$.

Definition 2.4.5. A measure m on the foliated metric space M is *smooth* if, whenever B is a Borel set, $m(B) = 0$ if and only if $B \cap L$ has Riemann measure 0, for m-almost every leaf L (*i.e.*, for all leaves in a leaf-saturated set of full m-measure).

Remark. Smooth measures are the same as measures induced by quasi-invariant currents (see Section 1.6).

Corollary 2.4.6. *Every nontrivial harmonic measure m on M is smooth. Furthermore, the diffusion operators D_t, $t > 0$, convert arbitrary measures m into smooth measures. Finally, $[D_t m] = [m]$, $t > 0$, if and only if m is smooth.*

Proof. Let m be a measure on M. Then

(1) $$D_t m(B) = \int_M \chi_B(x) \cdot (D_t m)(x) = \int_M D_t \chi_B(x) \cdot m(x),$$

(2) $$D_t \chi_B(x) = \int_{L_x} \chi_B(y) p(x,y;t) \cdot dy.$$

Let $t > 0$, recalling that $p(x,y;t)$ is then strictly positive everywhere. It follows from (2) that $D_t \chi_B | L \equiv 0$ on a leaf L if and only if $\mathrm{vol}(B \cap L) = 0$. Similarly, $D_t \chi_B | L > 0$ everywhere on the leaf L if and only if $\mathrm{vol}(B \cap L) > 0$. Since $D_t \chi_B$ is measurable (Lemma 2.3.1), it follows that the union B_0 of leaves L such that $\mathrm{vol}_L(B \cap L) = 0$ is a Borel set, as is the union B_+ of leaves L such that $\mathrm{vol}(B \cap L) > 0$.

Let B be a Borel set. By (1) and the above observations,

$$D_t m(B) = \int_M D_t \chi_B(x) \cdot m(x) = \int_{B_+} D_t \chi_B(x) \cdot m(x),$$

implying that $D_t m(B) = 0$ if and only if $m(B_+) = 0$. But B_+ is saturated, so $D_t \chi_{B_+} = \chi_{B_+}$ and (1) implies that $m(B_+) = 0$ if and only if $D_t m(B_+) = 0$. This proves that $D_t m$ is smooth. If m is harmonic, then $m = D_t m$ is smooth. The final assertion also follows easily. □

Corollary 2.4.7. *Let m be a harmonic measure on M. Then an open set $U \subset M$ is m-null if and only if the saturation of U is m-null. Consequently,* $\mathrm{supp}(m)$ *is a saturated Borel set.*

The next exercise recalls standard facts about the space of measures on a compact metric space. The one following it leads the reader through another proof of the existence of harmonic measures.

Exercise 2.4.8. Let M be a compact metric space, $C(M)$ the space of continuous, real-valued functions on M with the uniform norm. By the Riesz representation theorem, the topological dual $C(M)^*$ is the space of finite signed measures on M with the so-called weak* topology: a sequence of measures μ_n converges to a measure μ if $\mu(f_n) \to \mu(f)$ for every continuous function on M.

(1) The name weak* topology comes from the fact that this is the weakest topology on $C(M)^*$ making all the evaluation maps

$$f \in C(M) \to \mu(f),$$

$\mu \in C(M)^*$, continuous. Prove this.

Let $\mathcal{P}(M) \subset C(M)^*$ be the subset of probability measures.

(2) Show that $\mathcal{P}(M)$ is metrizable. In fact, a distance function inducing the topology can be defined as follows. Let $Q = \{f_n\}_{n=1}^\infty$ be a countable dense subset of $C(M)$, and, for measures μ, ν in $\mathcal{P}(M)$, let

$$d(\mu, \nu) = \sum_{n=1}^{\infty} \frac{|\mu(f_n) - \nu(f_n)|}{2^n \|f_n\|}.$$

(3) Prove that $\mathcal{P}(M)$ is a closed, convex subset of $C(M)^*$.

For each $f \in Q$, let I_f denote the smallest interval containing the range of the function, a compact subset of $\mathbb{R}$. The product space

$$X = \prod_{f \in Q} I_f$$

is compact in the product topology and metrizable.

(4) Show that there is a topological embedding $\mathcal{P}(M) \to X$ with closed, hence compact, image. Thus, $\mathcal{P}(M)$ is a compact, convex set of measures.

Exercise 2.4.9. Assume that M is a compact foliated space, fix a leafwise metric tensor, and let D_t be the diffusion semigroup. Fix a probability measure μ on M and, for each $t \in (0, \infty)$, let $\mu_t \in C(M)^*$ be defined by

$$\int_M f \cdot \mu_t = \frac{1}{t} \int_0^t \left(\int_M D_s f \cdot \mu(x) \right) \cdot ds.$$

(1) Prove that $\mu_t \in \mathcal{P}(M)$.

By compactness of $\mathcal{P}(M)$, $\{\mu_t\}$ has a limit point. Define

$$\mu_\infty = \lim_{n \to \infty} \mu_{t_n} \in \mathcal{P}(M)$$

over a suitable sequence $t_n \uparrow \infty$.

(2) Show that μ_∞ is a harmonic probability measure for $(M, \mathcal{F})$.

The second characterization of harmonic measures is local and exhibits their analogy with invariant transverse measures. This was anticipated in Examples 2.1.2 and 2.1.3.

Proposition 2.4.10. *A measure m on M is harmonic if and only if, on any given foliated chart $U = D \times Z$, it admits a decomposition of the form*

$$m = h(x, z) \cdot dx \otimes \nu(z),$$

where dx is the measure induced by the metric tensor, ν is a measure on the transversal Z, and $h(\bullet, z)$ is a fixed, positive, harmonic function on $D \times \{z\}$, for ν-almost all $z \in Z$.

Since constant functions are harmonic, the following observation is immediate.

Corollary 2.4.11. *A transverse invariant measure, when combined with the volume density along the leaves, is a harmonic measure.*

Definition 2.4.12. A harmonic measure as in Corollary 2.4.11 is said to be *completely invariant.*

Before proving Proposition 2.4.10, another example of its use in constructing harmonic measures will be given. This example is analogous to Example 2.1.3, but on a quite different foliated manifold.

Example 2.4.13. Recall the foliation $\mathcal{H}$ of the unit tangent bundle $T^1(\mathbf{H})$ of the hyperbolic plane constructed in [I, Example 1.3.14]. This foliation is transverse to the circle fibers. The foliation $\mathcal{H}$ passes to a foliation $\mathcal{H}_\Gamma$ of a compact manifold M by quotienting out the action of a suitable discrete group Γ of hyperbolic isometries. In fact, Γ is the group of covering transformations for a closed, orientable surface Σ of genus > 1 and $M = T^1(\Sigma)$. As in Example 2.1.3, a harmonic measure for $\mathcal{H}$ is constructed and shown to be invariant by hyperbolic isometries, inducing thus a harmonic measure for $\mathcal{H}_\Gamma$.

The leaves of $\mathcal{H}$ are parametrized by the points $s \in S_\infty$, the circle at infinity. (In the upper half-plane model, S_∞ is the compactification $\mathbb{R} \cup \{\infty\}$ of the x–axis.) Indeed, the leaf L_s will be the unit tangent field to the pencil $\mathcal{P}_s$ of geodesics with target $s \in S_\infty$. The general point on this leaf can be written as (z, v_s), $z \in \mathbf{H}$ and v_s the unique unit tangent vector at z whose associated geodesic ray limits on s. This gives a canonical parametrization

$$(z, v_s) \leftrightarrow (z, s)$$

identifying $T^1(\mathbf{H})$ with $\mathbf{H} \times S_\infty$. The hyperbolic isometries $\gamma \in \mathrm{PSl}(2, \mathbb{R})$ act on $\mathbf{H}$ by linear fractional transformations and these extend by the same formula to S_∞. Evidently,

$$(\gamma z, \gamma_{*z} v_s) = (\gamma z, v_{\gamma s}).$$

Under the identifications $\mathrm{PSl}(2, \mathbb{R}) = T^1(\mathbf{H}) = \mathbf{H} \times S_\infty$, left translation by $\gamma \in \mathrm{PSl}(2, \mathbb{R})$ becomes $\gamma \cdot (z, s) = (\gamma z, \gamma s)$.

To each $z \in \mathbf{H}$ associate the "visual measure" ν_z on S_∞. This is the unique Borel measure that, on each arc $[a, b] \subset S_\infty$, has value $\nu_z([a, b])$ equal to the radian measure of the angle between (z, v_a) and (z, v_b) subtended by $[a, b]$. If $\gamma \in \mathrm{PSl}(2, \mathbb{R})$, it is evident that

$$\nu_{\gamma z}([\gamma a, \gamma b]) = \nu_z([a, b]).$$

These measures belong to the Lebesgue measure class on S_∞ (*i.e.*, they have exactly the same Borel null sets as Lebesgue measure), so the Radon-Nikodym derivatives

$$h_{z_0}(z,s) = \frac{d\nu_z}{d\nu_{z_0}}(s)$$

are defined for almost all $s \in S_\infty$. An easy calculation shows that

$$h_{z_0}(x+iy,\infty) = c(z_0)y,$$

a harmonic function of $z = x+iy$. If $s \in S_\infty$ and g is the hyperbolic rotation about z_0 carrying ∞ to s, it is easy to check that $h_{z_0}(gz,s) = h_{z_0}(z,\infty)$. That is, since g is an isometry,

$$h_{z_0}(w,s) = h_{z_0}(g^{-1}w,\infty)$$

is a harmonic function of w, for each fixed choice of $s \in S_\infty$.

Let μ be the measure on $\mathbf{H} \times S_\infty$ given by

$$\int f \cdot \mu = \int_{\mathbf{H}} \left(\int_{S_\infty} f(z,s) \cdot \nu_z(s) \right) \cdot dz,$$

for each compactly supported, continuous function f. This is the candidate for a harmonic measure for $\mathcal{H}$. It is evident that

$$\int (f \circ \gamma) \cdot \mu = \int f \cdot \mu,$$

for all $\gamma \in \mathrm{PSl}(2,\mathbb{R})$; hence this measure descends to a measure m on $(M, \mathcal{H}_\Gamma)$. Exactly as in Example 2.1.3, m will be harmonic if μ satisfies the criterion of Proposition 2.4.10.

An important property of the visual measure is that it has the mean value property. More precisely, if $B \subset \mathbf{H}$ is an open hyperbolic disk with center z_0 and $[a,b]$ is an arc in S_∞, then

$$\int_B \nu_z([c,d]) \cdot dz = \mathrm{vol}(B)\nu_{z_0}([a,b]).$$

This property allows us to determine the local disintegration of μ relative to the projection $p : B \times [a,b] \to [a,b]$. The push-forward measure $p_*\mu = \nu$ is determined by its values on arbitrary subarcs $[c,d] \subset [a,b]$, computed by

$$\nu([c,d]) = \mu(p^{-1}[c,d]) = \int_B \nu_z([c,d]) \cdot dz = \mathrm{vol}(B)\nu_{z_0}([c,d]).$$

That is, $\nu = \mathrm{vol}(B)\nu_{z_0}$. Thus, the disintegration of μ has the form

$$\mu(z,s) = \mathrm{vol}(B) \int \sigma_s(z) \cdot \nu_{z_0}(s),$$

where the probability measure σ_s on $B \times \{s\}$ is defined for ν_{z_0}–almost all $s \in [a,b]$. For any integrable function f supported in $B \times [a,b]$,

$$\int f \cdot \mu = \mathrm{vol}(B) \int_{[a,b]} \left(\int_B f(z,s) \cdot \sigma_s(z) \right) \cdot \nu_{z_0}(s)$$

and

$$\begin{aligned} \int f \cdot \mu &= \int_B \left(\int_{[a,b]} f(z,s) \cdot \nu_z(s) \right) \cdot dz \\ &= \int_B \left(\int_{[a,b]} f(z,s) h_{z_0}(z,s) \cdot \right) \cdot \nu_{z_0}(s) dz \\ &= \int_{[a,b]} \left(\int_B f(z,s) h_{z_0}(z,s) \cdot dz \right) \cdot \nu_{z_0}(s). \end{aligned}$$

It follows that

$$\int_B f(z,s) h_{z_0}(z,s) \cdot dz = \mathrm{vol}(B) \int_B f(z,s) \cdot \sigma_s(z),$$

for ν_{z_0}–almost all $s \in [a,b]$ and all $f(z,s)$ as above. By Proposition 2.4.10, μ is a harmonic measure.

Exercise 2.4.14. As an exercise in hyperbolic geometry, prove the properties of the visual measures on S_∞ that were used in Example 2.4.13.

Remark. Examples 2.1.3 and 2.4.13 are special cases of a quite general construction using a unimodular Lie group G, a closed, connected subgroup $H \subset G$ and a cocompact lattice $\Gamma \subset G$. The compact manifold $\Gamma \backslash G$ is foliated by the projections of the left cosets of H. The Haar measure on G projects to a measure on the coset space G/H, and this foliation has a transverse invariant measure if and only if H is unimodular; a proof, as well as other related references, can be found in [**13**].

Nothing can be said about the measure being harmonic, because that requires a choice of metric tensor on the leaves of G. A particular case where there is a natural measure is that in which G is a semisimple Lie group of noncompact type and K is a maximal compact subgroup, so that there are a decomposition $G = KH$ and a natural semimetric tensor on G that induces a metric tensor on the leaves $P \times \{h\}$. The Haar measure on the compact group $G/H = K$ induces a harmonic measure for the foliation of G by H-leaves.

Proof of Proposition 2.4.10. Let w denote the general point in M and, on the foliated chart $U = D \times Z$, write $w = (x,z)$. The local decomposition is provided by the disintegration of the measure with respect to the fibration $p : D \times Z \to Z$, which is constant on the leaves. The projection p pushes m

forward to a measure $\nu = p_*(m|U)$ on Z. There is a measurable assignment of a probability measure λ_z on $D \times \{z\}$, for ν-almost all $z \in Z$, such that

$$\int_M f(w) \cdot m(w) = \int_Z \left(\int_{D\times\{z\}} f(x,z) \cdot \lambda_z(x) \right) \cdot \nu(z),$$

for every smooth function f with compact support in $D \times Z$. Since the support of $\triangle f$ is contained in the support of f, a partition of unity argument implies that the measure m is harmonic if and only if

$$(*) \qquad \int_Z \left(\int_{D\times\{z\}} \triangle f(x,z) \cdot \lambda_z(x) \right) \cdot \nu(z) = 0,$$

for all foliated charts $D \times Z$ and all f compactly supported in $D \times Z$.

Consider leafwise smooth functions of the form $f(x,z) = f(x)$ that are constant in z and compactly supported in D. These are bounded on $D \times Z$ and are limits in the C^2 topology (the topology of uniform convergence on compact sets of all derivatives of order ≤ 2) of sequences of functions that are compactly supported in $D \times Z$, so the condition $(*)$ holds for these functions also. Let Q be a countable, C^2-dense subset of these functions. Applying $(*)$ to $f \in Q$, it follows that

$$(**) \qquad \int_{D\times\{z\}} \triangle f(x) \cdot \lambda_z(x) = 0,$$

where z ranges over a subset $Z_f \subset Z$ of full ν-measure. The set

$$Z_* = \bigcap_{f\in Q} Z_f$$

also has full ν-measure. Thus $(**)$ holds for all compactly supported, smooth functions on D and for ν-almost all $z \in Z$. By a basic regularity result, Proposition B.4.6, this is equivalent to the existence of a measurable function $h(x,z)$ on $D\times Z$, such that $h(\cdot,z)$ is harmonic on $D\times\{z\}$, for ν-almost every $z \in Z$, and such that

$$\lambda_z(x) = h(x,z)\,dx,$$

for ν-almost all $z \in Z$.

The converse is an obvious consequence of the Green-Stokes formula. □

Exercise 2.4.15. Show that, if f is a bounded measurable function on M that is leafwise smooth and $\triangle f$ is bounded, then

$$\int_M \triangle f(x) \cdot m(x) = 0,$$

for every harmonic measure m.

Remark. The local decomposition of a harmonic measure m is not generally holonomy invariant. If $U = D \times Z$ and $U' = D' \times Z'$ are foliated charts, the local decompositions of m will be of the form

$$m|U = h(x,z) \cdot dx \otimes \nu(z),$$
$$m|U' = h'(x',z') \cdot dx' \otimes \nu'(z').$$

If $U \cap U' \neq \emptyset$, then there is a partially defined holonomy homeomorphism $g : z \in Z \mapsto z' = z'(z) \in Z'$. Since $m|U \cap U'$ is independent of the coordinates, the measures $g_*\nu$ and ν' are in the same measure class. The Radon-Nikodym theorem provides a ν'-integrable function

$$\delta(z') = \frac{d(g_*\nu)}{d\nu'}(z'),$$

defined for $z' = z'(z) \in \operatorname{im} g$.

Exercise 2.4.16. Let $\{U_i, x_i, z_i\}_{i,j\in I}$ be a regular foliated atlas on M, let m be a harmonic measure and let

$$m(x_i, z_i) = h_i(x_i, z_i)\, dx_i \otimes \nu_i(z_i),$$

where h_i is harmonic along plaques. Finally, let

$$\delta_{ij} = \frac{d(g_{ij*}\nu_j)}{d\nu_i}.$$

(1) Prove that
$$h_i(x_i, z_i) = h_j(x_j(x_i, z_i), z_j(z_i))\delta_{ij}(z_i).$$
Conclude that h_i/h_j is constant along the leaves of $\mathcal{F}|U_i \cap U_j$.

(2) Let d denote the leafwise exterior derivative and show that the measurable, leafwise smooth 1-forms $d \log h_i$, $i \in I$, fit together to define such a 1-form ω on M that is d-closed.

(3) Prove that the harmonic measure m is completely invariant (Definition 2.4.12) if and only if ω is d-exact (in the measurable category).

Exercise 2.4.17. Let $(M, \mathcal{F})$ be a compact foliated space of dimension one.

(1) Show that there is a metric along the leaves of M that makes every leaf a homogeneous space of $\mathbb{R}$.

(2) Show that every harmonic measure (with respect to the metric in (1)) is completely invariant.

(3) What about harmonic measures with respect to an arbitrary metric tensor?

Exercise 2.4.18. All the examples of foliated spaces with not completely invariant harmonic measures described so far are such that their leaves are either (hyperbolic) planes or cylinders. Show that the Hirsch foliated space [**98**] (also see [I, Example 13.3.22]) has a not completely invariant harmonic measure such that, for almost every leaf L, the harmonic function $\{h_i\}$ (given by Exercise 2.4.16) has the property that the components of its level sets are compact.

2.5. The Ergodic Theorem

This section is devoted to proving the Ergodic Theorem for the semigroup of diffusion operators on a compact foliated space.

Theorem 2.5.1. *Let $(M, \mathcal{F})$ be a compact foliated space endowed with a harmonic probability measure m. If f is an m-integrable function on M, then*

$$f^*(x) = \lim_{N\to\infty} \frac{1}{N} \sum_{k=1}^{N-1} D_k f(x)$$

exists for m-almost all $x \in M$, and f^* is a diffusion-invariant, m-integrable function, constant along m-almost every leaf, such that

$$\int_M f(x) \cdot m(x) = \int_M f^*(x) \cdot m(x).$$

The proof of this theorem is by application of a standard ergodic theorem, to be stated shortly. The discussion that follows is to verify that the required hypotheses hold.

Implicit in Theorem 2.5.1 is the fact that diffusion of m-integrable functions is defined and yields m-integrable functions. This is proven in the next proposition.

Proposition 2.5.2. *If m is a harmonic probability measure on M and if $f \in L^1(M, m)$, then the diffusion $D_t f$ is defined, $D_t f \in L^1(M, m)$, for all $t \geq 0$, and $\|D_t f\|_1 \leq \|f\|_1$.*

Proof. Fix an arbitrary time t and denote D_t by D. First, let f be a nonnegative m-integrable function. Thus, $Df(x)$ is defined (but possibly infinite) at arbitrary points $x \in M$. Define $f_n = \min\{f, n\}$ for $n = 1, 2, \ldots$. Then $\{f_n\}$ is an increasing sequence of bounded measurable functions converging pointwise to f. Therefore $Df_n \leq Df_{n+1}$, because the operator D is positive, and the sequence $\{Df_n\}_{n=1}^{\infty}$ increases pointwise to Df. Moreover,

$$\int_M Df_n(x) \cdot m(x) = \int_M f_n(x) \cdot m(x),$$

by the D-invariance of m. Since $\lim_n Df_n = Df$, it follows by Fatou's lemma that

$$\begin{aligned}\int_M Df(x)\cdot m(x) &\leq \liminf_{n\to\infty}\int_M Df_n(x)\cdot m(x)\\ &= \lim_{n\to\infty}\int_M f_n(x)\cdot m(x)\\ &= \int_M f(x)\cdot m(x),\end{aligned}$$

so that Df is m-integrable and $\|Df\|_1 \leq \|f\|_1$ in $L^1(M,m)$. For arbitrary $f \in L^1(M,m)$, write $f = f_+ - f_-$, where $f_\pm(x) = \max\{\pm f(x), 0\}$ are nonnegative and m-integrable. Clearly, $\|f\|_1 = \|f_+\|_1 + \|f_-\|_1$ and $Df = Df_+ - Df_-$; hence an application of the triangle inequality implies that $\|Df\|_1 \leq \|f\|_1$. □

Exercise 2.5.3. Show that, if $f \in L^1(m)$, then $D_t f \to f$ in $L^1(m)$ as $t \to 0$.

This proposition and the exercise show that D_t is a semigroup of bounded linear operators on $L^1(M,m)$ with L^1-norm bounded by 1. Furthermore, since the measure m is finite, the space of essentially bounded functions, $L^\infty(M,m)$, is a subspace of $L^1(M,m)$, and so D_t is defined on $L^\infty(M,m)$.

Lemma 2.5.4. *The diffusion operators D_t map essentially bounded functions to essentially bounded functions, and $\|D_t f\|_\infty \leq \|f\|_\infty$, for every $f \in L^\infty(M,m)$.*

Proof. The essential supremum norm $\|f\|_\infty$ is defined by

$$\|f\|_\infty = \inf\{a \geq 0 \mid m(\{|f(x)| > a\}) = 0\}$$

(with the convention that the infimum of the empty set is ∞).

If $f \in L^\infty(M,m)$ and $a > \|f\|_\infty$, then the set $\{|f(x)| > a\}$ has m-measure 0. Because m is a smooth measure (Corollary 2.4.6), for every leaf L, except perhaps for those in a saturated m-null set, the set of $x \in L$ such that $|f(x)| > a$ has Lebesgue measure zero. Because D_t is given on each leaf as an integral operator with strictly positive kernel $p(x,y;t)$ and with $\int_L p(x,y;t)\cdot dy = 1$, it follows that, for those same leaves L, the set of points $x \in L$ such that $|D_t f(x)| > a$ also has Lebesgue measure zero. Thus $m(\{|D_t f(x)| > a\}) = 0$, and the statement follows. □

For operators having those properties that D_t has been shown to have, the following ergodic theorem is available. A proof can be found in [**53**, Ch. 3].

Theorem 2.5.5. *Let (X, ν) be a finite measure space. Let D be a linear operator acting on $L^1(X, \nu)$ that maps essentially bounded functions to essentially bounded ones, with $\|D\|_1 \leq 1$ and $\|D\|_\infty \leq 1$. Let D^i denote the i-fold composition $D \circ \cdots \circ D$, and D^0 the identity operator. Then, for every ν-integrable function f, the limit*

$$f^*(x) = \lim_{N \to \infty} \frac{1}{N} \sum_{k=0}^{N-1} D^k f(x)$$

exists for ν-almost all $x \in X$. Moreover, f^* is a D-invariant ν-integrable function and

$$\int_X f^*(x) \cdot \nu(x) = \int_X f(x) \cdot \nu(x).$$

The more familiar ergodic theorem arises in the context of a measure-preserving transformation S of a space (X, ν), and the operator D in this setting is the precomposition $f \to f \circ S$. The output is a function f^* that is S-invariant, hence constant along the T-orbits.

Theorem 2.5.5 can be applied to the diffusion at time $t = 1$, thus providing a function f^* that is invariant under the powers D_1^n. It thus remains to prove the assertion that this function f^* is actually constant on each leaf in a saturated set of full m-measure. The next proposition takes care of this last detail.

Proposition 2.5.6. *Let m be a harmonic probability measure on M and let f be a measurable function on M that is m-integrable and such that $D_{t_0} f = f$ for a positive time $t_0 > 0$. Then the class of f in $L^1(M, m)$ contains a function that is constant along each leaf of M.*

Proof. Let f be the function in the statement of the theorem. For each nonnegative rational number r, let $f_r = \min\{f, r\}$. Since r is nonnegative, $|f_r| \leq |f|$; so f_r is m-integrable. By Proposition 2.5.2, $D_t f_r$ is also m-integrable for all $t \geq 0$.

Write $D = D_{t_0}$. By part (4) of Proposition B.10.1, $Dr = r$. Thus, the inequalities $f_r \leq f$ and $f_r \leq r$ imply

$$D f_r \leq D f = f,$$
$$D f_r \leq D r = r,$$

proving that $D f_r \leq f_r$. Furthermore, if $f_r < r$ on a set of positive Riemannian measure in a leaf L, then $D f_r(x) < r$, for every $x \in L$.

By the diffusion-invariance of m,

$$\int_M D f_r(x) \cdot m(x) = \int_M f_r(x) \cdot m(x).$$

Because m is smooth (Corollary 2.4.6) and $Df_r \le f_r$, this implies that there exists a saturated set B_r of full measure such that, for each leaf L in B_r, the set $\{x \in L \mid f_r(x) > Df_r(x)\}$ has Riemannian measure zero in L. That is, $Df_r = f_r$ almost everywhere on each leaf $L \subset B_r$, with respect to the Riemannian measure of L. Furthermore, there is $x \in L$ such that $f(x) = Df(x) > r$ if and only if $f > r$ on a set of positive Riemannian measure in L, in which case f_r is equal to its maximum value r on that set. But $Df_r = f_r$ almost everywhere on L and, by the observation at the end of the previous paragraph, this implies that $f_r = r$ almost everywhere on L. Equivalently, $f \ge r$ almost everywhere on L. Thus, for each leaf $L \subset B_r$, either $f \le r$ everywhere on L or $f \ge r$ almost everywhere on L. It was assumed that r is a nonnegative rational, but applying this same reasoning to $-f$ shows that r may be allowed to be any rational number. The saturated set

$$Z = \bigcap_{r \in \mathbb{Q}} B_r$$

has full m-measure and, for each leaf $L \subset Z$, the function $f|L$ is almost everywhere constant. Using the local disintegrations of m, it follows that f agrees m-almost everywhere in M with a function f' that is identically constant along each leaf in Z. □

The following analogue to Theorem B.3.2 is a corollary to the above proof.

Corollary 2.5.7. *A noncompact manifold L of bounded geometry admits no positive integrable harmonic functions.*

Exercise 2.5.8. Work out a proof of Corollary 2.5.7 following the outline of the proof of Proposition 2.5.6.

Exercise 2.5.9. Let M be a compact foliated space with a harmonic measure m. Show that, if h is an m-integrable function on M that is harmonic on all but a null set of leaves, then h is constant on all but, perhaps, a null set of leaves.

2.6. Ergodic Decomposition of Harmonic Measures

The theme of this section is to show that a harmonic measure can be decomposed as an integral of simpler ones, the so-called ergodic measures.

Definition 2.6.1. A probability measure m on the foliated space M is said to be *ergodic* if every saturated measurable subset of M either has m-measure 0 or m-measure 1.

Example 2.6.2. If L is a compact leaf of the foliated space M, then a transverse, holonomy-invariant probability measure concentrated on L (a Dirac measure) is harmonic and ergodic.

Example 2.6.3. The completely invariant measure of the foliation of the 2-torus given by lines of irrational slope is ergodic.

The fact that ergodic measures are important for dynamical questions is indicated by the following exercises.

Exercise 2.6.4. Let M be a compact foliated space with metric tensor and let m be a harmonic measure that is ergodic and positive on open sets. Show that, if M has more than one leaf, then, for m-almost every $x \in M$, the leaf through x is a dense subset of M of zero measure.

Exercise 2.6.5. Let M be as in the previous exercise and let m be an ergodic harmonic measure on M. Let $f : M \to \mathbb{R}$ be a measurable map that is constant along each leaf. Show that there is a saturated set $B \subset M$ of full m-measure whose image under f is a single point.

In light of Theorem 2.5.1, this exercise has the following corollary.

Corollary 2.6.6. *If m is an ergodic harmonic probability measure for M and if f is m-integrable, then*

$$f^* = \int_M f(x) \cdot m(x),$$

m-almost everywhere.

Exercise 2.6.7. Show that the harmonic measure on the torus bundle over the circle, constructed in Example 2.1.3, is ergodic.

Exercise 2.6.8. Show that the harmonic measure on the foliated space of Exercise 2.1.5 is ergodic.

Exercise 2.6.9. Let M be the foliated space of Example 2.4.13. The manifold M is a quotient of $\mathbf{H} \times S^1$ by a discrete subgroup Γ acting on it by linear fractional transformations. Prove that the measure m constructed in that example is ergodic. You may proceed as follows.

(1) Show first that every measurable union of leaves corresponds to a measurable subset of the circle invariant under the action of Γ.

(2) Let A be such an invariant subset. Apply the solution to the Dirichlet problem on the unit disc model for $\mathbf{H}$ to obtain a harmonic function u on $\mathbf{H}$ that is invariant under Γ and nonconstant if neither A nor its complement have measure zero.

(3) The quotient $\Sigma = \mathbf{H}/\Gamma$ is compact, and u descends to a harmonic function on Σ. Show that u must be constant on $\mathbf{H}$.

For this exercise you need to know something about the asymptotic behavior of $u(z)$ as z approaches S_∞. What is needed can be found in any reasonable book on complex analysis or inferred from Appendix C.

The following exercise may not have a complete answer because, in a certain sense, it may admit several.

Exercise 2.6.10. Let M be a foliated space with a leafwise metric. An averaging sequence consists of a sequence of bounded submanifolds L_i of leaves (not necessarily connected or lying in the same leaf) with the property that

$$\lim_{i\to\infty} \frac{\mathrm{vol}(\partial L_i)}{\mathrm{vol}(L_i)} = 0.$$

It was shown in [I, Theorem 10.2.22] that such a sequence defines a holonomy-invariant measure and thus a completely invariant harmonic measure m. The question is: what geometric properties of the elements L_i of the sequence guarantee that m is ergodic? For example, the reader would find it easy to describe an example where all L_i are connected but μ is not ergodic. But what about one where all ∂L_i are also connected?

The decomposition of a harmonic measure into ergodic ones will be carried out via the integral representation on compact convex sets. This is a very efficient but abstract technique, thus somewhat impractical because it does not give any hints for constructing ergodic measures. The following examples will serve to illustrate the basic integral representation concepts.

Example 2.6.11. Let M be a compact metric space. Exercise 2.4.8 shows that the space $\mathcal{P}(M)$ of probability measures on M is compact and convex in the usual topology. For each $x \in M$, let $\delta_x \in \mathcal{P}(M)$ denote the measure given by evaluation at x. That is,

$$\int_M f(y) \cdot \delta_x(y) = f(x).$$

This defines a map $\delta : M \to \mathcal{P}(M)$ that is, in fact, continuous. The justification of the continuity of this map is left as an exercise for the reader.

The measures δ_x have the property of being extreme points of the compact, convex set $\mathcal{P}(M)$. That is, the only way that δ_x can be expressed as $\delta_x = tm_1 + (1-t)m_2$, $0 \le t \le 1$, $m_1, m_2 \in \mathcal{P}(M)$, is if $t = 0, 1$ or $m_1 = m_2 = \delta_x$. Conversely, the only extremal measures are the point mass measures. Indeed, if m is a probability measure on M whose support contains more than two points, it is always possible to find an open set $U \subset M$ containing one and such that $\overline{U}$ does not contain the other. Therefore, $0 < m(U) < 1$. For a Borel subset $B \subset M$, define $m_B(A) = m(B \cap A)$, for all Borel sets $A \subset M$. Then

$$m = m(U)\frac{m_U}{m(U)} + (1 - m(U))\frac{m_{(M \smallsetminus U)}}{m(M \smallsetminus U)}$$

is a nontrivial convex decomposition of m in $\mathcal{P}(M)$.

Consider M to be a foliated space with 0-dimensional leaves (the points of M). Every probability measure on M is a transverse, invariant measure for this foliation, hence is harmonic. The Dirac measures are clearly ergodic and the converse was actually pointed out in the previous paragraph. Thus, the extremal measures δ_x in $\mathcal{P}(M)$ are exactly the ergodic ones.

If m is a probability measure on M, then the measure $\mu = \delta_* m$, the push-forward of m by δ, is a probability measure on $\mathcal{P}(M)$. The integral of a continuous function f on M with respect to m can thus be written as

$$\int_M f(x) \cdot m(x) = \int_{\mathcal{P}(M)} \left(\int_M f(y) \cdot \delta_x(y) \right) \cdot \mu(\delta_x).$$

That is, an arbitrary probability measure on M can be obtained as a convex integral combination of ergodic probability measures.

In the above example, the ergodic decomposition is quite trivial and it actually seems to obscure the situation. At the opposite end of the dimensional spectrum lie the compact foliated spaces with just one leaf, and for these the ergodic decomposition is even more trivial.

Proposition 2.6.12. *Let $(M, \mathcal{F})$ be a compact foliated space with a fixed metric tensor along the leaves. Let $\mathcal{H}$ denote the subset of $\mathcal{P}(M)$ consisting of those probability measures on M that are harmonic. Then $\mathcal{H}$ is a compact convex metrizable set.*

Proof. The property of being harmonic is preserved by linear operations; hence $\mathcal{H}$ is convex. The space of probability measures on M, endowed with the topology described in Exercise 2.4.8, is compact and metrizable. Because of Theorem 2.2.13, the semigroup D_t acts continuously on this compact set. Proposition 2.4.2 characterizes harmonic measures as the fixed points of this action, and so $\mathcal{H}$ is also compact and metrizable. □

Exercise 2.6.13. Let M be a compact foliated space. Show that the (possibly empty) subspace of $\mathcal{P}(M)$ consisting of those probability measures that are completely invariant (Definition 2.4.12) is also a compact convex set.

The following series of results serves to clarify the relation between ergodic properties of harmonic measures and geometric properties of these measures as points in the convex set $\mathcal{H}$. The relevant geometric property is defined as follows.

Definition 2.6.14. A measure $m \in \mathcal{H}$ is an extreme point of $\mathcal{H}$ if, whenever $m = tm_1 + (1-t)m_2$ holds for $t \in [0,1]$ and $m_1, m_2 \in \mathcal{H}$, then $t = 0$, $t = 1$, or $m_1 = m_2$.

Lemma 2.6.15. *If $m_1, m_2 \in \mathcal{H}$ are ergodic, the following are equivalent.*

(1) $[m_1] = [m_2]$.

(2) $m_1 \ll m_2$.

(3) $m_2 \ll m_1$.

(4) $m_1 = m_2$.

(5) m_1 *and* m_2 *have exactly the same saturated null sets.*

Proof. It is apparent that (4) implies (5). That (5) implies (1) is an easy consequence of the smoothness of these measures (Corollary 2.4.6). It is also immediate to verify that (1) implies both (2) and (3). Smoothness of harmonic measures and ergodicity imply that (2) and (3) are equivalent. To show that (2) implies (4), first note that the fact that $m_1 \ll m_2$ implies that the Radon-Nikodym derivative

$$\frac{dm_1}{dm_2}(x) = h(x)$$

exists, for m_2-almost all $x \in M$. By the Radon–Nikodym theorem, h is a representative of the unique element of $L^1(M, m_2)$ such that

$$m_1(B) = \int_M \chi_B \cdot m_1 = \int_M h\,\chi_B \cdot m_2,$$

for every Borel set B. The local structure of a harmonic measure implies that h may be assumed to be smooth along the leaves and that $\triangle h$ is measurable.

By the characterization of harmonicity of a measure, $\int_M \triangle f \cdot m_1 = 0$ for suitable functions f on M. By the definition of Radon-Nikodym derivative,

$$\int_M \triangle f \cdot m_1 = \int_M h \triangle f \cdot m_2.$$

For smooth functions f with compact support in a foliated chart, the local structure of a harmonic measure and Green's identity give

$$\int_M h \triangle f \cdot m_2 = \int_M f \triangle h \cdot m_2.$$

Since linear combinations of such functions f are dense in $C(M)$, it follows that $\triangle h = 0$ almost everywhere. Thus $D_t h = h$ and, since m_2 is ergodic, Corollary 2.6.6 implies that $h \equiv c$ is constant m_2-almost everywhere. Thus, for every Borel set $B \subset M$,

$$m_1(B) = \int_M \chi_B \cdot m_1 = \int_M c\chi_B \cdot m_2 = cm_2(B).$$

Since m_1 and m_2 are probability measures, $c = 1$ and $m_1 = m_2$. That (3) implies (4) is proved similarly. □

It will be shown that the extreme points in $\mathcal{H}$ are exactly the ergodic measures in $\mathcal{H}$ (Proposition 2.6.18). The following fact will be needed.

Lemma 2.6.16. *Let m be a harmonic measure for M, and let $R \subset M$ be a saturated Borel set. Then the measure m_R on M defined by $m_R(A) = m(A \cap R)$ is also a harmonic measure.*

Proof. First, m_R is a Radon measure (*cf.* Exercise A.2.3). Because of the characterization of harmonic measures provided by Proposition 2.4.2, it suffices to show that $D_t m_R = m_R$. If f is a bounded measurable function on M, then

$$(D_t f)\, \chi_R = D_t(f \chi_R),$$

because R is a saturated subset. Therefore, using the fact that $D_t m = m$,

$$\begin{aligned} \int_M f \cdot (D_t m_R) &= \int_M (D_t f) \cdot m_R \\ &= \int_M D_t(f\chi_R) \cdot m \\ &= \int_M f\chi_R \cdot D_t m \\ &= \int_M f \cdot m_R, \end{aligned}$$

and, since m_R and $D_t m_R$ are Radon measures, $D_t m_R = m_R$. □

Exercise 2.6.17. Measures m and m' are said to be mutually singular (this will be written as $m \perp m'$) if there is a Borel set N such that

$$m(N) = 0 = m'(M \setminus N).$$

Let m and m' be nontrivial harmonic measures on the compact foliated space M and prove that $m = m_s + m_c$, where m_s and m_c are harmonic measures for M such that $m_s \perp m'$ and $m_c \ll m'$. (Hint. The Lebesgue decomposition theorem [**183**, p. 181] provides a Borel subset N of M such that $m'(N) = 0$, $m(M \setminus N) = 0$ and $m_{M \setminus N} \ll m'$. Let S denote the union of leaves L for which the intersection $L \cap N$ has positive Riemannian measure, set $R = M \setminus S$ and use Lemma 2.6.16.)

Proposition 2.6.18. *Let M be a compact foliated space. Then a harmonic probability measure m on M is ergodic if and only if m is an extreme point of $\mathcal{H}$.*

Proof. If m is not ergodic, then it cannot be an extreme point of $\mathcal{H}$. For if B is a leaf-saturated Borel set of measure $0 < m(B) < 1$, then m can be decomposed as

$$m = m_B + m_{M \smallsetminus B}.$$

Each of these measures is nontrivial and harmonic by Lemma 2.6.16, and they are mutually singular by construction. If $m_1 = m_B/m(B)$ and $m_2 = m_{M\smallsetminus B}/m(M \smallsetminus B)$, then these are inequivalent harmonic probability measures and

$$m = tm_1 + (1-t)m_2$$

with $t = m(B)$.

For the converse, suppose that m is an ergodic harmonic probability measure, and that $m = tm_1 + (1-t)m_2$, where m_1 and m_2 are also harmonic probability measures and $0 < t < 1$. Evidently m_1 and m_2 are also ergodic. If $m_1 \neq m_2$, the implication (5) $\Rightarrow$ (4) in Lemma 2.6.15 implies that there is an $\mathcal{F}$-saturated Borel set B such that $m_1(B) > 0$ and $m_2(B) = 0$ (or *vice versa*). By ergodicity of m_1, $m_1(B) = 1$, and so $m(B) = t$. Since $0 < t < 1$, this contradicts the ergodicity of m. □

Next, three fundamental theorems on integral representation (the Bauer maximum principle, the Krein-Milman theorem, and the Choquet theorem) will be applied to the study of harmonic measures. Each of these theorems is a corollary of the next, and in fact the last one is that which provides the integral representation of harmonic measures as convex combinations of ergodic ones. It is nevertheless interesting to see what can be obtained from each of them. Proofs of these theorems can be found in [**30**] or [**149**].

The first result is the fact that a compact convex set has an extreme point. In the particular situation of the compact convex set of probability harmonic measures, this says that ergodic harmonic measures actually exist. The following exercise outlines a proof of this fact.

Exercise 2.6.19. Let M be a compact foliated space with metric tensor, and let $\mathcal{H}$ denote the compact convex set of harmonic probability measures, which is a subset of $C(M)^*$. Establish the following.

(1) If f is a continuous function on M, then the map $\mu \in C(M)^* \to \mu(f)$ has a maximum in $\mathcal{H}$.

(2) Let C be a closed convex subset of $C(M)^*$, and let $K(C, f)$ denote those elements of C where f reaches its maximum. Then $K(C, f)$ is a closed convex subset of C.

(3) Let $Q = \{f_n; n \geq 1\}$ be a countable dense subset of $C(M)$ with $f_1 = \mathbf{1}$. Let $\mathcal{H}_0 = \mathcal{H}$ and, inductively, define

$$\mathcal{H}_{n+1} = K(\mathcal{H}_n, f_{n+1}).$$

Note that $\mathcal{H}_n$ is a nested sequence of compact convex subsets of $\mathcal{H}$. Then $\bigcap_n \mathcal{H}_n$ consists of a single point, which is, in fact, an ergodic measure.

The exercise is an adaptation of the Bauer maximum principle (see [**30**]). The Krein-Milman theorem is the second of the fundamental theorems on integral representation to be used here, and subsumes the Bauer maximum principle. The statement is as follows.

Theorem 2.6.20. *Every compact convex set in a locally convex vector space is the closure of the convex hull of its extreme points.*

The next result gives an application of the Ergodic Theorem and the Krein-Milman theorem.

Proposition 2.6.21. *Let m be a probability measure on M that is diffusion-invariant for a positive time $t_0 > 0$. That is, $D_{t_0}m = m$. Then m is harmonic.*

Proof. Let $\mathcal{P}_0$ be the space of probability measures on M that are diffusion-invariant for a positive time $t_0 > 0$. Clearly, $\mathcal{H} \subset \mathcal{P}_0$.

It is apparent that $\mathcal{P}_0$ is a compact convex set of probability measures. Furthermore, the semigroup property of the operators D_t implies that each D_t leaves $\mathcal{P}_0$ invariant. Also, for $m \in \mathcal{P}_0$ and $t > 0$, the measures m and $D_t m$ are mutually absolutely continuous.

The first part of the proof of Proposition 2.6.18 shows that, if m is an extreme point of $\mathcal{P}_0$, then m is an ergodic measure. The Ergodic Theorem applies to the operator D_{t_0} and to the measure m, and thus, for every continuous function f, the limit

$$\lim_{N\to\infty} \frac{1}{N} \sum_{k=1}^{N-1} D_{t_0}^k f(x)$$

exists and equals $\int_M f \cdot m$, for all x in a saturated set S of full m-measure.

Let $t > 0$ and apply the Ergodic Theorem to the operator D_{t_0} and the D_{t_0}-invariant measure $D_t m$ to obtain a set S_t of full $D_t m$-measure such that, for each continuous function f on M and for all $x \in S_t$,

$$\lim_{N\to\infty} \frac{1}{N} \sum_{k=1}^{N-1} D_{t_0}^k f(x) = \int_M f \cdot D_t m.$$

Since the measures m and $D_t m$ are equivalent, it must be that $\int f \cdot m = \int f \cdot D_t m$, and thus $D_t m = m$.

Therefore, the operators D_t leave fixed every extreme point of $\mathcal{P}_0$ and thus, because of Theorem 2.6.20, every point of $\mathcal{P}_0$. □

The Krein-Milman theorem gives, in fact, an integral representation for points in a compact convex set. Indeed, an equivalent way of stating it is to say that, if K is a compact convex set in a locally convex vector space E and $x \in K$, then there is a measure μ on the closure of the set of extreme points such that x is the barycenter of μ; that is to say, such that

$$f(x) = \int_K f(y) \cdot \mu(y),$$

for every continuous linear functional f on E. (To say that E is a locally convex vector space means that E has a neighborhood base of 0 consisting of convex sets.) However, this result is not sufficient for the integral representation of harmonic measures as convex combinations of ergodic ones, because, as the following example of G. Choquet [**30**] shows, the set of extreme points need not be closed.

Exercise 2.6.22. Let $T = S^1 \times S^1$ be the two-dimensional torus. Let '+' denote the usual group operation on S^1, viewed as $\mathbb{R}/\mathbb{Z}$. Let $f : T \to T$ be the homeomorphism defined by $f(x, y) = (x, x + y)$. Let μ_n denote the measure defined by

$$\mu_n = \frac{1}{n} \sum_{k=0}^{n-1} \delta_{(1/n, k/n)}.$$

(1) Show that each measure μ_n is an invariant and ergodic probability measure for f.

(2) Show that the sequence of measures μ_n converges to an invariant measure μ that is not ergodic.

The third and final theorem on integral representation is Choquet's theorem, which refines the Krein-Milman theorem and, in particular, takes care of the problem exhibited by the previous example. Its statement follows.

Theorem 2.6.23. *Let K be a compact convex metrizable subset of a locally convex topological vector space E. Then every point $x \in K$ is the barycenter of a probability measure μ on K that is concentrated on the closure of the set of extreme points of K.*

That a measure μ is concentrated in the set $\mathcal{E}$ of extreme points of K means that $\mu(K \setminus \mathcal{E}) = 0$. It cannot be said that the support of μ is $\mathcal{E}$, because the set of extreme points need not be closed (but, under the metrizability assumptions of the theorem, it is a Borel set).

Culminating this lengthy discussion of compact convex sets and integral representation, and as a corollary to Choquet's theorem, the ergodic decomposition of harmonic measures can now be proven.

Theorem 2.6.24. *Let $(M, \mathcal{F})$ be a compact foliated space with metric tensor. Let $\mathcal{E}$ denote the subspace of $\mathcal{P}(M)$ consisting of the harmonic probability measures that are ergodic. If m is a harmonic probability measure on M, then there is a unique probability measure ν on $\mathcal{P}(M)$, concentrated in $\mathcal{E}$, such that*

$$\int_M f(x) \cdot m(x) = \int_{\mathcal{E}} \left(\int_M f(x) \cdot e(x) \right) \cdot \nu(e).$$

Proof. As has already been said, this is a corollary of Choquet's theorem. The vector space $E = C(M)^*$ is the space of signed Radon measures on M. Every continuous function on M defines, by double duality, a linear functional on the space of measures on M. The compact convex set is the set of harmonic probability measures $\mathcal{H}$. Its extreme points are the ergodic harmonic measures, as was proven in Proposition 2.6.18. Thus, given $m \in \mathcal{H}$, there is a measure ν, concentrated on the set of ergodic harmonic measures, such that

$$\int_M f(x) \cdot m(x) = \int_{\mathcal{E}} \left(\int_M f(x) \cdot e(x) \right) \cdot \nu(e),$$

for every continuous function f on M. □

The rest of this section discusses a different approach to the ergodic decomposition of a harmonic measure. This approach originates with Kryloff and Bogoliuboff [**116**] (for the case of flows on compact metric spaces) and the version presented here follows Yosida's [**191**] generalization to Markov processes. It is also possible to carry out the Kryloff and Bogoliuboff approach for the shift transformations on the path space $\Omega(M)$, which, in fact, gives a canonical construction of the sets of regular points given below. The version presented here still depends on the results from the integral representation theory just described (specifically, on Proposition 2.6.21), but, as it has some applications to dynamics of foliated spaces, a small discussion seems appropriate.

Definition 2.6.25. Let M be a compact foliated space with metric tensor. A subset $B \subset M$ is of *harmonic measure zero* if $m(B) = 0$ for every harmonic probability measure on M. A subset $B \subset M$ is of *harmonic measure one* if its complement is of harmonic measure zero.

Definition 2.6.26. A point $x \in M$ is a *quasi-regular point* if the limit

$$\lim_{N\to\infty} \frac{1}{N} \sum_{k=0}^{N-1} D_k f(x)$$

exists, for each continuous function f on M.

The relevance of quasi-regular points is the point of the following observation. If x is such a point, then there is a well defined positive linear functional $C(M) \to \mathbb{R}$ that assigns to a continuous function $f \in C(M)$ the value $f^*(x)$ given by the limit

$$f^*(x) = \lim_{N\to\infty} \frac{1}{N} \sum_{k=0}^{N-1} D_k f(x).$$

This positive linear functional is therefore represented by a probability measure on M (because $f^*(x) \geq 0$, if $f \geq 0$, and $\mathbf{1}^*(x) = 1$), which is called the *diffused measure* of x and is denoted by δ_x^*.

Proposition 2.6.27. *The set Q of quasi-regular points of the foliated space $(M, \mathcal{F})$ has the following properties.*

(1) *Q is a Borel set of harmonic measure one.*

(2) *If $x \in Q$, then the measure δ_x^* is a harmonic probability measure.*

Proof. If f is a continuous function on M, then the set of points $x \in M$ for which the limit

$$\lim_{N\to\infty} \frac{1}{N} \sum_{k=0}^{N-1} D_k f(x)$$

exists is a Borel subset of M. (It is the set of points where a sequence of continuous functions converges.) Let Q_f denote this set of points. By the Ergodic Theorem, the complement of Q_f is of harmonic measure zero. Since M is separable, there exist a countable subset $F \subset C(M)$ that is dense for the supremum norm topology. Then the intersection $\bigcap_{f\in F} Q_f$ is a Borel set of harmonic measure one.

Next, we show that this intersection coincides with the set of quasi-regular points. Let $f \in C(M)$ and let $\varepsilon > 0$. Then there exists $f_0 \in F$ such that the norm $\|f - f_0\| < \varepsilon/3$; hence $\|D_t f - D_t f_0\| < \varepsilon/3$ by the contraction property of the diffusion operators. Since the limit defining $f_0^*(x)$ exists by construction, there is an index N such that, if $N_1, N_2 \geq N$, then

$$\left| \frac{1}{N_1} \sum_{k=0}^{N_1-1} D_k f_0(x) - \frac{1}{N_2} \sum_{k=0}^{N_2-1} D_k f_0(x) \right| < \varepsilon/3.$$

Therefore, by a standard $\varepsilon/3$-argument, the limit that defines $f^*(x)$ also exists.

Regarding (2), it will be shown that, if $x \in Q$ is a quasi-regular point, then the measure δ_x^* is invariant under the operators D_s, s a positive integer. By Proposition 2.6.21, this is equivalent to being harmonic. For a continuous function f,

$$f^*(x) = \lim_{N\to\infty} \frac{1}{N} \sum_{k=1}^{N-1} D_k f(x).$$

For any integer time $s > 0$,

$$\begin{aligned}
\int_M D_s f(y) \cdot \delta_x^*(y) &= \lim_{N\to\infty} \frac{1}{N} \sum_{k=0}^{N-1} D_{k+s} f(x) \\
&= \lim_{N\to\infty} \frac{1}{N} \sum_{k=s}^{N+s-1} D_k f(x) \\
&= \lim_{N\to\infty} \frac{1}{N} \sum_{k=0}^{N-1} D_k f(x) \\
&\quad + \lim_{N\to\infty} \frac{1}{N} \left(\sum_{k=N}^{N+s-1} - \sum_{k=0}^{s-1} \right) D_k f(x).
\end{aligned}$$

Since f is continuous and $\|D_t f\| \leq \|f\|$, the second limit is bounded in absolute value by

$$\lim_{N\to\infty} \frac{2s\|f\|}{N} = 0.$$

This shows that

$$\int_M D_s f(y) \cdot \delta_x^*(y) = \int_M f(y) \cdot \delta_x^*(y),$$

so that δ_x^* is D_s-invariant, as advertised. □

In general, there is no apparent reason why the set of quasi-regular points Q should be saturated. Nevertheless, given a harmonic measure m, the Ergodic Theorem guarantees that Q contains a Borel saturated subset $Q(m)$ of total m-measure on which the limit

$$f^*(x) = \lim_{N\to\infty} \frac{1}{N} \sum_{k=0}^{N-1} D_k f(x)$$

exists and is leafwise constant for each continuous function $f \in C(M)$. Indeed, let $F \subset C(M)$ be a countable dense subset. An application of the Ergodic Theorem gives, for each $f \in F$, a Borel saturated set $Q(f) \subset Q$ of full m-measure where f^* is leafwise constant. Then $Q(m) = \bigcap_{f\in F} Q(f)$ is a Borel saturated set of regular points and of full m-measure. Furthermore, given $g \in C(M)$ and $\varepsilon > 0$, let $f \in F$ be such that $\|g - f\| \leq \varepsilon$. If x and y are two points in the same leaf of $Q(m)$, then $f^*(x) = f^*(y)$, and so

$$|g^*(x) - g^*(y)| \leq |g^*(x) - f^*(x)| + |g^*(y) - f^*(y)| \leq 2\varepsilon,$$

implying that g is also constant on each leaf of $Q(m)$.

Because of the several choices made during its construction, this set $Q(m)$ is not canonically associated to m. However, any other Borel subset of M with the properties of $Q(m)$ differs from this by a set of m-measure 0.

It follows that, for every continuous function f on M,

$$\int_M f(x)\cdot m(x) = \int_{Q(m)} \left(\int f(y)\cdot \delta_x^*(y)\right)\cdot m(x),$$

so that the measure m is a convex combination of the harmonic measures δ_x^*, $x \in Q(m)$. This is not an ergodic decomposition of m because the measures δ_x^* need not all be ergodic.

Definition 2.6.28. A point $x \in M$ is called a *transitive point* if $x \in Q$ and

$$\int_Q \left(f^*(y) - f^*(x)\right)^2 \cdot \delta_x^*(y) = 0$$

for every continuous function f on M.

Lemma 2.6.29. *The set Q_T of transitive points is of harmonic measure one. If $x \in Q_T$ is a transitive point, then the harmonic measure δ_x^* is ergodic.*

This is left as an exercise. The proof is given in [**191**, Chapter XIII].

Definition 2.6.30. A point $x \in M$ is called a density point if it is a quasi-regular point and δ_x^* is a harmonic measure that contains x in its support. The set of density points is denoted by Q_D.

Thus a quasi-regular point x is a density point if and only if $f^*(x) > 0$ for every nonnegative continuous function f such that the set $\{f > 0\}$ is a neighborhood of x.

Lemma 2.6.31. *The set Q_D of density points is a Borel set of harmonic measure one.*

Proof. Let $\{U_n\}_{n=1}^{\infty}$ be a countable base for the topology of M, and let f_n be a nonnegative continuous function such that $f_n > 0$ on U_n. Let B_n be the set of points $x \in M$ for which

$$\lim_{N\to\infty} \frac{1}{N} \sum_{k=1}^{N-1} D_k f_n(x) > 0$$

or $f_n(x) = 0$. This is a Borel set that has measure 1 with respect to every harmonic probability measure. Then $Q \cap \left(\bigcap_{n=1}^{\infty} B_n\right)$ is Borel and of harmonic measure one. It clearly contains the set of density points and, conversely, if $x \in Q \cap \left(\bigcap_{n=1}^{\infty} B_n\right)$ and $f \geq 0$ is continuous with $f(x) > 0$, then there exists f_n in the family above such that $f_n \leq f$ and $f_n(x) > 0$. Since x is also a regular point, it follows at once that $f^*(x) > 0$, thus showing that x is a density point. □

Definition 2.6.32. The set of *regular points* on the foliated space M is the intersection $R = Q_D \cap Q_T$.

Thus the set of regular points is a set of harmonic measure one. It is not necessarily a saturated subset of M, but, given a harmonic measure m, there is a Borel saturated subset $R(m)$ of regular points and of total m-measure. If f is a continuous function on M, then the discussion above shows that

$$\int_M f(x) \cdot m(x) = \int_R \left(\int f(x) \cdot \delta_y^*(x) \right) \cdot m(x),$$

expressing m as a convex combination of ergodic harmonic measures. That is, we have given an ergodic decomposition of m into harmonic measures.

The set $R(m)$ is called a regular set for the measure m. Note that it is not canonically associated to m, but, by virtue of the Ergodic Theorem, any other such set differs from this one by a set of m-measure 0.

2.7. Recurrence

In this final section we turn to the phenomenon of recurrence in a foliated space, and offer two theorems. The first, Theorem 2.7.4, is, in fact, a corollary (due to Garnett) of the work done in the previous section and shows that harmonic measures are adequate to study this phenomenon. That is, they cannot give positive weight to the family of leaves that do not matter much (the so-called wandering leaves, defined below). The second, Theorem 2.7.12, is due to E. Ghys, and has a spectacular application to the topology of the leaves of a compact foliated space.

Definition 2.7.1. A leaf L of a foliated space is called a *wandering leaf* if it is proper and noncompact. A point of a foliated space is said to be a *wandering point* if it lies on a wandering leaf, and the set of all wandering points is the *wandering set*. The complement of the wandering set is called the *nonwandering set*.

Thus, the nonwandering set consists of leaves that recur arbitrarily near themselves. This is the dynamically interesting part of the foliation.

Exercise 2.7.2. The union of the compact leaves of a foliated space is a Borel saturated subset.

Exercise 2.7.3. The wandering set is a saturated Borel subset of the foliated space; hence so is the nonwandering set.

Theorem 2.7.4. *Let $(M, \mathcal{F})$ be a compact foliated space with metric tensor. Then the wandering set has measure zero with respect to any harmonic measure for M.*

Proof. Let m be a harmonic probability measure for M. By the arguments of the previuous section, there is a Borel saturated set of full m-measure consisting of density points. Thus, it suffices to show that the set of density points Q_D of M contains no nonwandering leaf. (It would have been more natural to use the regular points, had Lemma 2.6.29 been proven.)

Suppose, on the contrary, that there is a wandering leaf L contained in the set of density points, and let $x \in L$. Since L is wandering, there is a relatively compact, foliated chart $U = D \times Z$, containing x and such that $L \cap U$ is a single plaque. Let f be a continuous function, supported on $\overline{U}$, positive on $U \setminus L \cap U$ and identically zero on $L \cap U$. Then $f|L \equiv 0$. Since x is a density point, the probability measure δ_x^* is harmonic, and the fact that

$$\int_M f(y) \cdot \delta_x^*(y) = f^*(x) = 0$$

implies that $\delta_x^*(U \setminus L \cap U) = 0$. Since $x \in \operatorname{supp}(\delta_x^*)$, we see that $\delta_x^*(L \cap U)$ is strictly positive. The saturation W of U is open, $\delta_x^*(W) > 0$, and, by Corollary 2.4.7, $\delta_x^*(W \setminus L) = 0$. By the same corollary, every point of L is in the support of δ_x^*. Let $\{U_i = D_i \times Z_i\}_{i \in I}$ be a family of relatively compact, foliated charts contained in W, covering L and such that each $L \cap U_i$ is a single plaque $P_i = D_i \times \{z_i\}$. The measure $\delta_x^*|U_i$ is concentrated on P_i, where its density is of the form $h_i(y)dy$ with h_i positive and harmonic (Proposition 2.4.10). The fact that $\delta_x^*|U_i \cap U_j$ can be expressed relative to either chart implies that h_i agrees with h_j on $P_i \cap P_j$. These piece together to give a global, positive, harmonic function h on L. Thus

$$0 < \delta_x^*(W) = \delta_x^*(L) = \int_L h(y) \cdot dy \leq 1.$$

Therefore, hdy is a finite harmonic measure on L. Since L is not compact, it has infinite volume; hence h cannot be constant. Corollary 2.5.7 then gives the desired contradiction. □

Exercise 2.7.5. Let M be a compact foliated space every leaf of which is compact. Show that every harmonic measure on M is completely invariant.

The rest of this section presents a recurrence result of Ghys [**80**]. This theorem requires one result from the theory of measure-preserving transformations, due to M. Kac.

Theorem 2.7.6. *Let S be a measure-preserving transformation of the probability space (X, ν). Let B be a measurable subset of X, and let F_B denote* *the "first return time to B" function, defined by*

$$F_B(x) = \inf\{n > 0 \mid S^n(x) \in B\},$$

with the convention that the infimum of the empty set is ∞. Then F_B is ν-integrable on B and

$$\int_B F_B(x) \cdot \nu(x) = \nu\left(\bigcup_{n=0}^{\infty} S^{-n}B\right).$$

The statement is stronger than the well-known Poincaré recurrence theorem (a proof of which is sketched in the next exercise), which says that ν-almost every point of B returns to B. With the assumption that S is ergodic it was proven by Kac [**108**]; the version stated here is due to F. Wright [**190**]. K. Petersen [**147**] gives three proofs of this result.

Exercise 2.7.7. Let (X, ν), S and B be as above. The objective of this exercise is to prove that $F_B < \infty$ ν-almost everywhere on B or, equivalently, that $B \subset \bigcup_{n=1}^{\infty} S^{-n}B$, up to a set of zero ν-measure.

(1) If that is not the case, show that the set $A = \{x \in B \mid F_B(x) = \infty\}$ has measure $\nu(A) > 0$ (which includes showing that A is measurable).

(2) Show that A is disjoint from $S^{-n}A$ for all $n > 0$.

(3) Conclude that the sets $S^{-n}A$, $n > 0$, are pairwise disjoint.

(4) Using the fact that S is measure-preserving and that $\nu(X) < \infty$, arrive at a contradiction.

Proof of Theorem 2.7.6. First, F_B will be shown to be measurable. Let A_n denote the set of points $x \in B$ such that $F_B(x) = n$. Let $B_0 = B$ and, for $n \geq 1$, let B_n denote the set of points $x \in B^c = X \setminus B$ for which $F_B(x) = n$. Then measurability of F_B follows by writing

$$A_n = B \cap S^{-1}B^c \cap \cdots \cap S^{-n+1}B^c \cap S^{-n}B$$

and

$$B_n = B^c \cap S^{-1}B^c \cap \cdots \cap S^{-n+1}B^c \cap S^{-n}B.$$

The sets A_n are pairwise disjoint, so the finiteness of the measure and Poincaré's recurrence imply that

$$\nu(B_0) = \sum_{n=1}^{\infty} \nu(A_n) < \infty. \tag{1}$$

In particular, $\lim_{n\to\infty} \nu(A_n) = 0$.

The sets B_n are pairwise disjoint and

$$\bigcup_{n=0}^{\infty} S^{-n}B = \bigcup_{n=0}^{\infty} B_n.$$

Hence

$$\sum_{n=0}^{\infty} \nu(B_n) = \nu(\bigcup_{n=0}^{\infty} S^{-n}B) < \infty. \tag{2}$$

By construction,

$$B_{n+1} \cup A_{n+1} = S^{-1}B_n$$

for $n \geq 0$. Thus, since A_n and B_n are disjoint and the measure is invariant,

$$\nu(B_n) = \nu(B_{n+1}) + \nu(A_{n+1}).$$

Iteration of this identity, together with convergence of the series (1) and (2), provides

$$\nu(B_n) = \sum_{k=1}^{\infty} \nu(A_{n+k}).$$

Therefore

$$\begin{aligned}\int_B F_B(x)\,d\nu(x) &= \sum_{n=1}^{\infty} n\nu(A_n) \\ &= \sum_{k=0}^{\infty} \sum_{n=k+1}^{\infty} \nu(A_k) \\ &= \sum_{k=0}^{\infty} \nu(B_k) < \infty.\end{aligned}$$

□

Let L be a complete Riemannian manifold of bounded geometry (for example, a leaf of a compact, foliated space M). Let $D \subset L$ be an unbounded regular domain in L with compact boundary. Define the "first exit time" from D,

$$T_D : \Omega(L) \to [0, \infty],$$

by

$$T_D(\omega) = \inf\{t > 0 \mid \omega(t) \in L \setminus D\},$$

with the convention that the infimum of the empty set is ∞. Because of bounded geometry and the fact that ∂D is compact, the unbounded domain D has infinite volume. Thus, it is reasonable to expect the "average" continuous path in L, issuing from a point $x \in D$, to wander about for an arbitrarily long time before finding (if ever) the compact exit ∂D. The formal statement is the following.

Proposition 2.7.8. *Let D be an unbounded regular domain in L with compact boundary. Then, for every $x \in D$, the expectation*

$$E_x[T_D] = \int_{\Omega(L)} T_D(\omega) \cdot P_x(\omega) = \infty.$$

Proof. For each $x \in L$, set $f(x) = E_x[T_D]$. Then $f(x) > 0$ for every $x \in D$, because $T_D(\omega) > 0$ if $\omega(0) \in D$. Likewise, $f|(L \smallsetminus D) \equiv 0$. Theorem C.8.9 gives two possibilities for f: either $f \equiv +\infty$ on D, or else f is finite on L. It will be shown that the second possibility implies that D has finite volume, contradicting the hypothesis that D is an unbounded regular domain.

If f is finite, then it has the following properties:

(1) because of Theorem C.8.9, $\triangle f \equiv -1$ on D, and

(2) because of Proposition C.9.7, f is a proper function on $\overline{D}$.

The elliptic equation $\triangle f = -\mathbf{1}$ implies that f is of class C^∞ on D, and thus, because of Sard's theorem, the set of regular values of f is residual in $\mathbb{R}$. For $a < b$ regular values of f, let $C[a,b] = f^{-1}[a,b]$ be the region of D bounded by $C_a = f^{-1}(a)$ and $C_b = f^{-1}(b)$. By property (2), $C[a,b]$ is compact for $a > 0$, and by (1), its volume is

$$\text{vol } C[a,b] = \int_{C[a,b]} -\triangle f = \int_{\partial C[a,b]} -\langle \operatorname{grad} f, \mathbf{n} \rangle,$$

the last equality by Theorem B.1.5. On C_a (respectively, on C_b), $\operatorname{grad} f$ is a negative multiple of the outward normal vector $\mathbf{n}$ (respectively, a positive multiple). Therefore

$$\text{vol } C[a,b] = \int_{C_a} |\operatorname{grad} f| - \int_{C_b} |\operatorname{grad} f|.$$

Letting b increase to ∞ through regular values, we see that

$$\text{vol } C[a,\infty) \leq \int_{C_a} |\operatorname{grad} f| < \infty.$$

Property (2) implies that $C[a,\infty)$ differs from D by a relatively compact set, and thus that D has finite volume. □

The next result makes explicit the phenomenon of recurrence.

Proposition 2.7.9. *Let M have a harmonic measure m and let B be a Borel subset of M. Then, for m-almost all x in B, the intersection $B \cap L_x$ approaches all ends of L_x.*

Proof. Let $F_B : \Omega(M) \to \{1, 2, 3, \ldots, \infty\}$ denote the first positive integral return time to B,

$$F_B(w) = \inf\{n > 0 \mid \omega(n) \in B\},$$

understood to be ∞ if ω never visits B at time t a positive integer. By Theorems 2.3.7 and 2.7.6, if $\Omega(B) = \{\omega \mid \omega(0) \in B\}$, then

$$\int_{\Omega(B)} F_B(\omega) \cdot \mu(\omega) = \int_B E_x[F_B] \cdot m(x) < \infty;$$

hence the set $\{x \in B \mid E_x[F_B] = \infty\}$ has m-measure zero.

Suppose that x is a point of B for which there exists an unbounded domain D in L_x, with compact boundary, such that $D \cap B = \emptyset$. Let A be the set of paths $\omega \in \Omega(M)$ such that $\omega(1) \in D$. Then A is a measurable subset of $\Omega(M)$ and

$$P_x(A) = \int_D p(x, y; 1) \cdot dy > 0.$$

For paths $\omega \in A$,

$$F_B(\omega) \geq 1 + T_D \circ \theta_1(\omega),$$

because Brownian particles move continuously. That is, the positive integral first return time $F_B(\omega)$ cannot be 1, since $\omega \in A$ visits D at time 1 and $D \cap B = \emptyset$. After that visit, ω cannot return to B before exiting D.

Upon integration, we see that

$$\begin{aligned} E_x[F_B] &\geq \int_A F_B(\omega) \cdot P_x(\omega) \\ &\geq \int_A (1 + T_D(\theta_1(\omega))) \cdot P_x(\omega) \\ &\geq \int_A T_D(\theta_1(\omega)) \cdot P_x(\omega) \\ &= \int_D p(x, y; 1) E_y[T_D] \cdot dy, \end{aligned}$$

where the equality is on account of the strong Markov property C.5.14. Because of Proposition 2.7.8,

$$E_y[T_D] = \infty,$$

for all $y \in D$; thus $E_x[F_B] = \infty$. By the first part of the proof, this can only happen on an m-null set of points x in B. □

The "Proposition Fondamentale" of Ghys [**80**] will serve as colophon to this chapter. One technical detail needs to be taken care of first.

Definition 2.7.10. A Borel subset B of M is good if, for each foliated chart U for M, the saturation of $B \cap U$ in U is a Borel set.

The point of this definition is that the saturation of a good Borel set is Borel, while this fails for general Borel sets (*cf.* Exercise 1.6.12).

Exercise 2.7.11. Show that the collection of all good Borel subsets of M is a σ-field of subsets of M that contains the σ-ring of all Borel transversals.

Theorem 2.7.12 (Proposition Fondamentale of Ghys). *Let M be a compact foliated space with harmonic measure m, and let B be a good Borel subset of M. Then, for m-almost every $x \in M$, the intersection $L_x \cap B$ either is empty or approaches all ends of L_x.*

Proof. That $L_x \cap B$ approaches all ends of L_x means that, if $D \subset L_x$ is an unbounded domain with compact boundary, then $D \cap B \neq \emptyset$. Let $\mathfrak{U} = \{U_i\}_{i \in N}$ be a regular cover of M by foliated charts. It may be assumed that B is a good Borel subset of one of the flow boxes in $\mathfrak{U}$ (in general, it is a finite union of such sets). For each finite sequence $J = \{i_1, \ldots, i_j\}$ of indices in N of length $j \leq k$, let B_J be the set

$$B_J = r(G_B \cap (G(U_{i_1}) \cdot G(U_{i_2}) \cdots G(U_{i_j}))).$$

Here, we recall that $r : G(M) \to M$ is the range map of the graph of the foliated space M and $G_B \subset G(M)$ is the subset with source in B. Then $B_k = \bigcup_{|I| \leq k} B_I$ is a Borel set and the union $\bigcup_k B_k$ is exactly the saturation of B.

For each k, $B_k \cap L_x$ approaches all ends of L_x if and only if $B \cap L_x$ approaches all ends of L_x. By Proposition 2.7.9, the intersection $B_k \cap L_x$, and hence $B \cap L_x$, approaches all ends of L_x for m-almost all points x in B_k. Passing to the union of the B_k's, we see that, for m-almost every point x in the saturation of B, $B \cap L_x$ approaches all ends of L_x. □

Exercise 2.7.13. With the hypothesis of Theorem 2.7.12, verify that the set B_k introduced during the proof is a Borel subset of M.

Exercise 2.7.14. Refer to Exercise C.9.4. Let M be the 3-torus densely foliated by planes of irrational slope. Let V denote a linear vector field along the leaves of the foliation. Fix $\alpha \in [0, 2\pi]$, and for each $x \in M$ let $C(x, \alpha)$ denote the cone in the leaf through x spanned by the line segments between those issuing from x with tangents $V(x)$ and $e^{i\alpha}V(x)$ (counterclockwise). Let B be a good Borel subset of M. Show that if, $\alpha > 2\pi/3$, then the intersection $B \cap C(x, \alpha)$ is unbounded in the leaf through x, for almost all $x \in B$. Can you improve α?

Chapter 3

Generic Leaves

By a well known theorem of Hopf [**100**], a regular covering of a compact, connected manifold has 0, 1, 2 or a Cantor set of ends. It is also known that, in compact, C^2-foliated manifolds without holonomy, the common diffeomorphism type of the leaves has 0, 1 or 2 ends [**19**]. In this chapter we will show that, in a suitable sense, the "generic" leaf of a foliation has 0, 1, 2 or a Cantor set of ends. A profound theorem of this type, due to E. Ghys [**80**], takes "generic leaf" to mean "almost every leaf" relative to a harmonic measure on the foliated space. Related results [**24**] take "generic" in the topological sense of "residual" [I, page 65]. We will prove Ghys's result and a fairly easy case of its topological analogue.

3.1. The Main Results and Examples

Before stating the main theorems, we discuss some examples of foliated manifolds, illustrating all nine of the noncompact, two-dimensional generic leaf types. For the leaves with one or two ends, we make a point of choosing examples in which not every leaf is of the generic type. For each of the six types with finite endset, it is easy to construct examples of foliated 3-manifolds without holonomy in which all leaves are of that type.

Example 3.1.1. In [I, Example 13.3.22], we constructed a minimal foliation of a closed 3-manifold in which each leaf had a Cantor set of ends. This construction is due to M. W. Hirsch [**98**]. Countably many of these leaves have genus one, the rest have genus zero. Thus, the generic leaf has no handles and a Cantor set of ends. One also notes that these generic leaves have trivial holonomy. Deleting the normal neighborhood of a closed transversal and sewing in $H \times S^1$, where H is a handle or crosscap, produces the other two types of generic leaf with a Cantor set of ends, one with handles clustering on the endset, the other with crosscaps clustering on the endset.

Example 3.1.2. The one-ended surfaces occurring generically as leaves are the plane, the plane with infinitely many handles and the plane with infinitely many crosscaps. If we can produce a minimal, foliated 3-manifold with the plane as generic leaf, then the trick employed in the previous example produces the other two types. In [I, Example 1.1.9], we constructed the dense-leaved foliation of the unit tangent bundle $M = T^1(\Sigma)$ of the 2-holed torus Σ that lifts to the foliation of $T^1(\mathbf{H})$ by geodesic pencils. The group of covering transformations consists of hyperbolic elements of $\mathrm{Psl}(2, \mathbb{R})$, each having two fixed points in $\mathbb{R} \cup \{\infty\}$. The pencil issuing from such a fixed point descends to a cylindrical leaf in M, the remaining pencils descending to planar leaves. Since the group $\pi_1(\Sigma)$ is countable, there are only countably many cylindrical leaves, so the generic leaves are planes. These also are exactly the leaves with trivial holonomy.

Example 3.1.3. The three remaining noncompact surfaces that can occur generically as leaves are the cylinder, the 2-ended ladder [I, Figure 4.1.3] and the cylinder with crosscaps clustering at both ends.

Consider the unimodular matrix

$$A = \begin{bmatrix} 2 & 1 \\ 1 & 1 \end{bmatrix}.$$

As a linear transformation, this maps the integer lattice $\mathbb{Z}^2$ onto itself, thereby inducing a diffeomorphism $f : T^2 \to T^2$. This is an Anosov diffeomorphism with interesting and well understood dynamics (*cf.* [**111**, pp. 84–86]). One of the features of interest is that the set of periodic points is countable and dense. Let F be a closed, connected surface of genus at least 1, let $\varphi : \pi_1(F) \to \{f^n\}_{n\in\mathbb{Z}}$ be a group surjection and form the suspension of φ, a foliated torus bundle $\pi : M \to F$. There are countably many compact leaves, all finite coverings of F, corresponding to the f-periodic points. The remaining leaves form a generic family of 2-ended leaves. Some of these leaves have both ends dense in M, some only one, and some have neither end dense. Each nondense end is asymptotic to a minimal set that may or may not be a compact leaf. If $F = T^2$, the generic leaves are cylinders. If F has higher genus and is orientable, the generic leaves are 2-ended ladders. If F has higher genus and is nonorientable, φ is chosen to annihilate an orientation-reversing loop and the generic leaves have both ends nonorientable.

Observe that Lebesgue measure on T^2 is invariant under f (since A is unimodular), so there is a holonomy-invariant measure of full support, and hence a completely invariant harmonic measure m fully supported on

M. The countably many compact leaves have m-measure zero, so m-almost every leaf is 2-ended.

Examples 3.1.1 and 3.1.2 were minimal foliations. They will exemplify both Theorem 3.1.4 and Theorem 3.1.5 below. Example 3.1.3 is far from minimal and exemplifies only Theorem 3.1.4. A more general version of Theorem 3.1.5 [**24**] does, in fact, cover this example. It would be interesting to know whether the topological types detected by each theorem in a given minimal foliated space could be different.

Theorem 3.1.4 (Ghys). *Let $(M, \mathcal{F})$ be a compact, leafwise C^3 foliated space and let m be an ergodic harmonic probability measure for $(M, \mathcal{F})$. Then there is an $\mathcal{F}$-saturated set G_m with $m(G_m) = 1$ for which one of the following holds:*

(1) *G_m reduces to a single compact leaf;*

(2) *every leaf in G_m has one end;*

(3) *every leaf in G_m has two ends;*

(4) *every leaf in G_m has a Cantor set of ends.*

If, in addition, the leaf dimension is $p = 2$ and G_m does not reduce to a compact leaf, then either all leaves in G_m have genus zero, or all leaves in G_m are orientable and have only nonplanar ends, or all leaves in G_m have only nonorientable ends.

Remark. If the harmonic measure m is not ergodic, Theorem 2.6.24 implies that the union of the supports of its ergodic components has full m-measure. Thus, m-almost every leaf is of one or another of the types in Theorem 3.1.4, and several of these types can occur simultaneously in $\operatorname{supp} m$.

Theorem 3.1.5 (Cantwell-Conlon). *Let the compact foliated space $(M, \mathcal{F})$ be minimal. Then there is a residual family G of leaves without holonomy for which one of the following holds:*

(1) *$G = M$ reduces to a single compact leaf;*

(2) *every leaf in G has one end;*

(3) *every leaf in G has two ends;*

(4) *every leaf in G has a Cantor set of ends.*

If, in addition, the leaf dimension is $p = 2$ and G does not reduce to a compact leaf, then either all leaves in G have genus zero, or all leaves in G are orientable and have only nonplanar ends, or all leaves in G have only nonorientable ends.

In comparing these two theorems, observe that the support of an ergodic harmonic measure can be viewed as a measure-theoretic analogue of a minimal set, while the topological analogue of a set of full measure is a residual set. The key to proving Theorem 3.1.4 is the fundamental recurrence result, Theorem 2.7.12. Because of the ergodicity hypothesis, we will be able to formulate a proof somewhat different from the one in [**80**], but closely analogous to the proof of Theorem 3.1.5.

Exercise 3.1.6. A leaf L of a foliated space $(M, \mathcal{F})$ is *totally recurrent* if $\lim_e L = M$, $\forall e \in \mathcal{E}(L)$, where $\lim_e L$ is the asymptote of the end e [I, Definition 4.3.1]. It should be evident that, in a minimal foliation, every leaf is totally recurrent. If m is an ergodic harmonic measure for $(M, \mathcal{F})$ and $X = \operatorname{supp} m$, prove that m-almost every leaf of the foliated space $(X, \mathcal{F}|X)$ is totally recurrent (in X).

3.2. The Holonomy Graph

The investigation of the endset of the generic leaf L will be reduced to an investigation of the endset of an associated 1-complex L^*, called the *holonomy graph* of L. This is an analogue of the Cayley graph of a group and is not to be confused with the graph of a foliated space treated in Chapter 1.

Let $(M, \mathcal{F})$ be a compact foliated space of leaf dimension p, modeled transversely on a locally compact, complete, separable metric space (T, ρ), and let $\mathcal{U} = \{U_i, x_i, z_i\}_{i=1}^K$ be a regular foliated atlas. We will make no smoothness hypotheses. The coordinates define homeomorphisms

$$U_i \cong D \times Z_i,$$

where we can take D to be the open unit ball in $\mathbb{R}^p$, and $Z_i \subset T$ is an open, relatively compact subset. The associated holonomy cocycle $\{\gamma_{ij}\}_{1 \le i,j \le K}$ generates the holonomy pseudogroup Γ on

$$Z = \bigcup_{i=1}^{K} Z_i.$$

As usual, we can arrange that the transverse sets $Z_i = z_i(U_i)$ have pairwise disjoint closures in T, we can identify Z_i with the set of plaques in U_i, or we can view these sets as imbedded subspaces of M, transverse to $\mathcal{F}$ and pairwise disjoint. If L_x is the leaf of $\mathcal{F}$ through $x \in Z$, then L_x^* will designate the 1-complex with vertices the $\mathcal{U}$-plaques in L_x, plaques P and Q being joined by an edge if and only if $P \cap Q \neq \emptyset$. If $\gamma_{ij}(P) = Q$, the edge directed from P to Q is labeled by γ_{ij}, the same edge directed from Q to P being labeled by γ_{ji}. This is the Cayley graph of the holonomy orbit $\Gamma(x)$.

There is a complete metric d_{L^*} on $L^* = L_x^*$ defined by requiring that each edge be isometric to the unit interval. Observe that the definition and properties of the endset of a manifold [I, Section 4.2] extend readily to this connected metric space, giving us a compact, totally disconnected, metrizable space $\mathcal{E}(L^*)$ of ends of the graph.

In [I, page 299], we saw that the metric $d = d_M$ on M induces a distance d_L on any leaf L, based on paths and compatible with the manifold topology of L. Note that, relative to this metric, there is a uniform, finite upper bound to the diameters of plaques. Thus, a sequence $\{x_k\}_{k=1}^{\infty}$ converges to an end $e \in \mathcal{E}(L)$ if and only if $\{P_k\}_{k=1}^{\infty}$ also converges to e, for every sequence of plaques such that $x_k \in P_k$, $\forall\, k \geq 1$. Then, as a sequence of vertices of the graph L^*, $\{P_k\}_{k=1}^{\infty}$ also converges to an end $e^* \in \mathcal{E}(L^*)$.

Exercise 3.2.1. Carry out the details of the above remarks, proving that the map $e \mapsto e^*$ is a well defined homeomorphism, canonically identifying the spaces $\mathcal{E}(L)$ and $\mathcal{E}(L^*)$.

Note that $L_x \mapsto \Gamma(x) \mapsto L_x^*$ defines canonical one-to-one correspondences between the set of leaves of $\mathcal{F}$, the set of Γ-orbits, and the set of graphs of these Γ-orbits. The $\mathcal{F}$-saturated subsets $Y \subseteq M$ correspond exactly to the Γ-invariant subsets of Z by $Y \leftrightarrow Y \cap Z$.

3.2.A. Borel properties. For the proof of Theorem 3.1.4, it is going to be necessary to check that certain naturally defined subsets of Z are Borel. For Theorem 3.1.5, it will be necessary to know topological properties of analogously defined subsets. This subsection and the following are devoted to technical lemmas of these types.

Definition 3.2.2. For each integer $r \geq 1$, the r-graph $Z^{(r)}$ of Γ is the set of $(r+1)$-tuples $(x_0, x_1, \dots, x_r)$ with all coordinates in the orbit $\Gamma(x_0)$. This is topologized as a subspace of the $(r+1)$-fold Cartesian product $Z^{\times r} = Z \times Z \times \cdots \times Z$.

In the proof of the following lemma and hereafter, we let Γ' denote the set of elements $g \in \Gamma$ that can be written as pure compositions of elements of the holonomy cocycle and that have maximal possible domain. This subset of Γ is countable.

Lemma 3.2.3. *For each integer $r \geq 1$, $Z^{(r)}$ is a Borel subset of $Z^{\times r}$ and the projection $\pi_i : Z^{(r)} \to Z$ onto the ith coordinate carries Borel subsets to Borel subsets.*

Proof. For $g_1, \dots, g_r \in \Gamma'$, let

$$D(g_1, \dots, g_r) = \bigcap_{i=1}^{r} \operatorname{dom} g_i,$$
$$W(g_1, \dots, g_r) = \{(x, g_1(x), \dots, g_r(x)) \mid x \in D(g_1, \dots, g_r)\}.$$

Since Z is locally compact and separable, the open subset $D(g_1, \dots, g_r)$ is a countable union of compact sets. Thus, its image in $Z^{\times r}$ under the continuous map

$$x \mapsto (x, g_1(x), \dots, g_r(x))$$

is a Borel set. But this image is $W(g_1, \dots, g_r)$ and $Z^{(r)}$ is the countable union of all such sets. Finally, π_i restricts to a homeomorphism of $W(g_1, \dots, g_r)$ onto an open subset of Z, the inverse being

$$y \mapsto (g_i^{-1}(y), g_1(g_i^{-1}(y)), \dots, g_r(g_i^{-1}(y))),$$

and the last assertion follows. □

Let us agree that g_0 always denotes id_Z. Thus, we can write the general point of $W(g_1, \dots, g_n)$ as $(g_0(x), g_1(x), \dots, g_n(x))$. For $0 \le i < j \le r$, define

$$d_{ij} : Z^{(r)} \to \mathbb{Z}^+,$$
$$d_{ij}(x, g_1(x), \dots, g_r(x)) = d_{L_x^*}(g_i(x), g_j(x)).$$

Also, for $h \in \Gamma'$, write $\|h\| = n$ if the shortest word in the elements of the holonomy cocycle that is equal to h has n terms.

Lemma 3.2.4. *For $1 \le i < j \le r$, d_{ij} is a Borel map. That is, for every subset $S \subseteq \mathbb{Z}^+$, $d_{ij}^{-1}(S)$ is a Borel subset of $Z^{(r)}$.*

Proof. It will be enough to show that, for each integer $n \ge 0$, the set of $\zeta \in Z^{(r)}$ such that $d_{ij}(\zeta) \le n$ is a Borel set. For this, we show that the intersection of this set with each $W(g_1, \dots, g_r)$ is a Borel set. Since

$$W(g_1, \dots, g_r) = \{(g_i^{-1}(y), g_1(g_i^{-1}(y)), \dots, g_r(g_i^{-1}(y))) \mid y \in g_i(D(g_1, \dots, g_r))\}$$

and $D(g_1, \dots, g_r)$ is invariant under permutations of the g_k's, we lose no generality in restricting to the case $i = 0$ and $j = 1$. Suppose that

$$d_{L_x^*}(x, g_1(x)) = d_{01}(x, g_1(x), \dots, g_r(x)) \le n.$$

Then there is $h \in \Gamma'$ such that $g_1(x) = h(x)$ and $\|h\| \le n$. The set of points $y \in D(g_1, \dots, g_r)$ such that $g_1(y) = h(y)$ is relatively closed in this open set, hence is a Borel set B_h. The union of these sets as $h \in \Gamma'$ ranges over the elements with $\|h\| \le n$ is exactly the Borel set we are seeking. □

If $g \in \Gamma'$, write

$$g = \gamma_{i_N i_{N-1}} \circ \gamma_{i_{N-1} i_{N-2}} \circ \cdots \circ \gamma_{i_1 i_0}$$

and, for $1 \le k \le N$, set

$$g_k = \gamma_{i_k i_{k-1}} \circ \gamma_{i_{k-1} i_{k-2}} \circ \cdots \circ \gamma_{i_1 i_0}.$$

Also, set $g_0 = \text{id}$. If $x \in \operatorname{dom} g$, let $\sigma_g(x)$ denote the edgepath in L_x^* with edges labeled by $\gamma_{i_1 i_0}, \dots, \gamma_{i_N i_{N-1}}$ and having successive vertices

$$x = g_0(x), g_1(x), \dots, g_N(x) = g(x).$$

Let $K_g(x) \subset L_x^*$ denote the finite, connected subcomplex that is the union of the vertices and edges of $\sigma_g(x)$. Every finite connected subcomplex can be so represented, although nonuniquely. Since Γ' is countable, this will be a useful device for making countable choices of compact, connected subcomplexes uniformly for families of leaves of $\mathcal{F}$.

Let $K_g(x)^c$ denote the closure in L_x^* of $L_x^* \smallsetminus K_g(x)$. This is a subcomplex of L_x^* having finitely many components. In determining the endset $\mathcal{E}(L_x^*)$, one is interested in the number of unbounded components of $K_g(x)^c$.

Definition 3.2.5. A component V of $K_g(x)^c$ *abuts* on a vertex $g_i(x)$ of $K_g(x)$ if this vertex lies in V. The *abutment* of V is the union of the vertices of $K_g(x)$ on which V abuts.

Clearly, the set of components of $K_g(x)^c$ corresponds bijectively to the finite set of (disjoint) abutments.

Lemma 3.2.6. *Let g be as above and fix i, $0 \le i \le N$. Then the set of $x \in \operatorname{dom} g$ such that an unbounded component of $K_g(x)^c$ abuts on $g_i(x)$ is a Borel set.*

Proof. We want to consider edgepaths $\sigma_h(g_i(x))$, if any, originating at $g_i(x)$, staying in $K_g(x)^c$, and ending at points arbitrarily far from $g_i(x)$. Accordingly, choose $h \in \Gamma'$, set $y = g_i(x)$, and let x range over the open (possibly empty) subset of $\operatorname{dom} g$ such that the edgepath $\sigma_h(y)$ is defined. Say that the length of this edgepath is ℓ. There is a corresponding Borel subset of points

$$(x, g_1(x), \dots, g_N(x), h_1(y), \dots, h_\ell(y)) \in Z^{(N+\ell)}.$$

Repeated applications of Lemma 3.2.4 prove that the subset for which each $h_k(y)$, $1 \le k \le \ell$, is at distance ≥ 1 from every vertex of $K_g(x)$ is a Borel set B_h. For each integer $n \ge 1$, let $B_h(n)$ be the subset for which $d_{L_x^*}(y, h(y)) \ge n$, again a Borel set. For fixed n, the union of $B_h(n)$ over the countably many choices of $h \in \Gamma'$ is also Borel, as is its projection B_n by π_0 into Z (Lemma 3.2.3). Finally, the Borel set obtained by intersecting the sets B_n over all $n \ge 1$ is exactly the set we seek. □

Lemma 3.2.7. *Let g be as above and fix i, j, $0 \le i < j \le N$. Then the set of $x \in \operatorname{dom} g$ such that a component of $K_g(x)^c$ abuts on both $g_i(x)$ and $g_j(x)$ is a Borel set, as is the set of $x \in \operatorname{dom} g$ such that no component of $K_g(x)^c$ abuts on both $g_i(x)$ and $g_j(x)$.*

Proof. The second of these sets is the complement in $\operatorname{dom} g$ of the first. The proof that the first is a Borel set is similar to the proof of Lemma 3.2.6. As x ranges over $\operatorname{dom} g$, let $y = g_i(x)$ and $z = g_j(x)$. Let $h \in \Gamma'$ and let $\ell \geq 1$ be the length of σ_h. The subset of $\operatorname{dom} g$ such that $h(y)$ is defined is open and the subset of that on which $h(y) = z$ is relatively closed, hence is a countable union of compact sets. The corresponding set of points

$$(x, g_1(x), \dots, g_N(x), h_1(y), \dots, h_\ell(y) = z) \in Z^{(N+\ell)}$$

is Borel, as is the subset in which $h_k(y)$ is at distance ≥ 1 from $K_g(x)$, $1 \leq k < \ell$ (the case $\ell = 1$ is special, but easily accomodated). The union of these sets as h ranges over Γ' is Borel and its projection by π_0 into Z is the Borel set we seek. □

Exercise 3.2.8. Let $\{X_1, \dots, X_s\}$ be a family of subsets of $\{g_0, g_1, \dots, g_N\}$ and, for each $x \in \operatorname{dom} g$, let $X_1(x), \dots, X_s(x)$ be the corresponding family of sets of vertices of $K_g(x)$. Show that the set of $x \in \operatorname{dom} g$ such that $X_1(x), \dots, X_s(x)$ is exactly the set of abutments of the unbounded components of $K_g(x)^c$ is a Borel set. For each integer k, $0 \leq k \leq N$, conclude that the set B_g^k of points $x \in \operatorname{dom} g$ such that $K_g(x)^c$ has exactly k unbounded components is also a Borel set.

Corollary 3.2.9. *For each integer $k \geq 1$ and each $g \in \Gamma'$, the set Z_g^k of points $x \in \operatorname{dom} g$ such that $K_g(x)^c$ has at least k unbounded components is a Borel set.*

3.2.B. Topological properties. Recall that the union of leaves without holonomy is residual [I, Theorem 2.3.12]. This was proven in [I] for foliated manifolds, but the proof goes through without change for foliated spaces. The results in this subsection will be used to prove that various $\mathcal{F}$-saturated subsets of M are residual or meager.

Let G_0 denote the union of leaves without holonomy. The Γ-invariant set $Z_0 = Z \cap G_0$ is residual in Z. We will need to throw away the meager saturated subset $B \subset G_0$ described in the following lemma. By abuse of notation, the residual set $G_0 \setminus B$ will again be denoted by G_0 and its intersection with Z by Z_0.

Lemma 3.2.10. *The union B of leaves that meet $\partial(\operatorname{dom} \gamma_{ij}) \subset T$, for at least one element γ_{ij} of the holonomy cocycle, is meager.*

The easy proof is left to the reader.

Definition 3.2.11. The star of a vertex z of L_x^*, denoted by $\operatorname{star}(z)$, is the union of $\{z\}$ and the open edges emanating from z. The vertex $z \in K_g(x)$ is an interior point of $K_g(x)$ if $\operatorname{star}(z) \subset K_g(x)$, and otherwise, it is a boundary point. The complex $\operatorname{int} K_g(x)$ is the subcomplex of $K_g(x)$ spanned by the interior points and $\partial K_g(x)$ is the subcomplex spanned by the boundary points.

Lemma 3.2.12 (Local Reeb Stability). *If $g \in \Gamma'$ and $x \in Z_0 \cap \operatorname{dom} g$, there are a neighborhood $V_x \subseteq \operatorname{dom} g$ of x and a canonical isomorphism of 1-complexes*

$$\pi_y : K_g(y) \to K_g(x),$$

defined for all $y \in V_x$, which preserves the labels γ_{ij} of directed edges and, on vertices, is given by

$$\pi_y(g_i(y)) = g_i(x).$$

Furthermore, $\pi_y(\partial K_g(y)) = \partial K_g(x)$ and $\pi_y(\operatorname{int} K_g(y)) = \operatorname{int} K_g(x)$.

Proof. We continue to use the notational conventions established above. The conditions $g_i(y) \neq g_j(y)$, $0 \leq i < j \leq N$, are *open*, which is to say that there is an open neighborhood V_x of x in $\operatorname{dom} g$ such that

$$g_i(x) \neq g_j(x) \Rightarrow g_i(y) \neq g_j(y), \quad \forall\, y \in V_x, \quad 0 \leq i < j \leq N.$$

Thus, π_y is well defined on the vertices of $K_g(y)$, for each $y \in V_x$, and extends linearly to a surjection of 1-complexes preserving the labels on directed edges. This does not use the hypothesis that $x \in Z_0$, but that hypothesis enables us to choose a possibly smaller V_x so that π_y is injective. Indeed, the equality $g_i(x) = g_j(x)$ is equivalent to $x = h_{ij}(x)$, where $h_{ij} = g_i^{-1} \circ g_j$. Since h_{ij} has trivial germ at x, this equality extends to all points y in a sufficiently small choice of V_x. There are only finitely many indices i, j to consider. For the last assertion, we only need to show that V_x can be chosen so small that $\operatorname{star}(z)$ and $\operatorname{star}(\pi_y(z))$ have edges with exactly the same labels, $\forall\, y \in V_x$, $\forall\, z \in K_g(y)$. But this is an easy consequence of the hypothesis that $x \notin B \cap \operatorname{dom} g$, where B is the meager set of Lemma 3.2.10. □

Lemma 3.2.13. *If $k \geq 0$ is an integer and $g \in \Gamma'$, the set of $x \in Z_0 \cap \operatorname{dom} g$ such that $K_g(x)^c$ has at most k components is relatively open in $Z_0 \cap \operatorname{dom} g$.*

Proof. Let $\mathbf{P}_x$ be the stated condition on x. It is equivalent to the condition that there exists $f \in \Gamma'$ such that $K_g(x) \subseteq \operatorname{int} K_{f \circ g}(x)$ and that the space $K_{f \circ g}(x) \smallsetminus K_g(x)$ have at most k components. (For $k = 0$, $f = \mathrm{id}$.) By Lemma 3.2.12, applied to $K_{f \circ g}(x)$, there is a neighborhood V_x of x in $\operatorname{dom} g$ such that

$$\mathbf{P}_x \Rightarrow \mathbf{P}_y, \quad \forall\, y \in V_x,$$

and the assertion follows. □

Corollary 3.2.14. *If $k \geq 0$ is an integer and $g \in \Gamma'$, the set of points $x \in Z_0 \cap \operatorname{dom} g$ such that $K_g(x)^c$ has at most k unbounded components is relatively open in $Z_0 \cap \operatorname{dom} g$.*

Proof. Let $x \in Z_0 \cap \operatorname{dom} g$ be a point such that $K_g(x)^c$ has at most k unbounded components. Choose $f \in \Gamma'$ such that $K_{f \circ g}(x)$ is exactly the union of $K_g(x)$ and the bounded components of $K_g(x)^c$. Then $K_{f \circ g}(x)^c$ is exactly the union of the unbounded components of $K_g(x)^c$, the number of these being $n \leq k$. Applying Lemma 3.2.13 to $K_{f \circ g}(x)$, we find a neighborhood V_x of x in $\operatorname{dom} g$ such that $K_{f \circ g}(y)^c$ has at most n components, $\forall\, y \in V_x$, so the number of unbounded components is at most $n \leq k$. Thus, the number of unbounded components of $K_g(y)^c$ is also at most k. □

Corollary 3.2.15. *If $k \geq 1$ is an integer and $g \in \Gamma'$, let Z_g^k be the set of points $x \in \operatorname{dom} g$ such that $K_g(x)^c$ has at least k unbounded components. Then $Z_0 \cap Z_g^k$ is relatively closed in $Z_0 \cap \operatorname{dom} g$.*

Indeed, apply Corollary 3.2.14 to the integer $k - 1 \geq 0$.

3.3. Proof of the Theorems

In this section we will consider Theorem 3.1.4 and Theorem 3.1.5 simultaneously. Accordingly, in each of the propositions below, the term "generic leaf" refers equally to "m-almost every leaf" relative to an ergodic harmonic probability measure and to a "residual family of leaves" in a minimal foliation.

First we need a trivial lemma, the proof being left to the reader. If m is a harmonic probability measure on M, let m_Z denote the measure on Z that, on each Z_i, is given by the usual disintegration of $m|U_i$. In the following, recall that the complement of a residual set is called meager.

Lemma 3.3.1. *If Y is an $\mathcal{F}$-saturated set, then $m(Y) = 0$ if and only if $m_Z(Y \cap Z) = 0$. Similarly, for an ergodic, harmonic probability measure m, $m(Y) = 1$ if and only if $m_Z(Y \cap Z) > 0$. Likewise, Y is meager (respectively, residual) in M if and only if $Y \cap Z$ is meager (respectively, residual) in Z. Finally, if a not necessarily Γ-invariant subset $X \subseteq Z$ is meager in Z (respectively, m_Z-null), the $\mathcal{F}$-saturation $\mathcal{F}(X)$ is also meager (respectively, m-null), as is the Γ-saturation $\Gamma(X)$.*

Proposition 3.3.2. *Either the generic leaf has a Cantor set of ends or the generic leaf has at most two ends.*

Proof. For $g \in \Gamma'$, let Z_g^3 denote the set of points $x \in \operatorname{dom} g$ such that $K_g(x)^c$ has at least 3 unbounded components. Let Z^3 denote the union of these as g ranges over the countable set Γ'. By Corollary 3.2.9, Z^3 is a Borel set. If $m_Z(Z^3) = 0$, then Lemma 3.3.1 implies that m-almost every leaf has at most two ends. If the measure of this set is positive, then $m_Z(Z_g^3) > 0$ for some $g \in \Gamma'$ and, m being ergodic, the saturation $\mathcal{F}(Z_g^3)$ has full measure.

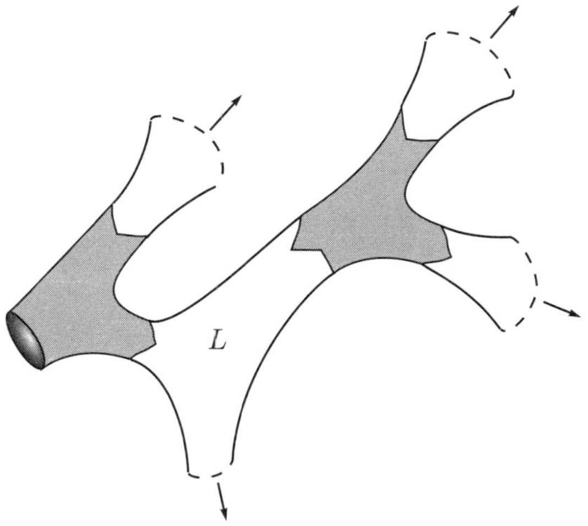

Figure 3.3.1. $L \cap Z_g^3$ splits off a Cantor set of ends

By Theorem 2.7.12, every neighborhood of every end of m-almost every leaf in this set meets Z_g^3. It follows that none of these ends are isolated (see Figure 3.3.1, where the shaded regions are the connected unions of plaques making up $L \cap Z_g^3$), so m-almost every leaf has a Cantor set of ends. In the topological context, Lemma 3.2.15 implies that either $Z_0 \cap Z^3$ is meager or some $Z_0 \cap Z_g^3$ has nonempty interior in Z_0. In the first case, Lemma 3.3.1 implies that a residual family $G \subseteq G_0$ consists of leaves with at most two ends, while, in the second case, the minimality of $(M, \mathcal{F})$ implies that every leaf is totally recurrent (*cf.* Exercise 3.1.6). Thus, every neighborhood of every end of every leaf $L \subset G_0$ meets Z_g^3; hence these leaves have a Cantor set of ends. □

In exactly the same way, using the sets Z_g^2 and Z_g^1, one proves the following results.

Proposition 3.3.3. *If the generic leaf has at most two ends, then either the generic leaf has exactly two ends or the generic leaf has at most one end.*

Proposition 3.3.4. *If the generic leaf has at most one end, then either the generic leaf has exactly one end or the generic leaf is compact.*

The only way the generic leaf of a minimal foliation can be compact is for the foliated space to be a single compact leaf. The corresponding property for $\operatorname{supp} m$ is the following exercise.

Exercise 3.3.5. If m is ergodic and the union of all compact leaves has positive measure, then $\operatorname{supp} m$ is a single compact leaf. (Hint. Use the full force of the hypothesis that Z is a locally compact, separable metric space).

Finally, we turn to the case in which the leaf dimension is 2. The key here is local Reeb stability, not the combinatorial version in Lemma 3.2.12, but the geometric version [I, Proposition 11.4.8]. That is, if L is a leaf and $F \subseteq L$ is a compact, connected submanifold such that every loop in F has trivial holonomy, then a neighborhood of F in M is trivially foliated as a product.

First, suppose that the foliation is minimal and that some leaf in G_0 is not orientable. This leaf contains an imbedded Möbius strip K; hence local Reeb stability provides a subset $K \times V \subset M$, where $V \subset Z$ is open and each factor $K \times \{x\}$ is imbedded in a leaf. By total recurrence, every leaf meets this set so as to pick up copies of K clustering at all ends. Alternatively, all leaves in G_0 are orientable. If one such leaf has genus > 0, local Reeb stability, together with total recurrence, implies that all ends of all leaves are cluster points of imbedded handles. The remaining alternative is that all leaves in G_0 have genus 0.

The case of an ergodic harmonic measure is only slightly more delicate. If some leaf contains an imbedded Möbius strip K, let $s \subset K$ be a core circle and find a plaque chain $x = P_0, P_1, \dots, P_m = x$ that covers s (see [I, Definition 2.3.1]). This chain can be viewed an edgeloop $\sigma_h(x)$, $h \in \Gamma'$. The set S_h of points $y \in \operatorname{dom} h$ such that $h(y) = y$ is closed. The saturation $\mathcal{F}(S_h)$, while generally not compact, is a foliated space in which local Reeb stability holds. Any imbedded Möbius strip with a core circle covered by $\sigma_h(y)$, $y \in S_h$, has trivial holonomy in $\mathcal{F}(S_h)$, since its fundamental group is generated by this core circle. Thus, local Reeb stability implies that the subset $S_h^* \subseteq S_h$ of such points y is relatively open, hence is a Borel set. Since Γ' is countable, the union S^* of these Borel sets is Borel. If $m_Z(S^*) = 0$, m-almost all leaves are orientable. Alternatively, for some $h \in \Gamma'$, $m_Z(S_h^*) > 0$ and, by the ergodicity of m and Theorem 2.7.12, m-almost every leaf has only nonorientable ends.

If m-almost every leaf is orientable and some leaf contains a handle H, let s and s' be loops in H generating $\pi_1(H, x)$, $\sigma_h(x)$ and $\sigma_g(x)$ edgeloops covering s and s', respectively. Now we consider the relatively closed subset $S_{h,g} \subseteq \operatorname{dom} h \cap \operatorname{dom} g$ consisting of the points y such that $h(y) = y = g(y)$. Let $S_{h,g}^* \subseteq S_{h,g}$ be the set of points y such that $\sigma_h(y)$ and $\sigma_g(y)$ cover generators of the fundamental group of a handle. By local Reeb stability, $S_{h,g}^*$ is relatively open in $S_{h,g}$, hence is a Borel set. Arguing as above, we conclude that either m-almost every leaf is orientable with only nonplanar ends, or m-almost every leaf has genus 0.

Exercise 3.3.6. Let $(M, \mathcal{F})$ be a compact, C^2-foliated manifold of codimension 1. Suppose that $X \subset M$ is an exceptional minimal set and show that the topologically generic leaf of the foliated space $(X, \mathcal{F}|X)$ cannot have two ends. (Hint. Show that the generic 2-ended leaf would have linear growth, hence that $(X, \mathcal{F}|X)$ would support a transverse, holonomy-invariant measure. This would contradict the existence of a resilient leaf in X [I, Corollary 8.2.5].)

Remark. This exercise raises the interesting and difficult problem of determining the generic endset of the leaves of an exceptional minimal set X as above. By a deep result of G. Duminy (unpublished, but see [**26**]), the semiproper leaves of X have a Cantor set of ends. For many examples, notably those that are real analytic (G. Hector, unpublished) and those of Markov type [**23**], *all* leaves in X have a Cantor set of ends. It is reasonable to conjecture this in general, but a good first step might be to prove that the generic leaf of X has a Cantor set of ends.

3.4. Generic Geometry of Leaves

In a series of papers, J. A. Álvarez López and the first author have studied the geometry of the generic leaf of a foliated space. This section describes some of these results; complete details can be found in the papers [**3**], [**4**] and [**5**].

One of the basic questions that anyone who has studied foliations asks is what the leaves of a foliation "look like". As noted in [I, Section 12.1], the formal meaning of "looking alike" is "having the same quasi-isometry type". A leaf of a compact foliated space has a well defined quasi-isometry type and it is a natural question to ask which quasi-isometry types of (intrinsic) metric spaces can appear as leaves of foliated spaces. There are two more or less related concepts of quasi-isometry. The first one is that which has been used before in this text, namely, two (Lipschitz) manifolds are quasi-isometric if there is a bi-Lipschitz homeomorphism $f : X \to Y$. The more general concept has also been used when discussing pseudogroups and the word metric given by a generating system. Two metric spaces X, Y are coarsely quasi-isometric if there is a mapping $f : X \to Y$ such that

(1) there are constants $K \geq 1$ and $A \geq 0$ so that

$$(1/K)d(x_1, x_2) - A \leq d(f(x_1), f(x_2)) \leq Kd(x_1, x_2) + A,$$

for all $x_1, x_2 \in X$, and

(2) $f(X)$ is B-dense in Y, for some constant $B \geq 0$.

If G is a finitely generated group, then a choice of generating system endows the Cayley graph of G with a right-invariant metric. If H is a subgroup of G, not necessarily normal, then the coset space G/H is a metric space with the induced distance. Such spaces G/H are called *discrete homogeneous spaces.* With this definition the problem of coarse quasi-isometry appearance of the leaves of a foliated space has the following answer.

Theorem 3.4.1. *Let $(M, \mathcal{F})$ be a compact foliated space. Then there exists a finitely generated group G such that every leaf of M is coarsely quasi-isometric to a discrete homogeneous space of G.*

Conversely, if G/H is a discrete homogeneous space, then there is a compact foliated space with a leaf coarsely quasi-isometric to G/H.

Sketch of the proof. The proof of the second part of the statement is by direct construction. In fact, given such a homogeneous space G/H, there is a two-dimensional foliated space with a leaf coarsely quasi-isometric to G/H.

The proof of the first part is based on two observations. One is that if a finitely generated group G acts on a topological space X, then the orbit of a point $x \in X$ is isometric to the homogeneous space G/G_x, where G_x is the stabilizer of x. The other is that the leaves of a compact foliated space are coarsely quasi-isometric to the orbits of a finitely generated holonomy pseudogroup.

Let $\{U_i\}_{i=1}^n$ be a finite cover of M by foliated charts $U_i = D_i \times Z_i$. For each pair of indices i, j such that $U_i \cap U_j \neq \emptyset$, there is a partially defined holonomy transformation $h_{ij} : Z_{ij} \subset Z_i \to Z_{ji} \subset Z_j$, where Z_{ij} is an open subset of Z_j. Let Z be the disjoint union of the transversals Z_i, $i = 1, \ldots, n$. For each pair of indices i, j, define a map $g_{ij} : Z \to Z$ by setting

$$g_{ij}(x) = \begin{cases} h_{ij}(x) & \text{if } x \in Z_{ij}, \\ x & \text{otherwise.} \end{cases}$$

Each g_{ij} is a Borel bijection of Z, with $g_{ji} = g_{ij}^{-1}$ (in fact, each g_{ij} is continuous on a residual subset of Z). The finite collection $\{g_{ij}\}$ defines an action of a finitely generated group G on Z whose orbits correspond to the orbits of the holonomy pseudogroup generated by $\{h_{ij}\}$ acting on Z. Moreover, it is not hard to check that the metric on an orbit induced by the group action is quasi-isometric to that induced by the action of the holonomy pseudogroup. □

The question also arises as to how many different quasi-isometry types can occur among the leaves of a given foliated space. A study of the relation of quasi-isometry among the leaves of a foliated space shows that the equivalence classes are Baire sets, and so basic topological dynamics provides the following answer.

Theorem 3.4.2. *Let $(M, \mathcal{F})$ be a transitive foliated space, either compact or with all the leaves of uniformly bounded geometry. Then either there are uncountable many quasi-isometry types of leaves, or else there exists a residual set of leaves that are all quasi-symmetric (see below).*

Roughly speaking, a metric space is *quasi-symmetric* if it has sufficiently many quasi-isometries of uniformly bounded distortion. Examples of quasi-symmetric spaces are the symmetric spaces of Lie groups.

A further study of the equivalence relation sheds more light on the problem. It happens that the second possibility of the previous statement is rather restrictive.

Theorem 3.4.3. *Let $(M, \mathcal{F})$ be a minimal foliated space, either compact or with all the leaves of uniformly bounded geometry. Then there is a residual set of quasi-isometric leaves if and only if there is a residual set of quasi-symmetric leaves.*

The following corollary is a basic consequence of this result.

Corollary 3.4.4. *Let $(M, \mathcal{F})$ be a two-dimensional compact foliated space that has all leaves dense and that admits no invariant transverse measure. If there is a simply connected leaf, then there is a residual set of leaves quasi-isometric to the Poincaré disk.*

Sketch of the proof. Given the hypothesis on the structure of M, there is, by [I, Theorem 12.6.3], a metric tensor on $(M, \mathcal{F})$ under which every leaf has constant curvature -1. The simply connected leaf then is isometric to the Poincaré disk, hence quasi-symmetric. A further argument implies that the holonomy cover of every leaf is quasi-isometric to the Poincaré disk. □

In a slightly different direction, another natural question to ask is what kind of quasi-isometry invariants of metric spaces are generic (*i.e.*, the same on a residual set) for the leaves of a foliated space. Well-known examples of quasi-isometry invariants appearing in foliation theory are the order of growth and the number of ends of leaves. A multitude of examples are the asymptotic invariants described in Gromov [**84**].

These invariants are best understood in terms of the Gromov-Hausdorff space $\mathcal{G}$. Points of this space are isometry classes $[L, x]$ of pointed metric spaces. This set is endowed with a topology in which a sequence $[L_n, x_n]$ converges to $[L, x]$ if, for each $R > 0$, the closed balls $B(x_n, R)$ in L_n converge, with respect to the Gromov-Hausdorff distance, to the closed ball

$B(x, R)$ in L. It is thus a sort of uniform convergence on compact sets for noncompact metric spaces. The Gromov-Hausdorff space does not have a foliated structure, but, given a foliated space M, there is a canonical mapping $M \to \mathcal{G}$ that sends the point $x \in M$ to the pointed metric space $[L_x, x]$, where L_x is the leaf containing x.

Theorem 3.4.5. *Let M be a foliated space. The canonical mapping*

$$x \in M \mapsto [L_x, x] \in \mathcal{G}$$

is continuous on the subfoliated space consisting of leaves without holonomy.

(As shown in [I, Theorem 2.3.12], the union of leaves with no holonomy is a dense G_δ-set in X, hence residual.)

A quasi-isometry invariant can be thought of a function on the Gromov-Hausdorff space $\mathcal{G}$ that is invariant under the obvious equivalence relation $[L, x] \sim [L, y]$. It turns out that, in all the known examples, such a function is, moreover, Borel measurable. Therefore, basic topological dynamics gives the result that for a transitive foliated space X and such a Borel invariant f with values in a complete separable metric space, there is a residual saturated subset of X so that all the leaves in this subset have the same invariant f.

This, for instance, applies to the two examples previously mentioned, the order of growth and the number of ends. In the latter case, a separate argument, having to do with recurrence, is needed to obtain the full statement of the results of E. Ghys [**80**] and J. Cantwell and the second author [**24**].

A large number of quasi-isometry invariants of metric spaces amenable to study are given by homotopy functors. Let F be a functor from the category of metric spaces to a category with limits $\mathfrak{A}$. If F is continuous (in a reasonable sense), then a quasi-isometry invariant of a space X can be defined as $F^\infty(X) = \lim_K F(X \setminus K)$, where K runs over all compact subsets K of X. The space of ends is related to one particular functor, namely, π_0.

Theorem 3.4.6. *Let F be a continuous functor with values in the category of vector spaces. Then the spaces $F^\infty(L)$ are isomorphic for a residual set of leaves L of a given foliated space.*

Other types of quasi-isometry invariants are given by compactifications of the leaves. A relevant compactification of a leaf, from the point of view of carrying quasi-isometric information, is the Higson-Roe compactification [**157**]. Its study has two aspects, one that relates to recurrence, and studies the limit sets of points in the Higson corona. The reader is referred to [**3**] for further details.

From a different point of view, it is also natural to try to find dynamical properties of a foliated space implying that all the leaves are quasi-isometric.

The example that comes to mind is the case of Riemannian foliation, for it follows from Molino's theory [**132**] that the holonomy covers of all the leaves are quasi-isometric via diffeomorphism. The topological analog of a Riemannian foliation is a foliated space whose holonomy pseudogroup is equicontinuous (see E. Ghys [**132**] and also M. Kellum [**112**]).

The analysis of the structure of these foliated spaces is the topic of [**4**]. Such analysis shows that the holonomy pseudogroup of an equicontinuous foliated space has properties similar to those of a group of isometries. However, due to the very general topological structure being studied, some further requirements are needed to realize quasi-isometries between leaves. One such particular requirement is the quasi-analyticity of the holonomy pseudogroup (which is essentially that given in Proposition 1.3.7).

Theorem 3.4.7. *Let $(M, \mathcal{F})$ be a compact, equicontinuous foliated space, with connected leaf space and whose holonomy pseudogroup is quasi-analytic. Then the holonomy covers of all the leaves of M are quasi-isometric.*

Disregarding the quasi-analytic condition, the following is available. The new tool needed is the concept of normal bundle to the leaves.

Theorem 3.4.8. *The universal covers of all leaves of an equicontinuous foliated space with connected leaf space are quasi-isometric.*

The reader is referred to [**4**] for further details and related results. The results of [**4**] permit, in fact, a purely topological characterization of Riemannian foliations, which is accomplished in [**5**].

Part 2

Characteristic Classes and Foliations

Foreword to Part 2

Characteristic classes are cohomology classes, defined on the base space B of a vector bundle and reflecting the "twisting" of the fibers over B. They have been enormously successful tools in the study of these bundles. An early application of this theory to foliations was a theorem of J. Milnor and J. Wood, using the Euler class to obstruct the existence of foliations transverse to the fibers of "overly twisted" circle bundles over surfaces. After treating this theorem, we will take up the general theory of characteristic classes of foliations.

Foliations are integrable subbundles of the tangent bundle, and a surprising theorem of R. Bott asserts that this integrability implies the vanishing of certain characteristic classes *at the level of differential forms.* We will establish this theorem using the classical Chern-Weil construction of characteristic classes and the Bott connection on the normal bundle of a foliation. As a consequence, we will define a system of "exotic" classes for foliations, of which the Godbillon-Vey invariant is the simplest. We will give a complete proof of Duminy's vanishing theorem for $\mathrm{gv}(\mathcal{F})$, as well as proving some partial extensions of that theorem to higher codimension that are due to S. Hurder.

Although the theory of characteristic classes pertains primarily to the quantitative theory of foliated manifolds, our emphasis is on the qualitative aspects. As elsewhere in this volume, our aim is to give a reasonably accessible introduction to this theory, together with some fairly deep qualitative applications. We will not totally ignore the quantitative theory, however, sketching the basic facts about the Haefliger classifying spaces where the exotic classes really live.

Chapter 4

The Euler Class of Circle Bundles

In this chapter, we give a hands-on construction of a characteristic class for oriented S^1-bundles E over finite simplicial complexes and cell complexes X, showing this class to be *the* obstruction to trivializing such a bundle. This is called the *Euler class* $e(E) \in H^2(X;\mathbb{Z})$. In particular, applying this to the unit circle bundle associated to an oriented, real 2-plane bundle V over X, we obtain the Euler class $e(V)$ of such a bundle. If the circle bundle is associated to a complex line bundle L over X, the class is denoted by $c_1(L)$ and called the first Chern class. We will use this class to prove the Poincaré-Hopf theorem and related theorems for surfaces, all of which will be needed in Part 3 of this volume. Finally, we will prove a quite nontrivial application of this class, the theorem of Milnor and Wood, concerning the existence of foliations transverse to the fibers of an oriented circle bundle over a compact, orientable surface of genus $g \geq 1$.

While the purely combinatorial approach in this chapter has the virtue of revealing exactly what is going on and is quite useful for studying 2-plane bundles and their associated circle bundles, such honesty becomes counterproductive in the general theory of characteristic classes. In the subsequent chapters, we will resort to more elegant methods (the Chern-Weil construction).

4.1. Generalities about Bundles

In order to work comfortably with bundles, certain basic results are needed. We will summarize these results, sometimes without proof. For a detailed treatment, see [**105**].

Let $\pi : E \to X$ be a locally trivial fiber bundle with fiber F. Here, we are working in the purely topological category. If $f : Y \to X$ is a continuous map of topological spaces, there is an important operation called the *pullback* of the bundle by f. This is a commutative diagram

$$\begin{array}{ccc} f^*(E) & \xrightarrow{\overline{f}} & E \\ {\scriptstyle \pi'}\downarrow & & \downarrow{\scriptstyle \pi} \\ Y & \xrightarrow[f]{} & X \end{array}$$

where $\pi' : f^*(E) \to Y$ is also a locally trivial bundle with fiber F (called the pullback of $\pi : E \to X$) and $\overline{f}$ is a continuous map taking the fiber over y homeomorphically onto the fiber over $f(y)$, $\forall\, y \in Y$. One easily checks uniqueness up to a canonical isomorphism. For existence, set

$$f^*(E) = \{(y, z) \in Y \times E \mid f(y) = \pi(z)\},$$

topologized as a subspace of $Y \times E$. Define $\overline{f}$ and π' by

$$\overline{f}(y, z) = z,$$
$$\pi'(y, z) = y.$$

It is easy to check that this construction performs as advertised. Furthermore, if the bundle has additional structure (*e.g.*, a principal G-bundle, a vector bundle, *etc.*), the pullback also has this additional structure. The pullback construction is also *functorial*:

$$\mathrm{id}^*(E) = E,$$
$$f^*(g^*(E)) = (g \circ f)^*(E),$$

where the equalities are canonical bundle isomorphisms. We note the following easy but useful fact (in which the critical word is "canonical").

Lemma 4.1.1. *If $E = X \times F$ and $f : Y \to X$ is continuous, then there is a canonical trivialization $f^*(E) = Y \times F$.*

Proof. Indeed, $f^*(E) = \{(y, (f(y), v)) \mid y \in Y, v \in F\}$. The canonical trivialization

$$\varphi : f^*(E) \to Y \times F$$

is the continuous map

$$\varphi(y, (f(y), v)) = (y, v),$$

with continuous inverse

$$\varphi^{-1}(y,v) = (y,(f(y),v)).$$

□

Finally, sections pull back to sections by the following lemma.

Lemma 4.1.2. *Let*

$$\begin{array}{ccc} f^*(E) & \xrightarrow{\overline{f}} & E \\ {\scriptstyle \pi'}\downarrow & & \downarrow{\scriptstyle \pi} \\ Y & \xrightarrow[f]{} & X \end{array}$$

be a pullback diagram for locally trivial bundles (with arbitrary fiber F). If

$$\sigma : X \to E$$

is a cross-section, there is a unique cross-section

$$\sigma' : Y \to f^*(E)$$

such that $\overline{f} \circ \sigma' = \sigma \circ f$.

Proof. In terms of our model of the pullback as a subspace of $Y \times E$, we set

$$\sigma'(y) = (y,\sigma(f(y))), \quad \forall\, y \in Y.$$

Since $\pi'(y,\sigma(f(y))) = y$, σ' is a section. Furthermore,

$$\overline{f}(\sigma'(y)) = \overline{f}(y,\sigma(f(y))) = \sigma(f(y)),$$

as required. These two equations also show that there is no other way to define σ'. □

If G is a topological group and $\pi : P \to X$ is a principal G-bundle, the pullback has a very important homotopy invariance property. For this, it is necessary that the spaces X and Y be *paracompact.*

Theorem 4.1.3. *Under the above hypotheses, let $f_t : Y \to X$, $0 \le t \le 1$, be a homotopy. Then the principal G-bundles*

$$\pi' : f_1^*(P) \to Y,$$
$$\pi' : f_0^*(P) \to Y$$

are isomorphic.

The general locally trivial F-bundle $\pi : E \to X$ is determined by a continuous *structure cocycle*

$$g_{\alpha\beta} : U_\alpha \cap U_\beta \to \mathrm{Homeo}(F),$$

where the group Homeo(F) is given the compact-open topology. This cocycle determines a principal Homeo(F)-bundle to which the original F-bundle is said to be associated. The pullback construction respects this "association", whence the above homotopy invariance for principal bundles extends to all locally trivial fiber bundles. In particular, taking functoriality into account, we have the following.

Corollary 4.1.4. *If $f : X \to Y$ is a homotopy equivalence between paracompact spaces, the pullback construction induces a one-to-one correspondence between the sets of isomorphism classes of F-bundles over X and Y, respectively.*

Corollary 4.1.5. *A locally trivial bundle over a contractible, paracompact space is trivial.*

4.2. Cell Complexes

Let X be a finite cell complex (also called a finite CW-complex) of dimension n. The 0-skeleton X_0 is a finite set of points called the *vertices* or 0-*cells*. The 1-skeleton X_1 is obtained by attaching finitely many 1-cells $D^1 = [-1, 1]$ (edges) to X_0 via maps

$$f_i : \{\pm 1\} \to X_0, \quad 1 \leq i \leq r.$$

These maps extend to maps (still denoted by f_i) of D^1 into X_1 that are homeomorphisms on $\operatorname{int} D^1$. The image of $f_i : D^1 \to X_1$ will be called a 1-cell of the complex, $1 \leq i \leq r$. (The case $r = 0$ is allowed, meaning that there are no 1-cells and $X_1 = X_0$.) Inductively, if the k-skeleton X_k has been defined for some $k < n$, X_{k+1} is obtained by attaching at most finitely many copies of the unit ball $D^{k+1} \subset \mathbb{R}^{k+1}$ to X_k via attaching maps $\partial D^{k+1} \to X_k$. Then $X_k \subseteq X_{k+1}$, and the components of $X_{k+1} \smallsetminus X_k$ are called the *open* $(k+1)$-cells of the complex, the closures in X_{k+1} of these open cells being called simply the $(k+1)$-cells of X. Finally, X is equal to its own n-skeleton X_n and, for $k > n$, we set $X_k = X$ also. The complex X is connected if and only if X_1 is connected. That is, every pair of vertices are connected by an edgepath in the 1-skeleton.

Subcomplexes $A \subseteq X$ are defined in the fairly obvious way, A being a closed subspace which, in its own right, is a cell complex, the set of cells of A being a subset of the set of cells of X.

The homology $H_*(X; \mathbb{Z})$ of a cell complex can be computed out of the above data. Briefly, $C_k(X; \mathbb{Z})$ is the free abelian group on the set $\{e_1^k, \dots, e_{r_k}^k\}$ of k-cells. Here, each k-cell e_i^k is given a fixed orientation. The oppositely oriented cell is denoted by $-e_i^k$, this being the honest negative of the generator e_i^k in $C_k(X;\mathbb{Z})$. The boundary operator

$$\partial : C_k(X;\mathbb{Z}) \to C_{k-1}(X;\mathbb{Z})$$

is defined via the attaching maps of the boundaries of the k-cells to X_{k-1} (the so-called "incidence relations"). Of course, it has the property that $\partial^2 = 0$ and that the homology of the chain complex $(C_*(X;\mathbb{Z}), \partial)$ is canonically isomorphic to the singular homology $H_*(X;\mathbb{Z})$. For details of this (as well as details of other facts sketched in this section), see [**179**, Chapter 2].

Dual to the cellular chain complex $(C_*(X;\mathbb{Z}), \partial)$ is the cellular cochain complex $(C^*(X;\mathbb{Z}), \delta)$, the coboundary operator δ being the adjoint of the boundary operator ∂. The characteristic class to be constructed in the next section will be an element of the cohomology group $H^2(X;\mathbb{Z})$.

If X and Y are finite cell complexes, a continuous map $f : X \to Y$ is called *cellular* if $f(X_k) \subseteq Y_k$, for all integers $k \geq 0$. Such a cellular map induces a canonical homomorphism $f_* : H_*(X;\mathbb{Z}) \to H_*(Y;\mathbb{Z})$. Of course, this is the same map induced by f in singular homology, but it is often helpful for computation to use the combinatorial nature of a cellular map. The following theorem, together with the homotopy invariance of singular homology, implies that no generality is lost by restricting to cellular maps.

Theorem 4.2.1 (Cellular Approximation Theorem). *If $f : X \to Y$ is a continuous map between finite cell complexes, then f is homotopic to a cellular map. If $A \subseteq X$ is a cellular subcomplex and $f|A$ is already cellular, the homotopy f_t can be chosen so that $f_t|A \equiv f|A$, $0 \leq t \leq 1$.*

This theorem is a consequence of the following, which is important in its own right.

Theorem 4.2.2 (Homotopy Extension Theorem). *Let $f : X \to Y$ be a continuous map, X a cell complex, $A \subseteq X$ a subcomplex, and $\widetilde{f}_t$ a homotopy of $f|A$. Then $\widetilde{f}_t$ extends to a homotopy f_t of f.*

Example 4.2.3. The sphere S^n has a particularly simple cell structure. We use one vertex x, no k-cells, $k = 1, 2, \ldots, n-1$, and one n-cell. That is, $X_{n-1} = \cdots = X_1 = X_0$. The n-cell is glued to $X_{n-1} = \{x\}$ by the unique map $\partial D^n \to \{x\}$ and the result is clearly the n-sphere. Since there are no k-cells, $1 \leq k \leq n-1$, $C_k(S^n;\mathbb{Z}) = 0$ in this range. Thus, if $n > 1$, *all* boundary operators are zero by default and we obtain the expected homology

$$H_k(S^2;\mathbb{Z}) = C_k(S^2;\mathbb{Z}) = \begin{cases} \mathbb{Z}, & k = 0, n, \\ 0, & \text{otherwise.} \end{cases}$$

For the case $n = 1$, one finds that $\partial : C_1(S^1;\mathbb{Z}) \to C_0(S^1;\mathbb{Z})$ vanishes, so the above values of homology also hold for this case.

A cell complex is said to be *regular* if each attaching map

$$f : \partial D^k \to X_{k-1}$$

is a homeomorphism onto a $(k-1)$-dimensional subcomplex, $1 \leq k \leq n$. Simplicial complexes are regular cell complexes, but there are many other examples. While useful for studying the Euler class, regularity is a bit too restrictive for our purposes. The following slightly weaker condition will be substituted.

Definition 4.2.4. A cell compex will be said to be *weakly* regular if each attaching map is a local homeomorphism onto a subcomplex.

Example 4.2.5. The cellular structure on S^n in Example 4.2.3 is neither regular nor weakly so. The simplest regular one has two cells of each dimension $k = 0, 1 \dots, n$, the k-skeleton being homeomorphic to S^k. Indeed, S^1 consists of two points (the vertices). Attaching two copies of D^1 by homeomorphisms $\partial D^1 \to S^0$ produces S^1. Two copies of D^2 are then attached by homeomorphisms $\partial D^2 \to S^1$ to obtain S^2, *etc.*

Example 4.2.6. Real projective space P^n can be assembled as a cell complex with one k-cell, $0 \leq k \leq n$. Thus, the 1-skeleton is forced to be a circle and the attaching map $g_1 : \partial D^2 \to X_1$ is a 2-fold covering map of the one circle onto the other. Inductively, if P^k has been defined, the 2-fold covering $g_k : \partial D^{k+1} = S^k \to P^k$, identifying antipodal points of S^k, attaches the cell D^{k+1} so as to produce P^{k+1}. This makes P^n into a weakly regular cell complex, but not a regular one.

The chain groups are

$$C_k(P^n; \mathbb{Z}) = \begin{cases} \mathbb{Z}, & k = 0, 1, \dots, n, \\ 0, & \text{otherwise.} \end{cases}$$

It can be shown that the boundary operator

$$\partial : C_k(P^n; \mathbb{Z}) \to C_{k-1}(P^n; \mathbb{Z}), \quad 1 \leq k \leq n,$$

as a map $\mathbb{Z} \to \mathbb{Z}$, is multiplication by 2 when k is even, and vanishes when k is odd. This reflects the fact that the attaching maps are the antipodal coverings. From this, the reader can easily compute the homology over $\mathbb{Z}$. If $\mathbb{Z}$ is replaced by the field $\mathbb{Z}_2$, all boundary operators vanish and

$$H_k(P^n; \mathbb{Z}_2) = \begin{cases} \mathbb{Z}_2, & 0 \leq k \leq n, \\ 0 & \text{otherwise.} \end{cases}$$

Over fields of characteristic $\neq 2$, the homology of P^{2n+1} agrees with that of S^{2n+1}, while that of P^{2n} agrees with the homology of a point.

Example 4.2.7. The torus T^2 is a compact, orientable surface obtained by identifying opposite edges of a square in an orientation-preserving way. This identifies all four vertices of the square to a single point, the unique vertex. The four edges are identified to two circles (the 1-cells) and the square itself (identified to T^2) is the 2-cell. This is a weakly regular, but not regular, cell complex. The next example generalizes this to closed, orientable surfaces of higher genus.

Example 4.2.8. The compact, orientable surface Σ_2 of genus 2 is the quotient of a hyperbolic octagon Δ under an equivalence relation that identifies all of the vertices to a single point x_0 and identifies edges pairwise. This octagon is depicted in Figure 4.2.1, the oriented edges e_i^+ and e_i^- being identified, $1 \le i \le 4$. The result can be viewed as a cell complex with one vertex, four 1-cells (glued to the vertex so as to form a bouquet $e_1 \vee e_2 \vee e_3 \vee e_4$ of four circles as in Figure 4.2.2) and one 2-cell. More generally, every compact, orientable surface Σ_g of genus $g \ge 1$ is a quotient of a $4g$-gon by a relation that identifies edges pairwise and identifies all vertices to a single point. Thus, Σ_g has a weakly regular cell structure with one vertex x_0, $2g$ 1-cells (X_1 is a bouquet of $2g$ circles $e_1 \vee \cdots \vee e_{2g}$) and one 2-cell Δ. Of course, a triangulation of Σ_g also provides a cell structure, but with more edges and lots of vertices and faces. The cell subdivision that we are using is much more efficient. In fact, one can show that the cellular boundary operator for this complex is trivial, so the homology is equal to the cellular chain group

$$H_k(\Sigma_g;\mathbb{Z}) = C_k(\Sigma_g;\mathbb{Z}) = \begin{cases} \mathbb{Z}, & k = 0,2, \\ \mathbb{Z}^{2g}, & k = 1, \\ 0, & \text{otherwise}. \end{cases}$$

Indeed, $\partial e_i = x_0 - x_0 = 0$, $1 \le i \le 2g$, and a glance at Figures 4.2.1 and 4.2.2 should convince the reader that

$$\partial\Delta = \sum_{i=1}^{2g} (e_i - e_i) = 0.$$

Exercise 4.2.9. If X is a cell complex and $f : S^k \to X$ is a local homeomorphism onto a subcomplex, show that there is a cell structure on S^k such that each open m-cell in S^k is carried homeomorphically onto an open m-cell of X, $0 \le m \le k$. In particular, for weakly regular cell complexes, the attaching maps $f : \partial D^{k+1} \to X_k$ are cellular with respect to this induced cell structure on ∂D^{k+1}.

In the remainder of this chapter, save mention to the contrary, all cell complexes will be weakly regular.

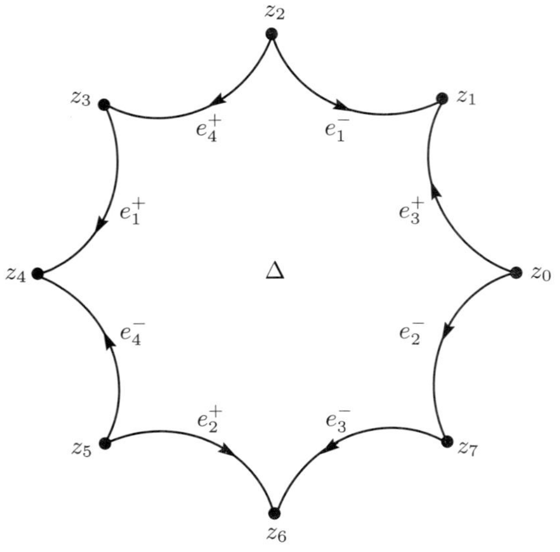

Figure 4.2.1. A hyperbolic octagon with pairwise edge identifications

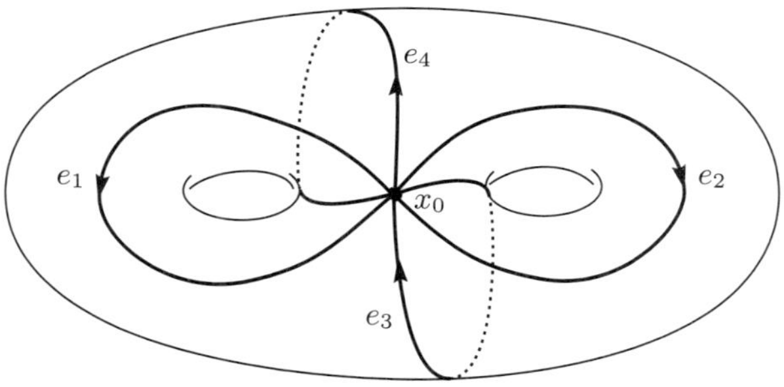

Figure 4.2.2. Σ_2 and its 1-skeleton

4.3. The First Obstruction

Let X be a weakly regular, finite, n-dimensional cell complex, $\pi : E \to X$ an oriented circle bundle. It is well known that this bundle is trivial if and only if it admits a global section. It is easy to construct a section over X_0, so we attempt to extend this to a section over X_1, thence to a section over X_2, *etc.* If all attempts are successful, we obtain a global extension at the nth step. It turns out that the only obstruction arises when we attempt to extend a section over X_1 to one over X_2.

Suppose that $\sigma : X_{k-1} \to E$ is a section of $E|X_{k-1}$, some $k \geq 1$. One tries to extend σ one k-cell at a time. Let

$$g_\alpha : \partial D^k \to X_{k-1}, \quad 1 \leq \alpha \leq r = r_k,$$

be the local imbeddings that are the attaching maps for the k-cells of X. The extension of g_α to a map (still denoted by g_α) of D^k into X is to be a homeomorphism on $\mathrm{int}(D^k)$. The images $e^k_\alpha = g_\alpha(D^k)$ are the k-cells of the complex, $1 \leq \alpha \leq r$, and their union with X_{k-1} is X_k. Fix α and consider the pullback

$$g^*_\alpha(E) \xrightarrow{\pi'} D^k,$$

and let σ_α be the unique section of the pullback over ∂D^k such that the diagram

$$\begin{array}{ccc} g^*_\alpha(E)|\partial D^k & \xrightarrow{\overline{g}_\alpha} & E|X_{k-1} \\ {\scriptstyle\sigma_\alpha}\uparrow & & \uparrow{\scriptstyle\sigma} \\ \partial D^k & \xrightarrow[g_\alpha]{} & X_{k-1} \end{array}$$

commutes (Lemma 4.1.2). Since D^k is contractible, $g^*_\alpha(E)$ is a trivial bundle and we fix an explicit trivialization

$$\begin{array}{ccc} g^*_\alpha(E) & \xrightarrow{\varphi_\alpha} & D^k \times S^1 \\ {\scriptstyle\pi'}\downarrow & & \downarrow{\scriptstyle p} \\ D^2 & \xrightarrow[\mathrm{id}]{} & D^k \end{array}$$

It is understood that the trivialization carries the oriented bundle $g^*_\alpha(E)$ to the standardly oriented bundle $D^k \times S^1$. Via the trivialization φ_α, we write the section σ_α as a continuous map

$$\widetilde{\sigma}_\alpha : \partial D^k \to S^1,$$
$$\varphi_\alpha(\sigma_\alpha(x)) = (x, \widetilde{\sigma}_\alpha(x)).$$

Lemma 4.3.1. *The section $\sigma : X_{k-1} \to E$ extends over X_k if and only if $\widetilde{\sigma}_\alpha$ extends over D^k, $1 \leq \alpha \leq r$.*

Proof. If σ extends to a section $s : X_k \to E$, this extension pulls back to a section

$$s_\alpha : D^k \to g^*_\alpha(E), \quad 1 \leq \alpha \leq r,$$

and we can write $\varphi_\alpha(s_\alpha(x)) = (x, \widetilde{s}_\alpha(x))$. Since s_α extends σ_α, $\widetilde{s}_\alpha$ extends $\widetilde{\sigma}_\alpha$.

For the converse, suppose that σ has been extended to a section

$$\sigma'_\alpha : X_{k-1} \cup e^k_1 \cup \cdots \cup e^k_\alpha \to E, \quad 0 \leq \alpha < r,$$

and that $\widetilde{\sigma}_{\alpha+1}$ extends to $\widetilde{s}_{\alpha+1}$. Let

$$s_{\alpha+1}(x) = \varphi_{\alpha+1}^{-1}(x, \widetilde{s}_{\alpha+1}(x)), \quad \forall x \in D^k,$$

an extension of $\sigma_{\alpha+1}$. When $\alpha = 0$, we take $\sigma_0' = \sigma$, so the induction gets started. From the commutative diagram

$$\begin{array}{ccccc} g_{\alpha+1}^*(E) & \xrightarrow{\text{id}} & g_{\alpha+1}^*(E) & \xrightarrow{\overline{g}_{\alpha+1}} & E \\ {\scriptstyle s_{\alpha+1}}\uparrow & & {\scriptstyle \pi'}\downarrow & & \downarrow{\scriptstyle \pi} \\ D^k & \underset{\text{id}}{\longrightarrow} & D^k & \underset{g_{\alpha+1}}{\longrightarrow} & X_k \end{array}$$

we see that

$$\pi \circ \overline{g}_{\alpha+1} \circ s_{\alpha+1}(x) = g_{\alpha+1}(x), \quad \forall x \in D^k.$$

Since $g_{\alpha+1}$ is a homeomorphism on $\operatorname{int} D^k$, it follows that σ_α' can be extended to a section $\sigma_{\alpha+1}'$ over $X_{k-1} \cup e_1^k \cup \cdots \cup e_{\alpha+1}^k$ by

$$\sigma_{\alpha+1}'(g_{\alpha+1}(x)) = \overline{g}_{\alpha+1} \circ s_{\alpha+1}(x), \quad \forall x \in \operatorname{int} D^k.$$

By finite induction, we reach the desired extension $s = \sigma_r'$. □

The extension $\widetilde{s}_\alpha$ of $\widetilde{\sigma}_\alpha : S^{k-1} \to S^1$ exists if and only if this map is homotopically trivial. Since

$$\pi_{k-1}(S^1) = \begin{cases} \mathbb{Z} & \text{if } k = 2, \\ 0 & \text{otherwise}, \end{cases}$$

we see that the only possible obstruction occurs when we try to extend a section $\sigma : X_1 \to E$ to one over X_2. This obstruction is $\deg \widetilde{\sigma}_\alpha$, $1 \le \alpha \le r$.

We will define the cochain c_σ on the basis $\{e_\alpha^2\}_{\alpha=1}^r$ of $C_2(X; \mathbb{Z})$ by

$$c_\sigma(e_\alpha^2) = \deg \widetilde{\sigma}_\alpha, \quad \alpha = 1, \ldots, r.$$

Here, each $e_\alpha^2 \in C_2(X; \mathbb{Z})$ is understood to be oriented. With the opposite orientation, this cell is $-e_\alpha^2 \in C_2(X; \mathbb{Z})$. The orientation pulls back to D^2 via g_α, hence induces an orientation on ∂D^2, and it is this orientation that is used in computing $\deg \widetilde{\sigma}_\alpha$. When the orientation is reversed, the degree changes sign and $c_\sigma(-e_\alpha^2) = -c_\sigma(e_\alpha^2)$, as it should. The other thing to check is that c_σ is independent of the choices of the orientation-respecting trivializations φ_α.

Lemma 4.3.2. *The cochain c_σ depends only on σ, not on the choices of trivializations.*

Proof. Let $\overline{\varphi}_\alpha$ be a second choice of trivialization of $g_\alpha^*(E)$ and write

$$\varphi_\alpha(\sigma_\alpha(x)) = (x, \widetilde{\sigma}_\alpha(x)),$$
$$\overline{\varphi}_\alpha(\sigma_\alpha(x)) = (x, \overline{\sigma}_\alpha(x)),$$

$\forall x \in \partial D^2$. We must show that $\deg \widetilde{\sigma}_\alpha = \deg \overline{\sigma}_\alpha$. The homeomorphism

$$\theta_\alpha = \overline{\varphi}_\alpha \circ \varphi_\alpha^{-1} : D^2 \times S^1 \to D^2 \times S^1$$

has the formula

$$\theta_\alpha(x, z) = (x, \psi_\alpha(x)(z)),$$

where

$$\psi_\alpha : D^2 \to \text{Homeo}_+(S^1)$$

is continuous. This homeomorphism, restricted to $\partial D^2 \times S^1$, gives

$$(x, \overline{\sigma}_\alpha(x)) = \theta_\alpha(x, \widetilde{\sigma}_\alpha(x)) = (x, \psi_\alpha(x)(\widetilde{\sigma}_\alpha(x))),$$

$\forall x \in \partial D^2$. But ψ_α can be viewed as a homotopy of $\psi_\alpha|\partial D^2$ to the constant $\psi_\alpha(0) \in \text{Homeo}_+(S^1)$, so

$$\overline{\sigma}_\alpha \sim \psi_\alpha(0) \circ \widetilde{\sigma}_\alpha.$$

Since degree is a homotopy invariant and since the degree of an orientation-preserving homeomorphism of S^1 is 1, we conclude that

$$\deg \overline{\sigma}_\alpha = (\deg \psi_\alpha(0)) \cdot (\deg \widetilde{\sigma}_\alpha) = \deg \widetilde{\sigma}_\alpha,$$

as desired. □

The discussion so far has the following consequence.

Corollary 4.3.3. *The cochain c_σ vanishes if and only if σ extends to a section $s : X \to E$.*

Definition 4.3.4. The cochain c_σ is called the first obstruction, or simply the obstruction cochain, to extending σ.

Relative to a fixed trivialization

$$\begin{array}{ccc} E|X_1 & \xrightarrow{\varphi} & X_1 \times S^1 \\ \pi \downarrow & & \downarrow p \\ X_1 & \xrightarrow[\text{id}]{} & X_1, \end{array}$$

sections can be written $\varphi(\sigma(x)) = (x, \widetilde{\sigma}(x))$. Fixing this trivialization, we abuse notation by writing $\sigma(x) = (x, \widetilde{\sigma}(x))$. Another section $\tau(x) = (x, \widetilde{\tau}(x))$ is homotopic to σ through sections if and only if $\widetilde{\sigma}$ and $\widetilde{\tau}$ are homotopic as maps of X_1 into S^1. The following is an obvious consequence of the fact that the degree of a map $S^1 \to S^1$ is homotopy-invariant.

Lemma 4.3.5. *If the sections σ and τ over X_1 are homotopic through sections, then $c_\sigma = c_\tau$.*

By Lemma 4.3.5 and Theorem 4.2.2, we can assume that

$$\widetilde{\sigma}|X_0 \equiv 1 \in S^1.$$

Thus, if e^1_β is an oriented 1-cell of X_1, $\widetilde{\sigma}|e^1_\beta$ takes the endpoint(s) of e^1_β to the basepoint $1 \in S^1$, hence has a well-defined degree $d_\sigma(e^1_\beta) \in \mathbb{Z}$. It is evident that $d_\sigma(-e^1_\beta) = -d_\sigma(e^1_\beta)$; hence d_σ extends to a 1-cochain in $C^1(X;\mathbb{Z})$. We emphasize that, unlike c_σ, *the cochain d_σ depends on the choice of trivialization φ.*

Relative to φ_α and φ, we can write the map $\overline{g}_\alpha$ in the pullback diagram

$$\begin{array}{ccc} g^*_\alpha(E)|\partial D^2 & \xrightarrow{\overline{g}_\alpha} & E|X_1 \\ {\scriptstyle \pi'}\downarrow & & \downarrow{\scriptstyle \pi} \\ \partial D^2 & \xrightarrow[g_\alpha]{} & X_1 \end{array}$$

as

$$\overline{g}_\alpha(x,z) = (g_\alpha(x), \widetilde{g}_\alpha(x)(z)), \quad \forall\, x \in \partial D^2, \quad \forall\, z \in S^1,$$

where

$$\widetilde{g}_\alpha : \partial D^2 \to \mathrm{Homeo}_+(S^1)$$

is continuous. These maps satisfy

$$\widetilde{\sigma}(g_\alpha(x)) = \widetilde{g}_\alpha(x)(\widetilde{\sigma}_\alpha(x)), \quad \forall\, x \in \partial D^2.$$

Exercise 4.3.6. Take $1 \in S^1$ as basepoint and let $n_\alpha \in \mathbb{Z}$ be the degree of the map

$$x \mapsto \widetilde{g}_\alpha(x)(1)$$

of ∂D^2 to S^1. Prove that the map

$$x \mapsto \widetilde{g}_\alpha(x)(\widetilde{\sigma}_\alpha(x))$$

has degree $n_\alpha + \deg \widetilde{\sigma}_\alpha$.

Exercise 4.3.7. By the previous exercise,

$$\deg(\widetilde{\sigma} \circ g_\alpha) = n_\alpha + c_\sigma(e^2_\alpha), \quad 1 \le \alpha \le r.$$

Use this to prove that

$$\delta d_\sigma(e^2_\alpha) = n_\alpha + c_\sigma(e^2_\alpha), \quad 1 \le \alpha \le r.$$

As a preliminary step in proving that c_σ is a cocycle, we have the following.

Lemma 4.3.8. *If E is the trivial circle bundle over X, then c_σ is a coboundary.*

Proof. We fix a trivialization, taking $E = X \times S^1$, and also use the restriction of this trivialization to $E|X_1$ as the trivialization φ in the above discussion. Furthermore, for each 2-cell $e^2_\alpha = g_\alpha(D^2)$, we choose the trivialization φ_α of $g^*_\alpha(E)$ to be the pull-back of the trivialization of E as in Lemma 4.1.1. With these choices, it is easy to see that the integers n_α in Exercise 4.3.6 all vanish; hence $c_\sigma = \delta d_\sigma$ by Exercise 4.3.7. □

Lemma 4.3.9. *For any oriented circle bundle E over X, the cochain c_σ is a cocycle, called the obstruction cocycle.*

Proof. If $\{e^3_1, \dots, e^3_m\}$ is the set of oriented 3-cells generating $C^3(X;\mathbb{Z})$, we must show that $c_\sigma(\partial e^3_i) = 0$, $1 \le i \le m$. Let $e^3_i = g_i(D^3)$, where $g_i : D^3 \to X_3$ is, as usual, the extension over D^3 of the attaching map for the cell. By weak regularity, we have a cell structure on D^3 as in Exercise 4.2.9, g_i being cellular with respect to this structure. Note that D^3 itself is the unique 3-cell and ∂D^3 is the 2-skeleton. Let D^3_1 denote the 1-skeleton.

Let σ_i be the unique section making the pull-back diagram

$$\begin{array}{ccc} g^*_i(E)|D^3_1 & \xrightarrow{\overline{g}_i} & E|X_1 \\ {\scriptstyle\sigma_i}\uparrow & & \uparrow{\scriptstyle\sigma} \\ D^3_1 & \xrightarrow[g_i]{} & X_1 \end{array}$$

commute. Note that, if

$$g^\#_i : C^*(X;\mathbb{Z}) \to C^*(D^3;\mathbb{Z})$$

is the induced cochain map (the adjoint of the chain map $g_{i\#}$), then

$$c_{\sigma_i} = g^\#_i(c_\sigma).$$

Since D^3 is contractible, $g^*_i(E)$ is trivial and Lemma 4.3.8 implies that c_{σ_i} is a coboundary. In particular, $\delta c_{\sigma_i} = 0$. Thus, if we let $b \in C^3(D^3;\mathbb{Z})$ denote D^3 when viewed as the generator, we obtain

$$\begin{aligned} c_\sigma(\partial e^3_i) &= c_\sigma(\partial g_{i\#}(b)) \\ &= c_\sigma(g_{i\#}(\partial b)) \\ &= g^\#_i(c_\sigma)(\partial b) \\ &= c_{\sigma_i}(\partial b) \\ &= 0. \end{aligned}$$

□

Even if the cocycle c_σ does not vanish, its class $[c_\sigma] \in H^2(X;\mathbb{Z})$ might be zero. We want to show that this will guarantee the existence of a global cross-section of E, even though the section σ itself may have been badly chosen so that *it* does not extend. The next two lemmas are critical.

Lemma 4.3.10. *If τ is another choice of section of $E|X_1$, then c_σ and c_τ are cohomologous.*

Proof. As usual, fix a trivialization of $E|X_1$ and write

$$\begin{aligned}\tau(x) &= (x, \widetilde{\tau}(x)),\\ \sigma(x) &= (x, \widetilde{\sigma}(x)),\end{aligned}$$

$\forall\, x \in X_1$. For each 2-cell e^2_α, $1 \le \alpha \le r$, write

$$\begin{aligned}\delta d_\sigma(e^2_\alpha) &= n_\alpha + c_\sigma(e^2_\alpha),\\ \delta d_\tau(e^2_\alpha) &= n_\alpha + c_\tau(e^2_\alpha),\end{aligned}$$

as in Exercise 4.3.7. It follows that $c_\sigma - c_\tau = \delta(d_\sigma - d_\tau)$. □

Lemma 4.3.11. *If $c \in C^2(X;\mathbb{Z}) = Z^2(X;\mathbb{Z})$ is cohomologous to c_σ, then there is a section $\tau : X_1 \to E$ such that $c = c_\tau$.*

Proof. There is a cochain $h \in C^1(X;\mathbb{Z})$ such that $c = c_\sigma - \delta h$. Let $\{e^1_\beta\}^s_{i=1}$ be the set of 1-cells generating $C^1(X;\mathbb{Z})$. We fix the trivializations φ and φ_α as before, and define $\tau : X_1 \to E$ by requiring that

$$d_\tau(e^1_\beta) = d_\sigma(e^1_\beta) - h(e^1_\beta), \quad 1 \le \beta \le s.$$

Then, as in the proof of the previous lemma, $c_\sigma - c_\tau = \delta(d_\sigma - d_\tau) = \delta h$, implying that $c = c_\sigma - \delta h = c_\tau$. □

At this point we have a well-defined cohomology class $[c_\sigma] \in H^2(X;\mathbb{Z})$ associated to the bundle E and (apparently) to the weakly regular cell structure $\mathcal{S}$ of X. Provisionally, let us write $(X, \mathcal{S})$ for the cell complex and denote the class $[c_\sigma]$ by $e(E, \mathcal{S})$.

Exercise 4.3.12. If $f : (X, \mathcal{S}) \to (Y, \mathcal{T})$ is a cellular map between weakly regular cell complexes, and if E is an oriented circle bundle over Y, prove that $f^*(e(E, \mathcal{T})) = e(f^*(E), \mathcal{S})$. We say that this class is *natural* with respect to cellular maps.

Lemma 4.3.13. *The class $e(E, \mathcal{S})$ does not depend on the choice of cellular structure $\mathcal{S}$ and is natural with respect to* continuous *maps.*

Proof. If $f : X \to Y$ is a continuous map between cell complexes, it is homotopic to a cellular map $\widetilde{f}$ (Theorem 4.2.1). By the canonical equality of cellular cohomology to singular cohomology, together with the homotopy invariance of singular theory, the induced cohomology maps f^* and $\widetilde{f}^*$ are the same. Thus, the naturality with respect to continuous maps follows from

Exercise 4.3.12. If $\mathcal{S}$ and $\mathcal{S}'$ are two weakly regular cellular structures on X, take $f = \mathrm{id}_X$ in this argument and conclude that $e(E, \mathcal{S}) = e(E, \mathcal{S}')$. □

The remainder of this chapter will be devoted to the geometric significance of this class. First, however, we discuss briefly a simple modification of this theory that defines the relative cohomology class of obstruction cocycles $c_{s,\sigma} \in C^2(X, A; \mathbb{Z})$ corresponding to a section $s : A \to E|A$ over a subcomplex A.

The section s, first restricted to the 1-skeleton of A and then extended to a section σ over X_1, gives rise to an obstruction cocycle $c_\sigma \in C^2(X; \mathbb{Z})$. By Corollary 4.3.3, this cocycle vanishes on the subgroup $C_2(A; \mathbb{Z}) \subseteq C_2(X; \mathbb{Z})$; hence it can be viewed as a cocycle in $C^2(X, A; \mathbb{Z})$. In order to remind us of the dependence of this relative cocycle on the section s, we will denote it by $c_{s,\sigma}$.

Exercise 4.3.14. Prove that a relative cocycle $c \in C^2(X, A; \mathbb{Z})$ is cohomologous to $c_{s,\sigma}$ if and only if it is of the form $c = c_{s,\tau}$, where τ is a section over X_1 extending $s|A_1$. (Hint. Choose the trivialization of $E|(A \cup X_1)$ so that d_σ is a 1-cochain vanishing in A.) Furthermore, show that $c_{s,\sigma} = 0$ if and only if the section $s \cup \sigma$ over $A \cup X_1$ extends to a section over X. Finally, discuss the independence of this class from the choice of weakly regular cellular decomposition.

4.4. The Euler Class

Our study of the first (and only) obstruction cocycle for oriented circle bundles produced a well defined characteristic class for such bundles.

Definition 4.4.1. The Euler class of the oriented circle bundle $\pi : E \to X$ is $e(E) = [c_\sigma] \in H^2(X; \mathbb{Z})$, where σ is any section of $E|X_1$. If $A \subseteq X$ is a subcomplex and s is a section of E over A, then the relative Euler class, modulo s, is $e(E, s) = [c_{s,\sigma}] \in H^2(X, A; \mathbb{Z})$, where σ is any section of $E|X_1$ that extends $s|A_1$. Finally, if E is the unit circle bundle associated to an oriented 2-plane bundle V, the Euler class of E is also called the Euler class of V and denoted by $e(V)$. In the relative case, a nonzero section v of V over $A \subseteq X$ can be normalized to a section s of $E|A$ and we obtain the relative Euler class $e(V, v) = e(E, s)$.

Theorem 4.4.2. *The Euler class $e(E)$ vanishes if and only if the bundle $\pi : E \to X$ is trivial. The relative Euler class $e(E, s)$ vanishes if and only if the section s over the subcomplex A extends to a section over X.*

Proof. If the class vanishes, any section σ over X_1 has $c_\sigma = \delta h$, for some 1-cochain h. By Lemma 4.3.11, there is a section τ over X_1 such that $c_\tau = 0$.

Such a section extends to a global section of E (Lemma 4.3.3) and, since E is orientable, it follows that the bundle is trivial. Conversely, if the bundle is trivial, it admits a global section. The restriction τ of that section to X_1 has obstruction cocycle $c_\tau = 0$ (Lemma 4.3.3), so $e(E) = [c_\tau] = 0$. The second assertion is similar and is left to the reader. □

Remark. It can be shown that every oriented circle bundle is isomorphic to the unit circle bundle of a complex line bundle that is unique up to isomorphism. The set $\mathcal{L}_{\mathbb{C}}(X)$ of isomorphism classes of complex line bundles over X is therefore in one-to-one correspondence with the set $\mathcal{S}(X)$ of isomorphism classes of oriented S^1-bundles over X. The Euler class of the circle bundle associated to the complex line bundle is called the first Chern class $c_1(L)$. The set $\mathcal{L}_{\mathbb{C}}(X)$ becomes an *abelian group* under tensor product (over $\mathbb{C}$) of line bundles and the Chern class

$$c_1 : \mathcal{L}_{\mathbb{C}}(X) \to H^2(X; \mathbb{Z})$$

is a group surjection (Exercise 4.4.3). By the above theorem, this homomorphism has trivial kernel, hence is an isomorphism. That is, the Euler class completely characterizes oriented circle bundles up to isomorphism.

Exercise 4.4.3. Prove the assertions in the above remark. Specifically,

(1) Prove that the set $\mathcal{L}_{\mathbb{C}}(X)$ of isomorphism classes $[L]$ of complex line bundles $\pi : L \to X$ is a group under tensor product. (Hint. Take as identity element 1 the isomorphism class of the trivial bundle $X \times \mathbb{C}$. Given a complex line bundle L, define the *dual bundle* $\overline{L}$ to be L itself, but with scalar multiplication $\mathbb{C} \times L \to L$ redefined by $(c, z) \mapsto \bar{c}z$. Prove that $[L \otimes \overline{L}] = 1$.)

(2) Show that the first Chern class

$$c_1 : \mathcal{L}_{\mathbb{C}}(X) \to H^2(X; \mathbb{Z})$$

is a surjective homomorphism of groups. Here, $c_1(L) = e(E)$, where $\pi : E \to X$ is the unit circle bundle of L.

Example 4.4.4. Realize $S^3 \subset \mathbb{C}^2$ as the sphere

$$\{(z, w) \mid |z|^2 + |w|^2 = 2\}.$$

The unit circle $S^1 \subset \mathbb{C}$ is a group and the action

$$S^1 \times S^3 \to S^3,$$
$$(e^{it}, (z, w)) \mapsto (e^{it}z, e^{it}w),$$

is free. The orbits of this action fiber S^3. These orbits are the intersections of S^3 with the complex, 1-dimensional subspaces of $\mathbb{C}^2$. This set of 1-dimensional subspaces is the complex projective line $\mathbb{C}P^1 \cong S^2$ and can be viewed as the base space of the fibration. We have constructed the *Hopf fibration*

$$\pi : S^3 \to S^2.$$

The Euler class of this bundle in $H^2(S^2;\mathbb{Z}) = \mathbb{Z}$ can be viewed as an integer, and we will see (when everything in sight is suitably oriented) that this integer is 1.

The inclusion $S^1 \times S^1 \subset S^3$ is the common interface of the familiar solid tori K_1 and K_2 into which S^3 is decomposed. Here, we take

$$K_1 = \{(z,w) \in S^3 \mid |z| \le |w|\},$$
$$K_2 = \{(z,w) \in S^3 \mid |z| \ge |w|\}.$$

The S^1-action leaves each of these solid tori invariant. The respective cores $\{0\} \times 2 \cdot S^1$ and $2 \cdot S^1 \times \{0\}$ are orbits and the remaining orbits in K_i wind once meridianally around the core while going once longitudinally along it, $i = 1, 2$. In particular, on the common boundary $S^1 \times S^1$, the action induces the linear foliation of slope 1. The fibers in K_i other than the core circle are said to be circles of type $(1,1)$. While the meanings of "meridian" and "longitude" depend on which solid torus, K_1 or K_2, one stands on, the circles on $S^1 \times S^1$ of type $(1,1)$ are the same from either point of view. Note that there is a cross-section of the foliation of $S^1 \times S^1$ that is a meridian from the point of view of K_1 and a longitude from the point of view of K_2.

Fix a weakly regular cellular structure on S^2 having one vertex on the equator, the equator itself being the sole 1-cell e^1. There are two 2-cells, the northern and southern hemispheres e_1^2 and e_2^2, sharing the equator as common boundary. These 2-cells are to be oriented compatibly with a fixed orientation of S^2; hence the induced orientations on the equator relative to the two hemispheres will be opposite. In terms of the cellular boundary operator, this turns into the equation

$$\partial(e_1^2 + e_2^2) = e^1 - e^1 = 0.$$

That is, $[e_1^2 + e_2^2] \in H^2(S^2;\mathbb{Z}) = \mathbb{Z}$ and, being nondivisible, this class can be identified as the generator 1.

This cell structure can be chosen so that the Hopf fibration fibers K_i over e_i^2, $i = 1, 2$, hence fibers $S^1 \times S^1$ over the equator e^1. Let $\sigma : e^1 \to S^1 \times S^1$ be a cross-section of this fibration that, from the point of view of K_1, is a meridian, hence, from the point of view of K_2, is a longitude. Our fibration restricts to trivial bundles

$$\pi_1 : K_1 \to e_1^2,$$
$$\pi_2 : K_2 \to e_2^2,$$

and σ defines a section σ_i of

$$\pi_i : K_i|\partial e_i^2 \to \partial e_i^2, \quad i = 1, 2.$$

Relative to explicit trivializations of these two bundles, however, these sections look very different. Indeed, in each case, the trivialization $K_i \cong e_i^2 \times S^1$ can be chosen to send meridians (circles of type $(1, 0)$) on $S^1 \times S^1$ to themselves and circles of type $(1, 1)$ to longitudes. A little thought shows that such a trivialization sends longitudes (circles of type $(0, 1)$) to circles of type $(-1, 1)$. Thus, $\deg \widetilde{\sigma}_1 = 0$ (σ_1 is a meridian, so $\widetilde{\sigma}_1$ is constant) and $\deg \widetilde{\sigma}_2 = \pm 1$ (σ_2 is a circle of type $(-1, 1)$, so $\widetilde{\sigma}_2$ is a homeomorphism). Thus, if we orient S^2 and the fibration so that the homeomorphism $\widetilde{\sigma}_2 : \partial e_2^2 \to S^1$ is orientation-preserving, then

$$c_\sigma(e_1^2 + e_2^2) = 0 + 1 = 1$$

and our assertion follows.

Example 4.4.5. The previous example provides a clue to analyzing all oriented circle bundles $\pi : E \to S^2$. The restrictions

$$\pi_i : E_i = E|e_i^2 \to e_i^2, \quad i = 1, 2,$$

must be trivial bundles because the 2-cells e_1^2 and e_2^2 are contractible. Thus, we can fix identifications $E_i \cong e_i^2 \times S^1$, $i = 1, 2$, and reassemble the bundle

$$E = E_1 \cup_\varphi E_2$$

via a gluing map

$$\varphi : e^1 \times S^1 \to e^1 \times S^1,$$
$$\varphi(x, z) = (x, \psi(x)(z)),$$

where

$$\psi : e^1 \to \mathrm{Homeo}_+(S^1)$$

is continuous. If the degree of the map

$$x \mapsto \psi(x)(1)$$

is k, then (up to a sign that depends on how everything in sight has been oriented), $e(E) = k$. Indeed, choose the section $\sigma : e^1 \to E$ that pulls back to the section

$$\sigma_1 : \partial e_1^2 \to \partial e_1^2 \times S^1,$$
$$\sigma_1(x) = (x, 1).$$

The section

$$\sigma_2 : \partial e_2^2 \to \partial e_2^2 \times S^1$$

that φ matches to σ_1 has formula

$$\sigma_2(x) = (x, \psi(x)^{-1}(1));$$

hence this is the pullback of σ relative to E_2. Thus,

$$c_\sigma(e_1^2) = 0,$$
$$c_\sigma(e_2^2) = \pm k \quad \text{(depending on orientations)}.$$

Remark. Careful attention to orientations is needed in this game. For instance, if we had chosen a section

$$\sigma' : e^1 \to E$$

in the above example so that $\sigma_2'(x) = (x, 1)$, then the formula for the other pullback would have to be $\sigma_1'(x) = (x, \psi(x)(1))$. The degrees of

$$x \mapsto \psi(x)(1),$$
$$x \mapsto \psi(x)^{-1}(1)$$

are the negatives of each other. But, taking into account the fact that ∂e_1^2 and ∂e_2^2 must be oppositely oriented versions of e^1, we see that

$$\deg \sigma_1' = \deg \sigma_2$$

and the cocycles c_σ and $c_{\sigma'}$ agree on the *cycle* $e_1^2 + e_2^2$.

Example 4.4.6. Let Σ_g be the closed, orientable surface of genus $g \geq 1$ (note that we are including the torus in this example). Realize this surface as the quotient of a $4g$-gon by pairwise identifications of edges in the usual way. Actually, it will be more convenient to replace the $4g$-gon with D^2, subdividing ∂D^2 into $4g$ arcs. The cell structure has one vertex x_0, $2g$ 1-cells $e_1, \dots, e_{2g}$ and one 2-cell D_0. Let $g_0 : D^2 \to \Sigma_g$ be the identification map. That is, $g_0|\partial D^2$ is the attaching map for the 2-cell. Let $\pi : E \to \Sigma_g$ be an oriented S^1-bundle and consider the pull-back diagram

$$\begin{array}{ccc} D^2 \times S^1 & \xrightarrow{\overline{g}_0} & E \\ {\scriptstyle p}\downarrow & & \downarrow{\scriptstyle \pi} \\ D^2 & \xrightarrow[g_0]{} & \Sigma_g \end{array}$$

where we have also fixed a trivialization of $g_0^*(E)$. In addition, fix a trivialization

$$\begin{array}{ccc} E|X_1 & \xrightarrow{\cong} & X_1 \times S^1 \\ {\scriptstyle \pi}\downarrow & & \downarrow{\scriptstyle p} \\ X_1 & \xrightarrow[\text{id}]{} & X_1 \end{array}$$

over the 1-skeleton $X_1 = e_1 \vee \cdots \vee e_{2g}$. Relative to these two trivializations, we can write

$$\begin{array}{ccc} \partial D^2 \times S^1 & \xrightarrow{\overline{g}_0} & X_1 \times S^1 \\ {\scriptstyle p}\downarrow & & \downarrow{\scriptstyle p} \\ \partial D^2 & \xrightarrow[g_0]{} & X_1 \end{array}$$

and the formula for $\overline{g}_0$ must have the form

$$\overline{g}_0(x, z) = (g_0(x), \psi(x)(z)), \quad \forall x \in \partial D^2,$$

where

$$\psi : \partial D^2 \to \mathrm{Homeo}_+(S^1)$$

is continuous. Let $m(\psi)$ be the degree of the map

$$x \mapsto \psi(x)(1).$$

Relative to the trivialization $E|X_1 \cong X_1 \times S^1$, let

$$\sigma : X_1 \to E|X_1$$

be the section $\sigma(y) = (y, 1)$. If σ_0 is the pullback of this section, commutativity of the diagram

$$\begin{array}{ccc} \partial D^2 \times S^1 & \xrightarrow{\overline{g}_0} & X_1 \times S^1 \\ {\scriptstyle \sigma_0}\uparrow & & \uparrow{\scriptstyle \sigma} \\ \partial D^2 & \xrightarrow[g_0]{} & X_1 \end{array}$$

implies that

$$\sigma_0(x) = (x, \psi(x)^{-1}(1)), \quad \forall x \in \partial D^2.$$

It follows that $e(E) = c_\sigma(D_0) = \deg \widetilde{\sigma}_0 = \pm m(\psi)$.

A slightly different way of looking at this picture is going to be useful. Let $\Delta \subset \Sigma_g \smallsetminus X_1$ be a compact, imbedded disk. Then X_1 is a deformation retract of the compact, bordered surface $S_g = \Sigma_g \smallsetminus \operatorname{int} \Delta$, and the bundle E restricts to trivial bundles over Δ and S_g. From the above analysis, it should be clear that the bundle E can be reassembled from $\Delta \times S^1$ and $S_g \times S^1$ by the attaching map

$$\varphi : \partial\Delta \times S^1 \to \partial S_g \times S^1,$$
$$\varphi(x, z) = (x, \psi(x)(z)),$$

where $\psi : \partial\Delta \to \mathrm{Homeo}_+(S^1)$ is essentially the same as above. To emphasize the role of ψ in determining the structure of the bundle, we will write $E = E_\psi$ hereafter. The formula

$$e(E_\psi) = \pm m(\psi)$$

will be very important shortly.

Exercise 4.4.7. A 2×2 matrix A with integer entries is said to be *unimodular* if $\det A = 1$. Such matrices induce orientation-preserving diffeomorphisms

$$f_A : T^2 \to T^2.$$

One can then fashion a closed 3-manifold N_A from $T^2 \times [0,1]$ by the identification $(z,1) \sim (f_A(z), 0)$. In an obvious way, this manifold fibers over S^1 with fiber T^2. If

$$A = \begin{bmatrix} 1 & p \\ 0 & 1 \end{bmatrix},$$

for $p \in \mathbb{Z}$, prove that N_A is also a circle bundle over T^2 with Euler class $\pm p$.

Example 4.4.8. The Euler class of the unit tangent bundle $T_1(M)$ of a compact, oriented surface M ($\partial M = \emptyset$) deserves special mention. Observe that the unit tangent bundle is canonically oriented by the orientation of the surface. As a consequence, $e(T_1(M)) \in \mathbb{Z}$ is determined without any sign ambiguities. It is always possible to find a vector field v on M with only isolated singularities, $x_1, \dots, x_r$. For $1 \le i \le r$, choose small, compact disks $D_i \subset M$ centered at x_i and pairwise disjoint. The restriction $v|\partial D_i$ can be normalized, defining a section σ_i of $T_1(M)|\partial D_i$ and an associated degree m_i of $\widetilde{\sigma}_i$. Here, the disks are oriented by the orientation of M. The number m_i is called the *index* of the vector field v at the isolated singularity x_i. It is possible to choose the cell decomposition of M so that the 1-skeleton is a deformation retract of $M \setminus \bigcup_{i=1}^r \operatorname{int} D_i$. Then, the disks D_i are essentially the 2-cells and the Euler class, as an integer, is

$$e(T_1(M)) = \sum_{i=1}^r m_i.$$

In particular, this integer, which is called the *index sum* of v, does not depend on the choice of the vector field v with isolated singularities.

Theorem 4.4.9 (Poincaré-Hopf Theorem). *The index sum of a vector field on a closed, oriented surface M is equal to the Euler characteristic $\chi(M)$.*

Since we have shown that this index sum is independent of the choice of v, the Poincaré-Hopf theorem can be proven by finding a particularly nice vector field whose index sum is readily seen to be $2 - 2g$, $g = \operatorname{genus} M$. For S^2, the vector field tangent to great circle arcs from the south pole x_1 to the north pole x_2 and having x_1 as a "source", x_2 as a "sink", has $m_1 = m_2 = 1$; hence the index sum is $2 = \chi(S^2)$. Alternatively, a vector field with only one singularity, a "magnetic monopole" with flowlines as in Figure 4.4.1, can be constructed on S^2, and the reader can ascertain that the index of this singularity is 2. On T^2, there is a nowhere zero vector field, so the index sum is $0 = \chi(T^2)$. For the surfaces Σ_g, $g \ge 2$, proceed as

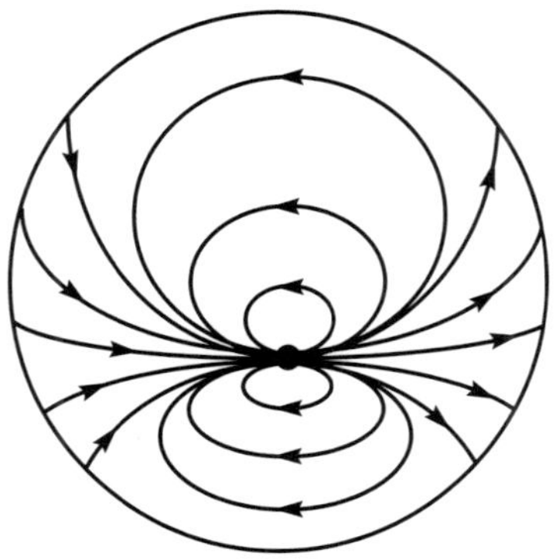

Figure 4.4.1. Flowlines at a monopole singularity on S^2

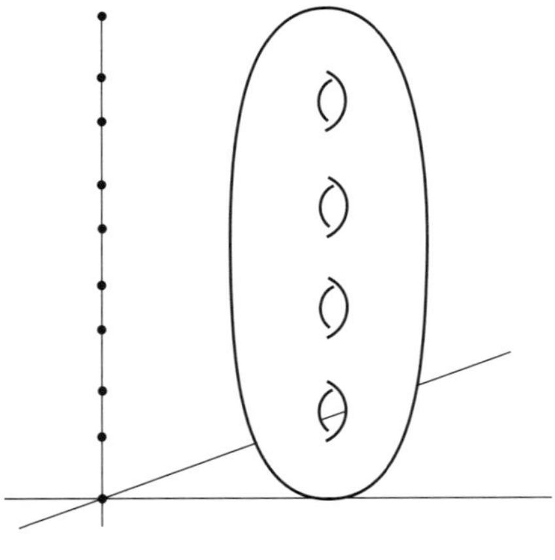

Figure 4.4.2. The height function for Σ_4

follows. Set Σ_g "on end" on the x, y-plane, as indicated in Figure 4.4.2, and consider the height function $f : \Sigma_g \to \mathbb{R}$. This function has $2g+2$ critical points (the corresponding critical values are indicated by dots on the z-axis in Figure 4.4.2). One is a minimum, $2g$ of them are saddle points, and one is a maximum. The gradient field ∇f has index $+1$ at its source (the minimum of f) and at its sink (the maximum of f) and has index -1 at each of the saddle points, as is easily seen. Everywhere else, ∇f is nonsingular. Thus, the index sum of this field is $2-2g=\chi(\Sigma_g)$. Observe that this also works when $g=1$, giving a vector field on T^2 with four singularities and index sum $2-2=0$, and when $g=0$, giving the vector field on S^2 having the north and south poles as sole singularities.

Thus, we have simultaneously verified the Poincaré-Hopf theorem and calculated

$$e(T_1(M))=\chi(M).$$

Example 4.4.10. The previous two examples can be modified for surfaces with boundary by using the relative Euler class. More precisely, let

$$s:\partial M\to E|\partial M$$

be a nowhere zero section. Such a section always exists. Then we can extend s to a section on the complement of finitely many points, say $x_1,\dots,x_r$. Let m_i denote the degree of this section about x_i, $1\le i\le r$, and argue as in Example 4.4.6 that the Euler class $e(E,s)\in H^2(M,\partial M;\mathbb{Z})=\mathbb{Z}$ is the integer $\sum_{i=1}^r m_i$.

In the case that $E=T_1(M)$, assume that $v\in\mathfrak{X}(M)$ is a vector field with isolated singularities in int M and that, on each component of ∂M, v is nonzero and either always tangent to that component or always transverse to it. In either case, the reader will quickly check that we can attach a disk to each component and extend v over that disk so that it has one singularity of index 1 in each of these disks. This produces a closed surface M' and a vector field with index sum $\chi(M')=e(T_1(M),v|\partial M)+c$, where c is the number of components of ∂M. We set $s=v/\|v\|$ on the complement of the singular set. Since $\chi(M)=\chi(M')-c$ and $e(T_1(M),s|\partial M)$ is equal to the index sum of v, we obtain the following relative version of Theorem 4.4.9.

Theorem 4.4.11 (Poincaré-Hopf II). *If M is a compact, connected, oriented surface, $\partial M\neq 0$, and if v is a vector field on M as above, then the index sum of v is equal to $\chi(M)$.*

Example 4.4.12. Let $(M,\mathcal{F})$ be an oriented and transversely oriented, C^2-foliated 3-manifold of codimension one, possibly with $\partial_\tau M\neq\emptyset\neq\partial_\pitchfork M$. Let

$$i:(N,\partial N)\hookrightarrow(M,\partial_\pitchfork M)$$

be a proper C^2 imbedding of a compact, oriented surface. If $\partial_\pitchfork M=\emptyset$, then, of course, $\partial N=\emptyset$. We assume that each component of ∂N either is transverse to $\mathcal{F}|\partial_\pitchfork M$ or is a leaf of that foliation. Let $[N,\partial N]\in H_2(M,\partial_\pitchfork M;\mathbb{Z})$ denote the corresponding homology class. Let w be a nowhere 0 section of $T(\mathcal{F})|\partial_\pitchfork M$ and assume that, on each component S of $\partial_\pitchfork M$, w is either everywhere tangent to S or points into M everywhere along S. We want to evaluate $e(T(\mathcal{F}),w)\in H^2(M,\partial_\pitchfork M;\mathbb{Z})$ on the class $[N,\partial N]$.

Consider first the case in which N is a leaf of $\mathcal{F}$. Then $i^*(T(\mathcal{F})) = T(N)$ and w restricts to a vector field s along ∂N that extends to a vector field v on N, satisfying the hypotheses in Example 4.4.10. Thus, in this case,

$$e(T(\mathcal{F}), w)([N, \partial N]) = \chi(N).$$

In particular, if $H^2(M, \partial_{\pitchfork} M; \mathbb{Z}) = 0$, we can conclude that every compact leaf of $\mathcal{F}$ is either a torus or an annulus.

If N is not a leaf, then a small isotopy makes the imbedding i transverse to $\mathcal{F}$ except at a finite number of interior points where it has Morse type tangencies to $\mathcal{F}$ (*cf.* [I, §7.1], where the C^2 hypothesis is used). That is, there is an induced singular foliation $\mathcal{F}' = i^*(\mathcal{F})$ on N, the singular points $\{x_1, \dots, x_r\}$ being of Morse type. This singular foliation is integral to a vector field v on N agreeing with $s = w|\partial N$ at the boundary. But this field should be viewed as a section of $i^*(T(\mathcal{F})) \neq T(N)$, and the index of v at x_i depends on whether the local orientations of N and $\mathcal{F}$ agree or not at the point x_i of tangency. If they agree, the index of v as a section of $i^*(T(\mathcal{F}))$ agrees with its index as a tangent field to N, but if the orientation is opposite, the signs of the two indices are opposite. That is, in the second case, the index of a center is -1 and of a saddle is $+1$. Adding these indices again gives the value of $e(T(\mathcal{F}), w)$ on $[N, \partial n]$, but this integer generally differs from $\chi(N)$. In the important case that all singularities are saddles, one obtains

$$|e(T(\mathcal{F}), w)([N, \partial N])| \leq |\chi(N)|,$$

an inequality that will be critical in Chapter 10.

4.5. Foliated Circle Bundles

Consider an oriented, $C^{1,0}$-foliated circle bundle $(E, \mathcal{F}, \pi)$ over Σ_g, $g \geq 1$, and use the same cell structure and notation as in Example 4.4.6. The 1-cells are actually loops based at x_0 and, if these loops are correctly oriented and indexed, we get the standard presentation of $\pi_1(\Sigma_g, x_0)$ as generated by $e_1, \dots, e_{2g}$ subject to the single relation

$$\prod_{i=1}^{g} [e_i, e_{2i}] = 1.$$

The foliated bundle is completely determined by the total holonomy homomorphism

$$h : \pi_1(\Sigma_g, x_0) \to \mathrm{Homeo}_+(S^1),$$

and we are going to demonstrate an algorithm of Milnor for determining $e(E)$ from the homomorphism h. A consequence of this algorithm is the following theorem.

Theorem 4.5.1 (Wood [**189**]). *Let $\pi : E \to \Sigma_g$ be an oriented circle bundle, $g \geq 1$. If there is a foliation transverse to the fibers, then*

$$|e(E)| \leq 2g - 2 = |\chi(\Sigma_g)|.$$

Thus, the "most twisted" circle bundle over Σ_g that can have a foliation transverse to the fibers is $T_1(\Sigma_g)$. In fact, the Roussarie example [I, Example 1.1.9] gives such a foliation of $T_1(\Sigma_g)$, for all $g \geq 2$. For $g = 1$, such foliations are easy to construct.

In an earlier theorem, Milnor [**129**] considered the circle bundle associated to a (smooth) 2-plane bundle over Σ_g. If the vector bundle admits a *flat connection* (curvature $\equiv 0$), then the circle bundle admits a smooth foliation transverse to the fibers. This is a stronger requirement on the bundle.

Theorem 4.5.2 (Milnor). *If the oriented, real 2-plane bundle $\pi : E \to \Sigma_g$ admits a connection with curvature zero, then*

$$|e(E)| \leq g - 1 = \frac{1}{2}|\chi(\Sigma_g)|.$$

Definition 4.5.3. If $(E, \mathcal{F}, \pi)$ is an oriented, foliated circle bundle over Σ_g with total holonomy homomorphism h, then $e(h)$ is the integer $e(E)$.

In order to give Milnor's algorithm for computing this number, we need some notation. If $a \in \mathbb{R}$, $T_a : \mathbb{R} \to \mathbb{R}$ will denote translation by a. The topological groups

$$G = \text{Homeo}_+(\mathbb{R}/\mathbb{Z}) = \text{Homeo}_+(S^1),$$
$$\widehat{G} = \{\widehat{f} \in \text{Homeo}_+(\mathbb{R}) \mid \widehat{f} \circ T_1 = T_1 \circ \widehat{f}\}$$

are related by the group surjection

$$\rho : \widehat{G} \to G,$$
$$\rho(\widehat{f})(x + \mathbb{Z}) = \widehat{f}(x) + \mathbb{Z}.$$

Clearly,

$$\ker \rho = \{T_n \mid n \in \mathbb{Z}\} = \mathbb{Z}$$

and we have an exact sequence

$$(*) \qquad 0 \to \mathbb{Z} \to \widehat{G} \xrightarrow{\rho} G \to 1.$$

The total holonomy homomorphism is determined by its values on the generators

$$h(e_i) = f_i \in G$$

and we choose $\widehat{f_i} \in \widehat{G}$ such that

$$\rho(\widehat{f_i}) = f_i, \quad 1 \leq i \leq 2g.$$

Since h is a group homomorphism, the relation on the generators e_i implies that

$$\prod_{i=1}^{g}[f_i, f_{2i}] = \mathrm{id} \in G.$$

We conclude from the exact sequence $(*)$ that

$$\prod_{i=1}^{g}[\widehat{f}_i, \widehat{f}_{2i}] = T_{m(\widehat{f}_1,\ldots,\widehat{f}_{2g})}$$

for some integer $m(\widehat{f}_1, \ldots, \widehat{f}_{2g})$.

Theorem 4.5.4 (Milnor). *If $(E, \mathcal{F}, \pi)$ is an oriented, foliated circle bundle over Σ_g with total holonomy homomorphism h, then $e(h) = \pm m(\widehat{f}_1, \ldots, \widehat{f}_{2g})$.*

Thus, $\pm m(\widehat{f}_1, \ldots, \widehat{f}_{2g})$ does not depend on the choice of the lifts $\widehat{f}_i$. Write

$$E = S_g \times S^1 \cup_\varphi \Delta \times S^1,$$
$$\varphi(x, z) = (x, \psi(x)(z)), \quad \forall\,(x, z) \in \Delta \times S^1,$$

as in Example 4.4.6, so $e(h) = \pm m(\psi)$ (the degree of the map $x \mapsto \psi(x)(1)$). We will show that,

$$m(\psi) = m(\widehat{f}_1, \ldots, \widehat{f}_{2g}),$$

thereby proving Theorem 4.5.4.

Since $\pi_1(S_g, x_0)$ is the free group on $\{e_i\}_{i=1}^{2g}$, we can define a group homomorphism

$$\widehat{h} : \pi_1(S_g, x_0) \to \widehat{G},$$
$$\widehat{h}(e_i) = \widehat{f}_i, \quad 1 \le i \le 2g,$$

which suspends to give a foliated $\mathbb{R}$-bundle $(S_g \times \mathbb{R}, \widehat{\mathcal{F}}, p)$ over S_g.

Claim 1. *There is a commutative diagram*

$$\begin{array}{ccc} S_g \times \mathbb{R} & \xrightarrow{\theta} & E|S_g \\ {\scriptstyle p}\downarrow & & \downarrow{\scriptstyle \pi} \\ S_g & \xrightarrow[\mathrm{id}]{} & S_g \end{array}$$

such that θ is an infinite cyclic covering and $\theta^{-1}(\mathcal{F}|(E|S_g)) = \widehat{\mathcal{F}}$.

Proof. Let $\Gamma = \pi_1(\Sigma_g; x_0)$ and let $\widetilde{S}_g$ denote the universal cover of S_g. The group actions

$$\Gamma \times (\widetilde{S}_g \times \mathbb{R}) \to \widetilde{S}_g \times \mathbb{R},$$
$$\Gamma \times (\widetilde{S}_g \times \mathbb{R}/\mathbb{Z}) \to \widetilde{S}_g \times \mathbb{R}/\mathbb{Z},$$

given respectively by

$$(\gamma, x, t) \mapsto (x \cdot \gamma^{-1}, \widehat{h}(\gamma)(t))$$
$$(\gamma, x, z) \mapsto (x \cdot \gamma^{-1}, h(\gamma)(z)),$$

quotient to give the foliated bundles $(S_g \times \mathbb{R}, \widehat{\mathcal{F}}, p)$ and $(E|S_g, \mathcal{F}|(E|S_g), \pi)$, respectively. Since $\rho(\widehat{h}(e_i)) = h(e_i)$, $1 \le i \le 2g$, these actions are equivariant with respect to the projection

$$\widetilde{S}_g \times \mathbb{R} \to \widetilde{S}_g \times \mathbb{R}/\mathbb{Z},$$

so this projection quotients to the map θ with the required properties. $\square$

We use Claim 1 to choose the trivialization $E|S_g = S_g \times S^1$ so that $\widetilde{\mathcal{F}} = \mathcal{F}|S_g \times S^1$ lifts to $\widehat{\mathcal{F}}$ under the projection $S_g \times \mathbb{R} \to S_g \times S^1$. Since $\pi_1(\Delta) = 0$, we can choose the trivialization $E|\Delta = \Delta \times S^1$ so that $\Delta \times \{z\}$ is a leaf of $\mathcal{F}|(E|\Delta)$, $\forall\, z \in S^1$. Note that the gluing map φ must match the foliation on $\partial S_g \times S^1$, induced by $\widetilde{\mathcal{F}}$, to the product foliation on $\partial\Delta \times S^1$.

Instead of choosing a section $\sigma : \partial\Delta = \partial S_g \to E$ that is the restriction of a section over S_g, choose σ to be the restriction to $\partial\Delta$ of a section over Δ whose image is a leaf of $\mathcal{F}|(E|\Delta)$. That is, relative to the trivialization of $E|\Delta$, we can take

$$\sigma(x) = (x, 1), \quad \forall\, x \in \partial\Delta.$$

Relative to the trivialization of $E|S_g$, write

$$\sigma(x) = (x, \widetilde{\sigma}(x)), \quad \forall\, x \in \partial S_g.$$

Since

$$(x, \widetilde{\sigma}(x)) = \varphi(x, 1) = (x, \psi(x)(1)), \quad \forall\, x \in \partial\Delta,$$

we see that $\widetilde{\sigma}(x) \equiv \psi(x)(1)$ on ∂S_g, so

$$|\deg \widetilde{\sigma}| = |m(\psi)| = |e(h)|.$$

We use the fact that $\sigma(x) = (x, \widetilde{\sigma}(x))$ lies on the boundary of a leaf of $\widetilde{\mathcal{F}}$ to compute this degree.

Claim 2. *Relative to the above choices,* $\deg \widetilde{\sigma} = m(\widehat{f}_1, \dots, \widehat{f}_{2g})$.

Proof. View the inclusion map

$$\iota : \partial S_g \hookrightarrow S_g$$

as a loop on S_g that is freely homotopic to $\prod_{i=1}^{g} [e_i, e_{2i}]$. It will be convenient to parametrize this loop as

$$\iota : [0,1] \to \partial S_g, \quad \iota(0) = \iota(1) = y_0.$$

Choose a path γ on S_g from x_0 to y_0, obtaining a loop $\tau = \gamma + \iota - \gamma$ based at x_0. We view $\tau \in \pi_1(S_g, x_0)$. Using the standard fact that the conjugacy class of an element of the fundamental group consists of all elements represented by loops that are *freely* homotopic to it, we see that τ belongs to the conjugacy class of $\prod_{i=1}^{g}[e_i, e_{2i}]$. Since the total holonomy homomorphism $\widetilde{h}$ of $\widetilde{\mathfrak{F}}$ carries this conjugacy class to the conjugacy class of $\prod_{i=1}^{g}[f_i, f_{2i}] = \mathrm{id}$, we see that $\widetilde{h}(\tau) = \mathrm{id}$ and τ lifts to a loop on any leaf of $\widetilde{\mathfrak{F}}$. If λ is a lift of γ to a path ending at $\sigma(y_0)$ and staying on the leaf through $\sigma(y_0)$, then we get a lift $\lambda + \sigma - \lambda$ of τ, where σ is our section. The parametrization of ι on $[0, 1]$ induces parametrizations

$$\sigma : [0,1] \to \partial S_g \times S^1,$$
$$\widetilde{\sigma} : [0,1] \to S^1,$$
$$\sigma(t) = (\iota(t), \widetilde{\sigma}(t)).$$

Since

$$\widehat{h}(\tau) = \widehat{h}\left(\prod_{i=1}^{g}[e_i, e_{2i}]\right) = T_{m(\widehat{f}_1,\dots,\widehat{f}_{2g})},$$

it is not generally true that $\iota : [0,1] \to \partial S_g$ lifts to a loop on a leaf of $\widehat{\mathfrak{F}}$. Indeed, ι lifts to

$$\widehat{\sigma} : [0,1] \to \partial S_g \times \mathbb{R},$$
$$\widehat{\sigma}(t) = (\iota(t), \overline{\sigma}(t)),$$

with $\operatorname{im}\widehat{\sigma}$ contained in a leaf of $\widehat{\mathfrak{F}}$. Hence

$$\overline{\sigma}(1) = T_{m(\widehat{f}_1,\dots,\widehat{f}_{2g})}(\overline{\sigma}(0)).$$

This leaf can be chosen to be the one through the point $\widehat{\sigma}(0)$ such that $\theta(\widehat{\sigma}(0)) = \sigma(0)$. That is, θ carries $\widehat{\sigma}$ to σ, so the projection $\mathbb{R} \to \mathbb{R}/\mathbb{Z}$ carries $\overline{\sigma}$ to $\widetilde{\sigma}$. Therefore, $\deg\widetilde{\sigma} = m(\widehat{f}_1, \dots, \widehat{f}_{2g})$. □

Since $|e(h)| = |m(\psi)| = |\deg\widetilde{\sigma}|$, Theorem 4.5.4 is proven.

If $\gamma \in \widehat{G}$, the function $\gamma(t) = \gamma(t) - t$ is periodic of period 1 on $\mathbb{R}$, so we can define

$$\underline{m}(\gamma) = \min_{t\in\mathbb{R}}(\gamma(t) - t),$$
$$\overline{m}(\gamma) = \max_{t\in\mathbb{R}}(\gamma(t) - t).$$

Two technical propositions about these functions are needed.

Proposition 4.5.5. *If $\gamma_i \in \widehat{G}$, $1 \le i \le n$, then*

$$\underline{m}\left(\prod_{i=1}^{n}\gamma_i\right) < \sum_{i=1}^{n}\underline{m}\,(\gamma_i) + n - 1,$$

$$\overline{m}\left(\prod_{i=1}^{n}\gamma_i\right) > \sum_{i=1}^{n}\overline{m}\,(\gamma_i) - n + 1.$$

Proposition 4.5.6. *If $\gamma \in \widehat{G}$ is the commutator of two elements of $\widehat{G}$, then*

$$\underline{m}\,(\gamma) < 1,$$
$$\overline{m}\,(\gamma) > -1.$$

Corollary 4.5.7. *If $T_a = \prod_{i=1}^{n}[\gamma_i, \gamma_{2i}]$, then $2 - 2n \le a \le 2n - 2$.*

Proof. Indeed,

$$a = \underline{m}\,(T_a) < \sum_{i=1}^{n}\underline{m}\,[\gamma_i, \gamma_{2i}] + n - 1 < 2n - 1,$$

by Propositions 4.5.5 and 4.5.6. The second inequality is entirely similar. □

Before proving the propositions, we use the corollary to prove Wood's theorem.

Proof of Theorem 4.5.1. By Theorem 4.5.4, we have

$$|e(E)| = |m(\widehat{f}_1, \dots, \widehat{f}_{2g})|,$$

where

$$\prod_{i=1}^{g}[\widehat{f}_i, \widehat{f}_{2i}] = T_{m(\widehat{f}_1, \dots, \widehat{f}_{2g})}.$$

By the above corollary, $2 - 2g \le m(\widehat{f}_1, \dots, \widehat{f}_{2g}) \le 2g - 2$. □

We verify the propositions via a series of claims. Fix $\gamma \in \widehat{G}$.

Claim 1. *If $x, y \in \mathbb{R}$, then*

(1) $x - y \in \mathbb{Z} \Rightarrow \gamma(x) - \gamma(y) = x - y$;
(2) $n \in \mathbb{Z}$ *and* $n < x - y < n + 1 \Rightarrow n < \gamma(x) - \gamma(y) < n + 1$.

Proof. Let $n \in \mathbb{Z}$. Since γ commutes with T_1, it commutes with T_n. Thus, if $x - y = n$,

$$\begin{aligned}\gamma(x) - \gamma(y) &= \gamma \circ T_{x-y}(y) - \gamma(y)\\ &= T_{x-y} \circ \gamma(y) - \gamma(y)\\ &= x - y + \gamma(y) - \gamma(y)\\ &= x - y.\end{aligned}$$

On the other hand,

$$\begin{aligned} n < x - y < n + 1 \Rightarrow y + n &< x < y + n - 1 \\ \Rightarrow \gamma(y+n) &< \gamma(x) < \gamma(y+n+1) \\ \Rightarrow \gamma(y) + n &< \gamma(x) < \gamma(y) + n + 1 \\ \Rightarrow n &< \gamma(x) - \gamma(y) < n + 1. \end{aligned}$$

□

Claim 2. $0 \leq \overline{m}(\gamma) - \underline{m}(\gamma) < 1$.

Proof. It is clear that $0 \leq \overline{m}(\gamma) - \underline{m}(\gamma)$. For the second inequality, choose $x, y \in \mathbb{R}$ such that $\gamma(x) - x = \underline{m}(\gamma)$ and $\gamma(y) - y = \overline{m}(\gamma)$. Since $\gamma(t) - t$ has period 1, we can assume that $|x - y| \leq 1$. If $1 \leq \overline{m}(\gamma) - \underline{m}(\gamma)$, then

$$\gamma(x) - \gamma(y) \leq x - y - 1.$$

If $x - y \in \mathbb{Z}$, this contradicts (1) of Claim 1. Otherwise, $n < x - y < n + 1$, for some integer n, and this contradicts (2) of that claim. □

Claim 3. $\underline{m}(\gamma^{-1}) = -\overline{m}(\gamma)$ *(hence,* $\overline{m}(\gamma^{-1}) = -\underline{m}(\gamma)$*).*

Proof. Indeed,

$$\begin{aligned} \underline{m}(\gamma^{-1}) &= \min_{t \in \mathbb{R}} (\gamma^{-1}(t) - t) \\ &= -\max_{t \in \mathbb{R}} (t - \gamma^{-1}(t)) \\ &= -\max_{s \in \mathbb{R}} (\gamma(s) - s) \\ &= -\overline{m}(\gamma), \end{aligned}$$

where we have made the substitution $t = \gamma(s)$. □

Proposition 4.5.5 is contained in the following.

Claim 4. *If* $\{\gamma_i\}_{i=1}^n \subset \widehat{G}$, *then*

$$\sum_{i=1}^n \underline{m}(\gamma_i) \leq \underline{m}\left(\prod_{i=1}^n \gamma_i\right) \leq \underline{m}(\gamma_n) + \sum_{i=1}^{n-1} \overline{m}(\gamma_i) < \sum_{i=1}^n \underline{m}(\gamma_i) + n - 1$$

and

$$\sum_{i=1}^n \overline{m}(\gamma_i) \geq \overline{m}\left(\prod_{i=1}^n \gamma_i\right) \geq \overline{m}(\gamma_n) + \sum_{i=1}^{n-1} \underline{m}(\gamma_i) > \sum_{i=1}^n \underline{m}(\gamma_i) - n + 1.$$

Proof. Note that the second set of inequalities follows from the first by replacing γ_i with γ_i^{-1}, $1 \leq i \leq n$, and applying Claim 3.

(1) If we set $\eta = \prod_{i=2}^{n} \gamma_i$, then

$$\begin{aligned} \underline{m}(\gamma_1) + \underline{m}(\eta) &= \min_{\eta(t)\in\mathbb{R}} (\gamma_1(\eta(t)) - \eta(t)) + \min_{t\in\mathbb{R}} (\eta(t) - t) \\ &\leq \min_{t\in\mathbb{R}} (\gamma_1(\eta(t)) - \eta(t) + \eta(t) - t) \\ &= \underline{m}\left(\prod_{i=1}^{n} \gamma_i\right). \end{aligned}$$

By induction on n, we get the first inequality,

$$\sum_{i=1}^{n} \underline{m}(\gamma_i) \leq \underline{m}\left(\prod_{i=1}^{n} \gamma_i\right).$$

(2) Choose $x \in \mathbb{R}$ with $\gamma_n(x) - x = \underline{m}(\gamma_n)$. Then

$$\begin{aligned} \underline{m}\left(\prod_{i=1}^{n} \gamma_i\right) &\leq \left(\prod_{i=1}^{n} \gamma_i\right)(x) - x \\ &= \left(\prod_{i=1}^{n} \gamma_i\right)(x) - \gamma_n(x) + \gamma_n(x) - x \\ &= \left(\prod_{i=1}^{n-1} \gamma_i\right)(\gamma_n(x)) - \gamma_n(x) + \underline{m}(\gamma_n) \\ &\leq \overline{m}\left(\prod_{i=1}^{n-1} \gamma_i\right) + \underline{m}(\gamma_n) \\ &\leq \sum_{i=1}^{n-1} \overline{m}(\gamma_i) + \underline{m}(\gamma_n) \\ &< \sum_{i=1}^{n} \underline{m}(\gamma_i) + n - 1. \end{aligned}$$

The last inequality is proven by $n-1$ applications of Claim 2. The next to last inequality results from step (1) and the remark preceding it. □

We will use the following notation:

$$\begin{aligned} [x]^- &= \text{greatest integer } \leq x \qquad \text{(the \textit{floor} of } x) \\ [x]^+ &= \text{least integer } \geq x \qquad \text{(the \textit{ceiling} of } x). \end{aligned}$$

Claim 5. *If $\gamma, \eta \in \widehat{G}$, then*

$$\begin{aligned} [\underline{m}(\eta^{-1}\gamma\eta)]^- &= [\underline{m}(\gamma)]^-, \\ [\underline{m}(\eta^{-1}\gamma\eta)]^+ &= [\underline{m}(\gamma)]^+, \end{aligned}$$

and the same for $\overline{m}$.

Proof. We give the proof for floors, the proof for ceilings being exactly the same. By Claim 1,

$$[x-y]^{-} = [\eta^{-1}(x) - \eta^{-1}(y)]^{-}, \quad \forall\, x, y \in \mathbb{R}.$$

Take $y = \eta(z)$ and $x = \gamma\eta(z)$, concluding that

$$[\gamma(\eta(z)) - \eta(z)]^{-} = [\eta^{-1}\gamma\eta(z) - z]^{-}, \quad \forall\, z \in \mathbb{R}.$$

Minimizing over $z \in \mathbb{R}$ (equivalently, over $\eta(z) \in \mathbb{R}$) gives the equality for $\underline{m}$ and maximizing gives it for $\overline{m}$. □

As we remarked above, Proposition 4.5.5 is contained in Claim 4.

Proof of Proposition 4.5.6. Consider an arbitrary commutator

$$\eta^{-1}\gamma^{-1}\eta\gamma \in \widehat{G}.$$

Then,

$$\begin{aligned}
\underline{m}\,(\eta^{-1}\gamma^{-1}\eta\gamma) &\leq \underline{m}\,(\gamma) + \overline{m}\,(\eta^{-1}\gamma\eta) && \text{(Claim 4)}\\
&= -\overline{m}\,(\gamma^{-1}) + \overline{m}\,(\eta^{-1}\gamma\eta) && \text{(Claim 3)}\\
&\leq -\overline{m}\,(\gamma^{-1}) + [\overline{m}\,(\eta^{-1}\gamma\eta)]^{+} &&\\
&= -\overline{m}\,(\gamma^{-1}) + [\overline{m}\,(\gamma^{-1})]^{+} && \text{(Claim 5)}\\
&< 1.
\end{aligned}$$

This proves the assertion for $\underline{m}$, and the one for $\overline{m}$ follows from this via Claim 3 and the fact that the inverse of a commutator is a commutator. □

The proof of Theorem 4.5.1 is complete and we turn to Theorem 4.5.2. Let $\alpha : S^1 \to S^1$ be the antipodal interchange map. In terms of the identification $S^1 = \mathbb{R}/\mathbb{Z}$, this is given by

$$\alpha(t + \mathbb{Z}) = \left(t + \tfrac{1}{2}\right) + \mathbb{Z}.$$

Define the groups

$$\begin{aligned}
G_\alpha &= \{f \in G \mid f \circ \alpha = \alpha \circ f\},\\
\widehat{G}_\alpha &= \left\{\widehat{f} \in \widehat{G} \mid \widehat{f}\left(t + \tfrac{1}{2}\right) = \widehat{f}(t) + \tfrac{1}{2}\right\}.
\end{aligned}$$

The exact sequence $(*)$ restricts to an exact sequence

$$(**) \qquad 0 \to \mathbb{Z} \to \widehat{G}_\alpha \xrightarrow{\rho} G_\alpha \to 1.$$

It is also easy to check that

$$\begin{aligned}
\lambda &: \widehat{G} \to \widehat{G}_\alpha,\\
\lambda(\widehat{f})(t) &= \tfrac{1}{2}\widehat{f}(2t),
\end{aligned}$$

defines a surjective group homomorphism.

Proposition 4.5.8. *If the oriented, foliated circle bundle $(E, \mathcal{F}, \pi)$ over Σ_g has total holonomy $h : \pi_1(\Sigma_g, x_0) \to G_\alpha$, then $|e(h)| \le g - 1$.*

Proof. If $a \in \mathbb{R}$ and $T_a = \prod_{i=1}^{g} [\widehat{f}_i, \widehat{f}_{2i}]$, where all $\widehat{f}_j \in \widehat{G}_\alpha$, choose $h_j \in \widehat{G}$ such that $\lambda(h_j) = \widehat{f}_j$, $1 \le j \le 2g$. Then

$$T_a = \lambda\left(\prod_{i=1}^{g} [h_i, h_{2i}]\right).$$

By the definition of λ, this becomes

$$\left(\prod_{i=1}^{g} [h_i, h_{2i}]\right)(2t) = 2t + 2a, \quad \forall\, t \in \mathbb{R}.$$

That is,

$$\prod_{i=1}^{g} [h_i, h_{2i}] = T_{2a}$$

and Corollary 4.5.7 implies that $|a| \le g-1$. As in the proof of Theorem 4.5.1, this applies to $a = \pm e(E)$. □

Proof of Theorem 4.5.2. Assume that $\pi : E \to \Sigma_g$ is a real, oriented 2-plane bundle. The associated frame bundle $\pi : P(E) \to \Sigma_g$ is a principal $\mathrm{Gl}_+(2, \mathbb{R})$-bundle. A flat connection on E is equivalently a flat connection on $P(E)$. The locally absolute parallelism induced by such a connection makes $P(E)$ a foliated $\mathrm{Gl}_+(2, \mathbb{R})$-bundle and E a foliated $\mathbb{R}^2$-bundle. Also, the set of rays out of the origin in $\mathbb{R}^2$ can be identified with S^1, so the set of rays out of the origin in E_x, $\forall\, x \in \Sigma_g$, defines an S^1-bundle $\pi : S^1(E) \to \Sigma_g$, associated to $P(E)$. (This bundle is isomorphic to the unit circle bundle of E relative to any Riemannian metric on E.) By definition, $e(E) = e(S^1(E))$. A flat connection on $P(E)$ induces a locally absolute parallelism on the associated circle bundle, making it a foliated S^1-bundle. But this foliated S^1-bundle has structure group $\mathrm{Gl}_+(2, \mathbb{R}) \subset G_\alpha$, and the total holonomy is a homomorphism

$$h : \pi_1(\Sigma_g, x_0) \to \mathrm{Gl}_+(2, \mathbb{R}).$$

Proposition 4.5.8 completes the proof. □

We know that $T(\Sigma_g)$ admits a flat *Riemannian* connection if and only if $g = 1$. Theorem 4.5.2 has a stronger corollary.

Corollary 4.5.9. *If $g \ge 2$, then $T(\Sigma_g)$ does not admit a flat connection.*

4.6. Further Developments

With little or no proof, we outline some of the further developments of this line of investigation.

Exercise 4.6.1. Let $(E, \mathcal{F}, \pi)$ be an oriented, foliated S^1-bundle over a compact manifold M. If L is a compact leaf of $\mathcal{F}$ and $p : L \to M$ is $\pi|L$, prove that the circle bundle $p^*(E)$ over L has a global section.

Corollary 4.6.2. *Let $(E, \mathcal{F}, \pi)$ be an oriented, foliated S^1-bundle over a compact, orientable surface M. If $\mathcal{F}$ has a compact leaf L, then the bundle $\pi : E \to M$ is trivial.*

Proof. The map $p = \pi|L : L \to M$ is a covering map, necessarily k-fold for some integer $k \geq 1$. The induced map

$$p^* : H^2(M; \mathbb{Z}) \to H^2(L; \mathbb{Z}),$$

being multiplication by k, is injective. Exercise 4.6.1 and the naturality of the Euler class imply that $p^*(e(E)) = 0$; hence $e(E) = 0$. That is, $\pi : E \to M$ is a trivial bundle. □

Corollary 4.6.3. *In a nontrivial, oriented, foliated S^1-bundle $\pi : E \to M$ over a compact, orientable surface M, either every leaf is dense in E or there is an exceptional minimal set.*

In fact, much sharper theorems are known. In the following, $(E, \mathcal{F}, \pi)$ is an oriented, foliated circle bundle over Σ_g, $g \geq 1$.

Theorem 4.6.4 (Ghys [**79**]). *If $\mathcal{F}$ is of class C^2 and has an exceptional minimal set, then $|e(E)| < 2g - 2$.*

In particular, over the torus, there is no $(E, \mathcal{F}, \pi)$ (of class C^2) with an exceptional minimal set, while, if $g \geq 2$, every C^2 foliation, transverse to the fibers of $T_1(\Sigma_g)$, has all leaves dense.

Theorem 4.6.5 (Ghys [**79**]). *If $\mathcal{F}$ is of class C^ω and has an exceptional minimal set, then $e(E) = 0$.*

Theorem 4.6.6 (Ghys and Sergiescu [**81**]). *If $\pi : E \to \Sigma_{12}$ is an oriented circle bundle with $|e(E)| = 1$, there is a C^∞ foliation transverse to the fibers that has an exceptional minimal set.*

It seems to be an open problem to determine the values of $|e(E)|$, strictly between 1 and $2g - 2$, for which there exists $(E, \mathcal{F}, \pi)$ (of class C^2) with an exceptional minimal set.

The following theorem of W. Thurston (doctoral dissertation), with a published proof and generalization by G. Levitt [**119**], combines with Theorem 4.5.1 to give an interesting compact leaf theorem.

Theorem 4.6.7 (Thurston, Levitt). *Let E be the total space of an orientable fibration by circles over Σ_g, $g > 1$, and let $\mathfrak{F}$ be a transversely orientable, C^2 foliation of E by surfaces, none of which is compact. Then $\mathfrak{F}$ is isotopic to a foliation everywhere transverse to the circle fibers.*

Corollary 4.6.8. *If E is the total space of an orientable fibration by circles over Σ_g, $g > 1$, and if $|e(E)| > 2g - 2$, then every transversely orientable, C^2 foliation of E by surfaces has a compact leaf.*

Chapter 5

The Chern-Weil Construction

Let $\mathbb{K} = \mathbb{C}$ or $\mathbb{R}$ and let $\mathfrak{gl}(n, \mathbb{K})$ denote the Lie algebra of $\mathrm{Gl}(n, \mathbb{K})$. We make the standard identification of $\mathfrak{gl}(n, \mathbb{K})$ with the Lie algebra of all $n \times n$ matrices over $\mathbb{K}$, the bracket operation being the commutator product of matrices. For each integer $q \geq 0$, denote by $I_n^{2q}(\mathbb{K})$ the space of homogeneous polynomial functions

$$P : \mathfrak{gl}(n, \mathbb{K}) \to \mathbb{K}$$

of degree q such that

$$P(AXA^{-1}) = P(X), \quad \forall\, A \in \mathrm{Gl}(n, \mathbb{K}) \text{ and } \forall\, X \in \mathfrak{gl}(n, \mathbb{K}).$$

For example, $\mathrm{tr} \in I_n^2(\mathbb{K})$ and $\det \in I_n^{2n}(\mathbb{K})$. Finally, set $I_n^{2q+1}(\mathbb{K}) = \{0\}$. This defines a graded commutative algebra $I_n^*(\mathbb{K})$ under multiplication of polynomial functions.

Remark. We use the terminology "graded commutative" for the old fashioned term "anticommutative". This means that

$$ab = (-1)^{\deg(a)\deg(b)} ba.$$

Actually, ordinary commutativity for graded algebras is not a very desirable property. Of course, multiplication of polynomials is commutative in the ordinary sense, but our indexing of $I_n^*(\mathbb{K})$ is designed to make this algebra graded commutative. The Chern-Weil homomorphism will be a homomorphism of this graded algebra into the graded commutative algebra $H^*(M; \mathbb{K})$.

5.1. The Chern-Weil Homomorphism

We consider smooth (class C^∞) n-plane bundles $\pi : E \to M$ over a manifold M, the field of scalars for the fibers being $\mathbb{K}$. We are going to define a canonical homomorphism

$$\varphi_E : I_n^*(\mathbb{K}) \to H^*(M; \mathbb{K})$$

of graded algebras. This will be the Chern-Weil homomorphism of the bundle E and its image will be the algebra of characteristic classes of E. When $\mathbb{K} = \mathbb{C}$, this is called the Chern algebra, and when $\mathbb{K} = \mathbb{R}$, it is the Pontryagin algebra.

In what follows, $C^\infty(M)$ is the algebra of real-valued, C^∞ functions on M and $C^\infty(M, \mathbb{K}) = C^\infty(M) \otimes \mathbb{K}$ is the algebra of $\mathbb{K}$-valued, C^∞ functions on M.

Let ∇ be a connection on E. Recall that the curvature tensor of ∇ is a $C^\infty(M)$-bilinear map

$$R : \mathfrak{X}(M) \times \mathfrak{X}(M) \to \mathrm{Hom}_{C^\infty(M,\mathbb{K})}(\Gamma(E), \Gamma(E)),$$

defined by

$$R(X, Y)(s) = \nabla_X \nabla_Y(s) - \nabla_Y \nabla_X(s) - \nabla_{[X,Y]}(s),$$

$\forall\, X, Y \in \mathfrak{X}(M)$ and $\forall\, s \in \Gamma(E)$. Choose a family $\{U_\alpha, s_1^\alpha, \dots, s_n^\alpha\}_{\alpha \in \mathfrak{A}}$ of local trivializations of E. Recall that the structure of the bundle is completely described by the *structure cocycle* $\{g_{\alpha\beta}\}_{\alpha,\beta \in \mathfrak{A}}$, where

$$g_{\alpha\beta} : U_\alpha \cap U_\beta \to \mathrm{Gl}(n, \mathbb{K})$$

is a continuous map such that

$$(s_1^\alpha(x), \dots, s_n^\alpha(x)) \cdot g_{\alpha\beta}(x) = (s_1^\beta(x), \dots, s_n^\beta(x)), \quad \forall\, x \in U_\alpha \cap U_\beta.$$

Here, "$\cdot$" indicates formal matrix multiplication.

Definition 5.1.1. The connection form on U_α relative to the trivialization $(s_1^\alpha, \dots, s_n^\alpha)$ is the $n \times n$ matrix $\theta^\alpha = \left(\theta_{ij}^\alpha\right)$ of forms in $A^1(U_\alpha, \mathbb{K})$ defined by

$$\nabla_X(s_j^\alpha) = \sum_{i=1}^n \theta_{ij}^\alpha(X) s_i^\alpha, \quad \forall\, X \in \mathfrak{X}(U_\alpha), \quad 1 \le j \le n.$$

Definition 5.1.2. The curvature form on U_α relative to the trivialization $(s_1^\alpha, \dots, s_n^\alpha)$ is the $n \times n$ matrix $\Omega^\alpha = \left(\Omega_{ij}^\alpha\right)$ of forms in $A^2(U_\alpha, \mathbb{K})$ defined by

$$R(X, Y)(s_j^\alpha) = \sum_{i=1}^n \Omega_{ij}^\alpha(X, Y) s_i^\alpha, \quad \forall\, X \in \mathfrak{X}(U_\alpha), \quad 1 \le j \le n.$$

The following is one of the Cartan structure equations. It is well known.

Lemma 5.1.3. *The curvature and connection forms satisfy the equation*

$$\Omega^\alpha = d\theta^\alpha + \theta^\alpha \wedge \theta^\alpha.$$

The following is an elementary consequence of the fact that curvature is a tensor. Its verification is left to the reader.

Lemma 5.1.4. *On* $U_\alpha \cap U_\beta$, $\Omega^\alpha = g_{\alpha\beta}\Omega^\beta g_{\alpha\beta}^{-1}$.

For $P \in I_n^{2p}(\mathbb{K})$, $P(\Omega^\alpha) \in A^{2p}(U_\alpha, \mathbb{K})$ makes sense. By Lemma 5.1.4,

$$P(\Omega^\alpha)|U_\alpha \cap U_\beta = P(\Omega^\beta)|U_\alpha \cap U_\beta,$$

so these local forms fit together to give a well defined global form

$$P(\Omega) \in A^{2p}(M, \mathbb{K}).$$

Furthermore, this is independent of the choices of systems of local trivializations, since the above reasoning can be applied to the union of two such systems. The form $P(\Omega)$ does depend on the choice of connection.

Lemma 5.1.5. *For each* $P \in I_n^{2p}(\mathbb{K})$, $P(\Omega)$ *is a closed form.*

Proof. Let $x_0 \in M$ and let $(U, x^1, \dots, x^n)$ be an open coordinate neighborhood of x_0 in which x_0 is the origin. We set up a special trivialization of $E|U$. Let σ_1 be the "x^1-axis" through x_0. Through each point of σ_1 there pass x^2-curves, sweeping out a smooth surface S; then through each point of S there pass x^3-curves, sweeping out a smooth 3-fold, *etc.* Next fix a basis $v_1, \dots, v_n$ of E_{x_0} and, using the connection ∇, parallel translate this frame along σ_1. Parallel translate the resulting frames along the x^2-curves, obtaining a smooth frame field along S, then parallel translate along the x^3-curves through S, *etc.* The result is a smooth frame field $(s_1, \dots, s_n)$ for $E|U$. We use this frame field to define the connection form θ and the curvature form Ω on U. For $e_i = \partial/\partial x^i_{x_0}$, our construction implies that the connection form θ_{x_0} vanishes on e_i, $1 \le i \le n$; hence $\theta_{x_0} = 0$. On U,

$$P(\Omega) = P(d\theta + \theta \wedge \theta),$$

and this is a sum of monomials of forms, in each of which either two of the factors are forms that vanish at x_0 or all of the factors are exact. Thus, $(d(P(\Omega)))_{x_0} = 0$. Since $x_0 \in M$ is arbitrary, $d(P(\Omega)) \equiv 0$. □

Thus, we obtain a homomorphism of graded algebras

$$\varphi_{(\nabla,E)} : I_n^*(\mathbb{K}) \to H^*(M; \mathbb{K}),$$
$$\varphi_{(\nabla,E)}(P) = [P(\Omega)].$$

Lemma 5.1.6. *The homomorphism* $\varphi_{(\nabla,E)}$ *is independent of the choice of connection* ∇.

Proof. Let ∇^0 and ∇^1 be two connections on E with respective curvature forms Ω^0 and Ω^1. We must show that

$$[P(\Omega^0)] = [P(\Omega^1)], \quad \forall P \in I_n^*(\mathbb{K}).$$

Consider the smooth n-plane bundle

$$E \times \mathbb{R} \xrightarrow{\pi \times \mathrm{id}} M \times \mathbb{R}.$$

We can define a connection ∇ on this bundle as follows. If $s \in \Gamma(E \times \mathbb{R})$ is constant in the coordinate $t \in \mathbb{R}$, define

$$\nabla_{\partial/\partial t}(s) \equiv 0$$

and, for $v \in T_{(x,t)}(M \times \{t\}) = T_x(M)$, set

$$\nabla_v(s) = (1-t)\nabla_v^0(s) + t\nabla_v^1(s).$$

Locally, arbitrary smooth sections of $E \times \mathbb{R}$ are $C^\infty(M \times \mathbb{R}, \mathbb{K})$-linear combinations of sections as above, so the formulas

$$\nabla_X\left(\sum_i f_i s_i\right) = \sum_i (X(f_i)s_i + f_i \nabla_X(s_i)),$$
$$\nabla_{X+Y}(s) = \nabla_X(s) + \nabla_Y(s)$$

define ∇ in general. It is elementary that, with this definition, ∇ is a connection on $E \times \mathbb{R}$. Clearly, ∇ restricts to ∇^0 on $M \times \{0\}$ and to ∇^1 on $M \times \{1\}$, so the curvature form Ω of ∇ restricts to Ω^i on $M \times \{i\}$, $i = 0, 1$. If we define

$$i_t : M \hookrightarrow M \times \mathbb{R},$$
$$i_t(x) = (x, t),$$

then $i_0 \sim i_1$ and, by the homotopy invariance of cohomology,

$$[P(\Omega^0)] = i_0^*[P(\Omega)] = i_1^*[P(\Omega)] = [P(\Omega^1)].$$

□

Definition 5.1.7. The Chern-Weil homomorphism

$$\varphi_E : I_n^*(\mathbb{K}) \to H^*(M; \mathbb{K})$$

for the bundle $\pi : E \to M$ is $\varphi_{(E,\nabla)}$, computed using any connection ∇ on the bundle.

Under the vague philosophy that the degree of "twistedness" of the bundle E can be measured by how "far" from being flat all connections on E must be, the Chern-Weil homomorphism is viewed as a measure of this twistedness. In the extreme case of a trivial bundle, there is a flat connection ($\Omega \equiv 0$) and, evidently,

$$\varphi_E : I_n^{2k}(\mathbb{K}) \to H^{2k}(M; \mathbb{K})$$

vanishes for all $k \geq 1$. (In all cases, φ_E is injective on $I_n^0(\mathbb{K}) = \mathbb{K}$.) Of course, a bundle can be nontrivial and admit a flat connection (hence be a foliated bundle), in which case this homomorphism is again trivial.

Definition 5.1.8. If $\mathbb{K} = \mathbb{C}$, the image of φ_E is called the Chern algebra $\mathrm{Chern}^*(E) \subseteq H^*(M;\mathbb{C})$. If $\mathbb{K} = \mathbb{R}$, the image of φ_E is the Pontryagin algebra $\mathrm{Pont}^*(E) \subseteq H^*(M;\mathbb{R})$.

5.2. The Structure of $I_n^*(\mathbb{K})$

In order to describe the graded algebras $\mathrm{Chern}^*(E)$ and $\mathrm{Pont}^*(E)$, we need to analyze the algebras $I_n^*(\mathbb{K})$. Let $S_n^{2k}(\mathbb{K})$ denote the space of *symmetric*, homogeneous polynomials p of degree k over $\mathbb{K}$ in n indeterminants $x_1, \dots, x_n$. Thus, if $\mathfrak{S}_n$ is the permutation group on n letters,

$$p(x_{\tau(1)}, x_{\tau(2)}, \dots, x_{\tau(n)}) = p(x_1, x_2, \dots, x_n), \quad \forall\, \tau \in \mathfrak{S}_n.$$

We set $S_n^{2k+1}(\mathbb{K}) = 0$.

The kth elementary symmetric function $\sigma_k \in S_n^{2k}(\mathbb{K})$ is given by

$$\sigma_k(x_1, \dots, x_n) = \sum_{1 \leq j_1 < \cdots < j_k \leq n} x_{j_1} \cdots x_{j_k}.$$

It is well known [**178**, p. 99] that

$$S_n^*(\mathbb{K}) = \mathbb{K}[\sigma_1, \dots, \sigma_n].$$

Consider the set of diagonal matrices

$$X = \begin{bmatrix} x_1 & 0 & \dots & 0 \\ 0 & x_2 & \dots & 0 \\ \vdots & \vdots & \vdots & \vdots \\ 0 & 0 & \dots & x_n \end{bmatrix} \in \mathfrak{gl}(n, \mathbb{K}).$$

Each permutation of the diagonal entries can be achieved by a transformation $X \mapsto AXA^{-1}$, for a suitable choice of $A \in \mathrm{Gl}(n, \mathbb{K})$. Therefore, we can define a homomorphism of graded algebras

$$\rho : I_n^*(\mathbb{K}) \to S_n^*(\mathbb{K})$$

by letting $\rho(P)$ be the restriction of P to the subspace of diagonal matrices, $\forall\, P \in I_n^*(\mathbb{K})$.

Lemma 5.2.1. *The homomorphism ρ is surjective.*

Proof. Define $c_i \in I_n^*$ by the formula

$$\det(I + tA) = 1 + \sum_{i=1}^{n} t^i c_i(A).$$

It is obvious that $\rho(c_i) = \sigma_i$, $1 \leq i \leq n$. □

Lemma 5.2.2. *The homomorphism $\rho : I_n^*(\mathbb{C}) \to S_n^*(\mathbb{C})$ is injective.*

Proof. Since $\mathbb{C}$ is algebraically closed, an arbitrary matrix $A \in \mathfrak{gl}(n, \mathbb{C})$ is conjugate under $\mathrm{Gl}(n, \mathbb{C})$ to an upper triangular matrix. Thus, a polynomial $P \in I_n^*(\mathbb{C})$ is completely determined by its restriction $P|\Delta_n(\mathbb{C})$ to the subspace of upper triangular matrices. Let $\Delta_n'(\mathbb{C}) \subset \Delta_n(\mathbb{C})$ be the dense subset of matrices having all diagonal entries distinct. Since the minimal polynomial of any matrix $A \in \Delta_n'(\mathbb{C})$ is a product of n nonrepeated linear factors, A is diagonalizable. That is, the set of diagonalizable, upper triangular matrices is dense in $\Delta_n(\mathbb{C})$. By continuity of polynomial functions, $P \in I_n^*(\mathbb{C})$ is uniquely determined by its values on the space of diagonal matrices, so ρ is one-to-one. □

Lemma 5.2.3. *The homomorphism $\rho : I_n^*(\mathbb{R}) \to S_n^*(\mathbb{R})$ is injective.*

Proof. Every $P \in I_n^*(\mathbb{R})$ extends uniquely to a polynomial function

$$P : \mathfrak{gl}(n, \mathbb{C}) \to \mathbb{C}$$

which, for purely formal reasons, is invariant under conjugation by elements of $\mathrm{Gl}(n, \mathbb{C})$. That is, we obtain an inclusion map $i : I_n^*(\mathbb{R}) \hookrightarrow I_n^*(\mathbb{C})$ as a real subalgebra. Similarly, there is an inclusion $i : S_n^*(\mathbb{R}) \hookrightarrow S_n^*(\mathbb{C})$ as a real subalgebra. Since the diagram

$$\begin{array}{ccc} I_n^*(\mathbb{R}) & \xrightarrow{i} & I_n^*(\mathbb{C}) \\ \rho \big\downarrow & & \big\downarrow \rho \\ S_n^*(\mathbb{R}) & \xrightarrow[i]{} & S_n^*(\mathbb{C}) \end{array}$$

is commutative, the assertion follows from the previous lemma. □

We summarize.

Theorem 5.2.4. *The homomorphism*

$$\rho : I_n^*(\mathbb{K}) \to S_n^*(\mathbb{K}) = \mathbb{K}[\sigma_1, \ldots, \sigma_n]$$

is an isomorphism of graded algebras. Furthermore, $\rho(c_i) = \sigma_i$, $1 \leq i \leq n$, where $c_i \in I_n^*(\mathbb{K})$ is defined by the formula

$$\det(I + tA) = 1 + \sum_{i=1}^{n} t^i c_i(A).$$

Hereafter, we write $I_n^*(\mathbb{K}) = \mathbb{K}[c_1, \ldots, c_n]$.

Definition 5.2.5. The elements $\Sigma_i \in I_n^{2i}(\mathbb{K})$, $i \geq 1$, are the functions

$$\Sigma_i(A) = \operatorname{tr} A^i.$$

Thus, $\rho(\Sigma_i)(x_1, \ldots, x_n) = x_1^i + \cdots + x_n^i$.

Lemma 5.2.6. *The algebra $I_n^*(\mathbb{K})$ is generated by $\{\Sigma_1, \ldots, \Sigma_n\}$.*

Proof. Since ρ is an isomorphism, it will be enough to show that the symmetric polynomials $\{\rho(\Sigma_1), \ldots, \rho(\Sigma_n)\}$ generate $S_n^*(\mathbb{K})$. By induction on k, the formula

$$(*) \qquad k\sigma_k + \sum_{i=1}^{k} (-1)^i \sigma_{k-i} \rho(\Sigma_i) = 0, \quad 1 \le k \le n$$

(where $\sigma_0 = 1$), implies that the elementary symmetric function σ_k can be written as a polynomial in $\{\rho(\Sigma_1), \ldots, \rho(\Sigma_k)\}$, $1 \le k \le n$. We must prove $(*)$. Let $f \in S_n^*(\mathbb{K})[t]$ be the polynomial

$$f = \sum_{j=0}^{n} \sigma_j t^j = (1 + x_1 t)(1 + x_2 t) \ldots (1 + x_n t).$$

By setting $\sigma_k = 0$, for all $k > n$, we view

$$f = \sum_{j=0}^{\infty} \sigma_j t^j$$

as a formal power series, $f \in S_n^*(\mathbb{K})[[t]]$. By purely formal properties of power series, we write

$$\begin{aligned} \frac{df}{dt} &= \sigma_1 + 2\sigma_2 t + \cdots + n\sigma_n t^{n-1} \\ &= \sum_{k=1}^{\infty} k\sigma_k t^{k-1} \end{aligned}$$

and

$$\begin{aligned} -\frac{t}{f}\frac{df}{dt} &= -t\frac{d}{dt} \log f(t) \\ &= -t\frac{d}{dt}\{\log(1 + x_1 t) + \cdots + \log(1 + x_n t)\} \\ &= \frac{-tx_1}{1 + tx_1} + \cdots + \frac{-tx_n}{1 + tx_n} \\ &= \sum_{i=1}^{\infty} (x_1^i + \cdots + x_n^i)(-t)^i \\ &= \sum_{i=1}^{\infty} (-1)^i \rho(\Sigma_i) t^i. \end{aligned}$$

Multiplying both sides of this last identity by f, we obtain

$$\begin{aligned}\Big(\sum_{j=0}^{\infty}\sigma_j t^j\Big)\Big(\sum_{i=1}^{\infty}(-1)^i\rho(\Sigma_i)t^i\Big) &= f\Big(-\frac{t}{f}\frac{df}{dt}\Big)\\ &= -t\frac{df}{dt}\\ &= -t\sum_{k=1}^{\infty}k\sigma_k t^{k-1}\\ &= \sum_{k=1}^{\infty}-k\sigma_k t^k.\end{aligned}$$

This gives

$$\sum_{k=1}^{\infty}\Big\{\Big(\sum_{i+j=k}(-1)^i\sigma_j\rho(\Sigma_i)\Big)+k\sigma_k\Big\}t^k=0,$$

implying equation $(*)$. □

5.3. Chern Classes and Pontryagin Classes

Let $\pi : E \to M$ be a smooth, complex n-plane bundle over the manifold M and let

$$\varphi_E : I_n^*(\mathbb{C}) \to H^*(M;\mathbb{C})$$

be the Chern-Weil homomorphism. We define the Chern classes

$$c_k(E)=\begin{cases}1\in H^0(M;\mathbb{C}), & k=0,\\ \left(\frac{\sqrt{-1}}{2\pi}\right)^k\varphi_E(c_k)\in H^{2k}(M;\mathbb{C}), & 1\le k\le n,\\ 0\in H^{2k}(M;\mathbb{C}), & k>n.\end{cases}$$

If $\pi : E \to M$ is a real n-plane bundle and

$$\varphi_E : I_n^*(\mathbb{R}) \to H^*(M;\mathbb{R})$$

is the Chern-Weil homomorphism, the Pontryagin classes are defined by

$$p_k(E)=\begin{cases}1\in H^0(M;\mathbb{R}), & k=0,\\ \left(\frac{-1}{4\pi^2}\right)^k\varphi_E(c_{2k})\in H^{4k}(M;\mathbb{R}), & 1\le k\le[n/2],\\ 0\in H^{4k}(M;\mathbb{R}), & k>[n/2].\end{cases}$$

Here, $[n/2]$ denotes the greatest integer $\le n/2$.

Proposition 5.3.1. *The Chern algebra* $\mathrm{Chern}^*(E)$ *of the complex n-plane bundle E over M is the subalgebra of $H^*(M;\mathbb{C})$ generated by $\{c_k(E)\}_{k=0}^n$.*

Since $\mathrm{Chern}^*(E) = \varphi_E(\mathbb{C}[c_1,\dots,c_n])$, this is evident. The following, however, needs proof.

Proposition 5.3.2. *The Pontryagin algebra* $\mathrm{Pont}^*(E)$ *of the real* n*-plane bundle* E *over* M *is generated by* $\{p_k(E)\}_{k=0}^{[n/2]}$.

Proof. This will be evident if we can show that $\mathrm{Pont}^j(E) = 0$ whenever j is not divisible by 4. Choose a smooth Riemann metric on the bundle and a connection ∇ which preserves this metric. In particular, if s_1 and s_2 are orthonormal sections of E over some open set U and $X \in \mathfrak{X}(U)$, then

$$\langle \nabla_X s_1, s_2 \rangle = \langle s_1, -\nabla_X s_2 \rangle.$$

It follows that, if $X, Y \in \mathfrak{X}(U)$ and R is the curvature tensor of ∇,

$$\langle R(X,Y)s_1, s_2 \rangle = \langle s_1, -R(X,Y)s_2 \rangle.$$

Thus, over a trivializing neighborhood U_α and relative to an orthonormal frame field $(s_1^\alpha, \dots, s_n^\alpha)$ there, the curvature matrix Ω^α is antisymmetric. Likewise, every odd power of Ω^α will be antisymmetric, so

$$\Sigma_{2k+1}(\Omega^\alpha) = \mathrm{tr}(\Omega^\alpha)^{2k+1} \equiv 0.$$

But every $P \in I_n^{2r}(\mathbb{R})$ is a linear combination of monomials $\Sigma_{i_1}\Sigma_{i_2}\cdots\Sigma_{i_m}$ (Lemma 5.2.6) and, if r is odd, some term Σ_{i_j} in each of these monomials has odd index i_j. That is, if r is odd and $P \in I_n^{2r}(\mathbb{R})$, then $P(\Omega) \equiv 0$. $\square$

Definition 5.3.3. The total Chern class of a complex vector bundle E is

$$c(E) = \sum_{k \geq 0} c_k(E)$$

and the total Pontryagin class of a real vector bundle E is

$$p(E) = \sum_{k \geq 0} p_k(E).$$

Corollary 5.3.4. *The total Chern and Pontryagin classes are given by*

$$c(E) = \left[\det\left(I + \frac{\sqrt{-1}}{2\pi}\Omega\right)\right],$$

$$p(E) = \left[\det\left(I + \frac{\sqrt{-1}}{2\pi}\Omega\right)\right],$$

respectively.

Theorem 5.3.5. *The total Chern and Pontryagin classes satisfy the "Whitney duality" formulas*

$$c(E_1 \oplus E_2) = c(E_1)c(E_2),$$
$$p(E_1 \oplus E_2) = p(E_1)p(E_2).$$

Proof. On $E_1 \oplus E_2$, use a connection of the form $\nabla_1 \oplus \nabla_2$, where ∇_i is a connection on E_i, $i = 1, 2$. Use neighborhoods U_α which trivialize both bundles simultaneously and let Ω_i^α be the respective curvature forms for $i = 1, 2$. Then,

$$\Omega^\alpha = \begin{bmatrix} \Omega_1^\alpha & 0 \\ 0 & \Omega_2^\alpha \end{bmatrix}$$

and

$$\det \begin{bmatrix} I + \frac{\sqrt{-1}}{2\pi}\Omega_1^\alpha & 0 \\ 0 & I + \frac{\sqrt{-1}}{2\pi}\Omega_2^\alpha \end{bmatrix} = \det\left(I + \frac{\sqrt{-1}}{2\pi}\Omega_1^\alpha\right) \wedge \det\left(I + \frac{\sqrt{-1}}{2\pi}\Omega_2^\alpha\right).$$

□

Suppose that $f : N \to M$ is a smooth map and that $\pi : E \to M$ is a smooth n-plane bundle with scalar field $\mathbb{K}$. Let $\pi' : f^*(E) \to N$ be the pull-back bundle. A connection ∇ on E pulls back to a connection $f^*(\nabla)$ on $f^*(E)$ in standard fashion and the pull-back of forms gives the curvature forms $f^*(\Omega^\alpha)$ on $f^{-1}(U_\alpha)$ of $f^*(\nabla)$. It should be clear that, for each $P \in I_n^*(\mathbb{K})$, $f^*(P(\Omega)) = P(f^*\Omega)$. This gives the *naturality* of the characteristic classes.

Theorem 5.3.6. *With all notation as above,*

$$f^*(c_k(E)) = c_k(f^*(E)),$$
$$f^*(p_k(E)) = p_k(f^*(E)),$$

$\forall\, k \geq 0$.

Exercise 5.3.7. For complex line bundles L_1 and L_2 over the same manifold M, prove that

$$c_1(L_1 \otimes L_2) = c_1(L_1) + c_1(L_2).$$

Chapter 6

Characteristic Classes and Integrability

Integrability of a p-plane field $E \subset T(M)$ forces the Pontryagin classes to vanish in a certain range of dimensions. This is the Bott vanishing theorem. It, in turn, will allow us to find *secondary* characteristic classes. We remark that this approach is what Bott has called the "low road" to these classes. We will not venture on the "high road", which arrives at the secondary classes via the Gelfand-Fuks cohomology of formal vector fields.

6.1. The Bott Vanishing Theorem

Let M be a smooth n-manifold, $E \subset T(M)$ a smooth p-plane distribution. The Frobenius theorem gives a necessary and sufficient condition that E be integrable, but gives no obvious way to determine whether E is *homotopic* to an integrable distribution. That is, we ask whether there is a subbundle

$$\widetilde{E} \subset T(M) \times \mathbb{R} \subset T(M \times \mathbb{R})$$

such that

$$\widetilde{E}_t = \widetilde{E}|(M \times \{t\}), \quad 0 \le t \le 1,$$

is a smooth homotopy of $E = \widetilde{E}_0$ to an integrable distribution $E' = \widetilde{E}_1$. By the homotopy invariance of bundles, the integrable bundle E' will be isomorphic to E, so we can ask the even weaker question: "Is E isomorphic as a bundle to an integrable distribution $E' \subset T(M)$?" An obstruction to this weak integrability condition is given by the "Bott vanishing theorem".

In order to state the theorem, we need to introduce the normal bundle

$$Q = T(M)/E$$

to E, a q-plane bundle over M, where $q = n - p$.

Theorem 6.1.1 (Bott [**9**]). *Let $E \subset T(M)$ be a p-plane distribution that is isomorphic to an integrable distribution on the n-manifold M and let $q = n - p$. Let $Q = T(M)/E$ be the normal bundle to E. Then* $\mathrm{Pont}^r(Q)$ *vanishes for $r > 2q$.*

Of course, this is only interesting when $q < n/2$. The proof of Theorem 6.1.1 will require several lemmas.

Let E and E' be q-plane distributions on M with normal bundles

$$Q = T(M)/E,$$
$$Q' = T(M)/E'.$$

If the bundles E and E' are isomorphic, it does *not* follow that $Q \cong Q'$. Nevertheless, Q and Q' have the same Pontryagin algebras. In order to prove this, we note the following little algebraic gem.

Lemma 6.1.2. *If A^* is a graded algebra with unity $1 \in A^0$ and such that $A^k = \{0\}$, $\forall\, k \geq n$, then*

$$G(A^*) = \{1 + a_1 + a_2 + \cdots + a_n \mid a_i \in A^i\}$$

is a group under multiplication.

Proof. It is clear that $G(A^*)$ is closed and associative under multiplication and that $1 = 1 + 0 + \cdots + 0$ is a multiplicative identity. In order to obtain an inverse $(1 + a_1 + \cdots + a_n)^{-1}$, we must solve the equation

$$(1 + a_1 + \cdots + a_n)(1 + b_1 + \cdots + b_n) = 1.$$

but this amounts to a system

$$\begin{aligned} a_1 + b_1 &= 0, \\ a_2 + a_1 b_1 + b_2 &= 0, \\ &\vdots \\ a_n + a_{n-1} b_1 + \cdots + a_1 b_{n-1} + b_n &= 0, \end{aligned}$$

which it is trivial to solve recursively for the b_i's. □

Lemma 6.1.3. *If E, E', Q and Q' are as above and $E \cong E'$, then*

$$\mathrm{Pont}^*(Q) = \mathrm{Pont}^*(Q').$$

Proof. It will be enough to show equality of the total Pontryagin classes $p(Q) = p(Q')$. Via a Riemannian metric on M, we obtain bundle isomorphisms

$$E \oplus Q \cong T(M) \cong E' \oplus Q' \cong E \oplus Q',$$

so Whitney duality gives

$$p(E)p(Q) = p(E)p(Q') = p(T(M)).$$

But total Pontryagin classes are elements of $G(H^*(M;\mathbb{R}))$, so

$$p(Q) = p(E)^{-1}p(T(M)) = p(Q').$$

□

Thus, to prove Theorem 6.1.1, we can assume that E is integrable and prove that $\operatorname{Pont}^r(Q) = 0$, $r > 2q$. The idea is to choose a connection ∇ on Q which restricts to a flat connection along each leaf. This is a *basic* connection (also called a Bott connection), and the Chern-Weil construction, using this connection, will yield Theorem 6.1.1 easily.

Let $\mathcal{F}$ be the codimension q foliation with normal bundle Q. If (U, x, y) is a foliated chart for $\mathcal{F}$, the transverse coordinate y defines a linear isomorphism

$$y_{*z} : Q_z \to T_{y(z)}(\mathbb{R}^q) = \mathbb{R}^q, \quad \forall\, z \in U.$$

Indeed, $y_{*z} : T_z(M) \to \mathbb{R}^q$ has kernel E_z, hence passes to the asserted isomorphism on $T_z(M)/E_z$.

Definition 6.1.4. Let $P(z) \subset U$ be the plaque $y^{-1}(y(z))$, $z_1, z_2 \in P(z)$. Let $v_i \in Q_{z_i}$, $i = 1, 2$. We say that v_1 is parallel to v_2 along $P(z)$ if $y_{*z_1}(v_1) = y_{*z_2}(v_2)$. If $Y \in \Gamma(Q|P(z))$, we say that Y is parallel if Y_w is parallel to Y_z along $P(z)$, $\forall\, w \in P(z)$. Finally, $Y \in \Gamma(U)$ is parallel if $Y|P(z)$ is parallel, $\forall\, z \in U$.

On overlapping foliated charts $(U_\alpha, x_\alpha, y_\alpha)$ and $(U_\beta, x_\beta, y_\beta)$, the transverse change of coordinates formula $y_\alpha = y_\alpha(y_\beta)$ implies that the notions of parallelism in U_α and U_β coincide in $U_\alpha \cap U_\beta$. Thus, one can parallel transport normal vectors $v \in Q$ along paths in leaves and the resulting parallelism is locally absolute. This is exactly how one would expect parallel transport of a connection to behave if its curvature R satisfies $R(X, X') \equiv 0$, $\forall\, X, X' \in \Gamma(E)$. We attempt to define such a connection as follows.

Let $\widetilde{\nabla}$ be an arbitrary connection on Q. By fixing a Riemannian metric on M, we can decompose the tangent bundle into an orthogonal direct sum $T(M) = Q' \oplus E$. The projection $T(M) \to T(M)/E = Q$ carries Q' isomorphically as a bundle onto Q. Having fixed the metric, we write $T(M) = Q \oplus E$. An arbitrary vector field $X \in \mathfrak{X}(M)$ decomposes into

$$X = X_Q + X_E, \quad X_Q \in \Gamma(Q) \text{ and } X_E \in \Gamma(E).$$

For $Y \in \Gamma(Q)$ and $X \in \mathfrak{X}(M)$, define

$$\nabla_X Y = \widetilde{\nabla}_{X_Q} Y + [X_E, Y]_Q.$$

Lemma 6.1.5. *The operator ∇, obtained as above, is a connection on Q satisfying*

$$\nabla_X Y = [X, Y]_Q,$$

for all $X \in \Gamma(E)$ and $Y \in \Gamma(Q)$. Every connection on Q having this property is called a (Bott) basic connection.

Exercise 6.1.6. Prove Lemma 6.1.5.

Lemma 6.1.5 makes no use of the requirement that E be an integrable distribution. Integrability is essential, however, for the following.

Lemma 6.1.7. *Let $E = T(\mathcal{F})$ be integrable and let ∇ be a basic connection on Q. If R is the curvature tensor of ∇ and $X, X' \in \Gamma(E) \subset \mathfrak{X}(M)$, then $R(X, X') \equiv 0$.*

Proof. Let $(U_\alpha, x_\alpha, y_\alpha)$ be a foliated chart. If $Y \in \Gamma(Q|U_\alpha)$ is parallel in the sense of Definition 6.1.4, then $y_{\alpha*}(Y)$ is a well defined vector field on $y_\alpha(U_\alpha) \subseteq \mathbb{R}^q$. If $X \in \Gamma(E|U_\alpha)$, then $y_{\alpha*}(X) \equiv 0$. Thus, by a basic property of the bracket,

$$y_{\alpha*}[X, Y] = [0, y_{\alpha*}(Y)] \equiv 0,$$

so

$$\nabla_X Y = [X, Y]_Q \equiv 0.$$

That is, Y is ∇-parallel along paths in plaques. In particular, given $z \in U_\alpha$ and $v \in Q_z$, one readily extends v to a section Y which is parallel in the sense of Definition 6.1.4. Hence the restriction of Y to an arbitrary path s in $P(z)$, beginning at z, produces a ∇-parallel field along s. Thus, the parallelism of ∇ is locally absolute along leaves, so the connection is flat along leaves. □

Exercise 6.1.8. Give a direct verification of Lemma 6.1.7, using only the formula for $R(X, X')$ and the property that

$$\nabla_X Y = [X, Y]_Q, \quad \forall X \in \Gamma(E) \text{ and } \forall Y \in \Gamma(Q).$$

(Hint. You must use the Frobenius theorem and the Bianchi identity.)

Example 6.1.9. Let M be an orientable 4-manifold and let $i : \Sigma_g \hookrightarrow \operatorname{int} M$ be a smooth imbedding, $g \geq 1$. Let $\nu(\Sigma_g)$ be the normal bundle of this imbedding, a 2-plane bundle oriented by choices of orientation for M and Σ_g. If there is a smooth foliation of M having $i(\Sigma_g)$ as a leaf, then $\nu(\Sigma_g)$ is just $i^*(Q)$. A Bott connection on Q pulls back to a flat connection on $\nu(\Sigma_g)$. By Theorem 4.5.2, it follows that $|e(\nu(\Sigma_g))| \leq g - 1$. That is, the Euler class of the normal bundle of the imbedding gives a topological obstruction to the extension of this imbedding to a smooth foliation. In particular, if $g \geq 2$, the diagonal imbedding

$$\Sigma_g \hookrightarrow \Sigma_g \times \Sigma_g,$$
$$x \mapsto (x, x),$$

cannot be a leaf of a foliation (Exercise 6.1.10).

Exercise 6.1.10. Let M be a manifold without boundary. Show that the normal bundle to the diagonal imbedding of M in the manifold $M \times M$ is isomorphic to the tangent bundle $T(\Sigma_g)$. Show that this completes the proof of the final assertion in Example 6.1.9.

A foliated chart (U, x, y) is simultaneously a trivializing neighborhood for Q and E. Let $A^*(U, E) \subseteq A^*(U)$ be the ideal of forms that vanish identically on $\Gamma(E|U)$. Clearly, $A^*(U, E)$ is generated as an ideal by the 1-forms $\{dy^1, \dots, dy^q\}$. Let Ω be the curvature form associated to a trivialization of $Q|U$ by a basic connection ∇. By Lemma 6.1.7, the following is immediate.

Lemma 6.1.11. *Under the above hypotheses, $\Omega_{ij} \in A^2(U, E)$, $1 \leq i, j \leq q$.*

Proof of Theorem 6.1.1. The elements of $A^*(U, E)$ are of the form

$$\omega = \sum_{i=1}^{q} \omega_i \wedge dy^i, \quad \text{for suitable } \omega_i \in A^*(U).$$

It follows that $(A^*(U, E))^{q+1} = \{0\}$. Thus, if $P \in I^{2k}(\mathbb{R})$, $k > q$, then $P(\Omega) = 0$. Since M is covered by foliated charts, Theorem 6.1.1 follows. □

It should be emphasized that, if Ω is defined by a *basic* connection, then we have proved the Bott vanishing theorem at the *form* level, not just for cohomology classes. This is key to the construction of the *secondary* or *exotic* classes for foliations.

Using Theorem 6.1.1, one can produce examples of p-plane distributions that are not homotopic to C^∞-integrable ones. For instance, there is a $2n$-plane distribution E on $M = P^n(\mathbb{C}) \times T^2$ which fails Bott's test. In this example, Q is a 2-plane bundle with $0 \neq p_1(Q)^2 \in H^8(M; \mathbb{R})$. Nevertheless, deep results of W. Thurston [**172**] and T. Tsuboi [**177**] show that there is no obstruction to homotopy of p-plane distributions to C^1-integrable ones. Thus, the Bott obstruction posits at least C^2-smoothness for the foliation.

Finally, we mention an interesting extension of the Bott vanishing range for Riemannian foliations.

Definition 6.1.12. A foliation is Riemannian if the normal bundle Q admits a Riemannian metric which is invariant under the natural parallelism along leaves.

Theorem 6.1.13 (Pasternack [**144**]). *If Q is the normal bundle of a Riemannian foliation, then* $\text{Pont}^r(Q) = 0$, *for* $r > q$.

Exercise 6.1.14. Prove Theorem 6.1.13.

6.2. The Godbillon-Vey Class in Arbitrary Codimension

We are going to give a construction of "Chern-Weil type" of the new characteristic classes that can be defined for (the normal bundle to) an integrable distribution. Before giving the general construction, we look at an important special case, the Godbillon-Vey class for a foliation of codimension q. Throughout, all foliated manifolds $(M, \mathcal{F})$ will be smooth of class C^∞ with $\partial M = \emptyset$.

Consider a foliated manifold $(M, \mathcal{F})$ of codimension q with normal bundle Q and tangent bundle E. For simplicity, we assume that $\mathcal{F}$ is transversely oriented. For the following remarks, the reader is referred to [I, Exercise 1.3.16]. There is a form $\omega \in A^q(M)$ such that E_z is exactly the null space of ω_z, for each $z \in M$. By the Frobenius integrability condition, there is a form $\eta \in A^1(M)$ such that $d\omega = \eta \wedge \omega$. The form $\eta \wedge (d\eta)^q \in A^{2q+1}(M)$ is closed and the cohomology class $[\eta \wedge (d\eta)^q]$ is independent of the allowable choices of ω and η.

Definition 6.2.1. The Godbillon-Vey class of $\mathcal{F}$ is $\text{gv}(\mathcal{F}) = [\eta \wedge (d\eta)^q]$.

A conjecture of R. Moussou, F. Pelletier [**136**] and D. Sullivan [**163**, Problem 17.3], formulated for the case $q = 1$, can be posited for the general case as well.

Conjecture. *If $(M, \mathcal{F})$ is a compact, smoothly foliated manifold and almost every leaf of $\mathcal{F}$ (relative to Lebesgue measure on the transverse space S) has nonexponential growth, then* $\text{gv}(\mathcal{F}) = 0$.

For the meaning of growth types of leaves and the particular types "exponential", "nonexponential", "quasi-polynomial", "polynomial", *etc.*, see [I, Chapter 12].

Duminy's theorem (Theorem 7.3.1) gives a stronger result when $q = 1$. In higher codimension, a somewhat weaker statement, with "nonexponential" replaced by "quasi-polynomial", is proven by S. Hurder [**102**] (who uses the term "subexponential" for such growth types). We will prove these results in subsequent chapters.

As a refresher on the Godbillon-Vey class in codimension one, here are a couple of exercises that could have been posed in [I, Section 3.6].

Exercise 6.2.2. Recall from [I, Exercise 3.6.18] that the Godbillon-Vey *number* $\int_M \eta \wedge d\eta$ for foliated 3-manifolds is a cobordism invariant. Let M be a closed manifold of arbitrary dimension n and suppose that $(M, \mathcal{F}_0)$ and $(M, \mathcal{F}_1)$ are cobordant foliated manifolds of codimension one. Since the underlying manifold M is the same, it makes sense to compare the Godbillon-Vey *classes* of the two foliations. Prove that $\mathrm{gv}(\mathcal{F}_0) = \mathrm{gv}(\mathcal{F}_1)$. (In the special case that the cobordism $W = M \times [0, 1]$, we say that the foliations are *concordant*.)

Exercise 6.2.3. If the compact, foliated manifold $(M, \mathcal{F}')$ is obtained from $(M, \mathcal{F})$ by turbulizing along a closed transversal [I, Example 3.3.11], then $\mathrm{gv}(\mathcal{F}') = \mathrm{gv}(\mathcal{F})$ [I, Proposition 3.6.14]. More generally, let $(M, \mathcal{F})$ be a foliated 3-manifold, turbulize along a finite family of disjoint closed transversals, then excise the Reeb components and glue them back in by new gluing maps. This produces a new manifold M' with smooth foliation $\mathcal{F}'$. Prove that $(M', \mathcal{F}')$ and $(M, \mathcal{F})$ have the same Godbillon-Vey number.

The following lemma brings Bott basic connections into the picture. Fix a decomposition $T(M) = Q \oplus E$ and a form $\omega \in A^q(M)$ that defines $\mathcal{F}$. Let $(U, Z_1, \dots, Z_q)$ be a local trivialization of Q such that $\omega(Z_1, \dots, Z_q) \equiv 1$ on U.

Lemma 6.2.4. *A form $\eta \in A^1(M)$ satisfies $d\omega = \eta \wedge \omega$ if and only if, for each local trivialization of Q as above, each basic connection ∇ and each $X \in \Gamma(E|U)$, $\eta(X) = -\operatorname{tr}(\theta)(X)$, where θ is the connection form in U of ∇ relative to the frame field $(Z_1, \dots, Z_q)$.*

Proof. Let $X \in \Gamma(E|U)$. Using a standard formula for the exterior derivative, we write

$$
\begin{aligned}
d\omega(X, Z_1, \dots, Z_q) = X\, &\underbrace{\omega(Z_1, \dots, Z_q)}_{\equiv 1} \\
&+ \sum_{i=1}^{q} (-1)^i Z_i \underbrace{\omega(X, Z_1, \dots, \widehat{Z}_i, \dots, Z_q)}_{\equiv 0} \\
&+ \sum_{1 \le i < j \le q} (-1)^{i+j} \underbrace{\omega([Z_i, Z_j], X, Z_1, \dots, \widehat{Z}_i, \dots, \widehat{Z}_j, \dots, Z_q)}_{\equiv 0} \\
&+ \sum_{i=1}^{q} (-1)^i \omega([X, Z_i], Z_1, \dots, \widehat{Z}_i, \dots, Z_q) \\
&= \sum_{i=1}^{q} (-1)^i \omega([X, Z_i]_Q, Z_1, \dots, \widehat{Z}_i, \dots, Z_q).
\end{aligned}
$$

The last equality uses the fact that E is the nullspace of ω. But, for each basic connection ∇,

$$[X, Z_i]_Q = \nabla_X Z_i = \sum_{j=1}^{q} \theta_{ji}(X) Z_j,$$

so

$$\begin{aligned} d\omega(X, Z_1, \dots, Z_q) &= \sum_{i=1}^{q} (-1)^i \omega(\theta_{ii}(X) Z_i, Z_1, \dots, \widehat{Z}_i, \dots, Z_q) \\ &= \sum_{i=1}^{q} -\theta_{ii}(X) \underbrace{\omega(Z_1, \dots, Z_q)}_{\equiv 1} \\ &= -\operatorname{tr}(\theta)(X). \end{aligned}$$

Thus, $d\omega = \eta \wedge \omega$ if and only if $\eta(X) = -\operatorname{tr}(\theta)(X)$, $\forall\, X \in \Gamma(E|U)$. □

Let $\{(U_\alpha, Z_1^\alpha, \dots, Z_q^\alpha)\}_{\alpha \in \mathfrak{A}}$ be a covering of M by local trivializations of Q such that $\omega(Z_1^\alpha, \dots, Z_q^\alpha) \equiv 1$. A corollary of the above lemma is the following.

Exercise 6.2.5. Prove that a form $\eta \in A^1(M)$ satisfies $d\omega = \eta \wedge \omega$ if and only if there exists a basic connection ∇ such that $\eta|U_\alpha = -\operatorname{tr}(\theta^\alpha)$, $\forall\, \alpha \in \mathfrak{A}$, where θ^α is the connection form of ∇ on U_α relative to the trivializing frame $(Z_1^\alpha, \dots, Z_q^\alpha)$. Furthermore, show that $d\eta = -\operatorname{tr}(\Omega) = -c_1(\Omega)$, where $\Omega = \{\Omega^\alpha\}_{\alpha \in \mathfrak{A}}$ is the curvature form of ∇.

In particular, the local forms $-\operatorname{tr}(\theta^\alpha)$ piece together to give a well defined global form $\eta = -\operatorname{tr}(\theta) = -c_1(\theta) \in A^1(M)$, for a suitably chosen basic connection ∇, and

$$\operatorname{gv}(\mathcal{F}) = [(-1)^{q+1} c_1(\theta) \wedge (c_1(\Omega))^q].$$

This suggests that there should be a Chern-Weil type construction of characteristic classes of $\mathcal{F}$.

6.3. Construction of the Exotic Classes

The construction uses two connections, one basic and one Riemannian. For the time being, however, let ∇^0 and ∇^1 be any two connections on the normal bundle Q of $\mathcal{F}$. Let

$$\lambda(\nabla^i) : \mathbb{R}[c_1, c_2, \dots, c_q] \to A^*(M)$$

denote the Chern-Weil map for Q and ∇^i, $i = 0, 1$. Then,

$$\lambda(\nabla^1)(c) - \lambda(\nabla^0)(c) = d\gamma$$

is exact, $\forall\, c \in \mathbb{R}[c_1, c_2, \dots, c_q]$. We need a precise formula for the form γ. This requires a version of the *Poincaré lemma.*

Lemma 6.3.1. *Let M be an n-manifold, $\partial M = \emptyset$, and let $I = [0,1]$. Let $i_t : M \hookrightarrow M \times I$ be the inclusion $i_t(x) = (x,t)$, $0 \le t \le 1$. Then there is a canonical $\mathbb{R}$-linear map*

$$\pi_* : A^k(M \times I) \to A^{k-1}(M)$$

which is additive on locally finite sums and satisfies

$$\pi_* \circ d + d \circ \pi_* = i_1^* - i_0^*.$$

Proof. This is a mild modification of the Poincaré lemma as presented in [**32**, Section 8.3]. Accordingly, we sketch the proof, relying on that reference for additional details. Forms in $A^k(M \times I)$ decompose into locally finite sums of forms of the following two types:

(a) $f(x,t)\pi^*(\eta)$, where $f \in C^\infty(M \times I)$ and $\eta \in A^k(M)$;
(b) $f(x,t)\,dt \wedge \pi^*(\eta)$, where $f \in C^\infty(M \times I)$ and $\eta \in A^{k-1}(M)$.

We define

$$\pi_*(f(x,t)\pi^*(\eta)) = 0,$$
$$\pi_*(f(x,t)\omega \wedge \pi^*(\eta)) = \left(\int_0^1 f(x,t)\,dt\right)\eta.$$

Clearly, there is at most one linear operator π_* which is additive on locally finite sums and satisfies these two equations. The construction of such an operator can be carried out using a fixed choice of local data to formulate an algorithm for decomposing arbitrary $\theta \in A^k(M \times I)$ into a locally finite sum of forms of types (a) and (b) (*cf.* [**32**, pp. 258-260]). One must check that this operator is additive on locally finite sums, a slightly tedious exercise. Finally, the exterior derivative d and the linear maps i_t^* are also additive on locally finite sums, so the desired identity need only be verified on forms of types (a) and (b). This is a straightforward exercise. □

The map π_* is sometimes referred to as "integration along the fiber". It has a version for general fiber bundles (*cf.* [**9**, p. 15]).

Proposition 6.3.2. *Given connections ∇^0 and ∇^1 on the normal bundle Q of $\mathcal{F}$, there is a canonical linear map*

$$\lambda(\nabla^0, \nabla^1) : \mathbb{R}[c_1, c_2, \dots, c_q] \to A^*(M)$$

which is of degree -1 relative to the graded structures of these spaces and satisfies

$$d(\lambda(\nabla^0, \nabla^1)(c)) = \lambda(\nabla^1)(c) - \lambda(\nabla^0)(c),$$

$\forall\, c \in \mathbb{R}[c_1, c_2, \dots, c_q]$.

Proof. Let $\nabla^{0,1}$ be the connection on the bundle

$$Q \times I \xrightarrow{\pi \times \mathrm{id}} M \times I,$$

constructed from ∇^0 and ∇^1 exactly as in the proof of Lemma 5.1.6, and let $\Omega^{0,1}$ be the locally defined curvature form of $\nabla^{0,1}$. For $c \in \mathbb{R}[c_1, c_2, \dots, c_q]$, we define

$$\lambda(\nabla^0, \nabla^1)(c) = \pi_*(c(\Omega^{0,1})).$$

This is $\mathbb{R}$-linear of degree -1 and, since $c(\Omega^{0,1})$ is a closed form,

$$\begin{aligned}
\lambda(\nabla^1)(c) - \lambda(\nabla^0)(c) &= c(\Omega^1) - c(\Omega^0) \\
&= i_1^*(c(\Omega^{0,1})) - i_0^*(c(\Omega^{0,1})) \\
&= \pi_*\underbrace{(d(c(\Omega^{0,1})))}_{\equiv 0} + d(\pi_*(c(\Omega^{0,1}))) \\
&= d(\lambda(\nabla^0, \nabla^1)(c)).
\end{aligned}$$

The third equality is by Lemma 6.3.1. □

Definition 6.3.3. Let $\mathcal{I}_q \subset \mathbb{R}[c_1, c_2, \dots, c_q]$ be the ideal consisting of all elements of degree $> 2q$. Then

$$\mathbb{R}_q[c_1, c_2, \dots, c_q] = \mathbb{R}[c_1, c_2, \dots, c_q]/\mathcal{I}_q$$

is the associated truncated polynomial algebra.

Lemma 6.3.4. *Let ∇^0 be a connection on Q that preserves some Riemannian metric on that bundle and let ∇^1 be a basic connection on Q. The maps $\lambda(\nabla^0, \nabla^1)$ and $\lambda(\nabla^1)$ satisfy*

(1) $d(\lambda(\nabla^0, \nabla^1)(c_{2j-1})) = \lambda(\nabla^1)(c_{2j-1})$;

(2) $\lambda(\nabla^1) : \mathbb{R}_q[c_1, c_2, \dots, c_q] \to A^*(M)$ *is a well defined homomorphism of graded algebras.*

Proof. As in the proof of Proposition 5.3.2, $\lambda(\nabla^0)(c_{2j-1}) = 0$. By Proposition 6.3.2, property (1) follows. By Bott vanishing at the *form* level, we see that

$$c \in \mathbb{R}[c_1, \dots, c_q] \text{ and } \deg c > 2q \Rightarrow \lambda(\nabla^1)(c) = 0,$$

and property (2) follows. □

We are going to form the tensor product $A^* \otimes_{\mathbb{R}} B^*$ of *graded commutative* $\mathbb{R}$-algebras. The reader is reminded that, in order that the result be graded commutative, the multiplication must be defined with the following sign convention:

$$(a \otimes b) \cdot (c \otimes d) = (-1)^{(\deg b)(\deg c)}(a \cdot c) \otimes (b \cdot d),$$

where "·" is used to denote the multiplication in all three algebras.

Let r be the largest odd integer $< q$ and consider the graded exterior algebra (over $\mathbb{R}$) $\Lambda(y_1, y_3, \ldots, y_r)$, with $\deg y_i = 2i - 1$. The tensor product of graded algebras

$$\mathrm{WO}_q = \mathbb{R}_q[c_1, c_2, \ldots, c_q] \otimes_{\mathbb{R}} \Lambda(y_1, y_3, \ldots, y_r)$$

is spanned by elements $c_J y_I$, where $J = (j_1, \ldots, j_q)$ and $I = (i_1, i_3, \ldots, i_r)$ are multi-indices and

$$\begin{aligned} c_J &= c_1^{j_1} c_2^{j_2} \cdots c_q^{j_q}, \\ y_I &= y_1^{i_1} y_3^{i_3} \cdots y_r^{i_r}. \end{aligned}$$

Exercise 6.3.5. Show that the graded algebra WO_q admits a unique linear operator

$$d : \mathrm{WO}_q \to \mathrm{WO}_q$$

with the properties

(1) d is of degree 1;
(2) $d(c_i) = 0$, $1 \le i \le q$;
(3) $d(y_i) = c_i$, $i = 1, 3, \ldots, r$;
(4) $d(a \cdot b) = d(a) \cdot b + (-1)^{\deg a} a \cdot d(b)$;
(5) $d^2 = 0$.

Thus, (WO_q, d) is a cochain complex giving rise to a graded commutative cohomology algebra $H^*(\mathrm{WO}_q)$. Since WO_q is generated by $c_J y_I$, subject only to the relations implied by graded commutativity and vanishing of terms c_J of degrees $> 2q$, property (2) of Lemma 6.3.4 implies that there is a well defined homomorphism

$$\lambda_{\mathcal{F},\nabla^0,\nabla^1} : \mathrm{WO}_q \to A^*(M)$$

of graded algebras, determined by requiring

$$\begin{aligned} \lambda_{\mathcal{F},\nabla^0,\nabla^1}(c_i) &= \lambda(\nabla^1)(c_i), \qquad && 1 \le i \le q, \\ \lambda_{\mathcal{F},\nabla^0,\nabla^1}(y_i) &= \lambda(\nabla^0, \nabla^1)(c_i), \qquad && i = 1, 3, \ldots, r. \end{aligned}$$

Lemma 6.3.6. *The graded algebra homomorphism $\lambda_{\mathcal{F},\nabla^0,\nabla^1}$ is a homomorphism of cochain complexes.*

Proof. Since $dc_i = 0$ and $\lambda(\nabla^1)(c_i)$ is a closed form,

$$\lambda_{\mathcal{F},\nabla^0,\nabla^1}(dc_i) = 0 = d\lambda_{\mathcal{F},\nabla^0,\nabla^1}(c_i), \quad 1 \le i \le q.$$

For odd values of $i \le r$,

$$\begin{aligned}\lambda_{\mathcal{F},\nabla^0,\nabla^1}(dy_i) &= \lambda_{\mathcal{F},\nabla^0,\nabla^1}(c_i)\\ &= \lambda(\nabla^1)(c_i)\\ &= d\lambda(\nabla^0,\nabla^1)(c_i)\\ &= d\lambda_{\mathcal{F},\nabla^0,\nabla^1}(y_i),\end{aligned}$$

where the next to last equality is by property (1) of Lemma 6.3.4. Since WO_q is spanned by elements $c_J y_I$, these observations and property (4) in Exercise 6.3.5 give the assertion. □

In this way, we obtain a homomorphism

$$\begin{aligned}\lambda_{\mathcal{F},\nabla^0,\nabla^1} &: H^*(\mathrm{WO}_q) \to H^*(M;\mathbb{R}),\\ \lambda_{\mathcal{F},\nabla^0,\nabla^1}[z] &= [\lambda_{\mathcal{F},\nabla^0,\nabla^1}(z)],\end{aligned}$$

of graded algebras.

Lemma 6.3.7. *At the level of cohomology, the homomorphism $\lambda_{\mathcal{F},\nabla^0,\nabla^1}$ is independent of the choices of Riemann metric, Riemann connection ∇^0 and basic connection ∇^1 on Q.*

Proof. Let $I = [0,1]$. The bundle $\pi \times \mathrm{id} : Q \times I \to M \times I$ is the normal bundle to the foliation $\mathcal{F} \times I$ of $M \times I$ (the foliation with leaves $L \times I$, where L ranges over the leaves of $\mathcal{F}$). If ∇^1 and $\widetilde{\nabla}^1$ are basic connections on Q, then the connection $\overline{\nabla}^1 = t\nabla^1 + (1-t)\widetilde{\nabla}^1$, defined by the usual conventions, is a basic connection on $Q \times I$. Similarly, the set of Riemannian metrics on Q is a convex set, implying that, if ∇^0 and $\widetilde{\nabla}^0$ are connections that preserve (possibly different) Riemann metrics, then $\overline{\nabla}^0 = t\nabla^0 + (1-t)\widetilde{\nabla}^0$ preserves a Riemann metric on $Q \times I$. Thus,

$$\lambda_{\mathcal{F},\widetilde{\nabla}^0,\widetilde{\nabla}^1} = i_0^* \circ \lambda_{\mathcal{F}\times I,\overline{\nabla}^0,\overline{\nabla}^1} = i_1^* \circ \lambda_{\mathcal{F}\times I,\overline{\nabla}^0,\overline{\nabla}^1} = \lambda_{\mathcal{F},\nabla^0,\nabla^1}$$

by the homotopy invariance of de Rham cohomology. □

Hereafter, we write the map in cohomology as

$$\lambda_{\mathcal{F}} : H^*(\mathrm{WO}_q) \to H^*(M;\mathbb{R}).$$

Since

$$\lambda_{\mathcal{F}}[c_{2j}] = \varphi_Q(c_{2j}) = (-4\pi)^2 p_j(Q),$$

we see that $\mathrm{Pont}^*(Q) \subseteq \operatorname{im} \lambda_{\mathcal{F}}$.

Definition 6.3.8. The elements of $\lambda_{\mathcal{F}}(H^*(\mathrm{WO}_q)) \smallsetminus \mathrm{Pont}^*(Q)$ are called the exotic characteristic classes of $\mathcal{F}$.

Example 6.3.9. If $\deg c_J = 2q$, then every element $y_I c_J$ of WO_q with $\deg y_I \geq 1$ is a cocycle. Indeed, monomials c_J are cocycles, so

$$d(y_I c_J) = d(y_I)c_J \pm y_I d(c_J) = d(y_I)c_J.$$

But $d(y_I)$ is a sum of monomials, each containing a factor c_j, $j \geq 1$, and $c_j c_J = 0$ in WO_q, so $d(y_I)c_J = 0$. The classes $\lambda_{\mathcal{F}}[y_I c_J]$, where $\deg c_J = 2q$ and $\deg y_I \geq 1$, are called *residual* classes of $\mathcal{F}$. If $y_I = y_1 y_{I'}$, the residual class is called a generalized Godbillon-Vey class. Of particular interest is the class with $y_I = y_1$ and $c_J = (c_1)^q$. We are going to show that, if $\mathcal{F}$ is transversely orientable, then

$$\lambda_{\mathcal{F}}[y_1(c_1)^q] = (-1)^{q+1}\,\mathrm{gv}(\mathcal{F}).$$

Since the left-hand side of this equation is defined whether or not $\mathcal{F}$ is transversely orientable, the Godbillon-Vey class extends to all smooth foliations.

Assume that $\mathcal{F}$ is transversely oriented and let $\omega \in A^q(M)$ define $\mathcal{F}$. We can view $\omega \in \Gamma(\Lambda^q(Q)^*)$ and choose a Riemann metric on Q (hence on $\Lambda^q(Q)$ and the dual $\Lambda^q(Q)^*$) such that $\|\omega\| \equiv 1$. Then, if $(Z_1^\alpha, \dots, Z_q^\alpha)$ is an orthonormal frame field for $Q|U_\alpha$, we have $\omega(Z_1^\alpha \wedge \cdots \wedge Z_q^\alpha) \equiv 1$. Let ∇^0 be a connection that preserves this metric and let θ_α^0 be the connection form on U_α relative to the orthonormal frame field. Thus, θ_α^0 is skew symmetric, so $\operatorname{tr}\theta_\alpha^0 = 0$. Since $\omega(Z_1^\alpha \wedge \cdots \wedge Z_q^\alpha) \equiv 1$, Corollary 6.2.5 allows us to choose a basic connection ∇^1 so that the traces $\operatorname{tr}(\theta_\alpha^1) \in A^1(U_\alpha)$ of its local connection forms piece together to define a form $-\eta \in A^1(M)$ such that $d\omega = \eta \wedge \omega$. Let $\nabla^{0,1}$ be the connection on $Q \times I$ formed out of ∇^0 and ∇^1 as usual, noting that the associated connection forms are

$$\theta_\alpha^{0,1} = t\theta_\alpha^1 + (1-t)\theta_\alpha^0$$

on $U_\alpha \times I$ (where we understand that these forms vanish on $\partial/\partial t$). Thus,

$$\begin{aligned}
\Omega_\alpha^{0,1} &= d\theta_\alpha^{0,1} + \theta_\alpha^{0,1} \wedge \theta_\alpha^{0,1} \\
&= dt \wedge (\theta_\alpha^1 - \theta_\alpha^0) + \zeta, \\
c_1(\Omega_\alpha^{0,1}) &= dt \wedge \operatorname{tr}\theta_\alpha^1 + \operatorname{tr}\zeta,
\end{aligned}$$

where ζ, hence $\operatorname{tr}\zeta$, has no terms involving dt. Thus,

$$\lambda(\nabla^0, \nabla^1)(c_1)|U_\alpha = \pi_*(c_1(\Omega_\alpha^{0,1})) = \left(\int_0^1 dt\right) \operatorname{tr}\theta_\alpha^1 = -\eta|U_\alpha.$$

Consequently,

$$\begin{aligned}
\lambda_{\mathcal{F},\nabla^0,\nabla^1}(y_1) &= \lambda(\nabla^0, \nabla^1)(c_1) = -\eta, \\
\lambda_{\mathcal{F},\nabla^0,\nabla^1}(c_1) &= \lambda(\nabla^1)(c_1) = d(\lambda(\nabla^0, \nabla^1)(c_1)) = -d\eta.
\end{aligned}$$

This proves that $\lambda_{\mathcal{F}}[y_1(c_1)^q] = (-1)^{q+1}[\eta \wedge (d\eta)^q]$, as claimed.

Exercise 6.3.10. Let $(M, \mathcal{F})$ be a foliated manifold and let $f : N \to M$ be a smooth map transverse to $\mathcal{F}$. Then the pullback construction [I, §3.2] produces a foliated manifold $(N, \mathcal{F}') = (N, f^*(\mathcal{F}))$. Prove that the diagram

$$\begin{array}{ccc} H^*(\mathrm{WO}_q) & \xrightarrow{\lambda_{\mathcal{F}}} & H^*(M;\mathbb{R}) \\ & \searrow^{\lambda_{\mathcal{F}'}} & \downarrow f^* \\ & & H^*(N;\mathbb{R}) \end{array}$$

commutes. This proves the *naturality* of the exotic classes.

6.4. Haefliger Structures and Classifying Spaces

This section will be an extended remark, sketching without proofs other ways of viewing the characteristic classes which we have constructed. Here we are dealing with the *quantitative* aspects of foliation theory (questions of existence and classification).

6.4.A. Vector bundles and G^q-cocycles. Let X be a topological space and consider real q-plane bundles $p : E \to X$. It is standard (for instance, see [**32**, §3.4]) that such a bundle is completely determined, up to isomorphism, by a G^q-cocycle, where $G^q = \mathrm{Gl}(q, \mathbb{R})$ denotes the group of nonsingular linear transformations of $\mathbb{R}^q$. Such a cocycle is given by a collection

$$g_{\mathfrak{A}} = \{U_\alpha, g_{\alpha\beta}\}_{\alpha,\beta\in\mathfrak{A}},$$

where the U_α's form an open cover of X and, whenever $U_\alpha \cap U_\beta \neq \emptyset$,

$$g_{\alpha\beta} : U_\alpha \cap U_\beta \to G^q$$

is a continuous map. The cocycle condition is that

$$g_{\alpha\beta}(w) g_{\beta\delta}(w) = g_{\alpha\delta}(w), \quad \forall\, w \in U_\alpha \cap U_\beta \cap U_\delta,$$

implying also that

$$g_{\alpha\alpha}(w) = I_q, \quad \forall\, w \in U_\alpha,$$
$$g_{\alpha\beta}(w) = g_{\beta\alpha}(w)^{-1}, \quad \forall\, w \in U_\alpha \cap U_\beta,$$

where I_q denotes the $q \times q$ identity matrix. Two such cocycles, $g_{\mathfrak{A}}$ and $g_{\mathfrak{B}}$, with respective open covers $\{U_\alpha\}_{\alpha\in\mathfrak{A}}$ and $\{U_\beta\}_{\beta\in\mathfrak{B}}$, are said to be *coherent* if there is a G^q-cocycle $g_{\mathfrak{L}}$, with open cover $\{U_\lambda\}_{\lambda\in\mathfrak{L}}$, where $\mathfrak{L}$ is the disjoint union of $\mathfrak{A}$ and $\mathfrak{B}$, which restricts to $g_{\mathfrak{A}}$ and to $g_{\mathfrak{B}}$ on the respective subcovers. This is exactly the equivalence relation that guarantees that $g_{\mathfrak{A}}$ and $g_{\mathfrak{B}}$ determine the same (isomorphism class of) q-plane bundles.

Remark. Because of a certain formal analogy with the first Čech cohomology $\check{H}^1(X;G)$ with coefficients in an abelian group G, one denotes the set of coherence classes of G^q-cocycles on X by $H^1(X;G^q)$. Since G^1 is an abelian group, one does obtain the Čech cohomology group in that case, but generally $H^1(X;G^q)$ is only a set. If one denotes by $\text{Vect}^q(X)$ the set of isomorphism classes of q-plane bundles over X, the above discussion claims that $H^1(X;G^q) = \text{Vect}^q(X)$ canonically.

Exercise 6.4.1. This exercise is for those familiar with Čech cohomology. As in Exercise 4.4.3, the set $\text{Vect}^1(X)$ is a group under tensor product of real line bundles. Prove this, and also prove that $\check{H}^1(X;\mathbb{Z}_2)$ is canonically isomorphic to the group $\text{Vect}^1(X)$. The isomorphism

$$w_1 : \text{Vect}^1(X) \to \check{H}^1(X;\mathbb{Z}_2)$$

is called the *first Whitney class.* The Whitney classes of real vector bundles are analogous to the Chern and Pontryagin classes, but take values in $\mathbb{Z}_2$-cohomology.

If $f : Y \to X$ is a continuous map and $g_{\mathfrak{A}}$, as above, determines the q-plane bundle E over X, there is a "pull-back" cocycle $g'_{\mathfrak{A}} = f^*(g_{\mathfrak{A}})$ on Y, given by

$$g'_A = \{f^{-1}(U_\alpha), g'_{\alpha\beta} = g_{\alpha\beta} \circ f\}_{\alpha,\beta\in\mathfrak{A}}.$$

The bundle over Y determined by $g'_{\mathfrak{A}}$ is exactly $f^*(E)$, the "pull-back" of E by f (Section 4.1). It is the unique bundle, up to isomorphism, that fits into a commutative diagram

$$\begin{array}{ccc} f^*(E) & \xrightarrow{\ \varphi\ } & E \\ {\scriptstyle p'}\big\downarrow & & \big\downarrow{\scriptstyle p} \\ Y & \xrightarrow[\ f\ ]{} & X \end{array}$$

where φ is continuous and carries the fiber over each $y \in Y$ linearly and bijectively to the fiber over $f(y)$. Bundle theory is homotopy invariant. That is, if $f_0, f_1 : Y \to X$ are homotopic, then $f_0^*(E) \cong f_1^*(E)$.

Remark. In the language of category theory, $X \mapsto H^1(X;G^q) = \text{Vect}^q(X)$ is a homotopy-invariant, contravariant functor from the topological category to the category of sets.

6.4.B. Milnor's classifying space for vector bundles.

For vector bundles

$$p : E \to X$$

over paracompact spaces X, there is a remarkable classification theory. Following J. Milnor (*cf.* [**130**, §5]), one builds a "universal" q-plane bundle

$$\pi : \mathcal{E}^q \to BG^q.$$

Here, BG^q is essentially the infinite dimensional Grassmann manifold of q-planes in $\mathbb{R}^\infty$ and the fiber $\mathcal{E}^q_v$ over $v \in BG^q$ is just the q-plane v. The "universal" property of this bundle is that, for each paracompact space X and each isomorphism class $[E]$ of q-plane bundles over X, there is a continuous map $f_E : X \to BG^q$, unique up to homotopy, such that there is an isomorphism of vector bundles

$$\begin{array}{ccc} f_E^*(\mathcal{E}^q) & \xrightarrow{\cong} & E \\ {\scriptstyle \pi'}\downarrow & & \downarrow{\scriptstyle p} \\ X & \xrightarrow[\text{id}]{} & X \end{array}$$

This sets up a bijective correspondence between $\mathrm{Vect}^q(X) = H^1(X; G^q)$ and the set of homotopy classes $[X, BG^q]$ of continuous maps of X into BG^q. For details, see [**105**].

Remark. We say that the contravariant functor $\mathrm{Vect}^q(\cdot) = H^1(\,\cdot\,; G^q)$ is "representable" on the category of paracompact spaces and that BG^q is a "classifying space" for this functor. Less formally, one can think of the universal q-plane bundle as a kind of "platonic ideal", all other q-plane bundles being "shadows" of this ideal bundle.

One shows that $H^*(BG^q; \mathbb{R})$ is a polynomial algebra. It has the form

$$H^*(BG^q; \mathbb{R}) = \mathbb{R}[p_1, p_2, \dots, p_{[q/2]}],$$

where $\deg p_i = 4i$, $1 \le i \le k$, and $[q/2]$ denotes the greatest integer in $q/2$. If X is a manifold and the bundle E over X is smooth, one proves that

$$f_E^*(p_i) = p_i(E), \quad 1 \le i \le [q/2].$$

For all of this, see [**130**, §15]. Thus, the Pontryagin classes actually live on the universal base space BG^q and induce, via the classifying map f_E, the characteristic classes constructed in Chapter 5. The homotopy invariance of cohomology guarantees that these classes are well defined. Indeed, we now have characteristic classes for vector bundles over general paracompact spaces, dispensing entirely with the differential geometry in the Chern-Weil construction.

There is a completely analogous theory for complex vector bundles and the Chern classes.

6.4.C. Foliations and Γ^q-cocycles. It was the idea of A. Haefliger to develop a classification theory for foliations [**86**] analogous to the theory just outlined for vector bundles. To do this, he generalized the notion of a foliation of codimension q to that of a "Γ^q-structure" (also called a "Haefliger structure") on a topological space. In this case, the role of the G^q-cocycle above will be played by a germinal version of the holonomy cocycle. Unlike bundles, such structures are not homotopy invariant, so the theory classifies them only up to homotopy, not up to isomorphism. The most interesting Haefliger structures, of course, are those corresponding to foliations on manifolds, and it becomes important to determine which these are.

Let Γ^q denote the pseudogroup of all local diffeomorphisms of $\mathbb{R}^q$. For simplicity, we work in the C^∞ category. The associated groupoid of germs, with its sheaf topology, will be denoted by $\mathcal{G}^q$. This topology is described as follows. Let $\widehat{f}_x$ denote the germ at $x \in \mathbb{R}^q$ of a diffeomorphism f defined on a neighborhood U of x. Then the set

$$\widehat{U}_f = \{\widehat{f}_y \mid y \in U\}$$

will be a basic neighborhood of $\widehat{f}_x$ in $\mathcal{G}^q$. It is easily seen that the intersection of two such basic neighborhoods is a union of such, so the basic neighborhoods form a base for a topology on $\mathcal{G}^q$.

An element $\gamma \in \mathcal{G}^q$ has *source* $s(\gamma) = x \in \mathbb{R}^q$ if $\gamma = \widehat{f}_x$, and it has *target* $t(\gamma) = y = f(x)$. Let

$$\mathcal{G}^q_{(x,y)} = \{\gamma \in \mathcal{G}^q \mid s(\gamma) = x, t(\gamma) = y\}.$$

Example 6.4.2. The familiar example of the holonomy cocycle γ associated to a (regular cover of a) smoothly foliated manifold $(M, \mathcal{F})$ [I, Definition 1.2.12] can be described in terms of $\mathcal{G}^q$. The transverse space S associated to a regular cover $\mathfrak{U} = \{U_\alpha, x_\alpha, y_\alpha\}_{\alpha,\beta\in\mathfrak{A}}$ can be realized as an open subset of $\mathbb{R}^q$. In this way the holonomy pseudogroup $\Gamma_{\mathfrak{U}}$ is realized as a subpseudogroup of Γ^q. The holonomy cocycle $\gamma = \{\gamma_{\alpha\beta}\}_{\alpha,\beta\in\mathfrak{A}}$ determines maps

$$w \mapsto \gamma^w_{\alpha\beta} = (\widehat{\gamma_{\alpha\beta}})_{y_\beta(w)} \in \mathcal{G}^q_{(y_\beta(w), y_\alpha(w))}$$

of $U_\alpha \cap U_\beta \to \mathcal{G}^q$. By the definition of the sheaf topology, it is clear that these maps are continuous. By the definition of the cocycle γ, one also has the crucial *cocycle property*

$$\gamma^w_{\alpha\beta} \circ \gamma^w_{\beta\delta} = \gamma^w_{\alpha\delta}, \quad \forall\, w \in U_\alpha \cap U_\beta \cap U_\delta.$$

Furthermore, if $w \in U_\alpha$ and if y^w_α denotes the germ of

$$y_\alpha : U_\alpha \to \mathbb{R}^q$$

at w, we have

$$y^w_\alpha = \gamma^w_{\alpha\beta} \circ y^w_\beta, \quad \forall\, w \in U_\alpha \cap U_\beta.$$

This will be the chief and motivating example of a Γ^q-cocycle, generally called a Haefliger cocycle. Two Haefliger cocycles will be equivalent if they arise from *coherent* foliated atlases [I, Definition 1.2.8]. The corresponding equivalence class will be called a Haefliger structure. This is just a slightly different way of looking at familiar concepts, but formulated in a way that will continue to make sense on topological spaces, where foliations and their holonomy pseudogroups no longer make sense.

Definition 6.4.3. Let X be a topological space and $\mathcal{U} = \{U_\alpha\}_{\alpha\in\mathfrak{A}}$ an open cover of X, together with continuous maps

$$y_\alpha : U_\alpha \to \mathbb{R}^q, \quad \forall\, \alpha \in \mathfrak{A}.$$

Denote by y_α^w the germ of y_α at $w \in U_\alpha$. For each $w \in U_\alpha \cap U_\beta$, let there be given $\gamma_{\alpha\beta}^w \in \mathcal{G}^q_{(y_\beta(w),y_\alpha(w))}$. This collection of data will be called a Haefliger cocycle if the following conditions are satisfied:

(1) the map $\gamma_{\alpha\beta} : U_\alpha \cap U_\beta \to \mathcal{G}^q$, defined by $w \mapsto \gamma_{\alpha\beta}^w$, is continuous, for all $\alpha, \beta \in \mathfrak{A}$;

(2) $y_\alpha^w = \gamma_{\alpha\beta}^w \circ y_\beta^w$, for all $w \in U_\alpha \cap U_\beta$ and all $\alpha, \beta \in \mathfrak{A}$;

(3) $\gamma_{\alpha\beta}^w \circ \gamma_{\beta\delta}^w = \gamma_{\alpha\delta}^w$, for all $w \in U_\alpha \cap U_\beta \cap U_\delta$ and all $\alpha, \beta, \delta \in \mathfrak{A}$.

Definition 6.4.4. Let

$$c_{\mathfrak{A}} = \{U_\alpha, y_\alpha, \gamma_{\alpha\beta}\}_{\alpha,\beta\in\mathfrak{A}},$$
$$c_{\mathfrak{B}} = \{U_\lambda, y_\lambda, \gamma_{\lambda\mu}\}_{\lambda,\mu\in\mathfrak{B}}$$

be two Haefliger cocycles on X. If $\mathfrak{L}$ denotes the disjoint union of $\mathfrak{A}$ and $\mathfrak{B}$ and there is a Haefliger cocycle $c_{\mathfrak{L}}$ corresponding to the cover $\{U_\delta\}_{\delta\in\mathfrak{L}}$ which restricts to $c_{\mathfrak{A}}$ and $c_{\mathfrak{B}}$, respectively, then we say that $c_{\mathfrak{A}}$ and $c_{\mathfrak{B}}$ are coherent.

Exercise 6.4.5. Show that coherence of Haefliger cocycles is an equivalence relation.

Definition 6.4.6. Let X be a topological space. A coherence class of Haefliger cocycles on X is called a Haefliger structure on X. The set of all Haefliger structures on X is denoted by $H^1(X;\Gamma^q)$.

Definition 6.4.7. If

$$c = \{U_\alpha, y_\alpha, \gamma_{\alpha\beta}\}_{\alpha,\beta\in\mathfrak{A}}$$

is a Γ^q-cocycle on X and $w \in U_\alpha$, the α-plaque of c through w is

$$P_\alpha^w = \{v \in U_\alpha \mid y_\alpha(v) = y_\alpha(w)\}.$$

Exercise 6.4.8. If $\{U_\alpha, y_\alpha, \gamma_{\alpha\beta}\}_{\alpha,\beta\in\mathfrak{A}}$ and $\{U_\lambda, y_\lambda, \gamma_{\lambda\mu}\}_{\lambda,\mu\in\mathfrak{L}}$ are coherent Γ^q-cocycles on a space X and $w \in U_\alpha \cap U_\lambda$, $\alpha \in \mathfrak{A}$ and $\lambda \in \mathfrak{L}$, prove that there is an open neighborhood V of w in X such that $P_\alpha^w \cap V = P_\lambda^w \cap V$.

Exercise 6.4.9. Show that $X \rightsquigarrow H^1(X;\Gamma^q)$ defines a contravariant functor from the category of topological spaces and continuous maps to the category of sets and set maps. That is, given a continuous map $g : X \to Y$, show how to define canonically a map $g^* : H^1(Y;\Gamma^q) \to H^1(X;\Gamma^q)$ and verify that $\mathrm{id}^* = \mathrm{id}$ and that $(f \circ g)^* = g^* \circ f^*$.

Definition 6.4.10. Two Haefliger structures $\zeta, \xi \in H^1(X;\Gamma^q)$ are homotopic if there is a Haefliger structure $\theta \in H^1(X \times I;\Gamma^q)$ such that

$$i_0^*(\theta) = \zeta,$$
$$i_1^*(\theta) = \xi,$$

where $i_t : X \hookrightarrow X \times I$ is the map $i_t(w) = (w,t)$, $0 \le t \le 1$. In this case, we write $\zeta \simeq \xi$.

Exercise 6.4.11. Prove the following properties of homotopy.

(1) The relation $\simeq$ is an equivalence relation on $H^1(X;\Gamma^q)$.
(2) If $f : X \to Y$ is continuous, $\zeta, \xi \in H^1(Y;\Gamma^q)$ and $\zeta \simeq \xi$, then $f^*(\zeta) \simeq f^*(\xi)$.
(3) If $f, g : X \to Y$ are homotopic and $\zeta \in H^1(Y;\Gamma^q)$, then $f^*(\zeta) \simeq g^*(\zeta)$.

Exercise 6.4.12. On $X = \mathbb{R}^q$, consider the Haefliger structure $[c]$, defined by the cocycle c with open cover $\mathfrak{U} = \{U_0 = \mathbb{R}^q\}$ a single open set, $y_0 = \mathrm{id}$ and γ_{00}^w equal to the germ at $y_0(w) = w$ of the identity, for all $w \in \mathbb{R}^q$. Let

$$g_t : \mathbb{R}^q \to \mathbb{R}^q, \quad g_t(w) = tw, \quad 0 \le t \le 1,$$

and prove that $g_0^*([c]) \ne g_1^*([c])$. (Hint. Use Exercise 6.4.8.) This proves that Haefliger structures are not homotopy invariant.

In light of these two exercises, we can define $\Gamma^q(X)$ to be the set of homotopy classes in $H^1(X;\Gamma^q)$, obtaining a contravariant functor which is homotopy invariant. If c is a cocycle, $\langle c \rangle$ will denote the homotopy class of $[c]$.

We turn to a description of the *normal bundle* of an element of $\Gamma^q(X)$. In light of the cocycle approach to bundle theory outlined above, this will be easy.

Let $[c] \in H^1(X;\Gamma^q)$ be represented by a cocycle

$$c = \{U_\alpha, y_\alpha, \gamma_{\alpha\beta}\}_{\alpha,\beta\in\mathfrak{A}}.$$

Define

$$g_{\alpha\beta} : U_\alpha \cap U_\beta \to G^q$$

by $g_{\alpha\beta}(w) = d(\gamma_{\alpha\beta}^w)$. This is a G^q-cocycle in the sense of bundle theory, giving a recipe for building an oriented, real q-plane bundle over X. We denote this cocycle by dc. The coherence relation between Γ^q-cocycles c and c'

induces the corresponding coherence relation between the associated G^q-cocycles dc and dc' which is necessary and sufficient for them to define isomorphic q-plane bundles over X. The (isomorphism class of the) bundle so obtained is called the normal bundle of the Haefliger structure $[c]$ and will be denoted by Q_c. Note that, by the homotopy invariance of vector bundles, Q_c depends only on the homotopy class $\zeta = \langle c \rangle \in \Gamma^q(X)$ of $[c]$, hence can be denoted by Q_ζ

Exercise 6.4.13. If $f : X \to Y$ is continuous and $\zeta \in \Gamma^q(Y)$ has normal bundle Q_ζ, prove that the normal bundle $Q_{f^*(\zeta)}$ of $f^*(\zeta)$ is the pull-back $f^*(Q_\zeta)$.

Exercise 6.4.14. If $X = M$ is a manifold and the cocycle c corresponds to a foliation $\mathcal{F}$ integral to $E \subset T(M)$, prove that the normal bundle Q_c is isomorphic to the normal bundle $Q = T(M)/E$ as already defined. In particular, if $f : N \to M$ is a smooth map, transverse to $\mathcal{F}$, note that the previous exercise agrees with the fact that the normal bundle Q' of $\mathcal{F}' = f^*(\mathcal{F})$ is the pull-back $f^*(Q)$ of the normal bundle Q of $\mathcal{F}$.

6.4.D. The Haefliger classifying space for $\Gamma^q(X)$. In analogy with Milnor's classification theory for vector bundles, Haefliger has shown [**86**] that the functor $\Gamma^q(X)$ is *representable*. That is, there are a topological space $B\Gamma^q$ and an element $\omega_q \in \Gamma^q(B\Gamma^q)$ such that, for each *paracompact* space X, the corresondence $[f] \leftrightarrow f^*(\omega_q)$ defines a canonical identification

$$[X, B\Gamma^q] = \Gamma^q(X).$$

Let Q^q denote the normal bundle of ω_q. Then there is a classifying map $\nu : B\Gamma^q \to BG^q$, unique up to homotopy, such that $\nu^*(\mathcal{E}^q) \cong Q^q$.

This structure has been used by Haefliger [**85**] to classify the set $\mathrm{Fol}^q(M)$ of integrable homotopy classes of codimension q foliations on an *open* manifold M (see [I, 3.6.B] for the definition of integrable homotopy). For this, he appeals to a very deep theorem of M. Gromov and A. Phillips. An account by V. Poénaru of this latter theorem will be found in [**151**].

We give an outline of this theory. Let

$$g : M \to BG^n$$

be the classifying map for the tangent bundle of M. This is also called the Gauss map of M. Let $\mathcal{F}$ be a foliation of M of codimension q and let $Q \subset T(M)$ be the normal bundle of $\mathcal{F}$. Corresponding to the splitting $T(M) = Q \oplus T\mathcal{F}$, there is a natural homotopy commutative diagram

$$\begin{array}{ccc} & & BG^q \times BG^{n-q} \\ & \overset{\overline{g}}{\nearrow} & \Big\downarrow p \\ M & \xrightarrow[g]{} & BG^n \end{array}$$

where p classifies the bundle $\mathcal{E}^q \times \mathcal{E}^{n-q}$.

Since Q is the normal bundle of $\mathcal{F}$, there is a homotopy commutative diagram

$$\begin{array}{ccc} & & B\Gamma^q \times BG^{n-q} \\ & \overset{\widetilde{g}}{\nearrow} & \Big\downarrow \nu \times \mathrm{id} \\ M & \xrightarrow[\overline{g}]{} & BG^q \times BG^{n-q} \end{array}$$

and composing these diagrams gives a homotopy commutative triangle

$$\begin{array}{ccc} & & B\Gamma^q \times BG^{n-q} \\ & \overset{\widetilde{g}}{\nearrow} & \Big\downarrow \rho \\ M & \xrightarrow[g]{} & BG^n \end{array}$$

Theorem 6.4.15 (Haefliger). *If the manifold M is open, the last of these diagrams sets up a one-to-one correspondence between $\mathrm{Fol}^q(M)$ and the set of homotopy classes of homotopy lifts $\widetilde{g}$ of the Gauss map g of M.*

The requirement that the manifold be open is needed for the application of the Phillips-Gromov theorem. By quite different methods, W. Thurston has obtained analogous results without this requirement [**174, 172**]. Here is a statement of his basic theorem.

Theorem 6.4.16 (Thurston). *Let M be a manifold, let $\zeta \in \Gamma^q(M)$ have normal bundle Q_ζ, and suppose that there is a bundle monomorphism*

$$i : Q_\zeta \to T(M).$$

Then there is a foliation $\mathcal{F}$ of codimension q, with normal bundle homotopic to $i(Q_\zeta)$, such that $\langle \mathcal{F} \rangle \in \zeta$.

In the case of codimension one, this leads to a particularly strong theorem. Indeed, it can be shown that every one-dimensional vector bundle is the normal bundle of a Haefliger structure, and a closed manifold M admits a one-dimensional subbundle of its tangent bundle if and only if $\chi(M) = 0$.

Corollary 6.4.17. *A closed manifold M admits a foliation of codimension one if and only if $\chi(M) = 0$.*

There are also relative versions of this theorem (the case of codimension 1 being special because of Reeb stability) and Thurston gives many beautiful corollaries.

Finally, it should be noted that there is a natural homomorphism

$$\lambda : H^*(\mathrm{WO}_q) \to H^*(B\Gamma^q),$$

and so there is a universal interpretation of the characteristic classes of foliations that extends to all elements of $\Gamma^q(X)$. It is an open question of long standing as to whether this map is injective.

Chapter 7

The Godbillon-Vey Classes

In this chapter we will prove the celebrated theorem of G. Duminy that codimension one foliations without resilient leaves have $\mathrm{gv}(\mathcal{F}) = 0$ (Theorem 7.3.1). We will also prove vanishing theorems of S. Hurder, including his partial extension of Duminy's result to all generalized Godbillon-Vey classes (Theorem 7.4.1). Hurder's ideas will also be useful for simplifying Duminy's proof. These results show that the exotic classes are sensitive to delicate qualitative properties of foliated manifolds.

It should be remarked that a new proof of Duminy's theorem has been obtained recently by Hurder and R. Langevin [**104**]. This proof does not use the theory of levels and gives considerable insight into the ergodic-theoretic basis of the theorem. Here, we will follow Duminy, giving the proof that uses the theory of levels, largely because we do not want this (unpublished) proof to be lost (but *cf.* [**20**]) and because it contains a wealth of geometric insight into the vanishing phenomenon. Of course, this proof also has ergodic-theoretic features.

7.1. The Godbillon Class and Measure Theory

We fix the hypotheses that M is a compact, connected, oriented n-manifold without boundary and that $\mathcal{F}$ is a smooth, transversely oriented foliation of codimension q. Recall that the generalized Godbillon-Vey classes $\lambda_{\mathcal{F}}[y_I c_J]$ are characterized by the property that y_1 is a factor of y_I and $\deg c_J = 2q$. It was a clever idea of Duminy to isolate the y_1 part of the class as a certain kind of (dual) cohomology class in its own right and then to "localize" this as a cohomology-valued measure on the σ-algebra of Lebesgue measurable, $\mathcal{F}$-saturated subsets of M. Duminy used this to analyze the Godbillon-Vey class for foliations of codimension one, but the same idea is useful for the generalized Godbillon-Vey classes.

7.1.A. The Godbillon Class. Let $\omega \in A^q(M)$ define $\mathcal{F}$, $d\omega = \eta \wedge \omega$. The principal ideal

$$A^*(M, \mathcal{F}) = A^*(M) \wedge \omega$$

depends only on the foliation, not otherwise on the choice of ω. Furthermore,

$$d(\gamma \wedge \omega) = (d\gamma \pm \gamma \wedge \eta) \wedge \omega,$$

so $(A^*(M, \mathcal{F}), d)$ is a cochain complex. The cohomology of this complex will be denoted by $H^*(M; \mathcal{F})$. Classes in this algebra will be denoted by $[\gamma]_{\mathcal{F}}$ and those in $H^*(M) = H^*(M; \mathbb{R})$ by $[\gamma]$.

Example 7.1.1. In a foliated chart (U, x, y), $\omega|U = f\, dy^1 \wedge \cdots \wedge dy^q$. Clearly

$$0 = d^2(\omega) = d(\eta \wedge \omega) = d\eta \wedge \omega,$$

and we conclude that

$$d\eta|U = \sum_{i=1}^{q} \eta_i \wedge dy^i,$$

for suitable 1-forms η_i on U. Consequently, the closed form $(d\eta)^q$ is in the space $A^{2q}(M, \mathcal{F})$. Following Duminy, we call the class

$$\mathrm{v}(\mathcal{F}) = [(d\eta)^q]_{\mathcal{F}} \in H^{2q}(M, \mathcal{F})$$

the *Vey class* of $\mathcal{F}$.

Lemma 7.1.2. *For each $k \geq 0$, a linear map*

$$\mathrm{g}(\mathcal{F}) : H^k(M, \mathcal{F}) \to H^{k+1}(M)$$

is well defined by

$$\mathrm{g}(\mathcal{F})[\alpha]_{\mathcal{F}} = [\eta \wedge \alpha]$$

and depends only on $\mathcal{F}$, not on the admissible choices of ω and η. The linear map $\mathrm{g}(\mathcal{F})$ is called the Godbillon class *of $\mathcal{F}$.*

Proof. First of all, we remark that $\eta \wedge \alpha$ is a closed form. Indeed, α is closed and $\alpha = \gamma \wedge \omega$, so

$$d(\eta \wedge \alpha) = d\eta \wedge \alpha = d\eta \wedge \gamma \wedge \omega = 0.$$

Suppose that $d\omega = \eta \wedge \omega = \widetilde{\eta} \wedge \omega$. If $\alpha = \gamma \wedge \omega \in A^k(M, \mathcal{F})$, then

$$\eta \wedge \alpha = \eta \wedge \gamma \wedge \omega = \widetilde{\eta} \wedge \gamma \wedge \omega = \widetilde{\eta} \wedge \alpha.$$

Thus, ω being fixed, the map is independent of the allowable choices of η. The form $\widetilde{\omega}$ also defines $\mathcal{F}$ if and only if $\widetilde{\omega} = f\omega$, where $f \in C^\infty(M)$ is nowhere zero, so

$$\begin{aligned} d\widetilde{\omega} &= \widetilde{\eta} \wedge \widetilde{\omega}, \\ \widetilde{\eta} &= \eta + d\log|f|. \end{aligned}$$

If $[\alpha]_\mathcal{F} \in H^k(M, \mathcal{F})$, then $d\alpha = 0$ and

$$\widetilde{\eta} \wedge \alpha = \eta \wedge \alpha + d(\log|f|\alpha),$$

so

$$[\widetilde{\eta} \wedge \alpha] = [\eta \wedge \alpha].$$

Finally, an arbitrary representative of $[\alpha]_\mathcal{F}$ is of the form $\widetilde{\alpha} = \alpha + d(\gamma \wedge \omega)$. Therefore,

$$\begin{aligned} \eta \wedge \widetilde{\alpha} &= \eta \wedge (\alpha + d(\gamma \wedge \omega)) \\ &= \eta \wedge \alpha + \eta \wedge (d\gamma \wedge \omega \pm \gamma \wedge \underbrace{\eta \wedge \omega}_{d\omega}) \\ &= \eta \wedge \alpha + \eta \wedge d\gamma \wedge \omega \\ &= \eta \wedge \alpha \pm d(\gamma \wedge d\omega), \end{aligned}$$

proving that $[\eta \wedge \widetilde{\alpha}] = [\eta \wedge \alpha]$. □

The term "Godbillon class" is also due to Duminy, who observes that

$$\mathrm{g}(\mathcal{F})\,\mathrm{v}(\mathcal{F}) = \mathrm{gv}(\mathcal{F}).$$

Lemma 7.1.3. *If* $\deg c_J = 2q$, *the closed form* $\lambda_{\mathcal{F},\nabla^0,\nabla^1}(y_I c_J)$ *is an element of* $A^*(M, \mathcal{F})$.

Proof. Since ∇^1 is basic, the $2q$-form $\lambda(\nabla^1)(c_J) = c_J(\Omega^1)$ has the property that

$$c_J(\Omega^1)(v_1 \wedge \ldots \wedge v_{2q}) = 0$$

whenever $q+1$ of the v_i's are tangent to $\mathcal{F}$. Therefore, $c_J(\Omega^1) \in A^{2q}(M, \mathcal{F})$ and $\lambda_{\mathcal{F},\nabla^0,\nabla^1}(y_I c_J) = \lambda_{\mathcal{F},\nabla^0,\nabla^1}(y_I) \wedge c_J(\Omega^1)$ is in the ideal $A^*(M, \mathcal{F})$. □

Corollary 7.1.4. *If* $\mathrm{g}(\mathcal{F}) \equiv 0$, *then all generalized Godbillon-Vey classes of* $\mathcal{F}$ *vanish.*

Indeed, from Example 6.3.9, recall that $\lambda_{\mathcal{F},\nabla^0,\nabla^1}(y_1) = -\eta$. Consequently,

$$\begin{aligned} \lambda_\mathcal{F}[y_1 y_I c_J] &= [-\eta \wedge \lambda_{\mathcal{F},\nabla^0,\nabla^1}(y_I c_J)] \\ &= -\mathrm{g}(\mathcal{F})[\lambda_{\mathcal{F},\nabla^0,\nabla^1}(y_I c_J)]_\mathcal{F} \\ &= 0. \end{aligned}$$

Definition 7.1.5. The dual class $\mathrm{g}(M,\mathcal{F}) \in H^{n-1}(M,\mathcal{F})^*$ is defined by

$$\mathrm{g}(M,\mathcal{F})[\alpha]_{\mathcal{F}} = \int_M \eta \wedge \alpha, \quad \forall\, [\alpha]_{\mathcal{F}} \in H^{n-1}(M,\mathcal{F}).$$

In other words, $\mathrm{g}(M,\mathcal{F})$ is the composition

$$H^{n-1}(M,\mathcal{F}) \xrightarrow{\mathrm{g}(\mathcal{F})} H^n(M) \xrightarrow{\int_M} \mathbb{R}.$$

Since $\int_M$ is an isomorphism, this contains exactly the same information as $\mathrm{g}(\mathcal{F})$ on $H^{n-1}(M,\mathcal{F})$. In fact,

Lemma 7.1.6. *The homomorphism* $\mathrm{g}(\mathcal{F}) : H^k(M,\mathcal{F}) \to H^{k+1}(M)$ *is completely determined by* $\mathrm{g}(M,\mathcal{F}) \in H^{n-1}(M,\mathcal{F})^*$, $\forall\, k \geq 0$.

Proof. By Poincaré duality, the class $[\gamma] \in H^{k+1}(M)$ is uniquely determined by the numbers

$$\int_M \gamma \wedge \beta, \quad \forall\, [\beta] \in H^{n-k-1}(M).$$

Thus, for $[\alpha]_{\mathcal{F}} \in H^k(M,\mathcal{F})$, $\mathrm{g}(\mathcal{F})[\alpha]_{\mathcal{F}} = [\eta \wedge \alpha]$ is uniquely determined by the numbers

$$\int_M \eta \wedge \alpha \wedge \beta = \mathrm{g}(M,\mathcal{F})[\alpha \wedge \beta]_{\mathcal{F}}, \quad \forall\, [\beta] \in H^{n-k-1}(M),$$

hence by $\mathrm{g}(M,\mathcal{F}) \in H^{n-1}(M,\mathcal{F})^*$. □

7.1.B. The Godbillon Measure. In order to prove vanishing theorems for the generalized Godbillon-Vey classes, it will be enough to prove them for the class $\mathrm{g}(M,\mathcal{F})$. This is the class that will be "localized" as a σ-additive, $H^{n-1}(M,\mathcal{F})^*$-valued measure on the σ-ring $\mathcal{B}(\mathcal{F})$ of $\mathcal{F}$-saturated, Lebesgue measurable sets.

Definition 7.1.7. If $B \in \mathcal{B}(\mathcal{F})$, the linear functional

$$\mathrm{g}(B,\mathcal{F}) : H^{n-1}(M,\mathcal{F}) \to \mathbb{R}$$

is defined by

$$\mathrm{g}(B,\mathcal{F})[\alpha]_{\mathcal{F}} = \int_B \eta \wedge \alpha, \quad \forall\, [\alpha]_{\mathcal{F}} \in H^{n-1}(M,\mathcal{F}).$$

Lemma 7.1.8. *For each* $B \in \mathcal{B}(\mathcal{F})$, *the class* $\mathrm{g}(B,\mathcal{F}) \in H^{n-1}(M,\mathcal{F})^*$ *is well defined by the above formula, independently of the allowed choices of* ω, η *and of the representatives* α *of classes* $[\alpha]_{\mathcal{F}} \in H^{n-1}(M,\mathcal{F})$.

Proof. As before, if $\omega, \widetilde{\omega}$ define $\mathcal{F}$, if

$$d\omega = \eta \wedge \omega,$$
$$d\widetilde{\omega} = \widetilde{\eta} \wedge \widetilde{\omega},$$

and if $\alpha, \widetilde{\alpha} \in [\alpha]_{\mathcal{F}} \in H^{n-1}(M, \mathcal{F})$, then

$$\widetilde{\eta} \wedge \widetilde{\alpha} = \eta \wedge \alpha + d(\gamma \wedge \omega).$$

Therefore, we must show that, if $\tau \in A^{n-1}(M, \mathcal{F})$, then $\int_B d\tau = 0$. Let $\{U_\alpha, x_\alpha, y_\alpha\}_{\alpha \in \mathfrak{A}}$ be a finite, regular, foliated atlas for $\mathcal{F}$, so coordinatized that $\overline{U}_\alpha = I^{n-q} \times I^q$. Let $\{\lambda_\alpha\}_{\alpha \in \mathfrak{A}}$ be a smooth partition of unity subordinate to this open cover. Then

$$\tau = \sum_{\alpha \in \mathfrak{A}} \tau_\alpha,$$

where $\tau_\alpha = \lambda_\alpha \tau$ is compactly supported in U_α. It is clear that we can write $\tau_\alpha = \gamma_\alpha \wedge dy_\alpha$, where γ_α is compactly supported in U_α and $dy_\alpha = dy_\alpha^1 \wedge \ldots \wedge dy_\alpha^q$. Therefore, $d\tau_\alpha = d\gamma_\alpha \wedge dy_\alpha$. Since $B \in \mathcal{B}(\mathcal{F})$,

$$B \cap \overline{U}_\alpha = I^{n-q} \times \widetilde{B},$$

where $\widetilde{B} \subseteq I^q$ is a Lebesgue measurable set. Therefore,

$$\begin{aligned} \int_B d\tau_\alpha &= \int_{\widetilde{B}} \left(\int_{I^{n-q} \times \{y\}} d\gamma_\alpha \right) dy_\alpha \\ &= \int_{\widetilde{B}} \left(\int_{\partial I^{n-q} \times \{y\}} \gamma_\alpha \right) dy_\alpha \\ &= 0, \end{aligned}$$

since $\gamma_\alpha | \partial I^{n-q} \times \{y\} \equiv 0$, $\forall\, y \in I^q$. Thus

$$\int_B d\tau = \sum_{\alpha \in \mathfrak{A}} \int_B d\tau_\alpha = 0,$$

by additivity of the integral. □

If

$$B = \bigcup_{k=1}^{\infty} B_k,$$

where the sets $B_k \in \mathcal{B}(\mathcal{F})$ are disjoint, then the σ-additivity of the integral gives

$$\mathrm{g}(B, \mathcal{F})[\alpha]_{\mathcal{F}} = \int_B \eta \wedge \alpha = \sum_{k=1}^{\infty} \int_{B_k} \eta \wedge \alpha = \sum_{k=1}^{\infty} \mathrm{g}(B_k, \mathcal{F})[\alpha]_{\mathcal{F}}.$$

That is, we can write

$$\mathrm{g}(B, \mathcal{F}) = \sum_{k=1}^{\infty} \mathrm{g}(B_k, \mathcal{F}),$$

obtaining a σ-additive function

$$\mathrm{g}(\cdot, \mathcal{F}) : \mathcal{B}(\mathcal{F}) \to H^{n-1}(M, \mathcal{F})^*.$$

Definition 7.1.9. The σ-additive function $\mathrm{g}(\cdot, \mathcal{F})$ is called the Godbillon measure associated to $\mathcal{F}$.

At this point, it is possible to outline the proof of Duminy's theorem. Here, the foliation is of codimension one, transversely orientable and smooth. The theory of levels [I, §8.3] for such foliations can be employed to produce a decomposition

$$M = B_1 \cup B_2 \cup \cdots \cup B_k \cup \cdots \cup B_\infty$$

into disjoint elements of $\mathcal{B}(\mathcal{F})$ such that B_k is either a local minimal set of $\mathcal{F}$ or empty, $1 \le k < \infty$. The set B_∞ will be "infinitely thin" in a sense (due to Duminy) to be defined later. This set may have positive Lebesgue measure, but we will see that $\mathrm{g}(B_\infty, \mathcal{F}) = 0$. In the absence of resilient leaves, no B_k can be exceptional and, if B_k is an open local minimal set, its leaves must have trivial holonomy. If B_k reduces to a single proper leaf, it has Lebesgue measure 0; hence $\mathrm{g}(B_k, \mathcal{F}) = 0$. Another basic result will be that, if B_k is open without holonomy, then $\mathrm{g}(B_k, \mathcal{F}) = 0$. We then obtain

$$\mathrm{g}(M, \mathcal{F}) = \mathrm{g}(B_\infty, \mathcal{F}) + \sum_{k \ge 1} \mathrm{g}(B_k, \mathcal{F}) = 0.$$

That is, the Godbillon class $\mathrm{g}(\mathcal{F}) = 0$, and hence

$$\mathrm{gv}(\mathcal{F}) = \mathrm{g}(\mathcal{F})\,\mathrm{v}(\mathcal{F}) = 0.$$

7.1.C. Infinitesimal Holonomy and the Godbillon Class. For foliations of codimension one, [I, Exercise 2.3.16] asserts that the form η, restricted to each leaf L, is a closed form representing the infinitesimal holonomy class of L. Following Hurder [**102**], we use infinitesimal holonomy in arbitrary codimension, encoded as a certain kind of cocycle, to give an explicit formula for η, hence for the Godbillon class. Hurder's methods amount to an elegant derivation of the Bott-Thurston formula for $\mathrm{gv}(\mathcal{F})$ [**10**, Proposition 2.25].

Fix a regular foliated atlas $\mathcal{U} = \{(U_\alpha, x_\alpha, y_\alpha)\}_{\alpha \in \mathfrak{A}}$ with associated holonomy cocycle $\{\gamma_{\alpha\beta}\}_{\alpha,\beta \in \mathfrak{A}}$. The transverse space S associated to this atlas [I, p. 57] can be viewed at our convenience as a submanifold of M or as an open subset of $\mathbb{R}^q$. The holonomy pseudogroup $\Gamma_{\mathcal{U}}$ will be denoted simply by Γ. Viewing $S \subset \mathbb{R}^q$, we denote by $J\gamma(y) \in \mathrm{Gl}(q, \mathbb{R})$ the Jacobian matrix of $\gamma \in \Gamma$ at $y \in \operatorname{dom} \gamma$. This is the infinitesimal holonomy of γ at y. Since we assume that $\mathcal{F}$ is transversely oriented, the matrices $J\gamma$ have positive determinants, so we can define

$$\nu(\gamma) = \log \det J\gamma, \quad \forall\, \gamma \in \Gamma,$$

a smooth, real-valued function on $\operatorname{dom}\gamma$. Observe that $\nu(\gamma \circ \theta) = \nu(\gamma) + \nu(\theta)$.

Definition 7.1.10. For $\alpha, \beta \in \mathfrak{A}$,

$$\zeta_{\beta\alpha} : U_\alpha \cap U_\beta \to \mathbb{R}$$

is the smooth map

$$\zeta_{\beta\alpha} = \nu(\gamma_{\beta\alpha}) \circ y_\alpha.$$

The cocycle property for $\{\gamma_{\beta\alpha}\}_{\alpha,\beta\in\mathfrak{A}}$ implies a cocycle property

$$(*) \qquad \zeta_{\beta\alpha} + \zeta_{\alpha\delta} = \zeta_{\beta\delta} \quad \text{on } U_\alpha \cap U_\beta \cap U_\delta,$$

as the reader will easily check. (This is a Čech 1-cocycle with values in the presheaf of smooth functions.) Fix a smooth partition of unity $\{\lambda_\alpha\}_{\alpha\in\mathfrak{A}}$, subordinate to $\mathcal{U}$ and, for each $\alpha \in \mathfrak{A}$, define

$$\eta_\alpha = \sum_{\delta\in\mathfrak{A}} \zeta_{\delta\alpha} d\lambda_\delta \in A^1(U_\alpha).$$

Lemma 7.1.11. *On $U_\alpha \cap U_\beta$, η_α and η_β agree, $\forall\, \alpha, \beta \in \mathfrak{A}$, so these forms fit together to define a 1-form η on M.*

Proof. On $U_\alpha \cap U_\beta$

$$\begin{aligned}
\eta_\alpha &= \sum_\delta \zeta_{\delta\alpha} d\lambda_\delta + \zeta_{\alpha\beta} \underbrace{\left(\sum_\delta d\lambda_\delta\right)}_{=d(1)\equiv 0} \\
&= \sum_\delta (\zeta_{\delta\alpha} + \zeta_{\alpha\beta})\, d\lambda_\delta \\
&= \sum_\delta \zeta_{\delta\beta}\, d\lambda_\delta \qquad (\text{by } (*)) \\
&= \eta_\beta.
\end{aligned}$$

□

On $S \subset \mathbb{R}^q$, let $dy = dy^1 \wedge \cdots \wedge dy^q$ and define

$$\omega_\alpha = \exp\left(\sum_\delta \zeta_{\delta\alpha}\lambda_\delta\right) y_\alpha^*(dy) \in A^q(U_\alpha).$$

Lemma 7.1.12. *On $U_\alpha \cap U_\beta$, ω_α and ω_β agree, so these forms fit together to yield $\omega \in A^q(M)$, a form which defines $\mathcal{F}$.*

Proof. On $U_\alpha \cap U_\beta$,

$$\begin{aligned}
\omega_\alpha &= \exp\Big(\sum_\delta \zeta_{\delta\alpha}\lambda_\delta\Big) y_\alpha^*(dy) \\
&= \exp\Big(\sum_\delta (\zeta_{\delta\alpha} + \zeta_{\alpha\beta})\lambda_\delta\Big) \exp(-\zeta_{\alpha\beta}) y_\alpha^*(dy) \\
&= \exp\Big(\sum_\delta \zeta_{\delta\beta}\lambda_\delta\Big) \exp(-\nu(\gamma_{\alpha\beta}) \circ y_\beta)(\gamma_{\alpha\beta} \circ y_\beta)^*(dy) \quad \text{(by } (*)\text{)} \\
&= \exp\Big(\sum_\delta \zeta_{\delta\beta}\lambda_\delta\Big) y_\beta^*(\exp(-\nu(\gamma_{\alpha\beta}))\gamma_{\alpha\beta}^*(dy)) \\
&= \exp\Big(\sum_\delta \zeta_{\delta\beta}\lambda_\delta\Big) y_\beta^*(dy) \\
&= \omega_\beta.
\end{aligned}$$

The next to last equality is seen by noting that

$$\exp(-\nu(\gamma_{\alpha\beta})) = \frac{1}{\det J\gamma_{\alpha\beta}}$$

and

$$\gamma_{\alpha\beta}^*(dy) = (\det J\gamma_{\alpha\beta})dy.$$

Since $y_\alpha^*(dy)$ defines $\mathcal{F}|U_\alpha$, $\forall\, \alpha \in \mathfrak{A}$, and exp is never zero, ω defines $\mathcal{F}$. □

Lemma 7.1.13. *The forms defined above satisfy $d\omega = \eta \wedge \omega$.*

Proof. For each $\alpha \in \mathfrak{A}$,

$$\begin{aligned}
d\omega_\alpha &= d\Big(\exp\Big(\sum_\delta \zeta_{\delta\alpha}\lambda_\delta\Big) y_\alpha^*(dy)\Big) \\
&= d\Big(\exp\Big(\sum_\delta \zeta_{\delta\alpha}\lambda_\delta\Big)\Big) \wedge y_\alpha^*(dy) \\
&= \exp\Big(\sum_\delta \zeta_{\delta\alpha}\lambda_\delta\Big) d\Big(\sum_\delta \zeta_{\delta\alpha}\lambda_\delta\Big) \wedge y_\alpha^*(dy) \\
&= \Big(\sum_\delta (\zeta_{\delta\alpha} d\lambda_\delta + d(\zeta_{\delta\alpha})\lambda_\delta)\Big) \wedge \omega_\alpha
\end{aligned}$$

on U_α. But $\zeta_{\delta\alpha} = \nu(\gamma_{\delta\alpha}) \circ y_\alpha$ is constant along each leaf of $\mathcal{F}|(U_\alpha \cap U_\delta)$; hence

$$d\zeta_{\delta\alpha} \wedge \omega_\alpha \equiv 0 \quad \text{on } U_\alpha \cap U_\delta$$

and

$$d(\zeta_{\delta\alpha})\lambda_\delta \wedge \omega_\alpha \equiv 0 \quad \text{on } U_\alpha,$$

so

$$d\omega_\alpha = \left(\sum_\delta \zeta_{\delta\alpha} d\lambda_\delta\right) \wedge \omega_\alpha = \eta_\alpha \wedge \omega_\alpha \quad \text{on } U_\alpha.$$

Since the formulas $d\omega_\alpha = \eta_\alpha \wedge \omega_\alpha$ agree on overlaps, the assertion follows. □

The moral is that all information about the Godbillon class and Godbillon measure is stored in ν.

7.1.D. Infinitesimal Holonomy and Groupoids. Recall that a groupoid $\mathcal{G}$ is a category with inverses [I, Definition 2.3.3]. A groupoid is assumed to be a "small" category, meaning that its objects form a set X in the Zermelo-Frankel sense. We say that $\mathcal{G}$ is a groupoid on X and write $\mathcal{G}_x^y$ for the set of morphisms in $\mathcal{G}$ from x to y. If $x \in X$, the orbit $\mathcal{G}(x)$ is the set of $y \in X$ such that $\mathcal{G}_x^y \neq \emptyset$. One way in which a groupoid arises is as the set of germs of elements of a pseudogroup. Another example is a group G, which can be viewed as a groupoid on a singleton $\{*\}$.

Definition 7.1.14. If $\mathcal{G}_1$ and $\mathcal{G}_2$ are groupoids with set of objects X_1 and X_2, respectively, a covariant functor $\varphi : \mathcal{G}_1 \to \mathcal{G}_2$ is called a homomorphism of groupoids. If

$$\varphi | X_1 = f : X_1 \to X_2,$$

the homomorphism φ is said to cover f.

Example 7.1.15. If Γ is the holonomy pseudogroup of $(M, \mathcal{F})$ and $\gamma \in \Gamma$, the Jacobian matrix $J\gamma(x)$ at $x \in \operatorname{dom}\gamma$ depends only on the germ $\widehat{\gamma}_x$ in the holonomy groupoid $\mathcal{G}$ of the foliation. If we view the general linear group as a groupoid, this defines a groupoid homomorphism

$$J : \mathcal{G} \to \mathrm{Gl}_+(q, \mathbb{R})$$

covering the constant map $S \to \{*\}$. Similarly, ν passes to a groupoid homomorphism

$$\nu : \mathcal{G} \to \mathbb{R}$$

covering the constant map.

An equivalence relation $\mathcal{R}$ on a set X can also be viewed as a groupoid on X. Formally, $\mathcal{R} \subset X \times X$ is identified with the set of ordered pairs (x, y) such that $x\mathcal{R}y$. The morphisms are

$$\mathcal{R}_y^x = \begin{cases} \{(x,y)\}, & \text{if } x\mathcal{R}y, \\ \emptyset, & \text{otherwise.} \end{cases}$$

The composition law in this groupoid is

$$\mathcal{R}_y^x \times \mathcal{R}_z^y \xrightarrow{\circ} \mathcal{R}_z^x,$$
$$(x,y) \circ (y,z) = (x,z),$$

each (x, x) being an identity and $(x, y)^{-1} = (y, x)$. The orbit $\mathcal{R}(x)$ is the equivalence class of $x \in X$.

A groupoid $\mathcal{G}$ on X induces an equivalence relation $\mathcal{R}$ on X, the equivalence classes $\mathcal{R}(x)$ being the orbits $\mathcal{G}(x)$, $\forall\, x \in X$. There is a "forgetful" functor

$$\begin{aligned} &\varphi : \mathcal{G} \to \mathcal{R}, \\ &\varphi(\gamma) = (y, x) \Leftrightarrow \gamma \in \mathcal{G}_x^y. \end{aligned}$$

Clearly, this is a groupoid homomorphism covering $\mathrm{id} : X \to X$.

We fix the assumption that $\mathcal{G}$ is the holonomy groupoid of $(M, \mathcal{F})$ on the transverse space S and $\mathcal{R}$ is the associated equivalence relation. The above abstract nonsense gives the setting for a fairly marvelous fact.

Proposition 7.1.16. *For Lebesgue almost every $x \in S$, all $y \in \mathcal{R}(x)$ and all $\widehat{\gamma}_y \in \mathcal{G}$, $J\gamma(y)$ depends only on $\varphi(\widehat{\gamma}_y) = (\gamma(y), y) \in \mathcal{R}$. In particular, almost every leaf of $\mathcal{F}$ has trivial infinitesimal holonomy group.*

That is, for almost every leaf L and every pair of points $y, z \in L \cap S$, there is a Jacobian matrix $J(z, y) \in \mathrm{Gl}_+(q, \mathbb{R})$ well defined as $J\gamma_s(y)$ for every path s in L from y to z. In particular, if the notion of "groupoid homomorphism" is weakened to allow homomorphisms that are only well defined on almost every orbit, then J and ν descend to groupoid homomorphisms

$$\begin{aligned} &J : \mathcal{R} \to \mathrm{Gl}_+(q, \mathbb{R}), \\ &\nu : \mathcal{R} \to \mathbb{R}. \end{aligned}$$

As we will see, this result does not require that M be compact nor that $(M, \mathcal{F})$ satisfy any orientation requirements.

In what follows, we use $|X|$ to denote the Lebesgue measure of a set $X \subset \mathbb{R}^q$ and denote by $D_x(r)$ the ball in $\mathbb{R}^q$ of radius r centered at x. The proof of Proposition 7.1.16 uses the following Lebesgue density theorem.

Theorem 7.1.17 (Lebesgue). *Let $X \subseteq \mathbb{R}^q$ with $|X| > 0$. Then almost every $x \in X$ is a point of density 1. That is, for almost every $x \in X$ and all sequences $r_k \downarrow 0$,*

$$\lim_{k \to \infty} \frac{|X \cap D_x(r_k)|}{|D_x(r_k)|} = 1.$$

Proof of Proposition 7.1.16. We emphasize that, for the moment, M is not assumed to be compact. First, we prove that almost every infinitesimal holonomy group vanishes. Let

$$X = \{x \in S \mid J(\mathcal{G}_x^x) \neq \{\mathrm{id}\}\}.$$

We assume that $|X| > 0$ and deduce a contradiction. The holonomy cocycle $\{\gamma_{\alpha\beta}\}_{\alpha,\beta\in\mathfrak{A}}$ is at most countably infinite. Thus the set of elements $\gamma \in \Gamma$ that can be written

$$\gamma = \gamma_{\alpha_r\alpha_{r-1}} \circ \cdots \circ \gamma_{\alpha_1\alpha_0}$$

and have maximal domain is at most countably infinite. Enumerate this set as $\{\gamma_k\}_{k=1}^{\infty}$. Then every germ $\widehat{\gamma}_x \in \mathcal{G}$ is the germ at x of some γ_k, $k \geq 1$. Define

$$X_k = \{x \in S \mid \gamma_k(x) = x \text{ and } J\gamma_k(x) \neq \mathrm{id}\}.$$

Evidently,

$$X = \bigcup_{k=1}^{\infty} X_k,$$

so $|X_k| > 0$ for some $k \geq 1$. Fix such a value of k and choose a point $x \in X_k$ of density 1. It follows that, for almost every $v \in S^{q-1}$, $\{x + tv\}_{t>0} \cap X$ clusters at x. In particular, we can choose a basis $\{v_1, \ldots, v_q\} \subset S^{q-1}$ of $\mathbb{R}^q$ so that the set $Y_i = X_k \cap \{x + tv_i\}_{t>0}$ clusters at x, $1 \leq i \leq q$. Since γ_k fixes $Y_i \subset X_k$ pointwise, it follows that $J\gamma_k(x) \cdot v_i = v_i$, $1 \leq i \leq k$, contradicting the fact that $x \in X_k$.

At this point, we have proven that the infinitesimal holonomy group $J(\mathcal{G}_x^x)$ is trivial for almost every $x \in S$. Since the infinitesimal holonomy groups at two points of $\mathcal{G}(x)$ are conjugate, this triviality holds for *all* points in almost every orbit. Finally, if $\mathcal{G}(x)$ is such an orbit, $y \in \mathcal{G}(x)$, and if $\widehat{\gamma}_y$ and $\widehat{\theta}_y \in \mathcal{G}$ have the same target $z = \gamma(y) = \theta(y)$, it follows that

$$J\theta(y)^{-1} \cdot J\gamma(y) = J(\theta^{-1})(z) \cdot J\gamma(y) = J(\theta^{-1} \circ \gamma)(y) = \mathrm{id};$$

hence $J\theta(y) = J\gamma(y)$, as asserted. □

7.1.E. The Ergodic Theory of ν. We will replace the smooth data

$$\nu : \mathcal{G} \to \mathbb{R}$$

with merely measurable data. If $X \subseteq S$ is Lebesgue measurable, then

$$|X| = \int_X dy,$$

so $dy = d|\cdot|$ is the differential of Lebesgue measure. Given an arbitrary measurable function $h : S \to \mathbb{R}$, another measure μ on S is defined by

$$\mu(X) = \int_X e^h dy.$$

Thus, $d\mu = e^h dy$. Evidently, μ has exactly the same sets of measure zero as does Lebesgue measure.

Definition 7.1.18. Two measures on a σ-algebra have the same measure class if they have exactly the same sets of measure zero.

This defines an equivalence relation. We restrict ourselves to measures on $X \subseteq S$ that are finite-valued. Here, X will be measurable and Γ-invariant. The Lebesgue measure class $\mathfrak{L}_X$ of such measures is also called the class of absolutely continuous measures. By the Radon-Nikodym theorem, it is characterized as follows.

Proposition 7.1.19. *A finite-valued measure μ on $X \subseteq S$ belongs to the Lebesgue measure class $\mathfrak{L}_X$ if and only if there is a function $h : X \to \mathbb{R}$, measurable and defined Lebesgue almost everywhere, such that $d\mu = e^h dy$.*

To motivate the technical discussion in this subsection, we state the following vanishing theorem [**102**]. It will be proven in the following subsection.

Theorem 7.1.20 (Hurder). *If the Lebesgue measure class $\mathfrak{L} = \mathfrak{L}_S$ on S contains a Γ-invariant measure, then all generalized Godbillon-Vey classes of $\mathcal{F}$ vanish.*

Consider an arbitrary measure μ on the σ-algebra of (Lebesgue) measurable subsets of X. Given $\gamma \in \Gamma$, we obtain a measure μ_γ on $X \cap \operatorname{dom} \gamma$ by setting

$$\mu_\gamma(Y) = \int_{\gamma(Y)} d\mu,$$

for every measurable subset $Y \subseteq X \cap \operatorname{dom} \gamma$. We write $d\mu_\gamma = \gamma^* d\mu$; hence

$$\int_{\gamma(Y)} d\mu = \int_Y \gamma^* d\mu.$$

In particular, it is obvious that

$$\gamma^* dy = e^{\nu(\gamma)} dy,$$

and we have the following more general formula for the way that Γ preserves the Lebesgue measure class $\mathfrak{L}_X$.

Lemma 7.1.21. *Let $d\mu = e^h dy$, where $h : X \to \mathbb{R}$ is defined Lebesgue almost everywhere and is measurable, and let $\gamma \in \Gamma$. Then*

$$\gamma^* d\mu = e^{\nu(\gamma) + h\circ\gamma - h} d\mu$$

on $Y = X \cap \operatorname{dom} \gamma$.

Proof. Indeed,

$$\begin{aligned}\int_{\gamma(Y)} d\mu &= \int_{\gamma(Y)} e^h dy \\ &= \int_Y e^{h\circ\gamma}\gamma^* dy \\ &= \int_Y e^{\nu(\gamma)+h\circ\gamma} dy \\ &= \int_Y e^{\nu(\gamma)+h\circ\gamma-h} e^h dy \\ &= \int_Y e^{\nu(\gamma)+h\circ\gamma-h} d\mu.\end{aligned}$$

□

Note. In what follows, we will let $\mathcal{G}_X$ and $\mathcal{R}_X$ denote the respective restrictions of the groupoids $\mathcal{G}$ and $\mathcal{R}$ to a measurable, Γ-invariant subset $X \subseteq S$. Likewise, we will write $\Gamma_X = \Gamma|X$.

For $x \in X$, set

$$\mathcal{G}_x = \bigcup_{y\in X} \mathcal{G}_x^y,$$

$$\mathcal{R}_x = \bigcup_{y\in X} \mathcal{R}_x^y.$$

We will say that a function φ defined on $\mathcal{G}_x$ (respectively, on $\mathcal{R}_x$), for (Lebesgue) almost every $x \in X$, is defined almost everywhere on $\mathcal{G}_X$ (respectively, on $\mathcal{R}_X$).

Definition 7.1.22. If $\varphi : \mathcal{G}_X \to \mathbb{R}$ (respectively, $\varphi : \mathcal{R}_X \to \mathbb{R}$) is defined almost everywhere and if $\gamma \in \Gamma$, then

$$\varphi(\gamma) : \operatorname{dom}\gamma \to \mathbb{R}$$

is the function defined almost everywhere by $\varphi(\gamma)(x) = \varphi(\widehat{\gamma}_x)$ (respectively, $\varphi(\gamma(x), x)$).

Definition 7.1.23. A cocycle on $\mathcal{G}_X$ is a groupoid homomorphism

$$a : \mathcal{G}_X \to \mathbb{R}$$

which is defined almost everywhere. A cocycle on $\mathcal{R}_X$ is defined similarly. Such cocycles are measurable (respectively, integrable) if

$$a(\gamma) : \operatorname{dom}\gamma \to \mathbb{R}$$

is measurable (respectively, integrable), $\forall\, \gamma \in \Gamma_X$.

For example, ν is a measurable cocycle both on $\mathcal{G}$ and on $\mathcal{R}$.

Remark. The cocycle condition

$$a(\widehat{\gamma}_{\theta(x)}\widehat{\theta}_x) = a(\widehat{\gamma}_{\theta(x)}) + a(\widehat{\theta}_x)$$

translates to the formula

$$a(\gamma \circ \theta) = a(\gamma) \circ \theta + a(\theta), \quad \text{on } \theta^{-1}(\operatorname{dom}\gamma) \cap \operatorname{dom}\theta,$$

where $\gamma, \theta \in \Gamma_X$.

Lemma 7.1.24. *The cocycle ν is integrable.*

Proof. We may as well prove this on $X = S$. Recall that $S \subset \mathbb{R}^q$ is relatively compact. Also, each element $\gamma_{\alpha\beta}$ of the holonomy cocycle has relatively compact domain $S_{\alpha\beta}$ and extends to a diffeomorphism on $\overline{S}_{\alpha\beta}$. Thus, $\nu(\gamma_{\alpha\beta})$ is bounded and continuous on $S_{\alpha\beta}$, hence integrable. Since the holonomy cocycle generates Γ, it follows that $\nu(\gamma)$ is integrable, $\forall\, \gamma \in \Gamma$. □

If a is a measurable cocycle on $\mathcal{G}_X$ and $h : X \to \mathbb{R}$ is a measurable function, one readily checks that the formula

$$b(\widehat{\gamma}_x) = a(\widehat{\gamma}_x) + h(\gamma(x)) - h(x), \quad \forall\, \widehat{\gamma}_x \in \mathcal{G}_X,$$

defines a measurable cocycle b on $\mathcal{G}_X$. A similar remark holds for cocycles on $\mathcal{R}_X$, assuring us that the following definition is sensible.

Definition 7.1.25. We say that measurable cocycles a and b on $\mathcal{G}_X$ are cohomologous, and write $a \sim b$, if there is a measurable function $h : X \to \mathbb{R}$, defined Lebesgue almost everywhere, such that

$$a(\widehat{\gamma}_x) = b(\widehat{\gamma}_x) + h(\gamma(x)) - h(x), \quad \text{for almost every } x \in X, \quad \forall\, \widehat{\gamma}_x \in \mathcal{G}_x.$$

Similarly, the measurable cocycles on $\mathcal{R}_X$ are cohomologous if there is such an h and

$$a(y,x) = b(y,x) + h(y) - h(x), \quad \text{for almost every } x \in X, \forall\, y \in \mathcal{R}_X(x).$$

If we wish to call attention to the function h, we will write $a \sim_h b$. Cohomology is an equivalence relation. The set of measurable cocycles on $\mathcal{G}_X$ or $\mathcal{R}_X$ forms an abelian group under addition, as do the integrable cocycles, and the cohomology relation respects this group structure (but does not preserve integrability).

Definition 7.1.26. The abelian groups $H(\mathcal{G}_X)$ and $H(\mathcal{R}_X)$ consist of the cohomology classes $[a]$ of measurable cocycles on $\mathcal{G}_X$ and $\mathcal{R}_X$, respectively. The subgroups $H'(\mathcal{G}_X) \subseteq H(\mathcal{G}_X)$ and $H'(\mathcal{R}_X) \subseteq H(\mathcal{R}_X)$ consist of those classes $[a]$ such that there exists an integrable cocycle $b \in [a]$.

Thus, $[\nu] \in H'(\mathcal{G})$ (and $H'(\mathcal{R})$), but not every cocycle $a \in [\nu]$ is integrable.

Remark. The standard construction of cohomology $H^*(K)$ of a group K can be extended to groupoids $\mathcal{K}$. Our cocycles are 1-cocycles in this cohomology theory, with the added requirement that the cocycles be measurable. The corresponding cohomology is denoted by $H^*_{\text{meas}}(\mathcal{K})$, $\mathcal{K} = \mathcal{G}_X$ or $\mathcal{R}_X$. The (measurable) function h is a 0-cochain and its coboundary $\delta(h)$ is the 1-cochain defined by

$$\delta(h)(\widehat{\gamma}_x) = h(\gamma(x)) - h(x)$$

when $\mathcal{K} = \mathcal{G}_X$ is the holonomy groupoid. For the groupoid $\mathcal{K} = \mathcal{R}_X$,

$$\delta(h)(y, x) = h(y) - h(x).$$

Thus, $H(\mathcal{G}_X) = H^1_{\text{meas}}(\mathcal{G}_X)$ and $H(\mathcal{R}_X) = H^1_{\text{meas}}(\mathcal{R}_X)$. One can define cohomology using *integrable* data instead of measurable. This cohomology has a natural (functorial) map to the measurable theory, and our groups $H'(\mathcal{G}_X)$ and $H'(\mathcal{R}_X)$ are the images of this map at the H^1 level.

As for cohomology of groups, groupoid homomorphisms induce homomorphisms (in a contravariant way) in cohomology. In our case, the measurable homomorphism $\varphi : \mathcal{G}_X \to \mathcal{R}_X$ (the forgetful functor) induces homomorphisms

$$\varphi^* : H(\mathcal{R}_X) \to H(\mathcal{G}_X),$$
$$\varphi^* : H'(\mathcal{R}_X) \to H'(\mathcal{G}_X).$$

Evidently, $[\nu] \in H'(\mathcal{G})$ is the image of $[\nu] \in H'(\mathcal{R})$ under φ^*. The inclusion maps $\mathcal{G}_X \hookrightarrow \mathcal{G}$ and $\mathcal{R}_X \hookrightarrow \mathcal{R}$ are also groupoid homomorphisms, pulling the class $[\nu]$ back to a class $[\nu]_X$.

Proposition 7.1.27. *For every cocycle $a \in [\nu]_X$ there is a measure $\mu_a \in \mathfrak{L}_X$ such that*

$$\gamma^*(d\mu_a) = e^{a(\gamma)} d\mu_a, \quad \forall \gamma \in \Gamma_X.$$

While this measure is not uniquely determined by $a \in [\nu]_X$, every measure in the Lebesgue measure class $\mathfrak{L}_X$ is of the form μ_a, for some $a \in [\nu]_X$.

Proof. As $h : X \to \mathbb{R}$ ranges over the measurable functions, $e^h dy$ ranges over $d\mu$ in the Lebesgue measure class $\mathfrak{L}_X$. Set $d\mu_a = e^h dy$, where $a \sim_h \nu$. By Lemma 7.1.21, this correspondence has the required property. Since h is not uniquely determined by a, μ_a is not unique, but every $\mu \in \mathfrak{L}$ is of the form $e^h dy$, h measurable, so $\mu = \mu_a$ for $a \sim_h \nu$. □

In our earlier construction of the Godbillon measure, we used the cocycle ν. We mimic this construction, replacing ν with an arbitrary measurable cocycle $a : \mathcal{G}_X \to \mathbb{R}$ (or, equally well, $a : \mathcal{R}_X \to \mathbb{R}$). It is *not* assumed that $a \in [\nu]$. In the following discussion, we may as well take $X = S$. The general case follows by extending all measures, functions, *etc.*, to be trivial outside of X.

Define

$$\zeta^a_{\beta\alpha} : U_\alpha \cap U_\beta \to \mathbb{R},$$
$$\zeta^a_{\beta\alpha} = a(\gamma_{\beta\alpha}) \circ y_\alpha.$$

The cocycle property for $\{\gamma_{\beta\alpha}\}_{\alpha,\beta\in\mathfrak{A}}$ and the fact that a is a homomorphism imply the cocycle property

$$\zeta^a_{\beta\alpha} + \zeta^a_{\alpha\delta} = \zeta^a_{\beta\delta}, \quad \text{on } U_\alpha \cap U_\beta \cap U_\delta.$$

The functions $\zeta^a_{\beta\alpha}$ are measurable and they are constant along the local leaves of $\mathcal{F}|U_\alpha \cap U_\beta$. If the cocycle a is integrable, so are all $\zeta^a_{\beta\alpha}$. As before, using a smooth partition of unity subordinate to the regular atlas, we define a "1-form"

$$\eta^a_\alpha = \sum_{\beta\in\mathfrak{A}} \zeta^a_{\beta\alpha} d\lambda_\beta$$

on U_α, $\forall\, \alpha \in \mathfrak{A}$. This 1-form is measurable (respectively, integrable) on U_α because a is measurable (respectively, integrable). Since the forms $d\lambda_\beta$ are smooth and the functions $\zeta^a_{\beta\alpha}$ are constant along local leaves, η^a_α is smooth along plaques. The proof of Lemma 7.1.11 goes through without change to prove

Lemma 7.1.28. *The forms η^a_α and η^a_β agree on $U_\alpha \cap U_\beta$, $\forall\, \alpha, \beta \in \mathfrak{A}$. Hence they define a measurable form η^a on M which is smooth along the leaves of $\mathcal{F}$.*

Remark. The notion of measurable k-form makes sense on a smooth manifold, being defined by measurability of coefficients in local coordinate charts. The notion is *well* defined, since the local coordinate changes, being smooth, preserve measurability. On compact, oriented n-manifolds, the notion of *integrability* makes sense for measurable n-forms ψ, the requirement being that the Lebesgue integral $\int_M \psi$ be defined and finite. For measurable k-forms ψ, integrability will mean that $\int_M \psi \wedge \theta$ is defined and finite, $\forall\, \theta \in A^{n-k}(M)$. The notion of a k-form that is smooth along leaves also makes sense in foliated charts and is invariant under changes of coordinates between two charts in a regular, foliated atlas. Since two such atlases have a common refinement which is regular, smoothness along leaves is a notion depending only on the foliation.

Definition 7.1.29. The space of integrable forms that are smooth along the leaves of $\mathcal{F}$ is denoted by $A^*_{\mathcal{F}}(M)$. We set

$$A^*_{\mathcal{F}}(M, \mathcal{F}) = A^*_{\mathcal{F}}(M) \wedge \omega = A^*_{\mathcal{F}}(M) \wedge A^*(M, \mathcal{F}).$$

Let (U, x, y) be a foliated chart, $\varphi \in A^k(M)$, and write

$$\varphi|U = \sum_{I,J} f_{IJ} dx^I \wedge dy^J$$

for suitable multi-indices I and J. In the local formula for $d\varphi \wedge \omega$, only the terms in $d\varphi$ of the form $f_I dx^I$ $(J = \emptyset)$ are involved. Observe also that the local formula for $d\varphi \wedge \omega$ involves only derivatives $\partial f_I / \partial x^i$, $1 \leq i \leq n - q$. Thus, if $\varphi \in A^k_{\mathcal{F}}(M)$, the local formula for $d\varphi \wedge \omega$ is defined, even though $d\varphi|U$ itself is *not* defined! On overlapping charts of a regular foliated atlas, the reader can check that the two local formulas for $d\varphi \wedge \omega$ agree, so $d\varphi \wedge \omega$ is well defined globally, $\forall\, \varphi \in A^k_{\mathcal{F}}(M)$. This is the key point in the proof of the following lemma.

Lemma 7.1.30. *There is a canonical extension of the exterior derivative on $A^*(M, \mathcal{F})$ to*

$$d : A^k_{\mathcal{F}}(M, \mathcal{F}) \to A^{k+1}_{\mathcal{F}}(M, \mathcal{F}), \quad \forall\, k \geq 0,$$

with the properties that $d^2 = 0$ and $d(\psi \wedge \theta) = d\psi \wedge \theta + (-1)^{\deg \psi} \psi \wedge d\theta$, for all $\psi \in A^*_{\mathcal{F}}(M)$ and all $\theta \in A^*(M, \mathcal{F})$.

Exercise 7.1.31. Prove Lemma 7.1.30. Show that this proof has as a corollary the "leafwise Stokes' theorem" of Heitsch and Hurder [**92**], stated below as Lemma 7.1.32. (Hint. You will need to formulate and prove a plaquewise version of Stokes' theorem. Then proceed as in the proof of Lemma 7.1.8).

Lemma 7.1.32. *If $\tau \in A^{n-1}_{\mathcal{F}}(M, \mathcal{F})$ and $B \in \mathcal{B}(\mathcal{F})$, then*

$$\int_B d\tau = 0.$$

Given an *integrable* cocycle $a : \mathcal{G} \to \mathbb{R}$ (or $a : \mathcal{R} \to \mathbb{R}$), we define a σ-additive function

$$\mathrm{g}^a(\cdot, \mathcal{F}) : \mathcal{B}(\mathcal{F}) \to H^{n-1}(M, \mathcal{F})^*$$

by

$$\mathrm{g}^a(B, \mathcal{F})[\psi]_{\mathcal{F}} = \int_B \eta^a \wedge \psi, \quad \forall\, [\psi]_{\mathcal{F}} \in H^{n-1}(M, \mathcal{F}), \quad \forall\, B \in \mathcal{B}(\mathcal{F}).$$

Lemma 7.1.33. *The above formula well defines $\mathrm{g}^a(B, \mathcal{F})$, $\forall\, B \in \mathcal{B}(\mathcal{F})$.*

Proof. The elements $\widetilde{\psi} \in [\psi]_{\mathcal{F}}$ are exactly the 1-forms

$$\widetilde{\psi} = \psi + d\rho, \quad \rho \in A^{n-2}(M, \mathcal{F}).$$

Thus,

$$\int_B \eta^a \wedge \widetilde{\psi} = \int_B \eta^a \wedge \psi + \int_B \eta^a \wedge d\rho,$$

and we must show that $\int_B \eta^a \wedge d\rho = 0$. By property (2) in Lemma 7.1.30,

$$d(\eta^a \wedge \rho) = d\eta^a \wedge \rho - \eta^a \wedge d\rho$$

and

$$d\eta^a \wedge \rho = d\Big(\sum_{\beta \in \mathfrak{A}} \zeta^a_{\beta\alpha} d\lambda_\beta\Big) \wedge \rho = \Big(\sum_{\beta \in \mathfrak{A}} \zeta^a_{\beta\alpha} d^2\lambda_\beta\Big) \wedge \rho = 0.$$

Here, we use the fact that $\zeta^a_{\beta\alpha}$ is constant along plaques. Therefore,

$$\int_B \eta^a \wedge d\rho = -\int_B d(\eta^a \wedge \rho) = 0$$

by Lemma 7.1.32. □

Thus, we have a σ-additive $H^{n-1}(M, \mathcal{F})^*$-valued measure g^a on $\mathcal{B}(\mathcal{F})$, depending only on the integrable cocycle a. The following key result is due to Hurder [**102**].

Theorem 7.1.34. *The measure* g^a *depends only on the class* $[a] \in H'(\mathcal{G})$ *(respectively,* $[a] \in H'(\mathcal{R})$*).*

Proof. Let $b \sim_h a$, where h is measurable, but a and b are both integrable. We must prove $\mathrm{g}^a = \mathrm{g}^b$. We will prove this first for the case that h is bounded, hence integrable. The general case will be deduced from this via a sequence $\{h^N\}_{N=1}^\infty$ of bounded, measurable functions converging pointwise to h.

Assume that h is bounded and let $[\psi]_{\mathcal{F}} \in H^{n-1}(M, \mathcal{F})$. On U_α,

$$\begin{aligned}(\eta^b - \eta^a) \wedge \psi &= \Big(\sum_{\beta \in \mathfrak{A}} (\zeta^b_{\beta\alpha} - \zeta^a_{\beta\alpha})\, d\lambda_\beta\Big) \wedge \psi \\ &= \sum_{\beta \in \mathfrak{A}} (b(\gamma_{\beta\alpha}) \circ y_\alpha - a(\gamma_{\beta\alpha}) \circ y_\alpha)\, d\lambda_\beta \wedge \psi \\ &= \sum_{\beta \in \mathfrak{A}} (h \circ \gamma_{\beta\alpha} \circ y_\alpha - h \circ y_\alpha)\, d\lambda_\beta \wedge \psi.\end{aligned}$$

But $y_\beta = \gamma_{\beta\alpha} \circ y_\alpha$, so

$$\begin{aligned}(\eta^b - \eta^a) \wedge \psi &= \sum_{\beta \in \mathfrak{A}} (h \circ y_\beta)\, d\lambda_\beta \wedge \psi - \Big(\sum_{\beta \in \mathfrak{A}} d\lambda_\beta\Big) \wedge (h \circ y_\alpha)\psi \\ &= \sum_{\beta \in \mathfrak{A}} (h \circ y_\beta)\, d\lambda_\beta \wedge \psi \\ &= d\Big(\sum_{\beta \in \mathfrak{A}} \lambda_\beta \cdot (h \circ y_\beta)\psi\Big),\end{aligned}$$

since $h \circ y_\beta$ is constant along plaques and $d\psi = 0$. Since h is bounded, it follows that

$$\rho = \sum_{\beta \in \mathfrak{A}} \lambda_\beta \cdot (h \circ y_\beta)\psi$$

is integrable and Lemma 7.1.32 implies

$$\int_B (\eta^b - \eta^a) \wedge \psi = \int_B d\rho = 0, \quad \forall\, B \in \mathcal{B}(\mathcal{F}).$$

Since $[\psi]_\mathcal{F} \in H^{n-1}(M, \mathcal{F})$ is arbitrary, $\mathrm{g}^b = \mathrm{g}^a$.

Allow h to be unbounded. For each positive integer N, define h^N on S by

$$h^N(x) = \begin{cases} h(x), & |h(x)| \leq N, \\ \operatorname{sign}(h(x))N, & |h(x)| > N. \end{cases}$$

Note that $|h^N(x) - h^N(y)| \leq |h(x) - h(y)|$, $\forall\, x, y \in S$. Furthermore, h^N is integrable and bounded and we can define an integrable cocycle $b^N \sim_{h^N} a$. By the previous paragraph, $\mathrm{g}^{b^N} = \mathrm{g}^a$, $\forall\, N \in \mathbb{Z}^+$. Thus,

$$\int_B \eta^{b^N} \wedge \psi$$

is independent of N and we can write

$$\int_B \eta^a \wedge \psi = \lim_{N \to \infty} \int_B \eta^{b^N} \wedge \psi, \quad \forall\, B \in \mathcal{B}(\mathcal{F}), \quad \forall\, [\psi]_\mathcal{F} \in H^{n-1}(M, \mathcal{F}).$$

In order to prove that

$$\int_B \eta^b \wedge \psi = \lim_{N \to \infty} \int_B \eta^{b^N} \wedge \psi,$$

we will use the Lebesgue dominated convergence theorem.

Write $(U_\alpha, x_\alpha, y_\alpha) = (U_\alpha, z_\alpha^1, \dots, z_\alpha^n)$, $\forall\, \alpha \in \mathfrak{A}$. Given $\varphi \in A^1_\mathcal{F}(M)$, write

$$\varphi | U_\alpha = \sum_{i=1}^n f_i^\alpha dz_\alpha^i$$

and define a measurable function $|\varphi| : M \to \mathbb{R}$ by

$$|\varphi|(z) = \max\{|f_i^\alpha(z)| \mid 1 \leq i \leq n,\ \alpha \in \mathfrak{A} \text{ such that } z \in U_\alpha\}.$$

Similarly, given an integrable cocycle $c : \mathcal{G} \to \mathbb{R}$, define the integrable function $|c| : M \to \mathbb{R}$ by

$$|c|(z) = \max\{|c(\gamma_{\beta\alpha})(y_\alpha(z))| \mid \alpha, \beta \in \mathfrak{A} \text{ such that } z \in U_\alpha \cap U_\beta\}.$$

For an integrable cocycle $c : \mathcal{R} \to \mathbb{R}$, the definition is

$$|c|(z) = \max\{|c(y_\beta(z), y_\alpha(z))| \mid \alpha, \beta \in \mathfrak{A} \text{ such that } z \in U_\alpha \cap U_\beta\}.$$

Finally, define

$$K = \max_{z \in M} \sum_{\beta \in \mathfrak{A}} |d\lambda_\beta|(z).$$

We use these quantities to define an integrable upper bound for $\left|\eta^{b^N}\right|$ which is independent of N. For $\alpha \in \mathfrak{A}$ and $z \in U_\alpha$,

$$\begin{aligned}
\left|\eta^{b^N}\right|(z) &= \left|\sum_{\beta \in \mathfrak{A}} \zeta V^{b^N}_{\beta\alpha} \, d\lambda_\beta\right|(z) \\
&\leq K \max_{\alpha,\beta} \left|b^N(\gamma_{\beta\alpha})(y_\alpha(z))\right| \\
&= K \max_{\alpha,\beta} \left|a(\gamma_{\beta\alpha})(y_\alpha(z)) + h^N(\gamma_{\beta\alpha}(y_\alpha(z))) - h^N(y_\alpha(z))\right| \\
&\leq K|a|(y_\alpha(z)) + K \max_{\alpha,\beta} \left|h^N(\gamma_{\beta\alpha}(y_\alpha(z))) - h^N(y_\alpha(z))\right|.
\end{aligned}$$

But

$$\begin{aligned}
\left|h^N(\gamma_{\beta\alpha}(y_\alpha(z))) - h^N(y_\alpha(z))\right| &\leq \left|h(\gamma_{\beta\alpha}(y_\alpha(z))) - h(y_\alpha(z))\right| \\
&= \left|b(\gamma_{\beta\alpha})(y_\alpha(z)) - a(\gamma_{\beta\alpha})(y_\alpha(z))\right| \\
&\leq |b|(z) + |a|(z).
\end{aligned}$$

Thus,

$$\left|\eta^{b^N}\right|(z) \leq 2K|a|(z) + K|b|(z).$$

This holds for all $z \in U_\alpha$ and all $\alpha \in \mathfrak{A}$, so $2K|a| + K|b|$ is the desired integrable upper bound.

For each fixed $\psi \in A^{n-1}(M, \mathcal{F})$ with $d\psi = 0$ we have

$$\lim_{N \to \infty} \eta^{b^N} \wedge \psi = \eta^b \wedge \psi,$$

pointwise on M, so the dominated convergence theorem gives

$$\lim_{N \to \infty} \int_B \eta^{b^N} \wedge \psi = \int_B \eta^b \wedge \psi,$$

completing the proof that $\int_B \eta^a \wedge \psi = \int_B \eta^b \wedge \psi$. □

7.1.F. Some Vanishing Theorems. The cohomology class

$$[\nu]_X \in H'(\mathcal{G}_X) \text{ (respectively, } \in H'(\mathcal{R}_X))$$

determines the Godbillon measure $\mathrm{g}\,|B = \mathrm{g}^\nu\,|B$ on the $\mathcal{F}$-saturation B of the Γ-invariant set X. When $X = S$, we get the Godbillon measure on M. Vanishing theorems for the generalized Godbillon-Vey classes are proven by showing that $[\nu]$ is "trying to be 0". The easiest case is that in which $[\nu] = 0$.

Proposition 7.1.35. *There is a Γ_X-invariant measure μ in the Lebesgue measure class $\mathfrak{L}_X$ on X if and only if $[\nu]_X = 0$.*

Proof. By Proposition 7.1.27, $[\nu]_X = 0$ (that is, $0 \in [\nu]_X$) if and only if there exists a measure $\mu_0 \in \mathfrak{L}_X$ such that

$$\gamma^* d\mu_0 = e^0 d\mu_0 = d\mu_0, \quad \forall \gamma \in \Gamma_X.$$

Equivalently, μ_0 is Γ_X-invariant. □

Proof of Theorem 7.1.20. If there is a Γ-invariant measure in $\mathfrak{L} = \mathfrak{L}_S$, we have $0 \in [\nu]$. By Theorem 7.1.34, $\mathrm{g}^\nu = \mathrm{g}^0 = 0$ and, in particular,

$$\mathrm{g}(M, \mathcal{F}) = \mathrm{g}^\nu(M, \mathcal{F}) = 0.$$

By Lemma 7.1.6, $\mathrm{g}(\mathcal{F}) = 0$; hence all generalized Godbillon-Vey classes vanish (Corollary 7.1.4). □

In fact, if there is a Γ_X-invariant measure in $\mathfrak{L}_X$, we can conclude that the restriction $\mathrm{g}^{\nu_X} = \mathrm{g}\,|B$ of the Godbillon measure to the $\mathcal{F}$-saturation B of X is trivial.

Conjecture. *If there is a Γ_U-invariant measure μ on an open, Γ-invariant subset $U \subseteq S$, not necessarily in the Lebesgue class, but finite on compact sets, nonatomic, and strictly positive on open sets, then the Godbillon measure g vanishes on the $\mathcal{F}$-saturation of U.*

A crucial step in the proof of Duminy's theorem will be to verify this conjecture in codimension one.

Definition 7.1.36. Let $B \in \mathcal{B}(\mathcal{F})$ and $X = B \cap S$. A measurable subset $Z \subseteq X$ is a section of $\mathcal{F}|B$ if almost every leaf of $\mathcal{F}|B$ meets Z in exactly one point.

Equivalently, Z is a section if $\mathcal{R}(x) \cap Z$ is a singleton, for almost every $x \in X$. Suppose that Z is a section of $\mathcal{F}|B$ and define $\rho : X \to Z$ almost everywhere by setting $\rho(x) = \mathcal{R}(x) \cap Z$ whenever this is a singleton. Define $h : X \to \mathbb{R}$ almost everywhere by setting

$$h(x) = \nu(\rho(x), x), \quad \text{for almost every } x \in X.$$

Lemma 7.1.37. *The function h is measurable.*

Proof. Consider all elements of Γ_X that are (restrictions to X of) pure compositions of the $\gamma_{\alpha\beta}$'s and have maximal possible domain. This is a countable set and we enumerate it as $\{\gamma_j\}_{j=1}^\infty$. Let $D_j = \operatorname{dom}\gamma_j$. Since Z is a section, the sets $X_j = \gamma_j(Z \cap D_j)$ unite to form a set that is almost all of X. On the measurable set X_j, h is defined almost everywhere by the measurable function $\nu(\gamma_j^{-1})$, so h is measurable. □

Theorem 7.1.38 (Hurder). *If $\mathcal{F}|B$ has a section, then $\mathrm{g}(B, \mathcal{F}) = 0$.*

Proof. Let h be as in the lemma and define $a \in [\nu]_X$ by $a \sim_h \nu$. Then, for almost every $y \in X$ and all $x \in \mathcal{R}(y)$, $\rho(x) = \rho(y)$ and

$$\begin{aligned} a(x,y) &= \nu(x,y) + h(x) - h(y) \\ &= \nu(x,y) + \nu(\rho(x),x) - \nu(\rho(y),y) \\ &= \nu(x,y) + \nu(y,\rho(y)) + \nu(\rho(y),x) \\ &= \nu(x,y) + \nu(y,x) \\ &= 0. \end{aligned}$$

Therefore, $0 = a \in [\nu]_X$ and the assertion follows. □

Corollary 7.1.39. *If the foliation $\mathcal{F}$ has a cross-section, all generalized Godbillon-Vey classes of $\mathcal{F}$ vanish.*

This corollary will be applied in the next section to prove that foliations with all leaves proper have trivial generalized Godbillon-Vey classes.

Finally, we turn to the notion of an ε-tempered cocycle [**103**], obtaining a sufficient condition for $\mathrm{g}(B,\mathcal{F}) = 0$ even when $[\nu]_X \neq 0$.

Recall the topological metric $d(y,z)$ on $\mathcal{R}(x)$ which we defined in [I, Subsection 12.2.C] in order to compute the growth of the leaf through x. This was the smallest integer N for which $y = \gamma(z)$ and

$$\gamma = \gamma_{\alpha_N \alpha_{N-1}} \circ \cdots \circ \gamma_{\alpha_1 \alpha_0}.$$

We view this as a function $d : \mathcal{R} \to \mathbb{R}$ and, by composition with the forgetful functor $\varphi : \mathcal{G} \to \mathcal{R}$, as a function $d : \mathcal{G} \to \mathbb{R}$. It will be notationally convenient to set $|\widehat{\gamma}_x| = d(\widehat{\gamma}_x)$ and $|y,x| = d(y,x)$.

Definition 7.1.40. Let $B \in \mathcal{B}(\mathcal{F})$, $X = B \cap S$, and let $\varepsilon > 0$. A cocycle $a : \mathcal{G}_X \to \mathbb{R}$ is ε-tempered if $|a(\widehat{\gamma}_x)| \leq \varepsilon |\widehat{\gamma}_x|$, for almost every $x \in X$ and $\forall\, \widehat{\gamma}_x \in \mathcal{G}_x$. Similarly, a cocycle $a : \mathcal{R}_X \to \mathbb{R}$ is ε-tempered if $|a(z,x)| \leq \varepsilon |z,x|$, for almost every $x \in X$ and $\forall\, z \in \mathcal{R}(x)$. If the ε-tempered cocycle a is obtained as the restriction of a cocycle b on $\mathcal{G}$ or $\mathcal{R}$, we say that b is ε-tempered on X.

In particular, a is ε-tempered if and only if $|a(\rho)| \leq \varepsilon$ whenever ρ is a germ of $\gamma_{\beta\alpha}$, $\forall\, \alpha, \beta \in \mathfrak{A}$. The following, therefore, is clear.

Lemma 7.1.41. *The cocycle $a : \mathcal{G}_X \to \mathbb{R}$ or $a : \mathcal{R}_X \to \mathbb{R}$ is ε-tempered if and only if*

$$\left|\zeta^a_{\beta\alpha}\right| = |a(\gamma_{\beta\alpha}) \circ y_\alpha| \leq \varepsilon,$$

uniformly (almost everywhere) on $B \cap U_\alpha \cap U_\beta$.

Note also that an ε-tempered cocycle is integrable.

Remark. If ν is ε-tempered on X, then

$$e^{-\varepsilon}|Y| \le |\gamma_{\beta\alpha}(Y)| \le e^{\varepsilon}|Y|, \quad \forall\, \beta, \alpha \in \mathfrak{A},$$

where $Y \subseteq X \cap \operatorname{dom}\gamma_{\beta\alpha}$ is measurable. Thus, the closer ε can be chosen to 0, the closer Lebesgue measure on X is to being Γ-invariant. Similarly, if $a \in [\nu]$ is ε-tempered on X, then $\mu_a \in \mathfrak{L}_X$ is $\gamma_{\beta\alpha}$-invariant up to a factor $e^{\pm\varepsilon}$, $\forall\, \alpha, \beta \in \mathfrak{A}$.

In the following argument, we define measurable functions $|\varphi| : M \to \mathbb{R}$ out of forms $\varphi \in A^*_{\mathcal{F}}(M)$ in analogy with similar definitions in the proof of Theorem 7.1.34.

Proposition 7.1.42. *Let $B \in \mathcal{B}(\mathcal{F})$, $X = B \cap S$, and $[a] \in H'(\mathcal{G})$ or $H'(\mathcal{R})$. If, for each $\varepsilon > 0$, there is a cocycle $a(\varepsilon) \in [a]$ which is ε-tempered on X, then $g^a(B, \mathcal{F}) = 0$.*

Proof. Let $[\psi]_{\mathcal{F}} \in H^{n-1}(M, \mathcal{F})$. Then

$$\left|\int_B \eta^a \wedge \psi\right| = \left|\int_B \eta^{a(\varepsilon)} \wedge \psi\right| \le \int_B \left|\eta^{a(\varepsilon)} \wedge \psi\right|.$$

On U_α,

$$\left|\eta^{a(\varepsilon)} \wedge \psi\right| \le \sum_{\beta\in\mathfrak{A}} |d\lambda_\beta| |\zeta^{a(\varepsilon)}_{\beta\alpha}| |\psi| \le \varepsilon \sum_{\beta\in\mathfrak{A}} |d\lambda_\beta| |\psi|.$$

Since $\alpha \in \mathfrak{A}$ is arbitrary,

$$\left|\eta^{a(\varepsilon)} \wedge \psi\right| \le \varepsilon \sum_{\beta\in\mathfrak{A}} |d\lambda_\beta| |\psi|,$$

almost everywhere on M, so

$$|\,\mathrm{g}^a(B, \mathcal{F})[\psi]_{\mathcal{F}}| \le \varepsilon \int_B \sum_{\beta\in\mathfrak{A}} |d\lambda_\beta| |\psi|.$$

As ε approaches 0, so does this quantity. Since $[\psi]_{\mathcal{F}} \in H^{n-1}(M, \mathcal{F})$ is arbitrary, $\mathrm{g}^a(B, \mathcal{F}) = 0$. □

Corollary 7.1.43. *If $[\nu]$ contains an ε-tempered cocycle, $\forall\, \varepsilon > 0$, then all generalized Godbillon-Vey classes vanish.*

Proposition 7.1.42 is critical for proving the deeper vanishing theorems of Hurder and simplifies the proof of Duminy's theorem.

7.2. Proper Foliations

Recall that a leaf L of $\mathcal{F}$ is proper if it is a properly imbedded submanifold of M. Equivalently, L is not asymptotic to itself. A foliation will be called proper if all of its leaves are proper. The following is proven in [**103**, Corollary 3.14], where proper foliations are called "type I" (terminology having to do with an associated C^*-algebra). The proof that we will give is modeled after Hurder's proof for the case of compact foliations [**101**].

Theorem 7.2.1 (Hurder and Katok). *If the foliation $\mathcal{F}$ is proper, then all of its generalized Godbillon-Vey classes vanish.*

By Corollary 7.1.39, we only need to prove the following.

Theorem 7.2.2. *A proper foliation admits a cross-section.*

The proof of Theorem 7.2.2 uses the Epstein-Millett filtration. This is the generalization to proper foliations, due to K. Millett [**128**], of a filtration invented by D. B. A. Epstein [**61**] to study foliations with all leaves compact. In order to describe this filtration, we need a definition.

Definition 7.2.3. Let $Y \subseteq M$ be an $\mathcal{F}$-saturated set and let $L \subseteq Y$ be a leaf. Then L is said to have locally trivial holonomy pseudogroup relative to Y if, $\forall x \in L \cap S$, there is a neighborhood of x in $Y \cap S$ that meets each leaf of $\mathcal{F}$ at most once.

Recall that $\mathcal{O}(\mathcal{F})$ denotes the family of open, $\mathcal{F}$-saturated subsets of M.

Theorem 7.2.4 (Millett [**128**]). *There are a countable ordinal $\gamma = \gamma(\mathcal{F})$ and a filtration*

$$\emptyset = U_0 \subset U_1 \subset \dots \subset U_\alpha \subset \dots \subset U_\gamma = M$$

by elements $U_\alpha \in \mathcal{O}(\mathcal{F})$, order-indexed by the ordinals $0 \le \alpha \le \gamma$, such that

(1) *U_α is dense in M, $0 < \alpha \le \gamma$;*

(2) *if $\beta \le \gamma$ is a limit ordinal, then*

$$U_\beta = \bigcup_{0<\alpha<\beta} U_\alpha;$$

(3) *for $0 \le \alpha < \gamma$, $U_{\alpha+1} \smallsetminus U_\alpha \neq \emptyset$ and is the union of all leaves of $\mathcal{F}$ that have locally trivial holonomy pseudogroup relative to $M \smallsetminus U_\alpha$.*

Proof of Theorem 7.2.2 using Theorem 7.2.4. Let $Y \in \mathcal{B}(\mathcal{F})$ and suppose that each leaf in Y has locally trivial holonomy pseudogroup relative to Y. Since $Y \cap S$ is second countable, there is a cover $\{V_k\}_{k=1}^\infty$ of $Y \cap S$

by relatively open subsets with the property that V_k meets each leaf of $\mathcal{F}$ at most once. Set $W_1 = V_1$ and, inductively, define

$$W_k = V_k \smallsetminus \Gamma\left(\bigcup_{i=1}^{k-1} W_i\right).$$

It is clear that $\bigcup_{k=1}^{\infty} W_k$ is a section of $\mathcal{F}|Y$. Since $Y_{\alpha+1} = U_{\alpha+1} \smallsetminus U_\alpha \in \mathcal{B}(\mathcal{F})$ and $\mathcal{F}|Y_{\alpha+1}$ has locally trivial holonomy pseudogroup relative to $Y_{\alpha+1}$, there is a section $X_{\alpha+1}$ of $\mathcal{F}|Y_{\alpha+1}$, $0 \le \alpha < \gamma$. But M is the disjoint union of the countable family $\{Y_{\alpha+1}\}_{0\le\alpha<\gamma}$, so $\bigcup_{0\le\alpha<\gamma} X_{\alpha+1}$ is a section of $\mathcal{F}$. □

Proof of Theorem 7.2.4. Let $d : S \times S \to [0, 1]$ be a metric, compatible with the topology of S and such that $d(x, y) = 1$ if and only if x and y lie in distinct components of S. We can also assume, by an appropriate choice of the regular foliated atlas, that S meets each leaf of $\mathcal{F}$ at least twice. For each $y \in S$, let L_y denote the leaf of $\mathcal{F}$ through y and define

$$\sigma(y) = \inf\{d(x, y) \mid x \neq y \text{ and } x \in L_y \cap S\}.$$

The function $\sigma : S \to [0, 1]$ is upper semicontinuous (Exercise 7.2.5). Furthermore, the fact that each leaf is proper implies that $\sigma > 0$ on all of S.

Let $x \in X \subseteq S$ and suppose that $\sigma|X$ is continuous at x. By continuity at x and the fact that $\sigma(x) > 0$, there is $\varepsilon > 0$ such that

$$y \in X \text{ and } d(x, y) < \varepsilon \Rightarrow \sigma(y) > 2\varepsilon.$$

Thus, the ε-neighborhood of x in X meets each leaf of $\mathcal{F}$ at most once.

Let $Y \subseteq M$ be a closed, $\mathcal{F}$-saturated set. By [**68**, Lemma 1.28, p. 39], the set of points of continuity of the upper semicontinuous function $\sigma|Y \cap S$ is a dense G_δ in $Y \cap S$. By the previous paragraph, it follows that the union of leaves of $\mathcal{F}|Y$ that have locally trivial holonomy pseudogroup relative to Y is relatively open and dense in Y.

By transfinite induction, we construct U_α, for each ordinal $\alpha \ge 0$. It is given that $U_0 = \emptyset$. If U_β has been constructed, $0 \le \beta < \alpha$, we proceed as follows.

(1) If α is a limit ordinal, set

$$U_\alpha = \bigcup_{0\le\beta<\alpha} U_\beta.$$

(2) If $\alpha = \beta + 1$ has an immediate predecessor β, we have shown that the union of leaves in $Y = M \smallsetminus U_\beta$ that have locally trivial holonomy pseudogroup relative to Y is relatively open and dense, so we define $U_{\beta+1} \smallsetminus U_\beta$ to be this set. It is obvious that $U_\alpha = U_{\beta+1}$ is as required.

Finally, we define γ to be the least ordinal such that $U_\gamma = U_{\gamma+1}$. This ordinal exists and is countable. Otherwise, M would be a strictly increasing union of uncountably many open subsets, contradicting second countability. □

Exercise 7.2.5. Prove that the function $\sigma : S \to (0,1]$, defined in the proof of Theorem 7.2.4, is upper semicontinuous.

Exercise 7.2.6. Prove that the leaf space $M/\mathcal{F}$ is a T_0 space if and only if $\mathcal{F}$ is a proper foliation.

Exercise 7.2.7. Prove that the existence of a filtration as in Theorem 7.2.4 implies that $\mathcal{F}$ is a proper foliation.

7.3. Codimension One

In this section, we assume that $\operatorname{codim}\mathcal{F} = 1$ and prove Duminy's theorem. In order to use the theory of levels and the results of Section 7.1, we assume $\mathcal{F}$ to be transversely orientable and M to be compact and orientable.

Theorem 7.3.1 (Duminy). *If* $\operatorname{codim}\mathcal{F} = 1$ *and no leaf is resilient, then* $\operatorname{gv}(\mathcal{F}) = 0$.

There are two intermediate propositions that must be proven. The first, Proposition 7.3.13, concerns "infinitely thin sets" or, in Duminy's terminology, sets $B \in \mathcal{B}(\mathcal{F})$ of "thickness" zero. The second, Proposition 7.3.14, concerns open, saturated sets with all leaves dense and without holonomy.

In what follows, we freely use the theory of levels and its associated notation, all as developed in [I, Chapter 8].

Recall that the union M_* of leaves at finite level is everywhere dense in M [I, Theorem 8.3.7]. Select the (finite) biregular atlas $\{U_\alpha, x_\alpha, y_\alpha\}_{\alpha\in\mathfrak{A}}$ so that the border plaques of $\overline{U}_\alpha = P_\alpha \times J_\alpha$ lie in M_*, $\forall\, \alpha \in \mathfrak{A}$. Thus, for some integer $k \geq 0$,

$$\bigcup_{\alpha\in\mathfrak{A}} P_\alpha \times \partial J_\alpha \subseteq M_k.$$

As usual, we select the y_α-coordinates so that $\overline{S}$ is a disjoint union of compact subintervals $J_\alpha = [a_\alpha, b_\alpha]$ of $\mathbb{R}$.

Definition 7.3.2. An element $W \in \mathcal{O}(\mathcal{F})$ is said to be thin if $W\cap\overline{S} = W\cap S$.

That is, W is thin precisely if none of the plaques $P_\alpha\times\{a_\alpha\}$ or $P_\alpha\times\{b_\alpha\}$ are contained in W. Thus, $\emptyset$ is thin, but M is not. If W is thin, let

$$W \cap \overline{S} = \bigsqcup_{i\in\mathcal{I}} (u_i, v_i) \quad \text{(disjoint union)},$$

where each (u_i, v_i) is an open interval in S and the index set $\mathcal{J}$ is at most countably infinite. The following assertions should be evident.

Lemma 7.3.3. *If W is thin and* $\operatorname{dom}\gamma_{\alpha\beta} \cap (u_i, v_i) \neq \emptyset$*, for some $i \in \mathcal{J}$, then* $(u_i, v_i) \subseteq \operatorname{dom}\gamma_{\alpha\beta}$.

Corollary 7.3.4. *If W is thin and γ is a pure composition of $\gamma_{\alpha\beta}$'s with maximal domain and if $(u_i, v_i) \cap \operatorname{dom}\gamma \neq \emptyset$, then $(u_i, v_i) \subseteq \operatorname{dom}\gamma$.*

Lemma 7.3.5. *For k sufficiently large, the $\mathcal{F}$-saturated set $M \smallsetminus M_k$ is thin.*

Definition 7.3.6. If $W \in \mathcal{O}(\mathcal{F})$, the thickness $\tau(W) \in [0, \infty]$ is defined by

(1) $\tau(W) = \infty$, if W is not thin;

(2) $\tau(\emptyset) = 0$;

(3) if W is thin and $W \cap \overline{S} = \bigsqcup_{i \in \mathcal{J}} (u_i, v_i)$, then

$$\tau(W) = \max_{i \in \mathcal{J}} (v_i - u_i).$$

Definition 7.3.7. The thickness of $X \in \mathcal{B}(\mathcal{F})$ is

$$\tau(X) = \inf_{X \subseteq W \in \mathcal{O}(\mathcal{F})} \tau(W).$$

Example 7.3.8. It is clear that $\tau(M \smallsetminus M_{k+1}) \leq \tau(M \smallsetminus M_k)$ and, since M_* is dense in M, that $\lim_{k \to \infty} \tau(M \smallsetminus M_k) = 0$. Consequently, the union M_∞ of leaves at infinite level has $\tau(M_\infty) = 0$. On the other hand, one easily constructs examples showing that the Lebesgue measure $|M_\infty|$ may be positive (Exercise 7.3.9). Furthermore, by [I, Corollary 8.3.24], a leaf at infinite level cannot be resilient and has trivial infinitesimal holonomy group.

Exercise 7.3.9. Use suspension of two diffeomorphisms of S^1 over the 2-holed torus Σ_2 to show that the Lebesgue measure of M_∞ may be positive.

Definition 7.3.10. Let $M = B_1 \cup B_2 \cup \cdots \cup B_k \cup \cdots \cup B_\infty$ be a decomposition of M into disjoint elements of $\mathcal{B}(\mathcal{F})$. This is a Duminy decomposition if

(1) $\tau(B_\infty) = 0$, and

(2) each B_k is either a local minimal set or $\emptyset$, $k < \infty$.

The following is standard by the theory of levels.

Lemma 7.3.11. *If $\mathcal{F}$ has no resilient leaves and $M = \bigcup_{1 \leq i \leq \infty} B_i$ is a Duminy decomposition, then B_i is either empty, a proper leaf, or an element of $\mathcal{O}(\mathcal{F})$ such that $\mathcal{F}|B_i$ has no holonomy, $1 \leq i < \infty$.*

Proposition 7.3.12. *A Duminy decomposition exists.*

Proof. Choose a countable dense subset $Z = \{z_j\}_{j=1}^{\infty} \subset M_* \cap \overline{S}$, making sure that the finite set $\{a_\alpha, b_\alpha\}_{\alpha\in\mathfrak{A}}$ is contained in Z. Let X_j denote the local minimal set containing z_j. By deleting repetitions, assume that $X_1, X_2, \dots, X_k, \dots$ are disjoint. It is possible that there are only finitely many of these local minimal sets, say $X_1, \dots, X_k$, in which case we set $X_j = \emptyset$, $j > k$. At any rate, it is clear that $\bigcup_{1\le j<\infty} X_j$ is dense in M.

If $X_i \ne \emptyset$, then $\overline{X}_i = Y_i^1 \cup \dots \cup Y_i^{r(i)}$, a finite, disjoint union of local minimal sets [I, Corollary 8.3.12]. If $X_i = \emptyset$, let $r(i) = 1$ and $Y_i^1 = \emptyset$. Generally, when $r(i) = 1$, $Y_i^1 = X_i = \overline{X}_i$ is compact. When $r(i) > 1$, we can choose the indexing so that

$$1 \le j < r(i) \Rightarrow \operatorname{level} Y_i^j \le \operatorname{level} Y_i^{j+1}.$$

It follows that $Y_i^1 \cup Y_i^2 \cup \dots \cup Y_i^j$ is compact, $1 \le j \le r(i)$. Linearly order $\{Y_i^j \mid 1 \le j \le r(i) \text{ and } i \ge 1\}$ by lexicographic order. That is, $Y_i^j \le Y_k^r$ means that $i \le k$ and, if $i = k$, then $j \le r$. Inductively, delete each nonempty Y_i^j if the same local minimal set appears earlier in the list. Renumber the remaining list as $B_1 < B_2 < \dots B_j < \cdots$. Each B_i is a local minimal set or is empty and $B_i \cap B_j = \emptyset$ if $i \ne j$. The set $\bigcup_{1\le i<\infty} B_i$ is dense in M and $\bigcup_{1\le i\le k} B_i$ is compact, $\forall\, k$. Set $W_k = M \smallsetminus \bigcup_{1\le i\le k} B_i$. Finally, define

$$B_\infty = M \smallsetminus \bigcup_{1\le i<\infty} B_i = \bigcap_{1\le k<\infty} W_k \in \mathcal{B}(\mathcal{F}).$$

Clearly, $\tau(B_\infty) = \lim_{k\to\infty} \tau(W_k) = 0$. □

Proposition 7.3.13. *If $B \in \mathcal{B}(\mathcal{F})$ and $\tau(B) = 0$, then $\mathrm{g}(B, \mathcal{F}) = 0$.*

Proof. Choose thin elements $W_k \in \mathcal{O}(\mathcal{F})$ such that

$$W_1 \supset W_2 \supset \cdots \supset W_k \supset \cdots \supset B$$

and $\lim_{k\to\infty} \tau(W_k) = 0$. Since there is a smooth extension $\overline{\gamma}_{\alpha\beta}$ to the compact closure of $\operatorname{dom}\gamma_{\alpha\beta}$, $\forall\,\alpha,\beta \in \mathfrak{A}$, we can choose

$$K = \max\{(\log \gamma'_{\alpha\beta})'(y) \mid y \in \operatorname{dom}\gamma_{\alpha\beta},\ \alpha,\beta \in \mathfrak{A}\}.$$

If (a, b) is a component of $W_k \cap S$, $k \ge 1$, and $y \in (a, b)$, define

$$h_k(y) = -\frac{1}{2}\log((y-a)(b-y)).$$

On $S \smallsetminus W_k$, let $h_k = 0$. This is measurable and we can define a cocycle $\eta_k \sim_{h_k} \nu$. Since $\eta_k = \nu$ on $S \smallsetminus W_k$, it is an integrable cocycle on that set. We will show that η_k is ε_k-tempered on $W_k \cap S$, where $\varepsilon_k = K\tau(W_k)$; hence η_k will be integrable on S. If $y \in W_k \cap \operatorname{dom}\gamma_{\alpha\beta}$, let $(a, b) \subset \operatorname{dom}\gamma_{\alpha\beta}$ be the component of $W_k \cap S$ that contains y. Then $(\overline{\gamma}_{\alpha\beta}(a), \overline{\gamma}_{\alpha\beta}(b))$ is also a component of $W_k \cap S$. Since $\nu(\gamma_{\alpha\beta})(y) = \log \gamma'_{\alpha\beta}(y)$, we obtain

$$\begin{aligned}\eta_k(\gamma_{\alpha\beta})(y) &= \log \gamma'_{\alpha\beta}(y) + \frac{1}{2}\log\left(\frac{(y-a)(b-y)}{(\gamma_{\alpha\beta}(y) - \overline{\gamma}_{\alpha\beta}(a))(\overline{\gamma}_{\alpha\beta}(b) - \gamma_{\alpha\beta}(y))}\right)\\ &= \log \gamma'_{\alpha\beta}(y) - \log \sqrt{\gamma'_{\alpha\beta}(\zeta)\gamma'_{\alpha\beta}(\xi)},\end{aligned}$$

for suitable $\zeta \in (a, y)$ and $\xi \in (y, b)$. This is just the mean value theorem. By an application of the intermediate value theorem, we find ρ in the closed interval bounded by ξ and ζ, hence in (a, b), such that $(\gamma'_{\alpha\beta}(\rho))^2 = \gamma'_{\alpha\beta}(\xi)\gamma'_{\alpha\beta}(\zeta)$. Another application of the mean value theorem gives λ between y and ρ such that

$$\eta_k(\gamma_{\alpha\beta})(y) = \log \gamma'_{\alpha\beta}(y) - \log \gamma'_{\alpha\beta}(\rho) = \big((\log \gamma'_{\alpha\beta})'(\lambda)\big)(y - \rho).$$

We conclude that

$$|\eta_k(\gamma_{\alpha\beta})(y)| \leq K(b-a) \leq K\tau(W_k).$$

Since $B \subset W_k$, η_k is ε_k-tempered on $B \cap S$, $\forall\, k \geq 1$. But $\eta_k \in [\nu]$ is integrable and $\lim_{k\to\infty} \varepsilon_k = 0$, so $\mathrm{g}(B, \mathcal{F}) = \mathrm{g}^\nu(B, \mathcal{F}) = 0$, by Proposition 7.1.42. □

Proposition 7.3.14. *If $W \in \mathcal{O}(\mathcal{F})$ is a local minimal set such that $\mathcal{F}|W$ is without holonomy, then* $\mathrm{g}(W, \mathcal{F}) = 0$.

Proof. Let μ be the transverse, holonomy-invariant measure on $W \cap S$ constructed in [I, Section 9.2]. This measure is continuous and finite on compact sets, so we use it to define a topological metric d_μ on each component (a, b) of $W \cap S$, setting

$$d_\mu(x, y) = d_\mu(y, x) = \mu[x, y], \quad a < x \leq y < b.$$

This metric gives the correct topology on (a, b). It is Γ-invariant in the sense that

$$x, y \in (a, b) \cap \operatorname{dom} \gamma_{\alpha\beta} \Rightarrow d_\mu(x, y) = d_\mu(\gamma_{\alpha\beta}(x), \gamma_{\alpha\beta}(y)).$$

If (a, b) is a component of $W \cap S$ and $b \notin W$ (respectively, $a \notin W$), [I, Lemma 9.2.17] implies that $\mu[y, b) = \infty$ (respectively, $\mu(a, y] = \infty$). Thus the metric "blows up" near b (respectively, a). It follows that "uniformly close" in the d_μ metric implies "uniformly close" in the Euclidean metric.

Let $\{\widetilde{U}_\alpha, \widetilde{x}_\alpha, \widetilde{y}_\alpha\}_{\alpha \in \mathfrak{A}}$ be a regular foliated atlas such that $\overline{U}_\alpha \subset \widetilde{U}_\alpha$ and $\widetilde{y}_\alpha$ extends y_α, $\forall\, \alpha \in \mathfrak{A}$. The associated holonomy cocycle $\{\widetilde{\gamma}_{\alpha\beta}\}_{\alpha,\beta \in \mathfrak{A}}$ consists of extensions of the corresponding $\gamma_{\alpha\beta}$'s. This cocycle generates the holonomy pseudogroup $\widetilde{\Gamma}$ on the corresponding transverse space $\widetilde{S}$. Finally, μ and d_μ extend to $W \cap \widetilde{S}$, and the extensions are $\widetilde{\Gamma}$-invariant.

Let $\delta > 0$. For $y \in W \cap S$, define

$$U_\delta(y) = \{x \in W \cap \widetilde{S} \mid d_\mu(y, x) \text{ is defined and } \leq \delta\}.$$

By choosing δ sufficiently small, we guarantee that $U_\delta(y)$ is a compact interval in $\widetilde{S}$, $\forall\, y \in W \cap S$. In fact, making δ even smaller, if necessary, we guarantee that, whenever $y \in W \cap \operatorname{dom} \gamma_{\alpha\beta}$, then $U_\delta(y) \subset \operatorname{dom} \widetilde{\gamma}_{\alpha\beta}$. By the $\widetilde{\Gamma}$-invariance of d_μ, we obtain the following crucial property:

$$(*) \qquad y \in W \cap \operatorname{dom} \gamma_{\alpha\beta} \Rightarrow \widetilde{\gamma}_{\alpha\beta}(U_\delta(y)) = U_\delta(\gamma_{\alpha\beta}(y)).$$

Define a measurable function $h_\delta : S \to \mathbb{R}$ by

$$h_\delta(y) = \begin{cases} -\log |U_\delta(y)|, & y \in W \cap S, \\ 0 & y \notin W \cap S. \end{cases}$$

Then, for $y \in W \cap \operatorname{dom} \gamma_{\alpha\beta}$,

$$\begin{aligned} h_\delta(\gamma_{\alpha\beta}(y)) &= -\log |U_\delta(\gamma_{\alpha\beta}(y))| \\ &= -\log |\widetilde{\gamma}_{\alpha\beta}(U_\delta(y))| \quad (\text{by } (*)) \\ &= -\log(\widetilde{\gamma}'_{\alpha\beta}(y^*)|U_\delta(y)|), \end{aligned}$$

where $y^* \in U_\delta(y)$ is given by the mean value theorem. Thus,

$$h_\delta(\gamma_{\alpha\beta}(y)) = -\log \widetilde{\gamma}'_{\alpha\beta}(y^*) + h_\delta(y).$$

Let $a_\delta \sim_{h_\delta} \nu$. By the above,

$$\begin{aligned} a_\delta(\gamma_{\alpha\beta})(y) &= \log \gamma'_{\alpha\beta}(y) + h_\delta(\gamma_{\alpha\beta}(y)) - h_\delta(y) \\ &= \log\left(\frac{\gamma'_{\alpha\beta}(y)}{\widetilde{\gamma}'_{\alpha\beta}(y^*)}\right), \end{aligned}$$

$\forall\, y \in W \cap \operatorname{dom} \gamma_{\alpha\beta}$ and some $y^* \in U_\delta(y)$. As $\delta \to 0$, we see that $y^* \to y$, d_μ-uniformly, hence also uniformly in the usual metric. Therefore,

$$\lim_{\delta \to 0} a_\delta(\gamma_{\alpha\beta}) = \log 1 = 0,$$

uniformly on $W \cap \operatorname{dom} \gamma_{\alpha\beta}$, $\forall\, \alpha, \beta \in \mathfrak{A}$. Thus, given $\varepsilon > 0$, we can choose $\delta > 0$ so small that $|a_\delta(\gamma_{\alpha\beta})| < \varepsilon$, uniformly on their domains, $\forall\, \alpha, \beta \in \mathfrak{A}$. Thus, $a_\delta \in [\nu]$ is ε-tempered on $W \cap S$ and agrees with ν on $S \smallsetminus W$, hence is integrable. Since $\varepsilon > 0$ is arbitrary, Proposition 7.1.42 implies that $\mathrm{g}(W, \mathcal{F}) = 0$. □

Proof of Theorem 7.3.1. Assume there is no resilient leaf. By Proposition 7.3.12, fix a Duminy decomposition

$$M = B_1 \cup B_2 \cup \cdots \cup B_\infty.$$

By Lemma 7.3.11, each nonempty B_k, $1 \le k < \infty$, is either a proper leaf or an open local minimal set without holonomy. In the first case, the Lebesgue measure $|B_k| = 0$, forcing $\mathrm{g}(B_k, \mathcal{F}) = 0$. In the second case, $\mathrm{g}(B_k, \mathcal{F}) = 0$ by Proposition 7.3.14. Finally, $\mathrm{g}(B_\infty, \mathcal{F}) = 0$ by Proposition 7.3.13. By the σ-additivity of the Godbillon measure, $\mathrm{g}(M, \mathcal{F}) = 0$, so $\mathrm{gv}(\mathcal{F}) = 0$. □

We know that there are only two ways that a resilient leaf can occur, either in an open local minimal set with holonomy or in an exceptional minimal set. In the known examples of foliations with $\mathrm{gv}(\mathcal{F}) \neq 0$, there is always an open local minimal set with holonomy. It can be conjectured that this is the only kind of resilient leaf detected by $\mathrm{gv}(\mathcal{F})$.

Conjecture. *If X is an exceptional local minimal set, then* $\mathrm{g}(X, \mathcal{F}) = 0$.

In fact, the most natural examples of exceptional local minimal sets, those of Markov type, have Lebesgue measure zero [**22**], so the Godbillon measure must vanish. It can be conjectured that, for foliations of class at least C^2, all exceptional local minimal sets have zero Lebesgue measure. This is a good problem.

7.4. Quasi-polynomial Leaves

We return to foliated manifolds $(M, \mathcal{F})$ of arbitrary codimension q. Our goal is to prove the following theorem from [**102**].

Theorem 7.4.1 (Hurder). *If almost every leaf of $\mathcal{F}$ has quasi-polynomial growth, then all generalized Godbillon-Vey classes of $\mathcal{F}$ vanish.*

In fact this is an immediate corollary of a vanishing theorem for the Godbillon measure.

Theorem 7.4.2 (Hurder). *If $B \in \mathcal{B}(\mathcal{F})$ and almost every leaf of $\mathcal{F}|B$ has quasi-polynomial growth, then* $\mathrm{g}(B, \mathcal{F}) = 0$.

7.4.A. ε-Tempering. A process, due to Hurder and Katok [**103**], for producing ε-tempered cocycles is key to the proof of Theorem 7.4.2. Here is a heuristic preliminary. Given an integrable cocycle $a : \mathcal{R} \to \mathbb{R}$, one tries to define

$$f : S \to \mathbb{R},$$

$$f(y) = \sum_{x \in \mathcal{R}(y)} e^{a(x,y)}.$$

If this converges almost everywhere , the reader can check that it is measurable. It satisfies

$$e^{a(z,y)} f(z) = \sum_{x \in \mathcal{R}(z)} e^{a(z,y)} e^{a(x,z)} = \sum_{x \in \mathcal{R}(y)} e^{a(x,y)} = f(y).$$

Then $h = \log \circ f$ is measurable and

$$a(z, y) + h(z) - h(y) = 0 \quad \text{(almost everywhere)},$$

so $[a] = 0$ and $\mathrm{g}^a(\cdot, \mathcal{F}) = 0$. Of course, one cannot generally expect convergence.

Given $\varepsilon > 0$ and $a = \nu$, try

$$f_\varepsilon(y) = \sum_{x \in \mathcal{R}(y)} e^{\nu(x,y) - \varepsilon|x,y|}.$$

This has a chance of converging wherever ν is δ-tempered, $0 < \delta < \varepsilon$. Define

$$h_\varepsilon(y) = \begin{cases} \log(f_\varepsilon(y)), & f_\varepsilon(y) < \infty, \\ 0, & f_\varepsilon(y) = \infty. \end{cases}$$

The fact that f_ε and h_ε are measurable is left as an exercise.

Lemma 7.4.3. $f_\varepsilon(y) < \infty \Leftrightarrow f_\varepsilon(z) < \infty$, $\forall\, z \in \mathcal{R}(y)$.

Proof. The implication "$\Leftarrow$" is trivial. For the converse, assume that $f_\varepsilon(y) < \infty$ and let $z \in \mathcal{R}(y)$. Then

$$e^{\nu(z,y) - \varepsilon|z,y|} f_\varepsilon(z) = \sum_{x \in \mathcal{R}(z)} e^{\nu(x,z) + \nu(z,y) - \varepsilon(|x,z| + |z,y|)}.$$

But $|x, z| + |z, y| \geq |x, y|$, so

$$e^{\nu(z,y) - \varepsilon|z,y|} f_\varepsilon(z) \leq \sum_{x \in \mathcal{R}(y)} e^{\nu(x,y) - \varepsilon|x,y|} = f_\varepsilon(y) < \infty.$$

□

Corollary 7.4.4. *The set $X \subseteq S$ on which f_ε converges is of the form $X = B \cap S$, $B \in \mathcal{B}(\mathcal{F})$.*

Proof. Since f_ε is measurable, $X = f_\varepsilon^{-1}(0, \infty)$ is measurable. By the above lemma, X is a union of $\mathcal{R}$-orbits, hence is a Γ-invariant set. □

Lemma 7.4.5. *If $f_\varepsilon(x) < \infty$, $y \in \mathcal{R}(x)$ and $|x, y| = 1$, then*

$$h_\varepsilon(x) - \varepsilon \leq \nu(y, x) + h_\varepsilon(y) \leq h_\varepsilon(x) + \varepsilon.$$

Proof. Write

$$f_\varepsilon(y) = \sum_{z \in \mathcal{R}(y)} e^{\nu(z,y) - \varepsilon|z,y|} = \sum_{z \in \mathcal{R}(y)} e^{\nu(z,x) + \nu(x,y) - \varepsilon|z,y|}.$$

But $|z, y| - 1 \leq |z, x| \leq |z, y| + 1$, so

$$f_\varepsilon(y) e^{\nu(y,x)} = \sum_{z \in \mathcal{R}(y)} e^{\nu(z,x) - \varepsilon|z,y|} \leq \sum_{z \in \mathcal{R}(y)} e^{\nu(z,x) - \varepsilon(|z,x| - 1)} = f_\varepsilon(x) e^{\varepsilon}.$$

Similarly, $f_\varepsilon(y) e^{\nu(y,x)} \geq f_\varepsilon(x) e^{-\varepsilon}$. Now apply log to both inequalities. □

Define a_ε by $a_\varepsilon \sim_{h_\varepsilon} \nu$. Let $X = B \cap S$ be the set on which $f_\varepsilon(y) < \infty$, $B \in \mathcal{B}(\mathcal{F})$.

Corollary 7.4.6. *The cocycle a_ε is integrable and ε-tempered on X.*

Proof. On $S \smallsetminus X$, $a_\varepsilon = \nu$; hence it is integrable there. If it is ε-tempered on X, it will be integrable. If $y, x \in X$ and $|y, x| = 1$, the previous lemma gives

$$-\varepsilon \le \underbrace{\nu(y,x) + h_\varepsilon(y) - h_\varepsilon(x)}_{a_\varepsilon(y,x)} \le \varepsilon;$$

hence $|a_\varepsilon(y, x)| \le \varepsilon$. □

7.4.B. The Game Plan. If L is a leaf and $\mathcal{R}(y) = L \cap S$, recall that the growth type $\operatorname{gr}(L)$ is the growth type of the function

$$H_y(m) = |\{z \in \mathcal{R}(y) \mid |z, y| \le m\}|,$$

where the absolute value symbol denotes cardinality. This growth is *quasi-polynomial* (*cf.* [I, Definition 12.2.15]) if and only if

$$\lim_{m\to\infty} \frac{\log H_y(m)}{m} = 0.$$

This holds if and only if, for each $\varepsilon > 0$, there is a constant $c_\varepsilon > 0$ such that $H_y(m) \le c_\varepsilon e^{\varepsilon m}$, $\forall\, m \ge 0$.

Fix $B \in \mathcal{B}(\mathcal{F})$ and assume that almost every leaf $L \subseteq B$ has quasi-polynomial growth. Set $X = B \cap S$. We will use the fact that $\mathcal{R}(y) \subseteq X$ has quasi-polynomial growth, for almost every $y \in X$, to prove that $f_\varepsilon < \infty$ almost everywhere on X for *every* $\varepsilon > 0$. Here, f_ε is defined as in the previous subsection and the results there show that $[\nu]$ contains an integrable cocycle a_ε which is ε-tempered on X. Since $\varepsilon > 0$ is arbitrary, we will conclude that $\operatorname{g}(B, \mathcal{F}) = 0$.

Definition 7.4.7. For $\delta > 0$, X_δ is the set of all $x \in X$ such that, for every $N > 0$, there is $(y, x) \in \mathcal{R}$ with $|y, x| > N$ and $\nu(y, x) > \delta|y, x|$. The Γ-invariant set $\widetilde{X}_\delta$ is the union of the orbits $\mathcal{R}(x)$ for $x \in X_\delta$.

Lemma 7.4.8. *For almost every $x \in X \smallsetminus \widetilde{X}_{\varepsilon/2}$, $f_\varepsilon(x) < \infty$.*

Proof. By the quasi-polynomial growth condition, we can define a function $c : X \to (0, \infty)$ with the property that

$$H_z(N) \le c(z) e^{\varepsilon N/4},$$

for almost every $z \in X$, and for all $N \ge 0$. We will see later that c can be chosen to be measurable, but we do not need that here. If $x \in X \smallsetminus \widetilde{X}_{\varepsilon/2}$, there is $N_0 > 0$ such that, whenever $|y, x| > N_0$,

$$\nu(y, x) \le \frac{\varepsilon}{2}|y, x|.$$

Therefore,

$$\begin{aligned} f_\varepsilon(x) &= \underbrace{\sum_{|y,x|\le N_0} e^{\nu(y,x)-\varepsilon|y,x|}}_{K} + \sum_{|y,x|>N_0} e^{\nu(y,x)-\varepsilon|y,x|} \\ &\le K + \sum_{|y,x|>N_0} e^{-\varepsilon|y,x|/2} \\ &\le K + \sum_{N>N_0} e^{-\varepsilon N/2} H_x(N) \\ &\le K + c(x) \sum_{N>N_0} e^{-\varepsilon N/4} \\ &< \infty. \end{aligned}$$

□

Thus, our game plan reduces to proving the following.

Lemma 7.4.9 (Key Lemma). *For each $\delta > 0$, $|X_\delta| = 0$, and hence $|\widetilde{X}_\delta| = 0$.*

7.4.C. Proof of the Key Lemma. Let Γ_0 denote the countable subset of Γ consisting of pure compositions of the $\gamma_{\alpha\beta}$'s, each with maximal domain. If $\gamma \in \Gamma_0$, $|\gamma|$ denotes the smallest integer $r \ge 0$ such that

$$\gamma = \gamma_{\alpha_r\alpha_{r-1}} \circ \dots \circ \gamma_{\alpha_1\alpha_0}.$$

Here, $|\gamma| = 0$ if and only if $\gamma = \gamma_{\alpha_0\alpha_0}$ is an identity.

In all that follows, $B \in \mathcal{B}(\mathcal{F})$, almost every leaf of $\mathcal{F}|B$ has quasi-polynomial growth, and $X = B \cap S$.

Claim 1. *There is a measurable function $c : X \to [1, \infty)$ such that*

$$H_x(N) \le c(x) e^{\delta N/4},$$

for all $N \ge 0$ and for almost every $x \in X$.

Proof. Since $\operatorname{gr}(H_x)$ is quasi-polynomial, for almost every $x \in X$, we can define

$$c(x) = \inf\{t \in \mathbb{R} \mid H_x(N) \le t e^{\delta N/4},\ \forall\, N \ge 0\},$$

a function taking finite values almost everywhere. Since $H_x(0) = 1$, $c(x) \ge 1$, for almost every $x \in X$. In order to prove that c is a measurable function, it will be enough to show that $c^{-1}(d, \infty)$ is relatively open in X, for all $d \in \mathbb{R}$. Clearly, $x \in c^{-1}(d, \infty)$ if and only if there is an integer $N_x \ge 0$ such that $H_x(N_x) > d e^{\delta N_x/4}$. For such a point x, let $\gamma_1, \dots, \gamma_r \in \Gamma_0$ be the elements with $|\gamma_j| \le N_x$ and $x \in \operatorname{dom}\gamma_j$, $1 \le j \le r$. Let U be an open neighborhood of x in $\bigcap_{j=1}^r \operatorname{dom}\gamma_j$, small enough that

$$\gamma_i(x) \ne \gamma_j(x) \Rightarrow \gamma_i(y) \ne \gamma_j(y), \quad \forall\, y \in U, \quad 1 \le i,j \le r.$$

Thus, if $y \in U$, $H_y(N_x) \ge H_x(N_x) > de^{\delta N_x/4}$, so $y \in c^{-1}(d,\infty)$. □

For each integer $N \ge 1$, set

$$X(\delta,N) = \{x \in X \mid \exists\, y \in \mathcal{R}(x) \text{ with } |y,x| \le N \text{ and } \nu(y,x) > \delta N\}.$$

Claim 2. *For each integer $N_0 \ge 1$, $X_\delta \subseteq \bigcup_{N=N_0}^\infty X(\delta,N)$.*

Proof. If $x \in X_\delta$, there is $y \in \mathcal{R}(x)$ such that

$$|y,x| > N_0 \quad \text{and} \quad \nu(y,x) > \delta|y,x|.$$

Take $N_1 = |y,x|$. Then $N_1 > N_0$ and $\nu(y,x) > \delta N_1$, so

$$x \in X(\delta,N_1) \subseteq \bigcup_{N=N_0}^\infty X(\delta,N).$$

□

Claim 3. *For each integer $N \ge 1$, $X(\delta,N)$ is relatively open in X, hence measurable.*

Proof. Given $x \in X(\delta,N)$, let $y \in \mathcal{R}(x)$ be such that $|y,x| \le N$ and $\nu(y,x) > \delta N$. Write $y = \gamma(x)$ for suitable $\gamma \in \Gamma_0$ with $|\gamma| = |y,x|$. Let $U_{\gamma,x} \subset X \cap \operatorname{dom}\gamma$ be a relatively open neighborhood of x such that

$$\nu(\gamma(z),z) = \nu(\widehat{\gamma}_z) > \delta N, \quad \forall\, z \in U_{\gamma,x}.$$

Since $|\gamma(z),z| \le |\gamma| \le N$, we conclude that $U_{\gamma,x} \subseteq X(\delta,N)$. Since $x \in X(\delta,N)$ was arbitrary, this set is relatively open in X. □

Thus, if we can prove that $\sum_{N=1}^\infty |X(\delta,N)| < \infty$, Claim 2 will imply that $|X_\delta| = 0$, as desired.

In the proof of Claim 3, the requirements that $\gamma \in \Gamma_0$ and $|\gamma| \le N$ can be met by only finitely many γ. Thus, we can list the elements $\gamma_1, \dots, \gamma_s \in \Gamma_0$ that actually satisfy the requirements in the proof for at least one $x \in X(\delta,N)$. We then get a cover of $X_{\delta,N}$ by the relatively open sets

$$U_j = \{x \in X(\delta,N) \cap \operatorname{dom}\gamma_j \mid \nu(\gamma_j(x),x) > \delta N\}, \quad 1 \le j \le s.$$

We use this to form a measurable partition

$$X(\delta,N) = \bigsqcup_{j=1}^s E_j,$$

where

$$\begin{aligned}
E_1 &= U_1 \\
E_2 &= U_2 \smallsetminus E_1 \\
&\vdots \\
E_s &= U_s \smallsetminus (E_1 \cup \cdots \cup E_{s-1}).
\end{aligned}$$

Using this partition, we define a measurable map

$$\begin{aligned}
\widetilde{\gamma} &: X(\delta, N) \to S, \\
\widetilde{\gamma}|E_j &= \gamma_j|E_j, \quad 1 \le j \le s.
\end{aligned}$$

In particular, $\widetilde{\gamma}|E_j$ is one-to-one, $\forall\, j$.

Definition 7.4.10. Let $Z \subseteq X(\delta, N)$ be measurable. A $\widetilde{\gamma}$-partition of Z is a finite partition of Z into measurable subsets $Z_1, \dots, Z_r$ with the property that $\widetilde{\gamma}|Z_i$ is one-to-one, $1 \le i \le r$.

Thus, the E_j's themselves constitute a $\widetilde{\gamma}$-partition of E and the sets $Z_j = Z \cap E_j$ constitute a $\widetilde{\gamma}$-partition of any measurable subset $Z \subseteq X(\delta, N)$. The problem with these partitions is our inability to control the number s of elements. In light of the following claim, it will be useful to repartition Z in such a way as to get good control of the number of elements in the partition. Note that the map $\widetilde{\gamma}$ itself is to remain the same.

Claim 4. *If $Z = \bigsqcup_{i=1}^r Z_i$ is a $\widetilde{\gamma}$-partition, then $|Z| \le re^{-\delta N}\,|S|$.*

Proof. Set $Z_{ij} = Z_i \cap E_j$. Then,

$$\begin{aligned}
|\widetilde{\gamma}(Z_{ij})| = |\gamma_j(Z_{ij})| &= \int_{\gamma_j(Z_{ij})} dy \\
&= \int_{Z_{ij}} \gamma_j^* dy \ge \int_{Z_{ij}} e^{\delta N} dy = e^{\delta N}|Z_{ij}|.
\end{aligned}$$

Since $\widetilde{\gamma}$ is one-to-one on Z_i, this gives

$$|S| \ge |\widetilde{\gamma}(Z_i)| = \sum_{j=1}^{s} |\widetilde{\gamma}(Z_{ij})| \ge e^{\delta N} \sum_{j=1}^{s} |Z_{ij}| = e^{\delta N}\,|Z_i|\,.$$

Since the Z_i's partition Z,

$$|Z| = \sum_{i=1}^{r} |Z_i| \le \sum_{i=1}^{r} e^{-\delta N}\,|S| = re^{-\delta N}\,|S|\,.$$

□

Let $z \in Z$. Since $\widetilde{\gamma}$ is one-to-one on each E_j, we can write $\widetilde{\gamma}^{-1}(\widetilde{\gamma}(z)) \cap Z$ uniquely as $\{z_{i_1}, z_{i_2}, \dots, z_{i_{r(z)}}\}$, where

$$z_{i_k} \in E_{i_k} \cap Z \quad \text{and} \quad 1 \leq i_1 < \cdots < i_{r(z)} \leq s.$$

Since $z \in \widetilde{\gamma}^{-1}(\widetilde{\gamma}(z))$, there is a unique integer $\lambda(z)$, $1 \leq \lambda(z) \leq r(z) \leq s$, such that $z = z_{i_{\lambda(z)}}$. This defines

$$\lambda : Z \to \{1, 2, \dots, s\}.$$

We define

$$r(Z) = \max_{z \in Z} \lambda(z),$$

noting that $r(Z)$ is also the maximum value of $r(z)$ on Z. Also, let

$$\mathfrak{I}_z = (i_1, \dots, i_{r(z)}),$$

the set of indices for $\widetilde{\gamma}^{-1}(\widetilde{\gamma}(z)) \cap Z$, where $1 \leq i_1 < \cdots < i_r \leq s$. Partially order the set $\{\mathfrak{I}_z \mid z \in Z\}$ by inclusion.

Claim 5. *If $\mathfrak{I}' = \{i_1, \dots, i_r\}$ is maximal in the above partial order, then $\{z \in Z \mid \mathfrak{I}_z = \mathfrak{I}'\}$ is measurable.*

Proof. This set is

$$\bigsqcup_{\ell=1}^{r} \{z \in Z \cap E_{i_\ell} \mid \mathfrak{I}_z = \mathfrak{I}'\},$$

so we prove that $\{z \in Z \cap E_{i_\ell} \mid \mathfrak{I}_z = \mathfrak{I}'\}$ is measurable, $\ell = 1, 2, \dots, r$. We will prove that this set is exactly $\bigcap_{j=1}^{r} \gamma_{i_\ell}^{-1}(\gamma_{i_j}(Z \cap E_{i_j}))$. Since this set is measurable, we will be done.

If $z \in Z \cap E_{i_\ell}$ and $\mathfrak{I}_z = \mathfrak{I}'$, then there is a unique element

$$z_{i_j} \in \widetilde{\gamma}^{-1}(\widetilde{\gamma}(z)) \cap E_{i_j} \cap Z, \quad 1 \leq j \leq r.$$

This implies that $z = \gamma_{i_\ell}^{-1}(\gamma_{i_j}(z_{i_j}))$, $1 \leq j \leq r$, hence that

$$z \in \bigcap_{j=1}^{r} \gamma_{i_\ell}^{-1}(\gamma_{i_j}(Z \cap E_{i_j})),$$

proving the inclusion in one direction. For the reverse inclusion, note the implications

$$\begin{aligned}
z \in \bigcap_{j=1}^{r} \gamma_{i_\ell}^{-1}(\gamma_{i_j}(Z \cap E_{i_j})) &\Rightarrow z \in Z \cap E_{i_\ell} \text{ and } \gamma_{i_\ell}(z) \in \bigcap_{j=1}^{r} \gamma_{i_j}(Z \cap E_{i_j}) \\
&\Rightarrow z \in Z \cap E_{i_\ell} \text{ and } \{\gamma_{i_j}^{-1}(\gamma_{i_\ell}(z))\}_{j=1}^{r} \subseteq \widetilde{\gamma}^{-1}(\widetilde{\gamma}(z)) \\
&\Rightarrow z \in Z \cap E_{i_\ell} \\
&\qquad \text{and } \{\gamma_{i_j}^{-1}(\gamma_{i_\ell}(z))\}_{j=1}^{r} = Z \cap \widetilde{\gamma}^{-1}(\widetilde{\gamma}(z))
\end{aligned}$$

by the maximality of $\mathcal{J}'$. But this implies that $z \in Z \cap E_{i_\ell}$ and $\mathcal{J}_z = \mathcal{J}'$, proving the reverse inclusion. □

Claim 6. *The map $\lambda : Z \to \{1, 2, \dots, s\}$ is measurable.*

Proof. We are going to allow the measurable subset $Z \subseteq X(\delta, N)$ to vary, so we denote the corresponding function λ by

$$\lambda_Z : Z \to \{1, 2, \dots, r(Z)\}$$

and prove λ_Z measurable by induction on $r(Z)$. If $r(Z) = 1$, then $\lambda_Z \equiv 1$ is measurable. If $r(Z) > 1$, assume that $\lambda_{Z'}$ is measurable, $1 \le r(Z') < r(Z)$. Let Z' be obtained from Z by throwing away the finitely many subsets of the form $\{z \in Z \mid \mathcal{J}_z = \mathcal{J}'\}$, where $\mathcal{J}'$ is maximal. By Claim 5, Z' is measurable. It is also clear that $r(Z') < r(Z)$ and that $\lambda_Z|Z' = \lambda_{Z'}$. By the inductive hypothesis, $\lambda_Z|Z'$ is measurable. Fix any of the maximal index strings, say $\mathcal{J}' = (i_1, \dots, i_r)$. On $\{z \in Z \mid \mathcal{J}_z = \mathcal{J}'\}$, λ_Z takes the value ℓ, $1 \le \ell \le r$, exactly on the measurable set $\{z \in Z \cap E_{i_\ell} \mid \mathcal{J}_z = \mathcal{J}'\}$; hence $\lambda_Z|(Z \smallsetminus Z')$ is also measurable. □

Set $Z_\ell = \lambda^{-1}(\ell)$, $1 \le \ell \le r(Z)$. These sets are disjoint and, by Claim 6, measurable. That is, the sets Z_ℓ, $1 \le \ell \le r(Z)$, form a $\widetilde{\gamma}$-partition of Z.

For each integer $k \ge 1$, set

$$X(k) = \{x \in X \mid c(x) \le k\}.$$

Then $X(k)$ is measurable (Claim 1) and $X = \bigcup_{k=1}^{\infty} X(k)$, so the proof of the Key Lemma is reduced to showing that

$$|X_\delta \cap X(k)| = 0, \quad \forall\, k \ge 1.$$

Set $X(\delta, N, k) = X(\delta, N) \cap X(k)$.

Claim 7. $r(X(\delta, N, k)) \le k e^{\delta N/2}$.

Proof. Let $x \in X(\delta, N, k)$. If $y \in \widetilde{\gamma}^{-1}(\widetilde{\gamma}(x))$, then $\gamma_i(x) = \gamma_j(y)$, for some i, j between 1 and s. That is, $y = \gamma_j^{-1}(\gamma_i(x))$, implying that $|y, x| \le 2N$. The set of all such y has cardinality bounded by $H_x(2N) \le k e^{\delta N/2}$ (by the definition of $c(x)$), so

$$\operatorname{card}(\widetilde{\gamma}^{-1}(\widetilde{\gamma}(x))) \le k e^{\delta N/2}, \quad \forall\, x \in X(\delta, N, k).$$

Therefore,

$$r(X(\delta, N, k)) = \max_{x \in X(\delta, N, k)} \operatorname{card}(\widetilde{\gamma}^{-1}(\widetilde{\gamma}(x)) \cap X(\delta, N, k)) \le k e^{\delta N/2}.$$

□

Proof of the Key Lemma. As observed above, it will be enough to prove that $|X_\delta \cap X(k)| = 0$, $\forall\, k \geq 1$. By Claim 2,

$$X_\delta \cap X(k) \subseteq \bigcup_{N=N_0}^{\infty} X(\delta, N, k), \quad \forall\, N_0 \geq 1.$$

Thus

$$\begin{aligned} |X_\delta \cap X(k)| &\leq \sum_{N=N_0}^{\infty} |X(\delta, N, k)| \\ &\leq \sum_{N=N_0}^{\infty} k e^{\delta N/2} e^{-\delta N} \, |S| \\ &= k|S| \sum_{N=N_0}^{\infty} e^{-\delta N/2}. \end{aligned}$$

Since this converges and $N_0 \geq 1$ is arbitrary, $|X_\delta \cap X(k)| = 0$. □

We have proven Theorem 7.4.2, hence Theorem 7.4.1. To the best of our knowledge, it remains open whether the Godbillon measure vanishes when almost every leaf has *nonexponential* growth.

Part 3

Foliated 3-Manifolds

Foreword to Part 3

Foliations of compact 3-manifolds by surfaces have been studied extensively, from both the quantitative and qualitative points of view. Furthermore, by work of D. Gabai [**71, 70, 73, 74, 72**], W. Thurston [**175**] and others, Reebless foliations have proven to be powerful tools for exploring the topology of 3-manifolds. It is the purpose of this part of the book to introduce the reader to some of this wonderful interaction between low dimensional topology and foliation theory.

In order to avoid a distracting mass of technical details, we assume familiarity with a number of basic facts and constructions from 3-manifold topology. Accordingly, we sometimes resort to "proof by picture" when the picture seems sufficienty clear and can be made rigorous by standard techniques. The basic facts and constructions needed from foliation theory are, for the most part, treated in Volume I.

Chapter 8

Constructing Foliations

We will show that every closed 3-manifold has a foliation of codimension one. In 1952, G. Reeb published his construction of a foliation of the 3-sphere. About twelve years later, W. Lickorish [**123**] exhibited foliations of codimension one on every closed, orientable 3-manifold. This was also achieved independently by S. P. Novikov and H. Zieschang [**141**]. The construction of foliations of nonorientable 3-manifolds was achieved by J. Wood [**188**] about four years later. Actually, Wood showed that every transversely oriented 2-plane field on a closed 3-manifold is homotopic to an integrable one. The methods of construction use surgery representations of the manifolds. Finally, W. Thurston [**173**], using a different method, eliminated the requirement of transverse orientability, proving that every 2-plane field is homotopic to a foliation.

A device frequently used in these constructions is "turbulization" or "Reeb modification" along a closed transversal γ. This inserts a Reeb component R with γ as core transversal, spinning the pierced leaves of the original foliation along the toroidal boundary leaf of R. A detailed account of this operation will be found in [I, Example 3.3.11]. Also useful is the operation of spinning a certain type of foliation along a transverse boundary component [I, 3.3.B].

In this chapter, we construct foliations of all closed 3-manifolds and sketch in some detail Thurston's method of homotoping plane fields to integrable ones. We will also present B. Raymond's construction of foliations with exceptional minimal sets in all closed 3-manifolds. Because of the fundamental use of turbulization and spinning in these constructions, there are many Reeb components. The necessity of this, in general, is a consequence of Novikov's theorem, to be discussed in the next chapter. The principal moral to be drawn from this chapter is that, since foliations with Reeb components are ubiquitous, they carry absolutely no information about the topology of closed 3-manifolds.

8.1. Orientable 3-Manifolds

In this section we prove the following.

Theorem 8.1.1 (Lickorish, Novikov and Zieschang). *Every closed, orientable 3-manifold admits a smooth, transversely orientable foliation of codimension one.*

Theorem 8.1.1 will be a consequence, via turbulization, of the following theorem of A. H. Wallace [**180**].

Theorem 8.1.2. *Every closed, connected, orientable 3-manifold can be obtained from S^3 by removing a finite collection of disjoint solid tori from S^3 and sewing them back differently.*

The process of removing solid tori from a 3-manifold and gluing (or sewing) them back in by suitable boundary diffeomorphisms is called *surgery*. For example, S^3 is obtained by gluing together two solid tori by a boundary identification that matches meridians on one to longitudes on the other and *vice versa*. Gluing the two solid tori by the identity map between the boundaries produces $S^2 \times S^1$, so this manifold is obtained from S^3 by a surgery.

In proving Theorem 8.1.2, we follow the account in [**123**], using a theorem of M. Dehn [**46**] (rediscovered by Lickorish [**121**]) about surface diffeomorphisms. For this, we need some terminology.

Let S be a closed orientable surface and let c be a smooth, simple closed curve in S. Let $\varphi : A = [-1,1] \times S^1 \to S$ be a smooth imbedding of the annulus in S such that $c = \varphi(\{0\} \times S^1)$. Let $\lambda : [-1,1] \to [0,2\pi]$ be smooth, nondecreasing, identically 0 near -1 and identically 2π near 1. We write $S^1 = [0,2\pi]/\{0 \equiv 2\pi\}$, letting $t \in [0,2\pi]$ represent the corresponding point of S^1. The map $h : A \to A$ defined by $h(x,t) = (x, t + \lambda(x))$ is a diffeomorphism of A which is the identity near ∂A. Hence $\varphi h \varphi^{-1}$ can be extended by the identity in $S \setminus \varphi(A)$ to give a diffeomorphism h_c of S. We call $h_c^{\pm 1}$ a *Dehn twist* along c.

Let Σ_g be a closed, orientable surface of genus g and let C be the family of curves on Σ_g indicated in Figure 8.1.1.

Theorem 8.1.3 (Dehn and Lickorish). *Each orientation-preserving diffeomorphism of Σ_g is isotopic to a composition $h_1 h_2 \cdots h_n$, where h_i is a Dehn twist along a curve in C, $1 \leq i \leq n$.*

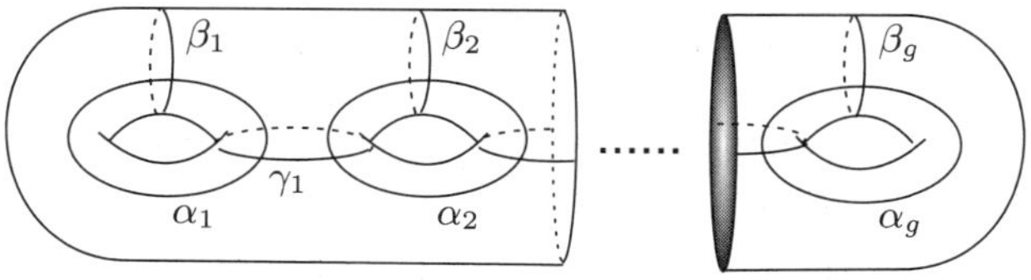

Figure 8.1.1. The family C of $3g-1$ circles for Dehn twists

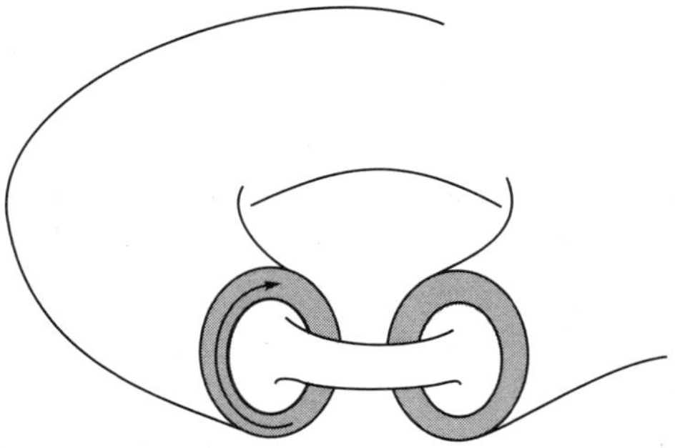

Figure 8.1.2. Extending Dehn twists

Theorem 8.1.3 is a standard tool of low dimensional topology, and will not be proven here.

Recall that the term "handlebody" (of genus g) refers to a solid g-holed torus M_g. More formally, one defines M_g by induction on g. We take $M_1 = S^1 \times D^2$. If M_g has been defined, we fix smoothly imbedded 2-disks $D \subset \partial M_g$ and $D' \subset \partial M_1$, choose a diffeomorphism $f : D' \to D$, and form the quotient space

$$M_{g+1} = M_g \cup_f M_1.$$

This can be done so that the resulting 3-manifold is smooth without corners. It is standard that, up to diffeomorphism, M_{g+1} is independent of the choices. Evidently, $\partial M_g = \Sigma_g$.

Lemma 8.1.4. *Let M_g be a handlebody and let $h : \partial M_g \to \partial M_g$ be a Dehn twist along one of the curves $c \in C$. Then there is a solid torus T in* $\operatorname{int} M_g$ *such that h extends to a diffeomorphism $\widetilde{h}$ of $M_g \setminus \operatorname{int} T$.*

Proof. Fix a collar neighborhood $\Sigma_g \times [0,1]$ of $\partial M_g = \Sigma_g \times \{0\}$. Let $A \subset \partial M_g$ denote the annulus such that $\operatorname{supp} h \subset \operatorname{int} A$ and work in the thickened annulus $A \times [0,1] \subset \Sigma_g \times [0,1]$. Fix choices $0 < a < b \leq 1$, and let $h' : A \times [0,a] \to A \times [0,a]$ be the homeomorphism defined by

$$h'(x,t) = (h(x),t), \quad 0 \le t \le a.$$

Since h' is the identity near $\partial A \times [0,a]$, we can excise the interior of the solid torus $T = A \times [a,b]$, round the corners and define $\widetilde{h}$ to be the extension of h' by the identity to all of $M_g \smallsetminus \operatorname{int} T$. (Figure 8.1.2 shows M_g with $T \cup A \times [0,a] = A \times [0,b]$ excised. The shaded annuli are the components of $\partial A \times [0,a]$ and the arrow is the direction of twist.) □

Observe that, among the $3g-1$ curves in C, exactly g are essential in M_g (not homotopically constant). As in Figure 8.1.1, enumerate the curves as

$$C = \{\alpha_1, \dots, \alpha_g, \beta_1, \dots, \beta_g, \gamma_1, \dots, \gamma_{g-1}\}$$

so that each α_i is essential in M_g, each β_i intersects α_i, but no other curve in C, and each γ_i intersects α_i and α_{i+1}, but no other curve in C. Let A_i be an annular neighborhood of α_i, B_i an annular neighborhood of β_i and G_i an annular neighborhood of γ_i, for each relevant value of i, chosen to intersect one another only in a 2-cell and to do so precisely when their core curves in C intersect. Denote the Dehn twist along the curve $c \in C$ by h_c, chosen so as to be supported well inside the corresponding annular neighborhood of c.

In the above proof, choose $a = 1/4$ and $b = 1/2$ for the curves $c = \alpha_i$, and choose $a = 3/4$ and $b = 1$ for the remaining curves in C. Corresponding to α_i, we designate the torus $A_i \times [1/4, 1/2]$ by T_i. The tori $B_i \times [3/4, 0]$ and $G_i \times [3/4, 0]$ will be denoted by S_i and N_i, respectively. Notice that these solid tori are all disjoint and that T_i is "lower" (closer to ∂M_g) than B_i, N_i and N_{i-1}, $1 \le i \le g$. It follows that $\widetilde{h}_{\alpha_i}$ is the identity on all other solid tori in this system, and hence it restricts to a diffeomorphism of the complement M_g' of the union of their interiors.

As for $\widetilde{h}_{\beta_i}$, this will be the identity on all the other tori except for T_i, which it twists as indicated in Figure 8.1.3. That figure also indicates a diffeomorphism φ_i of $M_g \smallsetminus \operatorname{int} S_i$ which is isotopic to the identity, is supported away from ∂M_g and from all the tori in the system except for T_i, and moves $\widetilde{h}_{\beta_i}(T_i)$ back to T_i. A careful proof of this can be written down by working in a thickened disk $D \times I \subset M_g$ such that $\partial D \times I = B_i$. Similarly, $\widetilde{h}_{\gamma_i}$ twists T_i and T_{i+1}, and an isotopy ψ_i of $M_g \smallsetminus \operatorname{int} N_i$ restores them to their original position. Set

$$\begin{aligned} \widehat{h}_{\alpha_i} &= \widetilde{h}_{\alpha_i}, \quad 1 \le i \le g, \\ \widehat{h}_{\beta_i} &= \varphi_i \circ \widetilde{h}_{\beta_i}, \quad 1 \le i \le g, \\ \widehat{h}_{\gamma_i} &= \psi_i \circ \widetilde{h}_{\gamma_i}, \quad 1 \le i \le g-1. \end{aligned}$$

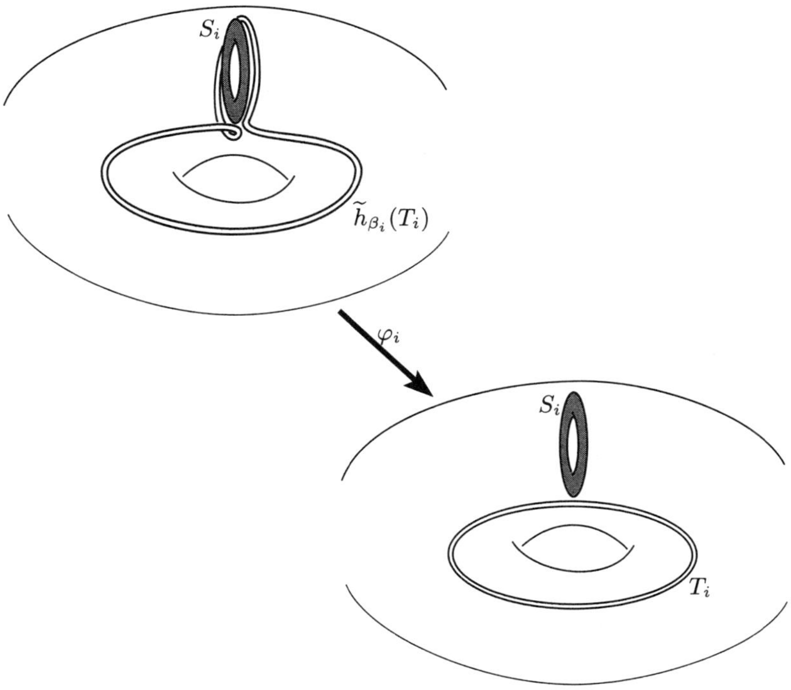

Figure 8.1.3. Untwisting $\widetilde{h}_{\beta_i}(T_i)$

If h is the inverse of any of the above Dehn twists, there is a similarly defined $\widehat{h}$. Finally, it should be evident that each $\widehat{h}^{\pm 1}$ restricts to a diffeomorphism of M_g' and extends the corresponding twist $h^{\pm 1}$. By a slight abuse, we denote this restriction by $\widehat{h}^{\pm 1}$ also.

Corollary 8.1.5. *Let M_g be a handlebody and let $f : \partial M_g \to \partial M_g$ be any orientation-preserving diffeomorphism. Then, after the interiors of $3g-1$ disjoint solid tori in $\operatorname{int} M_g$ have been excised, each parallel to one of the circles of $C \subset \partial M_g$ of Theorem 8.1.3, f extends to a diffeomorphism of the resulting manifold M_g'.*

Proof. Appealing to Theorem 8.1.3, we extend f over a collar neighborhood $\partial M_g \times [0,1]$ (where ∂M_g is identified with $\partial M_g \times \{0\}$) so that, on $\partial M_g \times \{1\}$, the extension is a product $h_1 h_2 \cdots h_n$ of Dehn twists as in that theorem. That is, without loss of generality, we assume that f is of this form. The desired extension is then $\widehat{h}_1 \widehat{h}_2 \cdots \widehat{h}_n$. □

The usefulness of this corollary for our purposes is rooted in the following two examples.

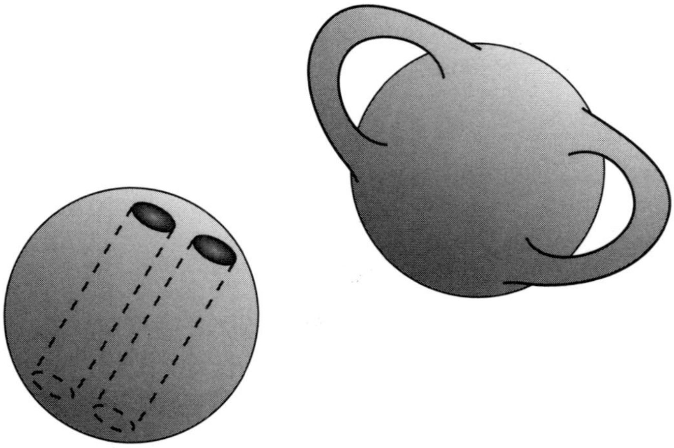

Figure 8.1.4. Decomposition of S^3 into handlebodies of genus 2

Example 8.1.6. The standard decomposition of S^3 into the union of two solid tori generalizes to a standard decomposition into two handlebodies of genus g, for each $g \geq 1$. Let $S^3 = \mathbb{R}^3 \cup \{\infty\}$ be the identification by stereographic projection. Taking ∞ as the north pole and 0 as the south pole, we let $B_0 = \{v \in \mathbb{R}^3 \mid \|v\| \leq 1\}$ denote the southern hemisphere and $B_\infty = \{v \in \mathbb{R}^3 \cup \{\infty\} \mid \|v\| \geq 1\}$ the northern one. Let $s_1, s_2, \dots, s_g$ be disjoint, properly imbedded arcs in B_0 and let $N_1, N_2, \dots, N_g$ be disjoint tubular neighborhoods of these arcs. Suppose that the northern hemisphere B_∞ "annexes" the tubes N_i, $1 \leq i \leq g$. What is left of the southern hemisphere is a handlebody of genus g, and the northern hemisphere with the tubes annexed is also a handlebody of genus g. This is indicated in Figure 8.1.4 for the case $g = 2$.

Example 8.1.7. Every closed, orientable 3-manifold M admits a decomposition into two handlebodies of the same genus g, although the minimum possible value of g will depend on the topological complexity of M. Such a decomposition is called a *Heegaard splitting*. We sketch this, referring the reader to [**93**, Chapter 2] for a more careful treatment.

A few preliminaries are needed. Let $K \subset M$ be a finite, connected, 1-dimensional, imbedded cell complex. If K has more than one vertex, it is homotopic to an imbedded complex $K' \subset M$ with one fewer vertex and one fewer edge. One just shrinks to a point an edge connecting two distinct vertices, this homotopy being an "isotopy up to the last moment". Iterating this process finitely often, we obtain a homotopic, imbedded cell complex

K^* which has just one vertex and (say) $g \geq 0$ edges. This is a bouquet of g circles. If the original complex K has no vertex from which only one edge issues, then $g \geq 1$.

Construct a compact, smoothly imbedded neighborhood N of K in M such that K is a deformation retract of N. For this, first produce disjoint ball neighborhoods of the vertices such that the edges each meet the union of the boundaries of these balls transversely and in two distinct points. Connecting the balls by disjoint tubular neighborhoods of the edges and rounding corners completes the construction of N. Since the union of two of the balls and a connecting solid tube is diffeomorphic to a ball, this neighborhood N is diffeomorphic to the analogously constructed neighborhood N^* of the bouquet of g circles K^*. Using the orientability of M, we note that no circle in this bouquet is an orientation-reversing loop in M, and it follows that N^*, hence N, is diffeomorphic to M_g.

We now use the fact that M can be smoothly triangulated [**137**, Chapter II]. If we take the 1-skeleton of a triangulation $\mathcal{T}$ as the complex K in the above discussion, we obtain an imbedded handlebody $N = M_g$ and, since $\partial M = \emptyset$, $g \geq 1$. Replacing $\mathcal{T}$ with its dual triangulation (the vertices are the barycenters of the 3-simplices of $\mathcal{T}$), we obtain a handlebody N' which is diffeomorphic to the complement of int N. The common interface of these two handlebodies is a closed surface whose genus is the genus of each handlebody. Thus, M is obtained by gluing together two copies of M_g by a diffeomorphism of the boundary.

Proof of Theorem 8.1.2. In order to distinguish two copies of M_g, we set $H_i = M_g \times \{i\}$, $i = 0, 1$. By Examples 8.1.6 and 8.1.7, fix decompositions

$$M = H_1 \cup_\varphi H_0,$$
$$S^3 = H_1 \cup_\psi H_0,$$

where $\varphi, \psi : \partial M_g \to \partial M_g$ are suitably chosen diffeomorphisms. More precisely, in M (respectively, S^3) the points are written as (x, i), $x \in M_g$ and $i = 0, 1$, the only nontrivial identifications being of the form

$$(x, 0) \equiv (\varphi(x), 1),$$
$$(x, 0) \equiv (\psi(x), 1)$$

in the respective cases, where $x \in \partial M_g$. Let h be the extension of $\varphi^{-1} \circ \psi$ to the complement M_g' in M_g of $3g - 1$ (open) solid tori as in Corollary 8.1.5. Let $H_0' = M_g' \times \{0\}$. Then the map

$$f : H_1 \cup_\psi H_0' \to H_1 \cup_\varphi H_0',$$

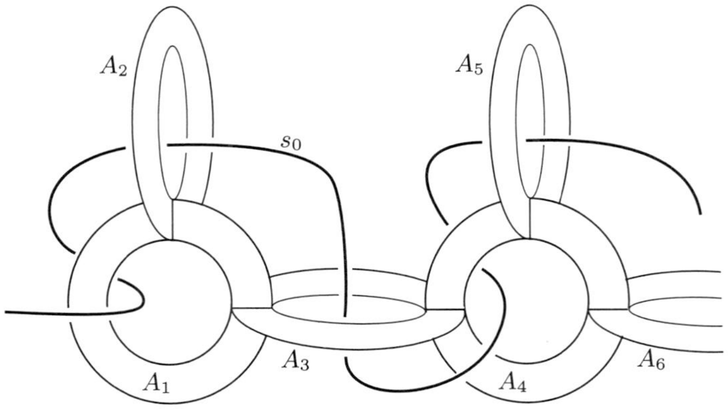

Figure 8.1.5. s_0 simply links the annuli A_i

defined by

$$f(y,i) = \begin{cases} (y,i), & i = 1, \\ (h(y),i), & i = 0, \end{cases}$$

is a diffeomorphism. But M and S^3 each result from suitably sewing the solid tori back into this manifold. □

Proof of Theorem 8.1.1. In order to use Theorem 8.1.2 to prove the existence of foliations on M, we show that the solid tori in S^3 on which the surgery is performed can be taken to be normal neighborhoods in S^3 of closed transversals to the Reeb foliation. A new foliation of S^3 is then produced by turbulization along these loops. The Reeb components introduced by this process are infinitesimally C^∞-trivial along their boundaries [I, Example 3.4.3]; hence excising them and sewing them back in by suitable diffeomorphisms produces a C^∞ foliation of M.

There is an isotopy of S^3 which moves the core circles of the surgery tori to positions transverse to the Reeb foliation $\mathcal{F}$. To see this, recall that the surgery tori in S^3 lie close to the $3g-1$ curves in Figure 8.1.1. Consequently, we can arrange that the cores of the tori are essential circles s_i in the annuli A_i, $1 \leq i \leq 3g-1$, indicated in Figure 8.1.5. Also recall that the surgery tori, hence their cores, were chosen to be unlinked in S^3. The curve s_0 in Figure 8.1.5 is an unknotted loop, passing through $\infty \in S^3$ and simply linking each of these circles. Thus, the link $\{s_0, s_1, \ldots, s_{3g-1}\}$ can be isotoped to the standard position in Figure 8.1.6. The Reeb foliation can be chosen

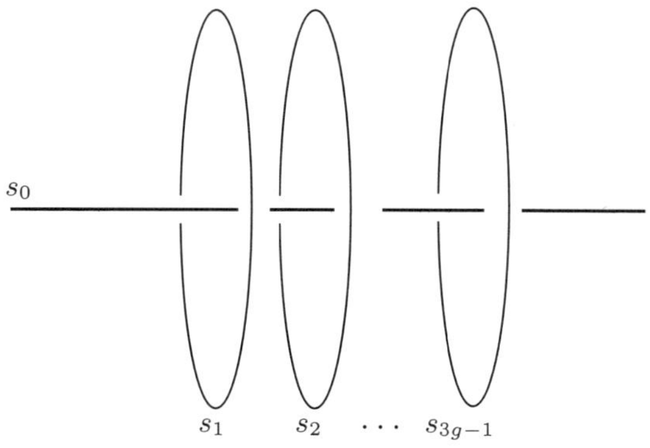

Figure 8.1.6. The link in standard position

so that s_0 is a closed tansversal in one Reeb component and $\{s_1, \ldots, s_{3g-1}\}$ are parallel closed transversals in the other Reeb component. □

8.2. Open Book Decompositions

Open book decompositions, also called Alexander decompositions or spinnable structures, were discovered by J. Alexander [**2**] in his investigations of knot theory. Their use in foliation theory, mainly for constructing foliations in higher dimensional manifolds, is due to H. B. Lawson and A. H. Durfee [**118, 54, 55**], I. Tamura [**170, 171**], *et al.* The name "open book" is due to E. Winkelnkemper [**185**].

Definition 8.2.1. A closed manifold M has an *open book decomposition* if there is a smooth map $f : M \to \mathbb{C}$ such that

(a) there is a neighborhood of $B = f^{-1}(0)$ in M on which f is a submersion, and

(b) the map
$$p = \frac{f}{|f|} : M \setminus B \to S^1$$
is a submersion.

The set B is a closed submanifold of codimension two, called the binding, and $p^{-1}(t)$, $t \in S^1$, are open submanifolds of codimension one, called the pages. For small $\varepsilon > 0$, the tubular neighborhood $N(B) = f^{-1}\{|z| < \varepsilon\}$ of B is diffeomorphic to $D^2 \times B$.

Example 8.2.2. Consider the polynomial $f(x, y) = x^p + y^q$ in two complex variables. Assume the integers p, q are relatively prime and > 1. The restriction of f to $\{(x, y) \mid |x|^2 + |y|^2 = 1\} \subset \mathbb{C}^2$ gives an open book decomposition of the 3-sphere with binding the (p, q)-torus knot. The case $p = 3$ and $q = 2$ played an important role in [I, Example 12.5.6]. In that case, the leaves of the open book are punctured tori.

The importance of open book decompositions for foliations is established in the following exercise.

Exercise 8.2.3. Let M be a closed manifold with an open book decomposition as above. If the manifold $D^2 \times B$ has a smooth, codimension one foliation which is infinitesimally C^∞-trivial at the boundary, proceed as follows to prove that M has a smooth codimension one foliation.

(1) Show that p fibers $M \smallsetminus B$ over S^1 and that each fiber F has closure $\overline{F} = F \cup B$.
(2) Use this and the "spinning" procedure of [I, 3.3.B] to produce a foliation of $M \smallsetminus \operatorname{int} N(B)$ which has the components of $\partial N(B)$ as leaves and is infinitesimally C^∞-trivial along these leaves.

Now use [I, Proposition 3.4.2].

If M as above is a 3-manifold, then B is the union of finitely many circles and $D^2 \times B$ always has a foliation trivial at the boundary (a collection of Reeb components). Hence, the next result gives another proof of the existence of codimension one foliations of orientable 3-manifolds.

Theorem 8.2.4 (Alexander). *Every closed orientable 3-manifold has an open book decomposition.*

This theorem was proven by Alexander in 1923. Lawson's applications of open books to foliations in higher dimension came forty-eight years later.

Exercise 8.2.5. Use the results of Section 8.1 to prove Alexander's Theorem.

8.3. Nonorientable 3-Manifolds

A procedure analogous to that in Section 8.1 was devised by J. Wood [**188**] for producing transversely orientable foliations in all closed, nonorientable 3-manifolds. Analogous to Theorem 8.1.2 is the following. A proof, similar to that of Theorem 8.1.2, will be found in [**122**, p. 317].

Theorem 8.3.1. *Let E be the total space of the nonorientable S^2-bundle over S^1 and let M be a closed, nonorientable, connected 3-manifold. Then* *there are disjoint, smoothly imbedded solid tori $T_i \subset E$, $1 \leq i \leq n$, such that M is diffeomorphic to a manifold obtained by suitable surgeries on these tori.*

Once again, it is necessary to show that the cores of these solid tori can be assumed to be transverse to a foliation, concluding that turbulization and surgery produce a foliation of M. In this case, the fibration $\pi : E \to S^1$ provides the foliation.

Theorem 8.3.2 (Wood). *By increasing the number of tori removed from E for the surgery that produces M, the cores of the tori can be taken to be transverse to the fibers of $\pi : E \to S^1$.*

Proof. Let θ denote the standard coordinate on S^1, well defined modulo 2π. Thus $d\theta$ and $\partial/\partial\theta$ are well defined and $\pi^*(d\theta)$ is a closed, nonsingular 1-form on E defining the foliation $\mathcal{F}$ by fibers. In particular, this foliation is transversely orientable.

Let $\sigma : S^1 \to E$ be a core of one of the surgery tori. By a small perturbation, we can assume that σ is in general position with respect to the fibers. That is, if $c : S^1 \to \mathbb{R}$ is defined by

$$(\pi \circ \sigma)_*(\partial/\partial\theta) = c(\theta)\partial/\partial\theta,$$

then c has only finitely many zeros. Evidently, these zeros correspond to the points where σ is not transverse to the foliation. We can suppose that c changes sign at each zero since, otherwise, it is obvious that a small isotopy will remove the tangency. Thus, these singular points occur in pairs. For each consecutive pair, we will show that, after removing two solid tori which are disjoint from the surgery tori and have cores transverse to $\mathcal{F}$, one can find a diffeomorphism which gets rid of these two singularities without introducing new ones.

Suppose $c(a) = c(b) = 0$ and $c(\theta) > 0$ for $a < \theta < b$. (If $c(\theta) < 0$ on that range, obvious modifications to the following constructions must be made.) Let $\sigma' = \sigma|[a, b]$, and let $j : I \to E$ be a path from $\sigma(b)$ to $\sigma(a)$ which is transverse to $\mathcal{F}$, directed by the transverse orientation and such that the $\sigma + j$ is a simple closed curve representing an even element of $\pi_1(E) \cong \mathbb{Z}$. By an arbitrarily small modification near $\sigma(a)$ and $\sigma(b)$, we produce a simple closed curve γ which is everywhere transverse to $\mathcal{F}$ and is not orientation-reversing in E. Thus, a normal neighborhood of γ will be a solid torus $T \cong [-1, 1] \times [-1, 1] \times S^1$ (rather than a solid Klein bottle). Here, to avoid confusion, we use z to denote the coordinate (modulo 2π) of the circle factor. By choosing T sufficiently thin and slightly displacing it, we assume that σ meets T exactly for $a - 2\varepsilon \leq \theta \leq a - \varepsilon$, for some small $\varepsilon > 0$, and that this arc is of the form $\{0\} \times [-1, 1] \times \{z\}$ in the coordinates

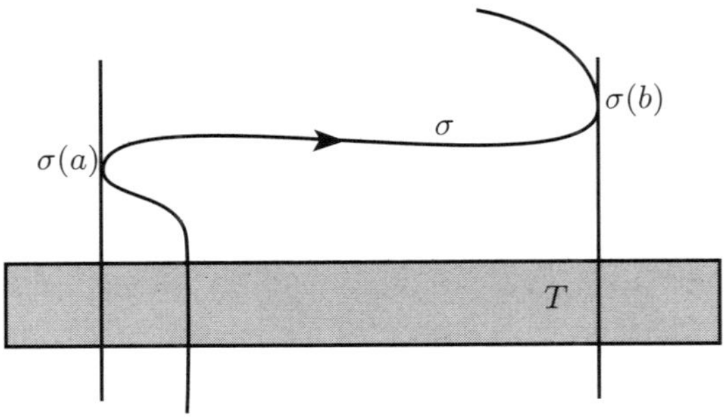

Figure 8.3.1. Relevant parts of T and σ

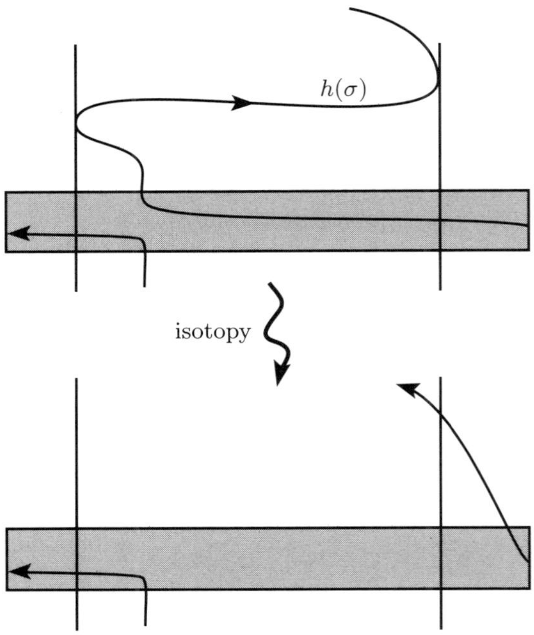

Figure 8.3.2. An isotopy removes two tangencies from $h(\sigma)$

of T. In Figure 8.3.1 we home in on the relevant parts of T (represented by the shaded rectangle) and σ. The vertical lines indicate leaves of $\mathcal{F}$.

Let $\lambda : [-1, 1] \to [0, 2\pi]$ be smooth, nondecreasing, constantly equal to 0 near -1 and to 2π near 1. If we remove from E (and from M) the interiors of two solid tori $T_1 = [-1, -1/3] \times I \times S^1$ and $T_2 = [2/3, 1] \times I \times S^1$, the Dehn twist,

$$h : [-1/3, 2/3] \times I \times S^1 \to [-1/3, 2/3] \times I \times S^1,$$
$$h(r, s, z) = (r, s, z - \lambda(s)),$$

of the solid torus extends smoothly over $E' = E \setminus \text{int}(T_1 \cup T_2)$ as the identity outside T. As indicated in Figure 8.3.2, $h(\sigma)$ can be isotoped so as to have two fewer tangencies with the fibers. One iterates this process until the cores of all the surgery tori are transverse to $\mathcal{F}$. □

Since $\mathcal{F}$ is transversely orientable, turbulization and surgery give the following.

Theorem 8.3.3 (Wood). *Every closed, nonorientable 3-manifold admits a smooth, transversely orientable foliation of codimension one.*

Exercise 8.3.4. Prove this theorem by showing that every closed, nonorientable 3-manifold has an open book decomposition. (Hint: You may have to extend the construction of open book decompositions. Look at the manifold E.)

8.4. Raymond's Theorem

The foliations constructed in the previous sections have rather simple structure: all leaves are proper, the compact leaves are tori and the others are each homeomorphic to a disc minus a finite set of points. It is natural, then, to ask how complicated a foliation and its leaves can be.

With a little knowledge of knot theory one can construct foliations having a leaf homeomorphic to a closed surface of higher genus minus a finite set of points. This is because many knots occur as bindings of open book decompositions of the 3-sphere. Realizing closed surfaces as leaves is not always possible. Indeed, Reeb stability, together with a result of Haefliger (*cf.* [I, Theorem 6.1.1]), implies that S^2 can only be a leaf in $S^1 \times S^2$ or in the nonorientable S^2-bundle E. Furthermore, in many closed 3-manifolds, the only possible closed leaf is T^2 [I, Corollary 6.3.8]. On the other hand, every open, orientable surface can be realized as a leaf of some foliation of an arbitrary 3-manifold and every open, nonorientable surface can be so realized in an arbitrary nonorientable 3-manifold. This is a fairly difficult theorem [**21**] having no analogue in higher dimensional manifolds [**78, 167**].

An interesting problem is that of realizing a minimal set in a given 3-manifold. The case of a closed surface was just mentioned. Furthermore, not every 3-manifold admits a foliation with all leaves dense. There remain exceptional minimal sets. Here, we will prove a surprising theorem of B. Raymond concerning the realizability of such sets.

Theorem 8.4.1 (Raymond). *There is a smooth, transversely orientable foliation of the solid torus, tangent to the boundary and infinitesimally C^∞-trivial there, which has an exceptional minimal set X.*

Of course, being minimal but not being a single compact leaf, X lies in the interior of the solid torus. In the proofs of Theorems 8.1.1 and 8.3.3, replace one of the Reeb components with this foliated torus, proving the following.

Corollary 8.4.2 (Raymond). *Every closed 3-manifold has a smooth, transversely orientable foliation having an exceptional minimal set.*

The construction uses a little knot theory. If $k \subset S^3$ is a knot (assumed to be smoothly imbedded), we denote a closed tubular neighborhood of k by $T(k)$. It is assumed that this is smoothly imbedded, so $M(k) = S^3 \setminus \operatorname{int} T(k)$ is a compact manifold with smooth boundary, called the knot complement.

Let $\kappa \subset S^3$ be the $(2,3)$-torus knot (the trefoil knot), κ^* its reflection in some hyperplane (the $(-2,3)$-torus knot). To avoid possible confusion later, recall that the torus knots of types (p,q), (q,p) and $(-p,-q)$ are all equivalent by an isotopy of S^3 [**15**, Proposition 3.27]. It is well known, however, that κ and κ^* are not equivalent.

The connected sum $\kappa \# \kappa^*$ is the square knot κ' (see Figure 8.4.1). We set $M = M(\kappa)$ and $M' = M(\kappa')$. Our basic plan is to foliate M' so that its boundary torus is a leaf and so that there is an exceptional minimal set in $\operatorname{int} M'$. One then fills in $T(\kappa')$ with a Reeb foliated solid torus to obtain a foliation of S^3 with an exceptional minimal set X. The final step is to find a closed, unknotted transversal σ to the foliation which does not meet X. Turbulizing along σ and removing the interior of the Reeb component gives a foliated solid torus having the boundary as a leaf and an exceptional minimal set in the interior.

In what follows, we identify S^1 as the coset space $\mathbb{R}/\mathbb{Z}$. Recall that κ is the binding of an open book decomposition (Example 8.2.2). This fibers M over the circle with fibers F meeting ∂M in longitudes of the knot. Let

$$V = S^1 \times [-1,1] \times S^1 \subset M$$

be a collar neighborhood of $\partial M = S^1 \times \{-1\} \times S^1$, so realized that, for arbitrary $z, w \in S^1$,

$$m_z = S^1 \times \{-1\} \times \{z\}$$

is a meridian of κ and

$$\ell_w = \{w\} \times \{-1\} \times S^1$$

is a longitude. In the coordinates (w,z), ∂M is naturally identified with $\mathbb{R}^2/\mathbb{Z}^2$, the meridians (respectively, the longitudes) being tangent to the

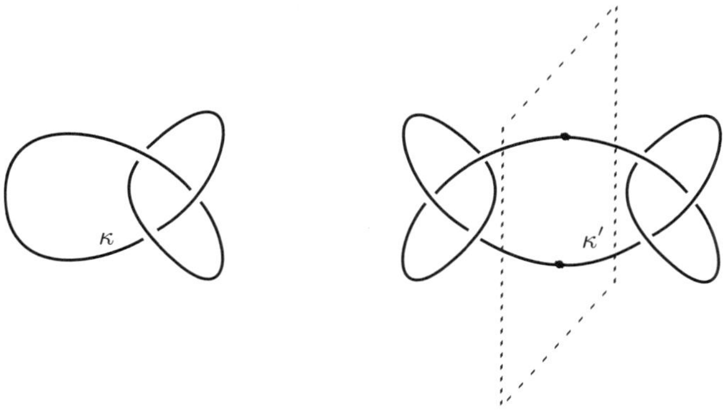

Figure 8.4.1. The trefoil and square knots

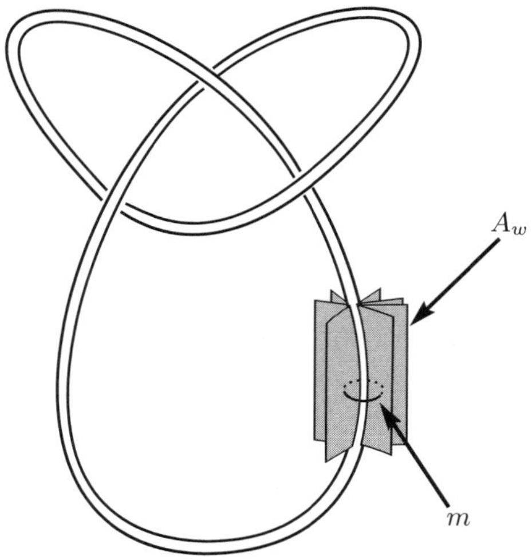

Figure 8.4.2. The open book fibration near ∂M

constant vector field

$$\mu = \begin{bmatrix} 1 \\ 0 \end{bmatrix} \quad \left(\text{respectively, } \lambda = \begin{bmatrix} 0 \\ 1 \end{bmatrix}\right).$$

Furthermore, each annulus $A_w = \{w\} \times [-1,1] \times S^1$ lies in a fiber F_w where it is a collar neighborhood of ∂F_w. This is indicated in Figure 8.4.2, where the small circle $m \subset V$ is $S^1 \times \{0\} \times \{1\}$ parallel to the meridians of κ.

Let N denote a closed, normal neighborhood of the curve m and let $W = M \setminus \operatorname{int} N$. This manifold is depicted in Figure 8.4.3 (it is the exterior of the tubes $T(\kappa) \cup N$, together with the point at infinity). In that figure,

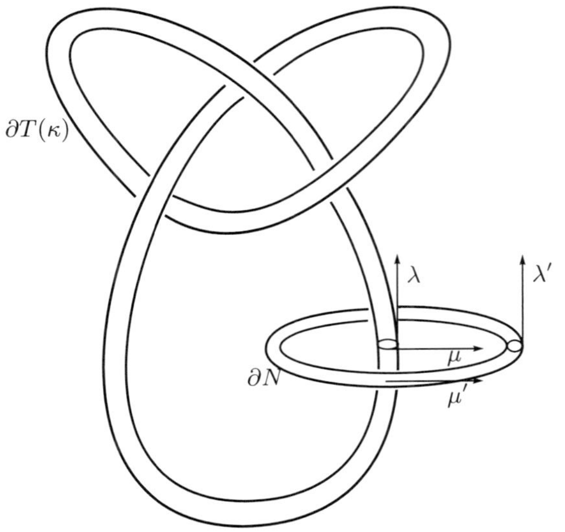

Figure 8.4.3. The manifold W

we indicate the tangent fields μ and λ on $\partial T(\kappa)$ corresponding to the coordinates (w, z). Coordinate fields μ' and λ' on ∂N are also indicated. The μ'-trajectories are circles isotopic in W to the meridians on $\partial T(\kappa)$ and the λ'-trajectories are homologous in W to the longitudes. These fields determine the standard coordinates (w', z') on ∂N.

Lemma 8.4.3. *There is a diffeomorphism of W to itself which interchanges the two boundary components and, relative to the respective coordinates (w, z) and (w', z'), is the identity map.*

Proof. The manifold $V \setminus \operatorname{int} N$ is naturally identified with $S^1 \times P$, where $\{w\} \times P$ is the pair of pants surface $A_w \setminus \operatorname{int} N$, for each $w \in S^1$. Let ℓ_w and ℓ'_w be the components of ∂A_w which are λ- and λ'-trajectories, respectively. There is a diffeomorphism of P which interchanges two boundary components and is the identity near the third. Applying this to each A_w so as to interchange ℓ_w and ℓ'_w defines a diffeomorphism of $V \setminus \operatorname{int} N$ which extends smoothly by the identity to W and has the asserted properties. □

It is intuitively helpful to picture W in a slightly different way. Using the homogeneity of S^3, one can move the point ∞ into the interior of N. This "blows up" the wormhole N, making it look like the outside of a solid torus, inside of which the trefoil-knotted wormhole looks more or less as

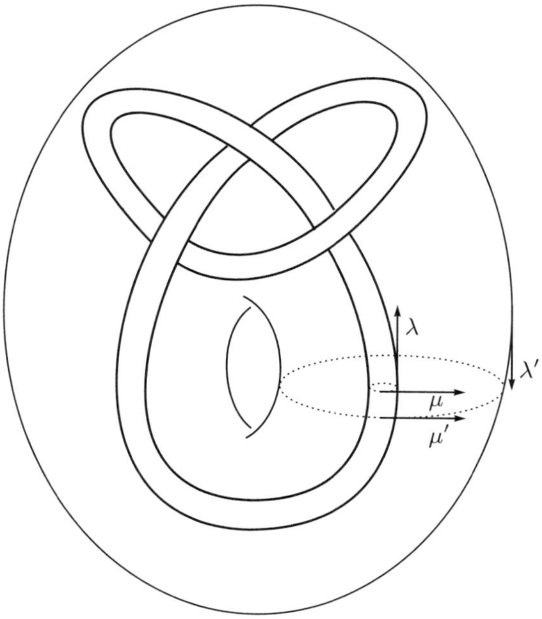

Figure 8.4.4. Another view of W

before. This is pictured in Figure 8.4.4, along with sample vectors from the coordinate fields on ∂W.

For $i = 1, 2$, let $M_i = M \times \{i\}$. In each of these copies of M, the objects defined above are likewise duplicated. That is, $N_i = N \times \{i\}$, $m_i = m \times \{i\}$, $\lambda_i = \lambda \times \{i\}$, *etc.* If $\iota : \partial M_1 \to \partial M_2$ is defined by

$$\iota(x, 1) = (x, 2),$$

the quotient space

$$DM = M_2 \cup_\iota M_1,$$

obtained by gluing M_1 to M_2 via ι, is called the *double* of M.

Lemma 8.4.4. *The manifold* $DM \smallsetminus \operatorname{int} N_1$ *is diffeomorphic to* M', *the complement of the square knot* $\kappa' = \kappa \# \kappa^*$, *and* m_2 *is isotopic in* M' *to a meridian of the square knot.*

Proof. Indeed,

$$DM \smallsetminus \operatorname{int} N_1 = M_2 \cup_\iota W_1,$$

and one applies Lemma 8.4.3 to W_1 so as to view the glued manifold as in Figure 8.4.5. The shading indicates the wormhole left by excising $\operatorname{int} N_1$. The figure also makes clear the assertion about m_2. □

The key construction is the following.

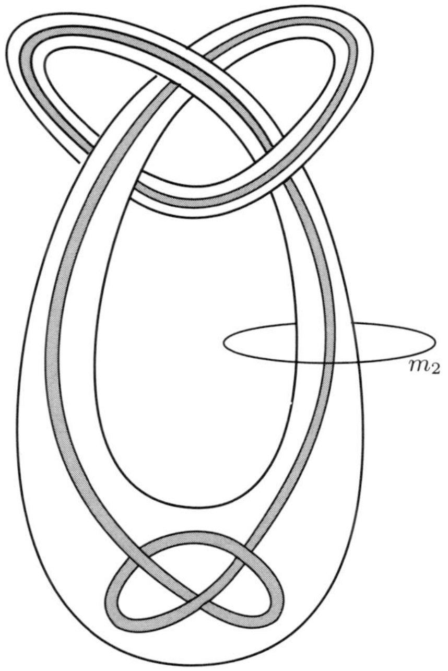

Figure 8.4.5. $M' = DM \smallsetminus \operatorname{int} N_1$

Proposition 8.4.5. *There is a smooth, transversely orientable foliation $\mathcal{F}$ of M, transverse to ∂M and having an exceptional minimal set X which meets ∂M. Furthermore, m is (isotopic to) a closed transversal to $\mathcal{F}$ missing X.*

Proof of Theorem 8.4.1 using Proposition 8.4.5. Let $(M_i, \mathcal{F}_i)$ be two copies of the foliated manifold of Proposition 8.4.5, $i = 1, 2$. Then the transverse gluing procedure developed in [I, Subsection 3.3.A] allows us to produce a doubled foliation $D\mathcal{F}$ of DM. Since $\mathcal{F}$ has an exceptional minimal set X meeting ∂M, the double DX of X is an exceptional minimal set of $D\mathcal{F}$. Since m_1 is transverse to $\mathcal{F}_1$ and does not meet X_1, $DM \smallsetminus \operatorname{int} N_1$ is foliated by the restriction of $D\mathcal{F}$ so that ∂N_1 is foliated by circles and does not meet the exceptional minimal set DX. Furthermore, m_2 is transverse to the foliation and does not meet DX. Since $\partial M'$ is foliated by circles, it is defined by a closed 1-form, and the spinning construction [I, Subsection 3.3.B] produces a smooth foliation. By [I, Exercise 3.4.5], this foliation is infinitesimally C^∞-trivial at $\partial M'$, and we can fill in the knotted wormhole with a Reeb component so as to obtain a smooth foliation of S^3 with an exceptional minimal set. The closed transversal m_2 is unknotted (it is a meridian of the square knot), so we can remove $\operatorname{int} N_2$ and spin so as to obtain a foliated solid torus as in Theorem 8.4.1. □

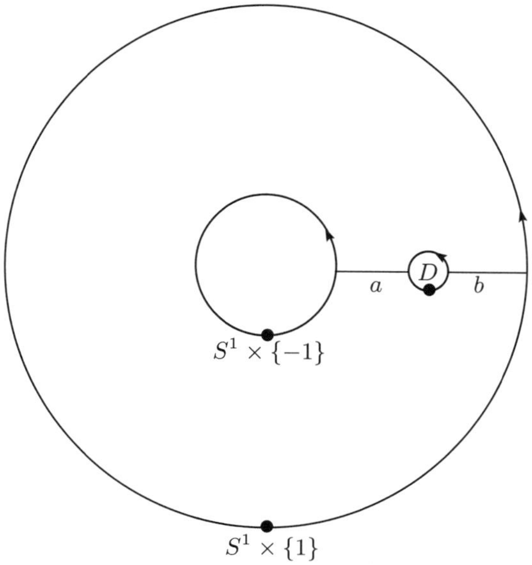

Figure 8.4.6. The pair of pants P

It remains for us to prove Proposition 8.4.5. We begin with a construction that will also be useful in the next section.

Let $A = S^1 \times [-1, 1]$, $D \subset \operatorname{int} A$ a small disk centered at $\{(0, 0)\}$ (where, as usual, we coordinatize S^1 as $\mathbb{R}/\mathbb{Z}$). Then $P = A \setminus \operatorname{int} D$ is a pair of pants as pictured in Figure 8.4.6, where we also indicate two properly imbedded arcs a and b. The three boundary tori of $P \times S^1$ will be denoted by

$$\begin{aligned} T_1 &= S^1 \times \{1\} \times S^1, \\ T_2 &= \partial D \times S^1, \\ T_3 &= S^1 \times \{-1\} \times S^1. \end{aligned}$$

Lemma 8.4.6. *Let $f, g \in \operatorname{Diff}^k_+(S^1)$, where $0 \le k \le \infty$ or $k = \omega$. Then there is a transversely oriented C^k foliation of $P \times S^1$, transverse to the circle factors $\{x\} \times S^1$, inducing foliations with first return maps $f, g^{-1} \circ f$ and g, respectively, on the boundary tori T_1, T_2 and T_3.*

Proof. In Figure 8.4.6, we indicate the meridional directions around the boundary tori relative to which we define the first return map. We also indicate by dots the circle fibers on which the respective first return maps are defined. Cut P along $a \cup b$, denoting the two resulting copies of each of these arcs by $a^\pm$ and $b^\pm$, respectively, the sign being chosen consistently with the meridional directions. Points $u \in a \cup b$ split into two corresponding points $u^{\pm}$ in the respective copies of $a \cup b$. If we cut $P \times S^1$ apart along the annuli $a \times S^1$ and $b \times S^1$ and reglue by

$$(u^-, y) \equiv (u^+, g(y)), \quad u \in a, \quad y \in S^1,$$
$$(u^-, y) \equiv (u^+, f(y)), \quad u \in b, \quad y \in S^1,$$

the foliation by the factors $P \times \{y\}$ is converted into a suspension foliation with the desired boundary behaviour. □

In $A \times S^1$, each torus $T_x = S^1 \times \{x\} \times S^1$, $-1 \leq x \leq 1$, has canonical coordinates $(w, z) \in \mathbb{R}^2/\mathbb{Z}^2$, enabling us to talk about (p, q)-torus knots in each of these concentric tori. We are particularly interested in the boundary tori $S^1 \times \{\pm 1\} \times S^1$ and the torus $S^1 \times \{0\} \times S^1$ through the center of D.

Lemma 8.4.7. *There is a transversely oriented C^ω foliation $\mathcal{F}'$ of $P \times S^1$, transverse to the three boundary tori, having an exceptional minimal set, inducing a foliation of $S^1 \times \{-1\} \times S^1$ by $(2,1)$-torus knots and inducing a foliation of $S^1 \times \{1\} \times S^1$ by $(3,1)$-torus knots.*

Proof. We adapt the construction of an exceptional minimal set in [I, Example 4.1.6]. Let $f, g \in \mathrm{Diff}^\omega_+(S^1)$ be defined as in that example. Recall that $f^3 = g^2 = \mathrm{id}$ and that the subgroup $G \subset \mathrm{Diff}^\omega_+(S^1)$ generated by f and g has a minimal, invariant Cantor set $C \subset S^1$. An application of Lemma 8.4.6 produces a suspension foliation with total holonomy group G. □

Exercise 8.4.8. Observe that the holonomy of $\mathcal{F}'$ around ∂D is

$$h = g^{-1} \circ f = g \circ F.$$

Show that h has exactly two fixed points, these being endpoints of a gap of the minimal Cantor set C, at one of which it is a 2-sided contraction and at the other of which it is a 2-sided expansion. Thus, the foliation induced on $\partial D \times S^1$ has two leaves which are circles, the remaining leaves all spiraling from one of these circles to the other. Use [I, Lemma 3.3.7] to prove that either of these circles can be isotoped to an arbitrarily nearby closed transversal to $\mathcal{F}'$ which does not meet the exceptional minimal set.

Proof of Proposition 8.4.5. Relative to the coordinates (w, x, z) of the manifold $A \times S^1$, $(w, z) \in \mathbb{R}^2/\mathbb{Z}^2$, $-1 \leq x \leq 1$, define the linear diffeomorphism θ by

$$\theta(w, x, z) = (w - 3z, x, w - 2z).$$

This is well defined, since

$$\det \begin{bmatrix} 1 & -3 \\ 1 & -2 \end{bmatrix} = 1.$$

This takes the $(2,1)$-circles on $S^1 \times \{-1\} \times S^1$ to $(-1,0)$-circles (meridians) and the $(3,1)$-circles on $S^1 \times \{1\} \times S^1$ to $(0,1)$-circles (longitudes). Furthermore, it takes the $(0,1)$-circle $\{(0,0)\} \times S^1$ to a $(-3,-2)$-torus knot. Thus, apply this diffeomorphism to $P \times S^1$ to obtain $A \times S^1 \smallsetminus V$, where V is a trefoil-knotted wormhole. The foliation $\mathcal{F}'$ is transformed into one meeting $S^1 \times \{-1\} \times S^1$ in meridians and $S^1 \times \{1\} \times S^1$ in longitudes. Gluing on two solid tori, foliated by disks meeting the boundary in meridians, we obtain a smoothly (in fact, analytically) foliated copy of $M = M(\kappa)$. The assertion in Proposition 8.4.5 about the loop m is an immediate consequence of Exercise 8.4.8. $\square$

Exercise 8.4.9. Investigate the holonomy of the exceptional minimal set X in Proposition 8.4.5. Observe that you may as well view X as the exceptional minimal set of Lemma 8.4.7 and study the dynamics of the total holonomy group G on $C \subset S^1$.

(1) If $y \in C$ and L is the leaf of X through y, the holonomy group $\mathcal{H}_y(L)$ is the group of germs at y of elements of G that fix y. Show that such a germ of $g \in G$ is nontrivial if and only if there is a neighborhood of y in S^1 on which g has no fixed point other than y.

(2) By the above, the germ $[g] \in \mathcal{H}_y(L)$ is nontrivial if and only if $g|C$ has nontrivial germ at y. Using this observation and symbolic dynamics (*cf.* [I, Exercise 4.1.8]), prove that $\mathcal{H}_y(L)$ is nontrivial for exactly a countable infinity of leaves in X and that, for these leaves, it is infinite cyclic.

(3) By [I, Exercise 4.1.8], there are exactly two semiproper leaves in X, and, by Exercise 8.4.8, each of these leaves has holonomy generated by the germ of a 2-sided contraction. Arguing as in the proof of [I, Theorem 8.1.26], show that this contraction corresponds to a compact juncture.

Exercise 8.4.10. Let L be a semiproper leaf of the exceptional minimal set X in Proposition 8.4.5. Let $\mathcal{E}(L)$ denote the endset of L, a compact, totally disconnected, metrizable space of ideal points at infinity (*cf.* [I, Sections 4.2 and 4.3]) and let $e \in \mathcal{E}(L)$. Using the final result of Exercise 8.4.9 and arguing in analogy with the proof of [I, Theorem 8.4.6], prove that every neighborhood of e in L contains a neighborhood B which spirals on L (see [I, Definition 8.4.2]). Deduce from this that $\mathcal{E}(L)$ has no isolated point, hence is a Cantor set. Finally, use this to prove that every leaf in X has a Cantor set of ends.

Remark. An unpublished theorem of G. Duminy asserts that, for codimension one foliations of class at least C^2, every semiproper leaf of an exceptional minimal set has a Cantor set of ends. It is unknown whether every leaf in the minimal set has a Cantor set of ends, although G. Hector has shown this for real analytic foliations. The idea behind these proofs is illustrated by Exercise 8.4.10.

8.5. Thurston's Construction

In addition to proving the existence of transversely orientable foliations by surfaces on all closed 3-manifolds, J. Wood proved that an arbitrary, transversely orientable 2-plane field on such a manifold is homotopic to an integrable one. We did not discuss this in Section 8.3, but here we give a detailed sketch of a very different approach, due to W. Thurston [**173**], proving the same result without the hypothesis of transverse orientability. A critical lemma (Lemma 8.5.2) was proven by Thurston using results of J. Mather [**126, 127**], of M. Herman and F. Sergeraert [**95**] and of D. B. A. Epstein [**60**] which, taken together, imply that $\mathrm{Diff}^\infty_+(S^1)$ is a simple group. We will give an alternative and somewhat simpler proof of this lemma, using instead an equally deep result of Herman [**94**] concerning rotation numbers of diffeomorphisms of S^1. We are grateful to Paul Schweitzer for considerable help in understanding Thurston's construction and especially for providing this proof of Lemma 8.5.2.

Theorem 8.5.1 (Thurston). *Every 2-plane field τ on a closed 3-manifold is homotopic to an integrable plane field.*

We emphasize that this theorem has no assumptions of orientability on M or τ, nor of transverse orientability on τ. A key step in the proof of Theorem 8.5.1 is the following result, which is interesting in its own right.

Lemma 8.5.2. *Let τ be a 2-plane field on $S^1 \times D^2$ transverse to the foliation by longitudinal circles $S^1 \times \{x\}$. Then τ is homotopic to an integrable plane field by a homotopy that keeps τ fixed on $S^1 \times \partial D^2$.*

The proof requires some preliminary discussion.

Recall that, to each $f \in \mathrm{Homeo}_+(S^1)$, there is assigned a Poincaré rotation number $\rho(f) \in \mathbb{R}/\mathbb{Z}$ [I, Exercise 9.2.19]. If $\alpha = \rho(f)$ is irrational and $f \in \mathrm{Diff}^2_+(S^1)$, Denjoy's theorem asserts that f is topologically conjugate to the rotation $R_{\rho(f)}$ of S^1 through $2\pi\rho(f)$ radians. In fact, for almost every irrational value $\alpha \in \mathbb{R}/\mathbb{Z}$, this can be improved immensely.

Theorem 8.5.3 (Herman [**94**]). *For a set of $\alpha \in \mathbb{R}/\mathbb{Z}$ of full Lebesgue measure, all diffeomorphisms $f \in \mathrm{Diff}^\infty_+(S^1)$ with $\rho(f) = \alpha$ are C^∞-conjugate to R_α.*

This is a consequence of the Fundamental Theorem of Herman [**94**, page 8], together with 10.3 on page 67 of the same reference.

The numbers α satisfying Theorem 8.5.3 will be called *Herman numbers.* We will not need the full force of this theorem – just the existence of a Herman number.

For the following corollary, we need the fact that the map

$$\lambda_f : \mathbb{R}/\mathbb{Z} \to \mathbb{R}/\mathbb{Z},$$
$$\lambda_f(\beta) = \rho(f \circ R_{-\beta}),$$

is a continuous surjection, for each $f \in \text{Homeo}_+(S^1)$. For this, see [**94**, 2.7 on p. 21 and 1.3 on p. 31].

Corollary 8.5.4. *Every* $f \in \text{Diff}^\infty_+(S^1)$ *can be written*

$$f = (\gamma \circ R_\alpha \circ \gamma^{-1}) \circ R_\beta,$$

for suitable $\gamma \in \text{Diff}^\infty_+(S^1)$ *and* $\alpha, \beta \in \mathbb{R}/\mathbb{Z}$.

Proof. Let α be any Herman number and consider $\lambda_f(\beta)$ as defined above. Since this takes all values in $\mathbb{R}/\mathbb{Z}$ as β varies, we can choose β so that $\rho(f \circ R_{-\beta}) = \alpha$. By Theorem 8.5.3, there is $\gamma \in \text{Diff}^\infty_+(S^1)$ such that

$$f \circ R_{-\beta} = \gamma \circ R_\alpha \circ \gamma^{-1}.$$

□

Proof of Lemma 8.5.2. Since the plane field τ is transverse to the circle fibers, it induces a nonsingular line field on $S^1 \times \partial D^2$, hence a one-dimensional foliation on this torus. Being transverse to the circle factors, this foliation is determined, up to smooth conjugacy, by its first return map

$$f : S^1 \to S^1,$$

an orientation-preserving C^∞ diffeomorphism. Applying Corollary 8.5.4 to this map, we write

$$(\gamma \circ R_\alpha \circ \gamma^{-1})^{-1} \circ f = R_\beta.$$

In Lemma 8.4.6, take $f = f$ and $g = \gamma \circ R_\alpha \circ \gamma^{-1}$. Applying that lemma, we drill two longitudinal open tubes out of the interior of $S^1 \times D^2$ and construct a suspension foliation on the resulting manifold $S^1 \times P$, transverse to the boundary and having first return map f on the "outer" boundary torus, first return map $\gamma \circ R_\alpha \circ \gamma^{-1}$ on another one, hence first return map

$$(\gamma \circ R_\alpha \circ \gamma^{-1})^{-1} \circ f = R_\beta$$

on the remaining boundary torus. Since the foliations on the inner boundary tori are the suspensions of rotations (up to conjugacy), the spinning construction in [I, 3.3.B] can be carried out smoothly along these tori. Note that the tangent plane field to this foliation is homotopic to $\tau|(S^1 \times P)$ by

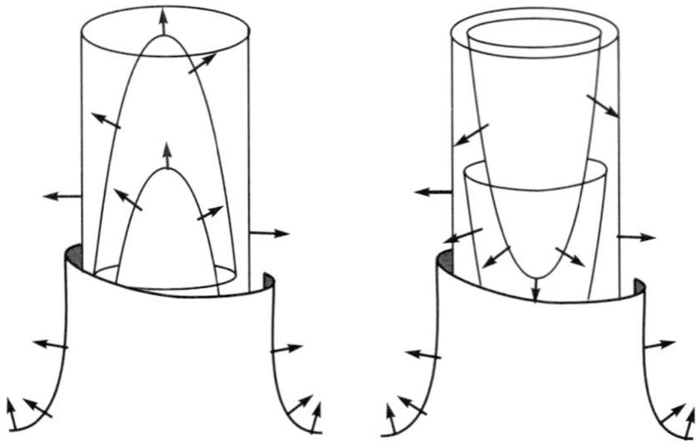

Figure 8.5.1. Right and wrong ways of inserting the Reeb components

a (large) homotopy that is constant on the outer boundary torus. If Reeb components, with leaves opening in the appropriate direction, are inserted into these inner wormholes, the resulting foliation has tangent plane field homotopic to τ. (See Figure 8.5.1 for illustrations of the right and wrong ways to direct the Reeb components.) □

The proof of Theorem 8.5.1 will proceed as follows. First, a way of subdividing a triangulation $\mathcal{D}(M)$ that does not produce overly misshapen simplices will be defined, the so-called "crystalline subdivision". After passing to the first barycentric subdivision of a suitably fine crystalline subdivision, we will be able to "jiggle" this to a triangulation $\mathcal{D}^*(M)$ such that τ is transverse to the 1- and 2-skeleton of $\mathcal{D}^*(M)$. After this, a small homotopy of τ, using a "counter-orientation" of the barycentric subdivision, will insure that the foliation on the 2-skeleton, induced by the transverse plane field, has certain nice properties. Another perturbation of the plane field makes it integrable in a neighborhood of the 2-skeleton. At this point, we have a partial foliation of M that has roughly spherical holes and we seem to be stuck. Indeed, Reeb stability will forbid an extension of the foliation across these holes. But the special properties of the partial foliation make it possible to find, for each 3-simplex σ, a transverse arc from one vertex of σ to another. Drilling out a neighborhood of these arcs replaces the spherical holes with toroidal wormholes, and Lemma 8.5.2 allows extension of the foliation across these wormholes. A more detailed account follows.

Fix a smooth triangulation $\mathcal{D}(M)$ of M. Such a triangulation can be viewed as a homeomorphism $\sigma : M_{\mathrm{PL}} \to M$, where M_{PL} is an affine simplicial complex in some Euclidean space $\mathbb{R}^n$, and σ is smooth on each simplex. If Δ is a simplex of M_{PL}, we can identify $\tau|\sigma(\Delta)$ with a smooth 2-plane field τ_Δ tangent to Δ. And we can interpret affine subdivisions of M_{PL} as smooth subdivisions of $\mathcal{D}(M)$. One should note that σ is not generally smooth at the interfaces of two simplices; hence the fields τ_Δ cannot be expected to fit together even continuously on M_{PL}.

One wants to perform small affine isotopies of M_{PL} in $\mathbb{R}^n$ by perturbing vertices along short line segments and extending the perturbation affinely to the simplicial complex M_{PL}. In order to interpret this as a smooth isotopy of $\mathcal{D}(M)$, one should be a bit careful. An elegant way to do this is attributed in [**172**] to A. Haefiger. One smoothly imbeds M in $\mathbb{R}^n$, n sufficiently large, and approximates M in a thin normal neighborhood $\nu(M)$ by an affine simplicial complex M_{PL}. Then σ is taken to be the restriction to this complex of the smooth bundle projection of $\nu(M)$ onto M. This projection can be composed with an affine isotopy of M_{PL} in $\nu(M)$, defined by small linear perturbations of the vertices, producing thereby a smooth isotopy of $\mathcal{D}(M)$. This is referred to by Thurston as a "jiggling" [**172**] of the triangulation.

Let Δ be a 3-simplex of M_{PL} and identify its tangent bundle as

$$T(\Delta) = \Delta \times \mathbb{R}^3.$$

Although τ is not assumed to be orientable nor transversely orientable, one can choose an orientation of τ_Δ. Since the Grassmann manifold $G_{2,3}$ of oriented 2-planes in 3-space is a 2-sphere, τ_Δ can then be viewed as a smooth map

$$\tau_\Delta : \Delta \to S^2.$$

Let Δ' denote the first barycentric subdivision of Δ and notice that the 2-planes in $T(\Delta)$ tangent to edges of Δ' determine finitely many circles $\Sigma_1, \dots, \Sigma_q \subset S^2$.

Definition 8.5.5. If $\operatorname{im} \tau_\Delta$ is disjoint from $\Sigma_1 \cup \cdots \cup \Sigma_q$, we will say that τ is in general position with respect to $\sigma(\Delta')$. If this holds for each 3-simplex Δ of M_{PL}, we will say that τ is in general position with respect to the first barycentric subdivision $\mathcal{D}'(M)$ of the triangulation $\mathcal{D}(M)$.

Note that this implies that τ is transverse not only to the 1-skeleton, but to the 2-skeleton of $\mathcal{D}'(M)$. Figure 8.5.2 indicates, via the induced 1-dimensional foliation on a 2-simplex, that general position is much stronger than transversality.

We will show that, after a suitably fine subdivision, the first barycentric subdivision of the triangulation can be slightly jiggled, as described above, so that τ is in general position. For this purpose, repeated barycentric subdivisions are bad, since they introduce more and more circles to be avoided by the plane field. Instead we use a *crystalline* subdivision.

Fix an integer N and intersect the affine 3-simplex Δ with a finite family of 2-planes parallel to a face of Δ and meeting each edge not on that face in

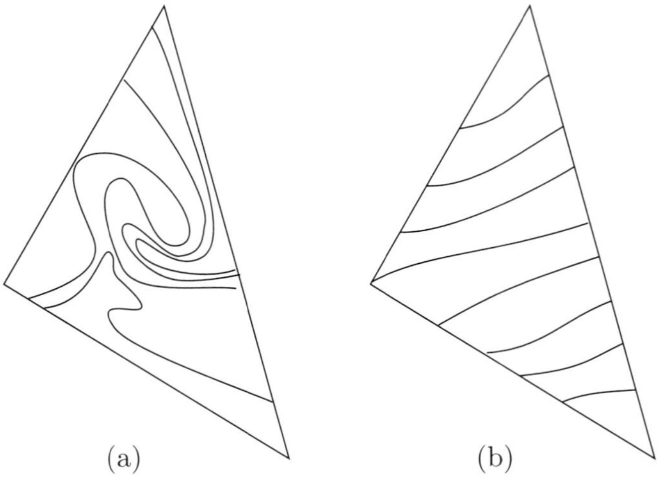

Figure 8.5.2. (a) The plane field is transverse to the 2-skeleton and (b) it is in general position.

N equally spaced points. Do this for each face. The result is a subdivision of Δ into simplices with faces and edges parallel to those of Δ itself. This will be called a crystalline subdivision of Δ. In Figure 8.5.3, we depict the induced subdivision on a face of Δ in the case $N = 3$. Notice that the simplices of the first barycentric subdivision, call it Δ^*, of this crystalline subdivision also consist of simplices with faces and edges parallel to those of the corresponding simplices of the first barycentric subdivision Δ' of Δ. Thus, the family of circles in S^2 to be missed by τ_Δ is unchanged for the simplices of Δ^*, but these simplices are made to be arbitrarily small by choosing N sufficiently large. By fixing a large value of N and carrying out the corresponding crystalline subdivision on each 3-simplex of M_{PL}, then passing to the first barycentric subdivision, we obtain a PL manifold M^*_{PL}. Let $\mathcal{D}^*(M)$ denote the smooth triangulation of M defined by $\sigma : M^*_{\text{PL}} \to M$.

Let v be a vertex of M^*_{PL} and let star(v) denote the star of v in this PL manifold. Let $\Delta_1, \dots, \Delta_r$ be the 3-simplices of this star, and set $\tau_i = \tau_{\Delta_i}$, $1 \le i \le r$. Note that there is a uniform bound on r, independent of the choice of crystalline subdivision. If the crystalline subdivision is fine enough, these plane fields can be thought of as essentially constant, and a small linear perturbation of v, fixing all of the other vertices of star(v), will perturb the edges so that the corresponding circles in the Grassmann manifold $G_{2,3} = S^2$ are moved off of the images of the τ_i, $1 \le i \le r$. In [**172**, Section 5], the details are given for n-dimensional complexes in $\mathbb{R}^n$, and these apply to our situation provided that star(v) can be viewed as an affine simplicial complex

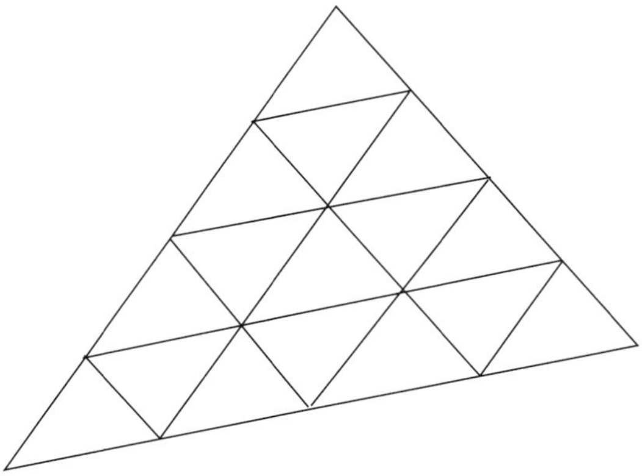

Figure 8.5.3. Crystalline subdivision of a 2-simplex

in $\mathbb{R}^3$. It is often the case that star(v) is contained in one of the original affine 3-simplices of M_{PL}, but this is not generally so. In general, an affine projection $p : \mathbb{R}^n \to \mathbb{R}^3$ can be found that imbeds star(v) as such a complex and carries a small neighborhood of v in $\mathbb{R}^n$ onto a small neighborhood of $p(v)$ in $\mathbb{R}^3$. One can then choose a linear preimage of the perturbation of $p(v)$ in $\mathbb{R}^3$.

Thus, we carry out the perturbation simplex-by-simplex in M^*_{PL}, guaranteeing that it is uniformly as small as desired on each simplex. Via the projection $\nu(M) \to M$, this is interpreted as a jiggling in M of the triangulation so that τ is in general position with respect to $\mathcal{D}^*(M)$. For notational simplicity, replace $\mathcal{D}(M)$ with a suitably fine crystalline subdivision and conclude the following.

Lemma 8.5.6. *There is a smooth triangulation $\mathcal{D}(M)$ such that τ is in general position with respect to the first barycentric subdivision $\mathcal{D}'(M)$. In particular, τ is transverse to the* 1- *and* 2-*skeleta of $\mathcal{D}'(M)$.*

By the definition of general position, the pull-back τ_Δ of τ to the affine preimage Δ of a 3-simplex σ_Δ of $\mathcal{D}'(M)$ has the property that no fiber is parallel to a face or an edge of Δ. Thus, each fiber of τ_Δ defines an affine projection of Δ onto a nondegenerate interval, giving a linear ordering (up to a reversal) of the vertices of σ_Δ. For a fine enough crystalline subdivision, the plane field τ_Δ is practically constant on Δ, and this linear ordering of the vertices is independent of the choice of fiber. There are two extreme vertices ("top" and "bottom", but these labels are interchangeable) and two intermediate ones. We will sometimes refer to the extreme vertices as the "poles". The one-dimensional, piecewise smooth foliation induced by τ on the boundary of $\sigma(\Delta)$ has no Reeb components, and it winds around this triangulated 2-sphere with the extreme vertices as sole singularities.

Construct a small foliated chart U_x about each vertex x in such a way that the fiber τ_x is tangent to the plaque through x. Making U_x sufficiently small, we see that an arbitrarily small homotopy makes $\tau|U_x$ tangent to the plaques without undoing the features of τ obtained thus far. If x is a pole of the 3-simplex $\sigma(\Delta)$ of $\mathcal{D}'(M)$, then the foliation on $\partial\sigma(\Delta)$ has closed leaves near the singularity x, and this singularity looks like a local maximum or minimum of a function. The closure of the union of x and these closed leaves is an imbedded disk D_x in $\partial\sigma(\Delta)$ (a "polar icecap").

Definition 8.5.7. A *counter-orientation* of a triangulation of an n-manifold M is an assignment of orientation to each n-simplex σ so that the induced orientations on a common $(n-1)$-dimensional face of two n-simplices always agree. A triangulation, together with a counter-orientation, is said to be a counter-oriented triangulation.

Another way to say this is that any two n-simplices that share a common $(n-1)$-dimensional face are oppositely oriented. While only orientable manifolds support an oriented triangulation, all manifolds support a counter-oriented triangulation. Indeed, we have the following.

Lemma 8.5.8. *If $\mathcal{D}(M)$ is a triangulation of an n-manifold M, then the first barycentric subdivision $\mathcal{D}'(M)$ has a canonical counter-orientation.*

Proof. Indeed, an n-simplex of $\mathcal{D}'(M)$ can be canonically written as an ordered n-tuple $\sigma = (x_0, x_1, \dots, x_n)$, where x_i is the unique vertex of σ that is the barycenter of an i-simplex of $\mathcal{D}(M)$, $0 \le i \le n$. This canonical ordering of the vertices, modulo even permutations, defines a canonical orientation of σ. The face σ_k opposite the vertex x_k, with orientation induced by the orientation of σ, can be written

$$\sigma_k = (-1)^k (x_0, \dots, \widehat{x}_k, \dots, x_n), \quad 0 \le k \le n.$$

There is one other n-simplex $\sigma' = (x_0, \dots, x'_k, \dots, x_n)$ sharing the face σ_k with σ and the orientation on σ_k induced from σ' is clearly the same. □

Of course, the reason we are using the first barycentric subdivision $\mathcal{D}'(M)$ is to take advantage of the canonical counter-orientation. This will be used to homotop the plane field τ so that it remains transverse to the 1- and 2-skeleta, but so that the induced foliation on the boundary of each 3-simplex $\sigma(\Delta)$ has only noncompact leaves outside small neighborhoods of the poles and these leaves spiral on closed leaves near the poles. The direction of spiraling will be dictated by the orientation assigned to $\sigma(\Delta)$ via the counter-orientation of $\mathcal{D}'(M)$. The fact that interfacing 3-simplices are

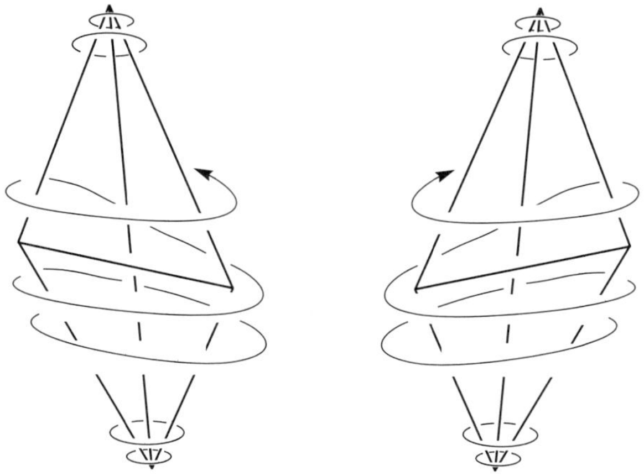

Figure 8.5.4. The spirals about interfacing 3-simplices σ and σ'

oppositely oriented makes the spirals compatible. More precisely, relative to a choice of local orientation of M in a neighborhood of two interfacing 3-simplices σ and σ', the spirals around one will be oriented as a right-handed screw and those around the other as a left-handed screw. (See Figure 8.5.4, where the simplices are separated and the leaves of the induced foliations are slightly displaced off of $\partial\sigma$ and $\partial\sigma'$.) We will now spell this out in greater detail.

Let $\sigma(\Delta)$ be a 3-simplex of $\mathcal{D}'(M)$, $\sigma(F)$ a face corresponding to the 2-face F of the affine simplex Δ. Let D_x and D_y be the polar icecaps on $\partial\sigma(\Delta)$ and let z be an interior point of $\sigma(F)$ outside the polar icecaps. The canonical orientation induced on $\sigma(F)$ gives a rotational direction in which we will rotate the fiber τ_z of τ. Indeed, an orientation of a 2-simplex is just a cyclic order of its vertices. Fixing a Riemannian metric, we decompose τ_z into the direct sum of the line ℓ_z tangent to $\sigma(F)$ and the line $\ell_z^\perp$ perpendicular to ℓ_z and rotate τ_z around the axis $\ell_z^\perp$. This extends to a rotational homotopy of the plane field that is supported in a small neighborhood of z in M. This rotational homotopy need not be arbitrarily small, but will be supported away from the 1-skeleton. In particular, this perturbed plane field will remain transverse to the 1- and 2-skeleta of $\mathcal{D}'(M)$ and will remain integrable near each vertex. The property that the planes (pulled back to the affine complex) are not parallel to an edge or face is no longer needed, the poles being determined now as the center singularities of the foliation on $\partial\sigma(\Delta)$.

A sufficiently large rotation at z guarantees that the leaf of the induced foliation of $\partial\sigma(\Delta)$ through z will spiral about $\sigma(\Delta)$ in the sense corresponding to the orientation of that simplex. This leaf will spiral on circle leaves that may not bound the respective polar icecaps but can be made as close as desired to these boundaries. Note that rotations on other faces of $\sigma(\Delta)$ only reinforce this spiraling effect, so the counter-orientation allows this to be carried out compatibly for all of the 3-simplices.

Lemma 8.5.9. *There is a homotopy of τ, keeping that plane field transverse to the 1- and 2-sleleta of $\mathcal{D}'(M)$, possibly shrinking the polar icecaps on the boundary of each 3-simplex, and such that, on each such boundary, a family of noncompact leaves spiral in on two circle leaves that are as close as desired to the respective poles.*

The next major move is to "inflate" the foliation on the 2-skeleton to a two-dimensional foliation in a thin normal neighborhood of the 2-skeleton and to do this compatibly with the foliation already present in the neighborhood of each vertex. Since the 2-skeleton is a neighborhood retract, it is intuitively transparent that this can be done and that the plane field τ can be homotoped in a smaller neighborhood to be tangent to this foliation. For more details, see [**172**, Section 6].

Lemma 8.5.10. *There is a homotopy of τ, supported in a neighborhood of the 2-skeleton, after which τ is integrable on a smaller neighborhood of the 2-skeleton and continues to induce the same foliation on the 2-skeleton.*

For the rest of the construction we will be working in the manifold M and will abuse notation, writing Δ for a simplex $\sigma(\Delta)$ of $\mathcal{D}'(M)$.

By Lemma 8.5.9, the noncompact leaves of the inflated foliation, restricted to a neighborhood of the boundary of each 3-simplex Δ, look like "spiral ramps" winding in on leaves near the poles. Denote by $(U, \mathcal{F}_U)$ the foliated neighborhood of the 2-skeleton. The complement of U in M consists of closed, disjoint topological balls, one in the interior of each 3-simplex. With a little care, each of these balls can be taken to be a smoothly imbedded cylinder $D^2 \times [-1,1]$ so that the disks $D^2 \times \{\pm 1\}$ lie in leaves of $\mathcal{F}_U$ and the lateral boundary $S^1 \times [-1,1]$ has foliation induced by $\mathcal{F}_U$ that is transverse to the fibers $\{z\} \times [-1,1]$. Since this induced foliation has spiral leaves, hence nontrivial holonomy, Reeb stability obstructs any attempt to extend the foliation over these cylinders. The key to resolving this impasse is the following.

Lemma 8.5.11. *For each 3-simplex Δ of $\mathcal{D}'(M)$, there is an $\mathcal{F}_U$-transverse arc in U from one pole of Δ to the other.*

Proof. Let x and y denote the poles. Since the triangulation is as fine as desired, we can assume that the connected subcomplex $X = \text{star}(x) \cup \text{star}(y)$ lies in a neighborhood in M on which τ is transversely orientable. Relative to this orientation, assume that x is the top vertex of Δ and y the bottom one. We can then choose a sequence $\Delta_0, \Delta_1, \dots, \Delta_m$ of 3-simplices of X with the following properties:

(1) A suitably oriented edge ℓ' of Δ_0 issues from x and is a positively directed transversal to $\mathcal{F}_U$.

(2) A suitably oriented edge ℓ'' of Δ_m is a positively directed transversal to $\mathcal{F}_U$ and terminates at y.

(3) The simplex Δ_{k-1} has a 2-face in common with Δ_k, $1 \le k \le m$.

Now construct a path ℓ that follows a segment of ℓ' at x out to one of the spiral ramps winding about Δ_0, then follows this ramp to one winding about Δ_1, continuing in this way until the path enters a spiral ramp about Δ_m. This ramp meets ℓ'', so we can complete the path to end with a segment of ℓ'' ending at y. At this point, a standard construction (see [I, 3.3.C]) allows us to perturb ℓ to the desired transverse arc. □

Proof of Theorem 8.5.1. We can assume that the transverse arcs given by Lemma 8.5.11 are all disjoint. Extend each end of each of these arcs into the interior of its associated 3-simplex Δ until it meets a boundary disk of the hole $D^2 \times [-1, 1] \subset \operatorname{int} \Delta$ described earlier. Drill out disjoint tubular neighborhoods of these arcs, rounding at the corners, to produce toroidal wormholes $D^2 \times S^1$. The foliation $\mathcal{F}_U$ induces on the boundary of each such wormhole a foliation transverse to the longitudinal circles $\{z\} \times S^1$. By Lemma 8.5.2, the plane field τ is homotopic in each of these solid tori to an integrable plane field and the homotopy fixes τ on the boundary. □

It should be noted that Thurston has general versions of this result for all q-plane fields on all closed n-manifolds [**172, 174**]. Generally the homotopy in question cannot be smooth, since the Bott vanishing theorem (Theorem 6.1.1) implies that many plane fields are not smoothly homotopic to foliations.

Chapter 9

Reebless Foliations

Much of this chapter will be devoted to conditions on a compact, orientable and transversely orientable foliated 3-manifold that force the presence of a leaf-saturated set homeomorphic to a Reeb component. Our main point, however, is the obverse principle that *Reebless foliations reflect interesting topological features of the ambient manifold.*

In a foundational paper [**141**], S. P. Novikov introduced the concept of a *vanishing cycle* for foliated 3-manifolds $(M, \mathfrak{F})$, proving that the presence of a vanishing cycle is a necessary and sufficient condition for the presence of a Reeb component. Roughly speaking, a vanishing cycle is a loop σ_0 on a leaf L_0 that is essential on L_0 but admits arbitrarily small displacements σ_t to nearby leaves L_t where σ_t is inessential. For example, a meridian on the boundary leaf of a Reeb component is a vanishing cycle. The hard part of Novikov's theorem is that, in compact, foliated 3-manifolds, this is the only way a vanishing cycle can occur.

Adapting an idea of Haefliger (*cf.* [I, Chapter 7]), Novikov uses the classical Poincaré-Bendixson theory to give sufficient conditions for the presence of a vanishing cycle. For instance, the existence of a closed, nullhomotopic transversal to $\mathfrak{F}$ implies that there is a vanishing cycle; hence closed manifolds with finite fundamental group only admit foliations with Reeb components. A similar argument shows that the existence of a loop on a leaf L that is inessential in M, but essential on L, implies the existence of a vanishing cycle. An immediate corollary is that the leaves of Reebless foliations are π_1-injective. These theorems are proven by applying Poincaré-Bendixson theory to the singular foliations induced by $\mathfrak{F}$ on immersed disks. Similar arguments, applied to immersed 2-spheres, prove that a Reebless foliation $\mathfrak{F}$ either forces $\pi_2(M) = 0$ or forces M to be diffeomorphic to $S^2 \times S^1$ (or to $S^2 \times I$ if $\partial M \neq \emptyset$) with the S^2 factors as leaves. This suggests that manifolds admitting Reebless foliations should be irreducible, although a counterexample to the Poincaré conjecture would cast doubt on this conclusion. Happily, a theorem of H. Rosenberg [**159**] does imply that these manifolds are irreducible.

In fact, Rosenberg's theorem can be deduced from a result of F. Palmeira [**143**] that simply connected n-manifolds, foliated by $(n-1)$-planes, are diffeomorphic to $\mathbb{R}^n$. This theorem implies that a closed 3-manifold with a Reebless foliation has universal cover $\mathbb{R}^3$. Palmeira's theorem will be proven in Appendix D.

As presented here, Novikov's theorems are only for foliated 3-manifolds. However, recent work by F. Alcalde Cuesta, G. Hector and P. Schweitzer [**42**] gives a remarkable generalization to arbitrary dimensions. In this work, Novikov's "homotopy vanishing cycle" is replaced by a homological analogue.

Throughout this chapter, we assume that $(M, \mathcal{F})$ is a compact, orientable, transversely orientable, connected, C^∞-foliated 3-manifold with 2-dimensional leaves. In fact, smoothness of class C^2 is all that will be needed and, with a great deal of care, even that degree of regularity can be weakened. Indeed, the theorems are true for C^0 foliations by work of V. Solodov [**165**]. While it is customary to prove these theorems only for closed 3-manifolds, they hold for cases in which $\partial M \neq \emptyset$. As usual, we will assume that $\partial M = \partial_\tau M \cup \partial_\pitchfork M$, allowing corners. We fix a transverse, 1-dimensional foliation $\mathcal{L}$, assumed to be tangent to $\partial_\pitchfork M$.

Remark. Even purely C^0 foliations admit transverse, 1-dimensional foliations. This is due essentially to L. C. Siebenmann [**164**] (*cf.* [**91**, 1.1]), and makes it possible to translate many local deformation arguments, usually given in the C^2 category, into the C^0 category. In particular, this simplifies Solodov's proof of the C^0 Novikov theorems.

The terms "essential" and "inessential" will be used frequently. Recall that a loop σ in a manifold N, based at a point $x \in N$, is homotopic to the constant x through loops based at x (*i.e.*, represents the identity $[\sigma] = 1 \in \pi_1(N, x)$) if and only if it is *freely* homotopic to a constant. Such loops are called *inessential*, those not homotopic to a constant being essential. Inessential loops are also said to be nullhomotopic.

9.1. Statements of Results

As remarked above, Novikov's work centers on the concept of a (homotopy) vanishing cycle. While the notion makes sense in higher dimensions, it is only useful for foliated 3-manifolds.

Definition 9.1.1. Let $\sigma_0 : S^1 \to L$ be a smooth loop on a leaf of $\mathcal{F}$. Let $\sigma : S^1 \times [0,1] \to M$ be a smooth map (and write $\sigma(z,t) = \sigma_t(z)$) such that:

(1) $\sigma_t : S^1 \to M$ has image in a leaf L_t of $\mathcal{F}$, $0 \le t \le 1$.

(2) For each fixed $z \in S^1$, $\sigma_t(z)$ describes an arc in a leaf of $\mathcal{L}$ as t varies over $[0,1]$.

(3) σ_0 is not homotopic to a constant in $L = L_0$.

(4) σ_t is nullhomotopic in L_t, $0 < t \le 1$.

Then we say that σ_0 is a vanishing cycle on the leaf L.

Example 9.1.2. In the Reeb foliation of S^3 or in any foliated 3-manifold having a Reeb component, the toroidal leaf bounding the Reeb component has a vanishing cycle. Indeed, the meridian σ_0 on this boundary leaf is essential on that leaf, but it pushes smoothly along the leaves of $\mathcal{L}$ to loops σ_t on all the nearby planar leaves L_t. Of course, σ_t is nullhomotopic on L_t, $0 < t \le 1$.

Here are our main goals in this chapter, most of which are due to Novikov [**141**]. All manifolds are assumed to be connected.

Theorem 9.1.3 (Novikov). *For compact, orientable and transversely orientable foliated* 3*-manifolds* $(M, \mathcal{F})$*, the following are equivalent.*

(1) *The foliation* $\mathcal{F}$ *has a Reeb component.*

(2) *There is a leaf* L *of* $\mathcal{F}$ *that is not* π_1*-injective. That is, the inclusion* $i : L \hookrightarrow M$ *induces a homomorphism* $i_* : \pi_1(L) \to \pi_1(M)$ *with nontrivial kernel.*

(3) *Some leaf of* $\mathcal{F}$ *contains a vanishing cycle.*

The implication (1) $\Rightarrow$ (2) is easy. The boundary torus of a Reeb foliated solid torus R has an essential loop (a meridian) that is nullhomotopic in R, hence in M. The fact that (2) $\Rightarrow$ (3) is fairly deep, using the Poincaré-Bendixson theory on a compressing disk that meets the foliation with only Morse type singularities. This will rely on results proven in [I, Sections 7.1 and 7.2]. Finally, the implication (3) $\Rightarrow$ (1) is the most difficult. The main step will be to prove that the vanishing cycle produces an "exploding disk", this being a special case of the exploding plateaus of [I, Definition 10.4.4]. The theory of foliation cycles, together with a theorem of S. Goodman, will then imply that the leaf L supporting the vanishing cycle is a torus. Here the fact that ∂M may be nonempty presents some problems that will be sidestepped by suitable doublings. One must do a bit more work to see that L bounds a Reeb component.

Theorem 9.1.4 (Novikov). *Let $(M, \mathcal{F})$ be a compact, orientable and transversely orientable foliated 3-manifold. Then each of the following implies that $\mathcal{F}$ has a Reeb component.*

(1) *There is a closed, nullhomotopic transversal to $\mathcal{F}$.*

(2) *The tangential boundary of M is empty and $\pi_1(M)$ is finite.*

It is easy to see that (2) $\Rightarrow$ (1). Indeed, since M is compact and $\partial_\tau M = \emptyset$, any transverse arc can be extended indefinitely; hence there is a transverse arc passing at least twice through some foliated chart. Appealing to transverse orientability, we see that a transverse arc with both endpoints in such a chart can be modified in that chart to be a closed transversal τ. If $\pi_1(M)$ is finite, a suitable iteration of τ is a closed, nullhomotopic transversal. Thus, Theorem 9.1.4 will be a consequence of Theorem 9.1.3 and the following.

Proposition 9.1.5. *If $\mathcal{F}$ admits a closed, nullhomotopic transversal, then some leaf of $\mathcal{F}$ has a vanishing cycle.*

The proof again uses Poincaré-Bendixson theory as in [I, Section 7.2], as does the proof of the following.

Proposition 9.1.6. *Let $\mathcal{F}$ be a foliation without vanishing cycles and let $f : S^2 \to M$. Then f is homotopic to a map that takes its image in a leaf of $\mathcal{F}$.*

Theorem 9.1.7 (Novikov). *Let $(M, \mathcal{F})$ be a compact, orientable and transversely orientable foliated 3-manifold with $\pi_2(M) \neq 0$. Then, either $\mathcal{F}$ has a Reeb component, or M is diffeomorphic to $S^2 \times S^1$ or to $S^2 \times [0,1]$ and $\mathcal{F}$ is the product foliation*

Proof. Indeed, if $\mathcal{F}$ has no Reeb component, it has no vanishing cycles by Theorem 9.1.3. If also $\pi_2(M) \neq 0$, let $f : S^2 \to M$ be homotopically nontrivial. By Proposition 9.1.6, f is homotopic to a map $g : S^2 \to L$, for some leaf L. Thus, $\pi_2(L) \neq 0$ and the only connected, orientable surface with this property is S^2. By [I, Theorem 6.1.5], it follows that $M \cong S^2 \times S^1$ or $M \cong S^2 \times [0,1]$, foliated as a product. □

Recall that a 3-manifold is *irreducible* if every tamely imbedded 2-sphere bounds a 3-ball. Theorem 9.1.7 suggests that manifolds supporting Reebless foliations, other than $S^2 \times S^1$ or $S^2 \times [0,1]$, are irreducible. Indeed, this would follow if we knew the Poincaré conjecture. Instead, the following result can be used.

Theorem 9.1.8 (Rosenberg). *If V is a 3-manifold without boundary, not necessarily compact, and if $\mathcal{F}$ is a foliation of V by planes, then V is irreducible.*

Corollary 9.1.9. *If $(M, \mathcal{F})$ is a compact, foliated 3-manifold without Reeb components and possibly with boundary, then either M is irreducible or it is homeomorphic to the product of S^2 and a compact 1-manifold.*

Proof. Indeed, the leaves are π_1-injective by Theorem 9.1.3. If we set $M_0 = M \setminus \partial M$ and $\mathcal{F}_0 = \mathcal{F}|M_0$, the leaves of $\mathcal{F}_0$ are also π_1-injective. Again, Theorem 9.1.7 allows us to assume that $\pi_2(M) = 0$ and it follows that the lifted foliation on the universal cover of M_0 has all leaves diffeomorphic to $\mathbb{R}^2$. Theorem 9.1.8 then implies that the universal cover of M_0 is irreducible. By Exercise 9.1.15, M_0 itself, hence also M, is irreducible. □

The fact that the lift of a Reebless foliation to the universal cover has planar leaves has a stronger consequence than Corollary 9.1.9. The following theorem, valid for n-manifolds with $n \geq 3$, will be proven in Appendix D.

Theorem 9.1.10 (Palmeira). *Let M be a simply connected $(n+1)$-manifold, $n \geq 2$, admitting a smooth foliation by leaves diffeomorphic to $\mathbb{R}^n$. Then M is diffeomorphic to Euclidean space $\mathbb{R}^{n+1}$.*

In fact, this theorem is true as stated (and rather easy) for $n = 1$. It will be a corollary of a stronger theorem of Palmeira that is valid only for $n \geq 2$. Another corollary will be that the foliation $\mathcal{F}$ is smoothly conjugate to a foliation $\mathbb{R}^{n-1} \times \mathcal{F}'$, where $\mathcal{F}'$ is a foliation of $\mathbb{R}^2$.

Corollary 9.1.11. *If $(M, \mathcal{F})$ is a compact, foliated 3-manifold without Reeb components, possibly with boundary, and not conjugate to $S^2 \times S^1$ nor to $S^2 \times [0, 1]$ with the product foliation, then the universal cover of* $\operatorname{int} M$ *is diffeomorphic to $\mathbb{R}^3$.*

Since $\mathbb{R}^3$ is irreducible (J. Alexander [**1**]), Theorem 9.1.8 and Corollary 9.1.9 follow.

We conclude with some elementary but useful exercises.

Exercise 9.1.12. Let $\sigma_i : S^1 \to L$ be piecewise smooth loops on a leaf L of $\mathcal{F}$, $i = 0, 1$. Fix a Riemannian metric g on M and let d_L denote the distance function on L defined by the metric $g|L$. Prove that there is a value $\delta > 0$ such that, if

$$d_L(\sigma_1(z), \sigma_0(z)) < \delta, \quad \forall z \in S^1,$$

then σ_1 and σ_0 are homotopic in L. Conclude that, if the requirement (2) in Definition 9.1.1 is weakened to read: "For each fixed $z \in S^1$, $\sigma_t(z)$ describes an $\mathcal{F}$-transverse arc as t varies over $[0, 1]$", then σ_0 is still a vanishing cycle.

Exercise 9.1.13. If σ_0 is a vanishing cycle on L, show that any loop σ_1 that is homotopic to σ_0 in L is also a vanishing cycle.

Exercise 9.1.14. If σ_0 is a loop on a leaf L_0 that is nullhomotopic in L_0, and if σ_t, $0 \le t \le 1$, is a homotopy of σ_0 through loops σ_t on leaves L_t, show that there is a value of $\delta > 0$ such that σ_t is nullhomotopic on L_t, $0 \le t < \delta$.

Exercise 9.1.15. Let M be a 3-manifold, not necessarily compact, with $\partial M = \emptyset$. If there is a regular covering $\pi : \widetilde{M} \to M$ with $\widetilde{M}$ irreducible, prove that M is irreducible.

9.2. Poincaré-Bendixson Theory and Vanishing Cycles

Here and throughout this chapter, we assume familiarity with the definitions, notation and results of [I, Chapter 7]. In particular, smooth imbeddings (respectively, immersions) $f : T \hookrightarrow M$ with suitable conditions on $f|\partial T$ can be uniformly well approximated in the C^k norm, $k \ge 2$, by a smooth imbedding (respectively, immersion) g that has only Morse type tangencies with a given foliation. As in [I, Exercise 7.1.13], g is homotopic to f and it is not hard to show that the homotopy F_t stays as close to f in the C^k norm as g. One can arrange, therefore, that each F_t is a smooth imbedding (respectively, immersion). That is, the homotopy is actually an *isotopy*. This observation will be important in Section 9.5.

Throughout this section, assertions that something "bounds a disk" are frequent and are justified, of course, by the Jordan curve theorem, although this is seldom made explicit.

9.2.A. Immersed disks. We leave it to the reader to modify the proof of [I, Theorem 7.1.10] to obtain the following.

Lemma 9.2.1. *Let $f : D^2 \to \operatorname{int} M$ be a smooth map such that $f(\partial D^2)$ lies in a leaf L of $\mathcal{F}$. Then, given $\varepsilon > 0$, there exists a smooth immersion $g : D^2 \to \operatorname{int} M$ that is ε-near f in the C^2-topology and satisfies the following*:

(1) *$g(\partial D^2) \subset L$.*

(2) *g is in general position with respect to $\mathcal{F}$.*

(3) *If $\{p_1, \dots, p_\ell\} \subset D^2$ is the set of points such that g is tangent to $\mathcal{F}$ at $g(p_i)$, $1 \le i \le \ell$, then $g(p_i)$ and $g(p_j)$ lie in distinct leaves of $\mathcal{F}$, whenever $i \ne j$.*

Lemma 9.2.2. *Let $f : D^2 \to \operatorname{int} M$ be a smooth immersion in general position with respect to $\mathcal{F}$ and such that $f|\partial D^2$ either is transverse to $\mathcal{F}$ or has image in a leaf L of $\mathcal{F}$. In the latter case, assume further that the loop $f(\partial D^2)$ is essential in L. Then there is a compact, $\mathcal{F}^*$-saturated disk $D \subseteq D^2$ that contains no limit cycles and is such that $f(\partial D)$ is an essential loop on a leaf of $\mathcal{F}$.*

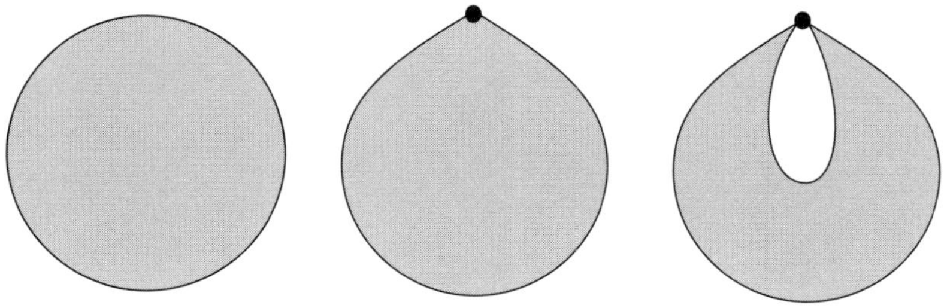

Figure 9.2.1. $\mathcal{D}$ consists of disks or pinched annuli

Proof. Consider first the case of transverse boundary. Let $(\widetilde{\mathcal{S}}, \preceq)$ be the partially ordered, inductive set of graphs as defined in [I, p. 161] and let $\mathcal{G} \in \widetilde{\mathcal{S}}$ be a minimal element. As in the proof of [I, Proposition 7.3.2], either $\mathcal{G}$ bounds a disk D with the asserted properties or $\mathcal{G}$ bounds two disks joined at a saddle, at least one of which is as desired. The case of tangential boundary is nearly the same. Indeed, if $\mathcal{F}^*$ has limit cycles, one defines $(\widetilde{\mathcal{S}}, \preceq)$ exactly as before, mimics the proof of [I, Lemma 7.2.7] to show that this is inductive, and argues as above. In the alternative case, there are no limit cycles and we can take $D = D^2$. □

We assume that $f : D^2 \to \operatorname{int} M$ satisfies the hypotheses of Lemma 9.2.2. If $\sigma \subset D^2$ is a closed orbit or a compact graph such that $f(\sigma)$ is nullhomotopic on its leaf, we will say that σ itself is inessential. Otherwise, σ is essential. If $f(\sigma)$ is a vanishing cycle, we will say that σ is a vanishing cycle.

Consider the set $\mathcal{D}$ of subsets $\Delta \subseteq D^2$ that are the closures of open, $\mathcal{F}^*$-saturated disks containing no limit cycles and such that $\partial\Delta$ is either an essential closed orbit or an essential graph. By Lemma 9.2.2, $\mathcal{D} \neq \emptyset$. A little thought shows that the only possibilities are that Δ is topologically a disk, with $\partial\Delta$ either a closed orbit or a graph consisting of a saddle point and one separatrix, or Δ is a pinched annulus with boundary a figure eight graph (see Figure 9.2.1).

Exercise 9.2.3. If $\Delta \in \mathcal{D}$ and there is a saddle point $q \in \partial\Delta$, show that the usual Poincaré-Hopf formula (Theorem 4.4.11) for $\chi(\Delta)$ generalizes so that q contributes -1 to $\chi(\Delta)$ if two separatrices lie in Δ and contributes 0 if only one does. (Hint. Modify $\mathcal{F}^*$ in a slightly larger disk or annulus.) Conclude that every $\Delta \in \mathcal{D}$ contains a center.

The following will be proven by induction on the number k of centers in $\Delta \in \mathcal{D}$.

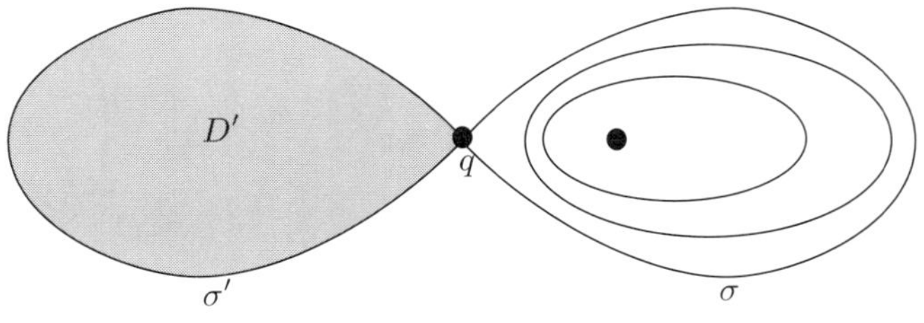

Figure 9.2.2. Case 1(a)

Lemma 9.2.4. *Each $\Delta \in \mathcal{D}$ contains a vanishing cycle.*

Proof. Let $\Delta \in \mathcal{D}$ with $k = 1$. Then there can be no saddles in $\operatorname{int} \Delta$ and each closed orbit near enough to the center p must be inessential. Indeed, let U be a biregular chart containing $f(p)$ and notice that the closed orbits near p are carried by f to loops on (simply connected) $\mathcal{F}$-plaques. Consider all open, $\mathcal{F}^*$-saturated disks U in Δ, centered at p and such that $U \smallsetminus \{p\}$ consists entirely of closed orbits that are inessential. The union of these is a maximal such open disk and its closure D is bounded either by a closed orbit $\sigma \subset \operatorname{int} D$ or by $\partial\Delta$. In the first case, if σ were inessential, Exercise 9.1.14 would contradict the maximality of $U = \operatorname{int} D$. That is, σ is a vanishing cycle. In the second case, $D = \Delta$ and it is immediate that $\partial\Delta$ is a vanishing cycle.

For the inductive step, let $k \geq 2$ and assume that the assertion has been verified for all cases in which the number of centers in $\Delta \in \mathcal{D}$ is at most $k - 1$. Beginning at a center p, consider, as before, the maximal open, $\mathcal{F}^*$-saturated disk U in Δ that is a union of p and closed, inessential orbits. If $D = \overline{U} \subset \operatorname{int} \Delta$ is a disk bounded by a closed orbit σ, then, as above, σ is a vanishing cycle. Suppose, therefore, that ∂D is a compact graph. Then D need not be a disk (Case 2, below) and we consider two cases.

Case 1. D is a disk and the graph ∂D is the union of a saddle point q and one separatrix. Since there are more centers in Δ than in D, we see that $D \subset \operatorname{int} \Delta$; hence ∂D is one loop σ of a figure eight graph $\mathcal{G}$. There are two subcases.

(a) The other loop σ' of $\mathcal{G}$ bounds a disk D' that meets D exactly at q (Figure 9.2.2). If ∂D is essential, we are done, so suppose that it is inessential. If σ' is essential, then it bounds a disk $D' \in \mathcal{D}$ with at most $k - 1$ centers. By the inductive hypothesis, D' contains a vanishing cycle. If σ' and σ are each inessential, then the entire figure eight is carried by f to a loop on a leaf L that defines trivial holonomy on L. Thus, the figure

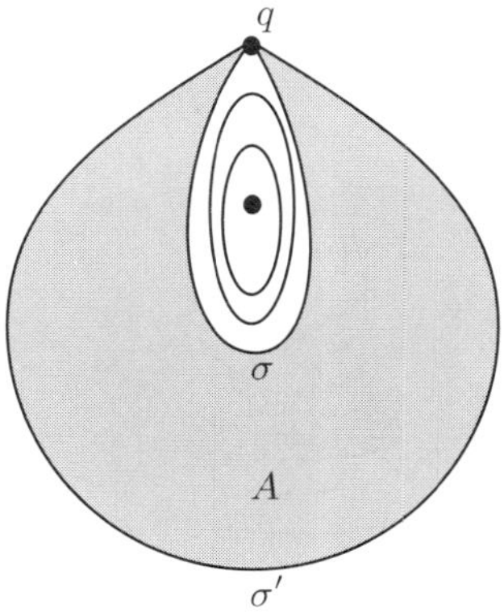

Figure 9.2.3. Case 1(b)

eight graph is encircled by a continuum of closed, inessential orbits, and we obtain an open disk that is the union of the two closed disks bounded by the figure eight and a band of closed, inessential orbits. As before, we can take D_1 to be the closure of an open disk that is maximal with respect to these properties and investigate whether it is a disk such that ∂D_1 is essential (hence, a vanishing cycle) or not. If its boundary is a closed orbit, D_1 is a disk and ∂D_1 is essential by the maximality of D_1. If D_1 contains all the centers and its boundary is not a closed orbit, then $D_1 = \Delta$ and $\partial\Delta$ is a vanishing cycle. Otherwise, we again encounter Case 1 or Case 2.

(b) The other loop σ' of $\mathcal{G}$ bounds a disk D' that contains D (Figure 9.2.3). Again, we suppose that $\sigma = \partial D$ is inessential (otherwise, we are done) and remark that $\mathcal{G}$ bounds the pinched annulus $A = D' \setminus \operatorname{int} D$. If σ' is essential, so is $\mathcal{G}$, and $A \in \mathcal{D}$. But A contains at most $k-1$ centers; hence, by the inductive hypothesis, it contains a vanishing cycle. If σ' is inessential, then it is encircled by a continuum of closed, inessential orbits and we obtain an open disk that is the union of D' and a band of closed, inessential orbits. Again take D_1 to be the closure of an open disk that is maximal with respect to these properties and investigate whether it is a disk such that ∂D_1 is essential (hence, a vanishing cycle) or not. If its boundary is a closed orbit, D_1 is a disk and ∂D_1 is essential by the maximality of D_1. If D_1 contains all the centers and its boundary is not a closed orbit, then $D_1 = \Delta$ and $\partial\Delta$ is a vanishing cycle. Otherwise, we again encounter Case 1 or Case 2 below.

Case 2. D is a pinched annulus and ∂D is the union of a saddle point and two separatrices. Let σ and σ' be the two closed loops consisting of the saddle and one of the separatrices as in Figure 9.2.4. If ∂D is essential, it is a

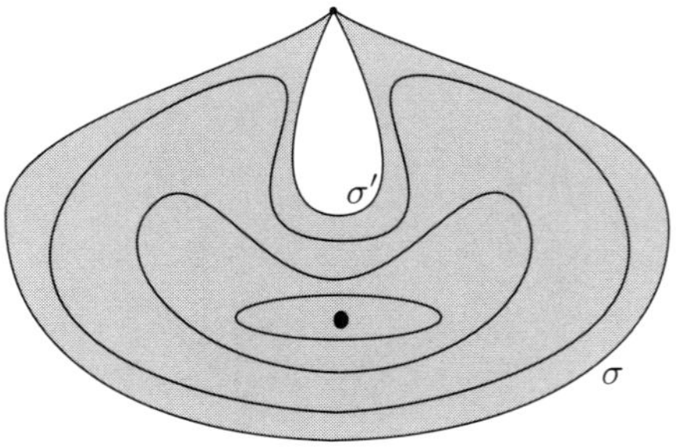

Figure 9.2.4. Case 2

vanishing cycle and we are done. Otherwise, $\partial D = \sigma + \sigma'$ is inessential; hence σ and σ' are either both inessential or both essential. If both are essential, then one of them, say σ', bounds a disk $D' \in \mathcal{D}$ that is disjoint from $\operatorname{int} A$ and so contains at most $k-1$ centers. By the inductive hypothesis, D' contains a vanishing cycle. If σ and σ' are both inessential, then there is an open, $\mathcal{F}^*$-saturated disk U that is the union of the disk $D'' \supset A$ bounded by σ and an annulus that is a continuum of closed, inessential orbits. We take D_1 to be the closure of the open disk that is maximal with these properties and investigate whether it is a disk such that ∂D_1 is essential (hence, a vanishing cycle) or not. If its boundary is a closed orbit, D_1 is a disk and ∂D_1 is essential by the maximality of D_1. If D_1 contains all the centers and its boundary is not a closed orbit, then $D_1 = \Delta$ and $\partial \Delta$ is a vanishing cycle. Otherwise, we again encounter Case 1 or Case 2 below.

Since there are only finitely many compact graphs, this process must stop after finitely many steps. But the only way it stops is by finding a vanishing cycle in D. □

Proof of Proposition 9.1.5. Suppose that $\sigma : S^1 \to M$ is a closed, nullhomotopic transversal. We can assume that σ is smooth and find a smooth nullhomotopy $g : D^2 \to M$, $\sigma = g|\partial D^2$. By [I, Theorem 7.1.10], an arbitrarily C^2-close approximation produces a smooth immersion $f : D^2 \to M$ in general position with respect to $\mathcal{F}$. Since the approximation is C^2-close (C^1-close would be enough), $f|\partial D^2$ is transverse to $\mathcal{F}$. By Lemmas 9.2.2 and 9.2.4, there is a vanishing cycle. □

The following is the implication (2) ⇒ (3) in Theorem 9.1.3.

Proposition 9.2.5. *Let $i : L \hookrightarrow M$ be the inclusion of a leaf, $x \in L$. If $i_* : \pi_1(L, x) \to \pi_1(M, x)$ is not one-to-one, then $\mathcal{F}$ has a vanishing cycle.*

Proof. Let $\sigma : S^1 \to L$ be a loop through x that is essential in L but not in M. We can assume that σ is smooth and define a smooth nullhomotopy $g : D^2 \to M$, $g|\partial D^2 = \sigma$. By Lemma 9.2.1, arbitrarily C^2-close to g there is a smooth immersion $f : D^2 \to M$, in general position with respect to $\mathcal{F}$ and having $f(\partial D^2) \subset L$. By Exercise 9.1.12, $f|\partial D^2$ is essential in L. By Lemmas 9.2.2 and 9.2.4, there is a vanishing cycle. □

9.2.B. Immersed spheres. We turn to the proof of Proposition 9.1.6. Thus, for an arbitrary continuous map $f : S^2 \to M$, we must prove that, in the absence of vanishing cycles, f is homotopic to a map into a leaf. We will use the following elementary lemma.

Lemma 9.2.6. *Let L be a leaf of $\mathcal{F}$ and let $h : K \to L$ be a continuous map, defined on a compact, path-connected space and nullhomotopic in L. Then, for every leaf L' of $\mathcal{F}$ that passes sufficiently near L, there is a continuous map $H : K \times [0, 1] \to M$ such that*

(1) $H(K \times \{t\}) = H_t(K) \subset L_t$, *where L_t is a leaf of $\mathcal{F}$, $0 \le t \le 1$;*

(2) $H(\{x\} \times [0, 1]) \subset T_x$, *where T_x is the leaf of $\mathcal{L}$ through $h(x)$, $\forall x \in K$;*

(3) $H_0 = h$;

(4) $H_1 : K \to L'$.

Proof. Fix a basepoint $x_0 \in K$ and let $y_0 = h(x_0)$. Let $[0, 1] \subset T_{x_0}$ be so parametrized that $0 = y_0$ and let L_t denote the leaf of $\mathcal{F}$ through $t \in [0, 1]$. For any $x \in K$, choose a path σ_x in K from x_0 to x and consider the path $h \circ \sigma_x$ on L from y_0 to $h(x)$. The holonomy lift of this path defines $\gamma_x : [0, \varepsilon] \to T_x$, for some $\varepsilon > 0$, so that $\gamma_x(0) = h(x)$ and $\gamma_x(\varepsilon) \in L_\varepsilon$. We attempt to define $H_t(x) = \gamma_x(t)$, $0 \le t \le \varepsilon$. This will have all the desired properties, provided that it is well defined. If τ_x is another path in K from x_0 to x, we consider the loop $\lambda_x = \sigma_x^{-1} * \tau_x$, based at x_0. We must show that $h \circ \lambda_x$ has trivial holonomy. Since h is nullhomotopic in L, this loop is freely nullhomotopic in L; hence it is also nullhomotopic relative to y_0. It follows that the two definitions of $H_t(x)$ will agree, $0 \le t \le \varepsilon'$, for a suitable choice of $\varepsilon' \in (0, \varepsilon]$. Since K is compact, we can restrict the set of paths σ_x that we use to correspond to only finitely many distinct plaque chains, in which case there is a minimal $\varepsilon' > 0$ for which everything is well defined. For each choice $L' = L_t$, $0 < t \le \varepsilon'$, we only need to reparametrize $[0, \varepsilon']$ as the unit interval to complete the proof. □

Let $f : S^2 \to M$ be given. We assume that $\mathcal{F}$ has no vanishing cycles and must prove that f is homotopic to a continuous map taking its image in some leaf L of $\mathcal{F}$.

Lemma 9.2.7. *If $\pi_2(L) \neq 0$, for some leaf L of $\mathcal{F}$, then f is homotopic to a map taking its image in L.*

Proof. The only connected, orientable surface with nontrivial second homotopy group is S^2. Thus, $L \cong S^2$ and M is homeomorphic either to $S^1 \times S^2$ or to $I \times S^2$ in such a way that the leaves of $\mathcal{F}$ are the factors $\{z\} \times S^2$ [I, Theorem 6.1.5]. Thus, $i : L \hookrightarrow M$ generates $\pi_2(M) \cong \mathbf{Z}$ and every map $f : S^2 \to M$ is homotopic to a map taking its image in L. □

Thus, we assume that every leaf L of $\mathcal{F}$ has $\pi_2(L) = 0$ and that no leaf has a vanishing cycle. By [I, Theorem 7.1.10], we can assume that f is in general position with respect to $\mathcal{F}$, no two distinct saddles in $\mathcal{F}^* = f^{-1}(\mathcal{F})$ being connected by a separatrix. Again, there is a vector field on S^2 whose orbits are the leaves or singular points of $\mathcal{F}^*$; hence we often call these the orbits of $\mathcal{F}^*$.

The basic idea is to start the deformation at the center singularities of $\mathcal{F}^*$, showing that the absence of vanishing cycles implies that there is no obstruction to continuing the deformation until S^2 is carried entirely into a single leaf. We will use the assumption that $\pi_2(L) = 0$ in two parts of the proof.

Lemma 9.2.8. *If $\mathcal{G} \subset S^2$ is either a closed orbit or a graph, then $f(\mathcal{G})$ is nullhomotopic in its leaf. Furthermore, every orbit of $\mathcal{F}^*$ is either a singular point, a closed orbit or a separatrix.*

Proof. If $\mathcal{G}$ is a closed orbit or a graph containing a single separatrix, the Jordan curve theorem on S^2 implies that $\mathcal{G}$ bounds a disk, hence is inessential in M. Thus, if it were essential in its leaf, there would be a vanishing cycle by Proposition 9.2.5. If $\mathcal{G}$ is a figure eight graph, the same argument shows that each loop of $\mathcal{G}$ is inessential in its leaf. If some orbit of $\mathcal{F}^*$ is neither singular, closed nor a separatrix, then there is a limit cycle. The Jordan curve theorem allows us to apply Lemmas 9.2.2 and 9.2.4 to prove there would be a vanishing cycle. □

Let $\mathcal{D}$ denote the family of $\mathcal{F}^*$-saturated disks $D \subset S^2$ such that ∂D is either a closed orbit or a graph. Let $\mathcal{D}'$ denote the union of $\mathcal{D}$ and the family of $\mathcal{F}^*$-saturated "pinched annuli" as in Figure 9.2.5. Note that $\Delta \in \mathcal{D}'$ implies that $\partial\Delta$ is either a closed orbit or a graph.

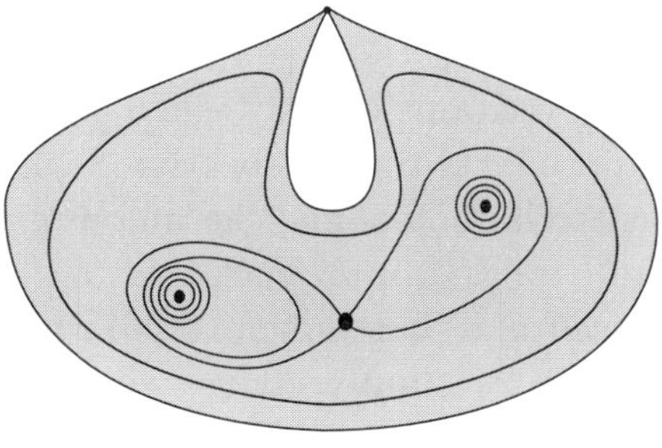

Figure 9.2.5. A pinched annulus in $\mathcal{F}^*$

Proposition 9.2.9. *If $D \in \mathcal{D}$, there is a homotopy $f \sim g$ such that g and f agree on $S^2 \smallsetminus \operatorname{int}(D)$ and $g(D) \subset L$, where L is the leaf of $\mathcal{F}$ such that $f(\partial D) \subset L$.*

Proof of Proposition 9.1.6 using Proposition 9.2.9. We can express S^2 as the union of disks D_1 and D_2 with common boundary $\mathcal{G}$ (a closed orbit or a closed graph). Let L be the leaf of $\mathcal{F}$ such that $f(\mathcal{G}) \subset L$. By Proposition 9.2.9, f is homotopic to g such that $g(D_1) \subset L$ and $f|D_2 = g|D_2$. Similarly, g is homotopic to h such that $h|D_1 = g|D_1$ and $h(D_2) \subset L$. That is, f is homotopic to h such that $h(S^2) \subset L$. □

We are reduced to proving Proposition 9.2.9. This will be accomplished by a series of lemmas.

Definition 9.2.10. The element $\Delta \in \mathcal{D}'$ is $\mathcal{F}$-flat if $f \sim g$, where f and g agree on $S^2 \smallsetminus \operatorname{int}(\Delta)$ and $g(\Delta) \subset L$, L being the leaf of $\mathcal{F}$ containing $f(\partial\Delta)$.

Lemma 9.2.11. *Let $p \in S^2$ be a center singularity. Then there is an $\mathcal{F}$-flat disk $D \in \mathcal{D}$ such that $p \in D$.*

Proof. A sufficiently small $D \in \mathcal{D}$ with $p \in D$ will be a union of p and closed orbits and will be carried by f entirely into a biregular neighborhood in M. Construction of the desired homotopy is completely elementary. □

Lemma 9.2.12. *Let $\mathcal{G} \subset S^2$ be a closed orbit or a maximal graph and let L be the leaf of $\mathcal{F}$ containing $f(\mathcal{G})$. Then there is an $\mathcal{F}^*$-saturated neighborhood N of $\mathcal{G}$ in S^2 such that $f_0 = f|N$ is homotopic in M to a map $f_1 : N \to L$. Furthermore, the homotopy f_t can be chosen so that $f_t|\mathcal{G} = f|\mathcal{G}$, $0 \le t \le 1$, and so that $f_t(x)$ stays in a plaque of $\mathcal{L}$, $0 \le t \le 1$, $\forall x \in N$.*

Proof. By Lemma 9.2.8, it should be clear that $\mathcal{G}$ admits $\mathcal{F}^*$-saturated neighborhoods that are uniformly close to $\mathcal{G}$. By the same lemma, every cycle in N is carried by f to a nullhomotopic cycle on a leaf. In particular, all have trivial holonomy. Cover $f(\mathcal{G})$ by a set of biregular neighborhoods in M, so chosen that each contains exactly one $\mathcal{F}$-plaque meeting $f(\mathcal{G})$ and so that two overlap if and only if their respective $\mathcal{F}$-plaques that meet $f(\mathcal{G})$ overlap. Since $f(\mathcal{G})$ has trivial holonomy, the union of these neighborhoods contains a neighborhood U of $f(\mathcal{G})$ in M such that $\mathcal{F}|U$ and $\mathcal{L}|U$ are product foliations. That is, U itself is a "biregular neighborhood", except that the "$\mathcal{F}$-plaques" in U need not be simply connected. Choosing N uniformly close enough to $\mathcal{G}$, we guarantee that $f(N) \subset U$. Using the biregularity of U, it is elementary to construct the desired homotopy. □

Lemma 9.2.13. *Let $D \in \mathcal{D}$ be $\mathcal{F}$-flat and suppose that all orbits outside D that pass sufficiently near ∂D are closed. Then there is an $\mathcal{F}$-flat disk $D' \in \mathcal{D}$ such that $D \subset \operatorname{int}(D')$. Similarly, if $D_i \in \mathcal{D}$ are $\mathcal{F}$-flat, $i = 1, 2$, and if $D_1 \cap D_2$ is a single point (necessarily a saddle), then there is an $\mathcal{F}$-flat disk $D' \in \mathcal{D}$ such that $D_1 \cup D_2 \subset \operatorname{int}(D')$.*

Proof. Consider both cases simultaneously by letting Δ represent either D or $D_1 \cup D_2$. By assumption, we have found g, homotopic to f and agreeing with f outside of Δ, such that $g(\Delta)$ lies in the leaf of $\mathcal{F}$ that contains $f(\partial\Delta)$. Since Δ is simply connected, Lemma 9.2.6 allows us to define a deformation of $g|\Delta$ along the leaves of $\mathcal{L}$ to a map of Δ into any nearby leaf of $\mathcal{F}$. Coupled with the deformation in Lemma 9.2.12, this allows us to choose $D' \in \mathcal{D}$ by adding a saturated neighborhood N of $\partial\Delta$ to Δ, and then to extend the deformation of $g|\Delta$ to a deformation of $g|D'$ into the leaf of $\mathcal{F}$ containing $g(\partial D')$. The deformation is constant on $g|\partial D' = f|\partial D'$; hence it extends by the constant deformation of $f|(S^2 \setminus \operatorname{int}(D'))$ to a deformation of f, as required. □

The following elementary exercise will be used in the proof of the next lemma.

Exercise 9.2.14. Let X be a path-connected space with $\pi_2(X) = 0$. Let $\alpha, \beta : D^2 \to X$ be continuous maps such that $\alpha|\partial D^2 = \beta|\partial D^2 = \lambda$. Show that α and β are homotopic through maps that restrict to λ on ∂D^2.

Lemma 9.2.15. *Let $\Delta \in \mathcal{D}'$ and suppose that $\operatorname{int}(\Delta) = \bigcup_{n=1}^{\infty} D_n$, an increasing union of $\mathcal{F}$-flat disks $D_n \in \mathcal{D}$. Then Δ is $\mathcal{F}$-flat.*

Proof. Let L be the leaf of $\mathcal{F}$ that contains $f(\partial\Delta)$. We consider three cases.

Case 1. Suppose that $\Delta \in \mathcal{D}$ and that $\partial\Delta$ is a closed orbit. Let N be a neighborhood of $\partial\Delta$ as in Lemma 9.2.12, chosen so that $N \cap \operatorname{int}(\Delta)$ is a union of closed orbits. Let $f_1 : N \to L$ be the map given by Lemma 9.2.12.

By Exercise 9.1.12, we can choose N uniformly so close to $\partial\Delta$ that, for every closed orbit $\sigma \subset N \cap \mathrm{int}(\Delta)$, $f_1(\sigma)$ is homotopic to $f(\partial\Delta)$ in L, hence nullhomotopic there. We can assume that $\sigma_n = \partial D_n \subset N \cap \mathrm{int}(\Delta)$. Thus, restrict f_1 to $N \cap D_n$ and extend it to a map $h : D_n \to L$. By Lemma 9.2.6, we can deform h along the leaves of $\mathcal{L}$ to $\tilde{h} : D_n \to L_n$, where L_n is the leaf of $\mathcal{F}$ containing the image $f(\sigma_n)$. Let $g : D_n \to L_n$ be as in Definition 9.2.10. Note that $\tilde{h}|\sigma_n = f|\sigma_n = g|\sigma_n$. That is, $g|D_n$ and $\tilde{h}|D_n$ agree on ∂D_n and, by Exercise 9.2.14, $g : D_n \to L_n$ is homotopic to $\tilde{h} : D_n \to L_n$ by a homotopy that is fixed on the boundary. Following this homotopy by the reverse deformation of $\tilde{h}|D_n$ back to $h|D_n$ and using Lemma 9.2.12 to extend this deformation by a deformation of $g|N \cap \Delta = f|N \cap \Delta$ into L, we obtain a deformation of $f|\Delta$ into L that is fixed on $\partial\Delta$, thereby proving that Δ is $\mathcal{F}$-flat.

Case 2. Suppose that $\Delta \in \mathcal{D}$ and that $\partial\Delta$ is a graph, hence the union of a saddle and one of the separatrices that issue from and return to that saddle. The assumption that $\mathrm{int}(\Delta) = \bigcup_{n=1}^{\infty} D_n$ implies that the other separatrix lies outside of Δ. Let N be the neighborhood of the maximal graph $\mathcal{G} \supset \partial\Delta$ as in Lemma 9.2.12, and note that $N \cap \mathrm{int}(\Delta)$ is exactly as in Case 1; hence the argument given there carries over, essentially without change.

Case 3. Suppose that $\Delta \in \mathcal{D}'$ is a pinched annulus. Thus, $\partial\Delta$ is a graph containing two separatrices. Let N be the neighborhood of $\partial\Delta$ as in Lemma 9.2.12. The loop $f(\partial\Delta)$ is nullhomotopic on its leaf L and, as in the proof of Case 1, we obtain a map $h : \Delta \to L$ representing this nullhomotopy and agreeing with the map $f_1 : N \to L$, N a suitable neighborhood of $\partial\Delta$. It is clear that h itself is nullhomotopic in L; hence Lemma 9.2.6 allows us to deform h along the leaves of $\mathcal{L}$ to a map $\tilde{h} : \Delta \to L_n$ and to complete the proof as in Case 1. □

Lemma 9.2.16. *Let $D \in \mathcal{D}$ and suppose that D contains only one center singularity. Then D is $\mathcal{F}$-flat.*

Proof. Let $p \in D$ be the center. There can be no saddles in $\mathrm{int}(D)$. By Lemma 9.2.11, one can find an $\mathcal{F}$-flat D', properly contained in D, with $p \in D'$. Let D_0 be the union of all such D'. Clearly D_0 is either an open or a closed disk, properly contained in D. If closed, it would be a maximal $\mathcal{F}$-flat disk, properly contained in D. But Lemma 9.2.13 would show that it is contained in a larger $\mathcal{F}$-flat disk, properly contained in D. Thus, D_0 is open. By Lemma 9.2.15, the closure of D_0 is $\mathcal{F}$-flat and contained in D. If it were properly contained in D, we would contradict the definition of D_0. Hence $D = \overline{D}_0$ is $\mathcal{F}$-flat. □

Lemma 9.2.17. *Let $k \geq 2$ and suppose that Proposition 9.2.9 is true for disks $D \in \mathcal{D}$ containing at most $k-1$ centers. Then it is true for disks $D \in \mathcal{D}$ containing k centers.*

Proof. Suppose that $D \in \mathcal{D}$ contains k centers, and let $p \in D$ be one of these centers. Let $\mathcal{D}_p$ be the set of disks $D' \in \mathcal{D}$ such that $p \in D' \subseteq D$. By Lemma 9.2.11, there is an $\mathcal{F}$-flat disk $D' \in \mathcal{D}_p$. Let D_0 be the union of all such disks. By Lemma 9.2.15, $\overline{D}_0$ is $\mathcal{F}$-flat. If $\overline{D}_0 = D$, we are done. If $\overline{D}_0$ is properly contained in D, there are several possibilities to be considered, each leading to a contradiction.

Case 1. The closure $\overline{D}_0 = \Delta \in \mathcal{D}_p$, and a neighborhood of $\partial\Delta$ on the outside of Δ is saturated by closed orbits. Since Δ is $\mathcal{F}$-flat, Lemma 9.2.13 provides a properly larger $\mathcal{F}$-flat disk $D'' \in \mathcal{D}_p$, contradicting the definition of D_0.

Case 2. The closure $\overline{D}_0 = D_1 \in \mathcal{D}_p$. This disk is $\mathcal{F}$-flat but is not as in Case 1. There are two possibilities.

Subcase 1. There is a disk $D_2 \subset D$ such that $D_1 \cup D_2$ is as in Lemma 9.2.13. Each of these disks contains at least one center; hence D_2 contains at most $k-1$ centers. Hence D_2 is also $\mathcal{F}$-flat, and Lemma 9.2.13 provides an $\mathcal{F}$-flat disk $D' \in \mathcal{D}_p$ containing $D_1 \cup D_2$, again contradicting the definition of D_0.

Subcase 2. There is a disk $D_2 \subset D$, $D_2 = D_1 \cup \Delta$, where $\Delta \in \mathcal{D}'$ is a pinched annulus. Both D_1 and Δ must contain at least one center; hence Δ contains at most $k-1$ centers. Thus, $\operatorname{int}(\Delta) = \bigcup_{n=1}^{\infty} \Delta_n$, an increasing union of elements of $\mathcal{D}$, each containing at most $k-1$ centers. By hypothesis, each Δ_n must be $\mathcal{F}$-flat; hence Δ is $\mathcal{F}$-flat by Lemma 9.2.15. Since D_1 is $\mathcal{F}$-flat, it follows that $D_2 \in \mathcal{D}_p$ is $\mathcal{F}$-flat, contradicting the definition of D_0.

Case 3. The closure $\overline{D}_0 = \Delta$ is a pinched annulus as in Figure 9.2.5 and is $\mathcal{F}$-flat. Let $D' \subset D$ be the disk bounded by one loop of $\partial\Delta$ as in that figure. Then, since Δ contains at least one center, D' has at most $k-1$ centers, hence is $\mathcal{F}$-flat. This shows that the disk $D'' = D' \cup \Delta \in \mathcal{D}_p$ is also $\mathcal{F}$-flat. But D'' properly contains $\overline{D}_0$, again a contradiction. □

By induction on the number of centers in D, these last two lemmas prove Proposition 9.2.9.

9.3. Novikov's Exploding Disk

Recall the concept of an "exploding plateau"

$$\varphi : P \times (0, 1] \to M$$

[I, Definition 10.4.4]. Here we are interested in producing an exploding plateau of the type

$$\varphi : D^2 \times (0,1] \to M$$

(an exploding disk). By way of a brief review, we note that this is to be a smooth family of leaf imbeddings $\varphi_t : D^2 \to L_t$ that extends to a smooth map of $\partial D^2 \times [0,1]$ carrying $\partial D^2 \times \{0\}$ into a leaf L_0 of $\mathcal{F}$, but does not extend even continuously to a map of $D^2 \times [0,1]$ carrying $D^2 \times \{0\}$ into a leaf. We will write $D_t = \varphi_t(D^2)$, $0 < t \leq 1$. As shown in [I, Section 10.4], the area of D_t becomes arbitrarily large as $t \downarrow 0$, while the length of ∂D_t remains bounded. It is this exploding phenomenon that forces the presence of a Reeb component. Observe that we have changed a convention used in [I, Section 10.4], where φ was defined on $D^2 \times [0,1)$ and the explosion occurred as $t \uparrow 1$. This harmless change is consistent with the conventional parametrization of vanishing cycles.

Our current goal is the proof of the following.

Proposition 9.3.1. *If some leaf L_0 of $\mathcal{F}$ contains a vanishing cycle σ_0, then there is an exploding disk.*

Definition 9.3.2. A loop $\sigma : S^1 \to L$ is in general position if it is an immersion, transverse to itself at every self-intersection and such that $\sigma^{-1}(z)$ contains at most 2 points, $\forall z \in L$.

Since $\dim(L) = 2$, it is a standard fact that every loop is homotopic to one in general position. The self-intersections of loops in general position are called *double points*. By Exercise 9.1.13, we obtain the following.

Lemma 9.3.3. *Without loss of generality, it can be assumed that every vanishing cycle is in general position.*

Let σ_0 be a vanishing cycle on $L = L_0$, assumed to be in general position, and let σ_t be as in Definition 9.1.1, $0 \leq t \leq 1$. Let n_t denote the number of double points of σ_t.

Lemma 9.3.4. *If σ_0 is a vanishing cycle, then it can be assumed without loss of generality that $n_t \leq n_0$, $0 \leq t \leq 1$. Indeed, if $\sigma_t(x) = \sigma_t(y)$, then $\sigma_0(x) = \sigma_0(y)$.*

Proof. For each $z \in S^1$, $\sigma_t(z)$ describes an arc in a leaf of $\mathcal{L}$ as t ranges over $[0,1]$. For small enough $\varepsilon > 0$, the arcs $J_z = \{\sigma_t(z) \mid 0 \leq t \leq \varepsilon\}$ are disjoint for distinct values of $\sigma_0(z)$. By reparametrizing, we can assume that $\varepsilon = 1$. Thus, if $x \neq y$ are points of S^1 and $\sigma_{t_0}(x) = \sigma_{t_0}(y)$ for some $t_0 \in [0,1]$, it follows that $J_x \cap J_y \neq \emptyset$, hence that $\sigma_0(x) = \sigma_0(y)$. $\square$

A small perturbation puts σ_0 in

$$M^\circ = M \smallsetminus \partial_{\pitchfork} M,$$

where it is still a vanishing cycle for $\mathcal{F}^\circ = \mathcal{F}|M^\circ$. We will work regularly in the possibly noncompact foliated manifold $(M^\circ, \mathcal{F}^\circ)$. Thus, all leaves will have empty boundary and we will continue to denote by L_t the leaf of $\mathcal{F}^\circ$ containing σ_t.

It is standard that the only connected and simply connected 2-manifolds with empty boundary are $\mathbb{R}^2$ and S^2. Since M is compact, connected and the leaves of $\mathcal{F}$ are orientable, [I, Theorem 6.1.5] implies that no leaf has universal cover diffeomorphic to S^2. Indeed, all the leaves would be diffeomorphic to S^2, hence simply connected, and so no leaf could have a vanishing cycle. Thus, no leaf of $\mathcal{F}^\circ$ can be covered by S^2 either. Let $\pi_t : \mathbb{R}^2 \to L_t$ be a realization of $\mathbb{R}^2$ as the universal cover, $0 \le t \le 1$. Since σ_t is nullhomotopic in L_t, $0 < t \le 1$, every lift $\widehat{\sigma}_t : [0,1] \to \mathbb{R}^2$ is a closed loop, $0 < t \le 1$; hence we will write $\widehat{\sigma}_t : S^1 \to \mathbb{R}^2$. Normally, we write $\widehat{\sigma}_t$ for an arbitrary lift, unless some choice of specific basepoints has been made.

Definition 9.3.5. The vanishing cycle σ_0 is simple if a lift $\widehat{\sigma}_t$ is a simple closed curve in $\mathbb{R}^2$, $0 < t \le \varepsilon$, for some $\varepsilon > 0$.

Of course, one lift is a simple closed curve if and only if every lift is.

Lemma 9.3.6. *If the vanishing cycle σ_0 is not simple, there is a sequence $\{t_k\}_{k=1}^\infty \subset [0,1]$, $\lim_{k\to\infty}(t_k) = 0$, such that L_{t_k} has a vanishing cycle with strictly fewer double points than σ_0.*

Proof. Let the double points of σ_0 be $\sigma_0(x_i) = \sigma_0(y_i)$, $x_i \ne y_i$, $1 \le i \le m$. Let $U_i = \{t \in (0,1] \mid \widehat{\sigma}_t(x_i) = \widehat{\sigma}_t(y_i)\}$, $1 \le i \le m$.

First observe that each U_i is open in $(0,1]$. Indeed, x_i and y_i divide S^1 into two arcs, call them α_i and β_i. Then $\widehat{\sigma}_{t_0}(x_i) = \widehat{\sigma}_{t_0}(y_i)$ is equivalent to $\sigma_{t_0} \circ \alpha_i$ and $\sigma_{t_0} \circ \beta_i$ being nullhomotopic loops on L_{t_0}. By Exercise 9.1.14, it follows that $\sigma_t \circ \alpha_i$ and $\sigma_t \circ \beta_i$ are nullhomotopic on L_t, $\forall t \in (t_0 - \delta, t_0 + \delta)$, for some $\delta > 0$.

The assertion that σ_0 is simple is equivalent to the assertion that, for some $\varepsilon > 0$, $U_i \cap (0,\varepsilon) = \emptyset$, $1 \le i \le m$. Suppose, then, that this fails for, say, $i = 1$. There are two possibilities.

Case 1. Assume that $U_1 \cap (0,1) = (0,\varepsilon)$, for some $\varepsilon > 0$. Then $\sigma_t \circ \alpha_1$ and $\sigma_t \circ \beta_1$ are both nullhomotopic on L_t, $0 < t < \varepsilon$. Writing

$$\sigma_0 = \sigma_0 \circ \alpha_1 + \sigma_0 \circ \beta_1,$$

we see that, since σ_0 is homotopically essential on L_0, so is at least one of $\sigma_0 \circ \alpha_1$ and $\sigma_0 \circ \beta_1$. Thus, one of these loops is a vanishing cycle and has one less double point than σ_0. (Actually, since vanishing cycles were defined to be smooth, one must smooth the corner on this loop by a small homotopy. Hereafter, such a move will be assumed without comment.) In this case, take the constant sequence $t_k = 0$.

Case 2. Assume that $U_1 \cap (0,1)$ has infinitely many components (t_k, s_k) such that $\lim_{k\to\infty}(t_k) = 0$. For every $t \in (t_k, s_k)$, $\sigma_t(x_1) = \sigma_t(y_1)$ is a fixed point for the holonomy of $\sigma_0 \circ \alpha_1$ as well as for $\sigma_0 \circ \beta_1$. By continuity, $\sigma_{t_k}(x_1) = \sigma_{t_k}(y_1)$ is also a fixed point for the holonomy of those loops. But $t_k \notin U_1$. Hence $\widehat{\sigma}_{t_k}(x_1) \neq \widehat{\sigma}_{t_k}(y_1)$ and it follows that neither of the loops $\sigma_{t_k} \circ \alpha_1$ and $\sigma_{t_k} \circ \beta_1$ is nullhomotopic on L_{t_k}; hence each of them is a vanishing cycle with one less double point than σ_0. □

Corollary 9.3.7. *If L_0 is a leaf with a vanishing cycle, there is a (possibly constant) sequence $s_k \downarrow 0$ such that L_{s_k} contains a simple vanishing cycle.*

Proof. If σ_0 is simple or if a cofinal subsequence of the sequence $\{t_k\}$ as in Lemma 9.3.6 has σ_{t_k} simple, we are done. Otherwise, apply the lemma to all but finitely many σ_{t_k}, obtaining $\{t_k^1\}_{k=1}^{\infty}$ such that $t_{k-1} > t_k^1 > t_k$ and a vanishing cycle $\sigma_{t_k^1}$ with strictly fewer double points than σ_{t_k}. If no cofinal subsequence consists of simple vanishing cycles, iterate this procedure. Since the number of double points of all these vanishing cycles remains bounded by n_0, the process must finally terminate, giving the desired sequence. □

Thus, replacing L_0 with L_t, where t is arbitrarily close to 0, we assume that σ_0 is simple and we try to define $F : D^2 \times (0,1] \to M$. We can reparametrize the family σ_t so that every $\widehat{\sigma}_t : S^1 \to \mathbb{R}^2$ is an imbedding, $0 < t \le 1$. In particular, $\widehat{\sigma}_1(S^1)$ is a smooth, simple closed curve in $\mathbb{R}^2$; hence it bounds a smoothly imbedded disk Δ_1 (by the Jordan curve theorem). Fix a diffeomorphism $\varphi : D^2 \to \Delta_1$ and define $F_1 : D^2 \to L_1$ by $F_1 = \pi_1 \circ \varphi$. Then, F_1 is a smooth immersion defining a nullhomotopy of σ_1 on L_1. The following is a close analogue to Lemma 9.2.6.

Lemma 9.3.8. *For some $t_0 \in [0,1)$, F_1 extends to a smooth immersion*

$$F : D^2 \times (t_0, 1] \to M$$

such that, for $t_0 < t \le 1$,

$$F_t(D^2) = F(D^2 \times \{t\}) \subset L_t,$$
$$F_t|\partial D^2 = \sigma_t,$$

and $F(\{x\} \times (t_0, 1])$ lies in a leaf of $\mathcal{L}$, $\forall x \in D^2$.

Proof. Covering the image of F_1 with biregular charts, we lift F_1 along the $\mathcal{L}$-plaques to produce $F_t : D^2 \to L_t$, $t_0 < t \le 1$, for some $t_0 \in (0,1)$. Indeed, since D^2 is simply connected, any two paths τ_1, τ_2 in D^2 from $1 \in S^1$ to $x \in D^2$ define homotopic paths $\widetilde{\tau}_i = F_1 \circ \tau_i$ in L_1, $i = 1, 2$. Thus, the holonomy lifts of these paths to nearby leaves L_t, starting at $\sigma_t(1)$, end at the same point, which we call $F_t(x)$. □

Another way of viewing this immersion uses the following.

Exercise 9.3.9. Let $H : D^2 \to \mathbb{R}^2$ be an immersion such that $H|\partial D^2$ is one-to-one. Let $\Delta \subset \mathbb{R}^2$ be the imbedded disk bounded by $H(\partial D^2)$ and prove that H maps D^2 diffeomorphically onto Δ.

For any $t \in (t_0, 1]$ and choice of lift $\widehat{\sigma}_t$, $F_t : D^2 \to L_t$ lifts canonically to an immersion $\widehat{F}_t$ of D^2 into $\mathbb{R}^2$ such that $\widehat{F}_t|\partial D^2 = \widehat{\sigma}_t$. Since $\widehat{\sigma}_t$ is one-to-one, the above exercise guarantees that $\widehat{F}_t$ is a diffeomorphism of D^2 onto the disk Δ_t bounded by the smooth simple closed curve $\widehat{\sigma}_t(S^1)$. We can define $\widetilde{F} : D^2 \times (t_0, 1] \to M$ by

$$\widetilde{F}(x, t) = \pi_t(\widehat{F}_t(x)).$$

We have the following trivial observation.

Lemma 9.3.10. *The map $\widetilde{F}$ and the immersion F are identical.*

Lemma 9.3.11. *In Lemma 9.3.8, one can take $t_0 = 0$, but F cannot be extended continuously to $D^2 \times [0, 1]$ so as to carry $D^2 \times \{0\}$ into a leaf of $\mathcal{F}$.*

Proof. Consider the set of all immersions $F : D^2 \times (t_F, 1] \to M$ as in Lemma 9.3.8. This is partially ordered by setting $F \le F'$ whenever $t_F \le t_{F'}$ and $F|D^2 \times (t_{F'}, 1] = F'$. This set is inductive; hence let $F : D^2 \times (t_F, 1] \to M$ be minimal. We must show that $t_F = 0$. Otherwise, $\widehat{\sigma}_{t_F}$ bounds a disk Δ_{t_F} and, using any diffeomorphism $D^2 \to \Delta_{t_F}$, we construct an immersion

$$F' : D^2 \times (t_F - \eta, t_F + \eta) \to M$$

as in the proof of Lemma 9.3.8. Let $t_F < t' < t_F + \eta$. By Lemma 9.3.10 and the properties of these immersions, F and F' are determined by $\widehat{F}_{t'}$ and $\widehat{F}'_{t'}$ respectively. But $\widehat{F}'_{t'} \circ \psi = \widehat{F}_{t'}$ for unique $\psi \in \mathrm{Diff}(\Delta_{t'})$, and we replace $\widehat{F}'_{t'}$ with $\widehat{F}'_{t'} \circ \psi$ to define

$$F'' : D^2 \times (t_F - \eta, t_F + \eta) \to M$$

with all the desired properties and agreeing with F on $D^2 \times (t_F, t_F + \eta)$. Together, these extend F to the subset $D^2 \times (t_F - \eta, 1]$, contradicting the minimality of F. Note that $F|\partial D^2 \times (0, 1]$ extends continuously by σ_0 to $\partial D^2 \times [0, 1]$. If F could be extended continuously to $D^2 \times [0, 1]$ so as tc carry $D^2 \times \{0\}$ into a leaf, this leaf would have to be L_0, contradicting the fact that σ_0 is essential in this leaf. □

Thus, to prove that F is an exploding disk, it remains for us to prove that $F_t : D^2 \to L_t$ is an imbedding, $0 < t \le 1$. In fact, it will be enough to prove this for $0 < t \leq \eta$, for suitable $\eta \in (0,1]$, since we only need to reparametrize $(0,\eta]$ as $(0,1]$ in order to get the exploding disk. What we will show is that the covering maps $\pi_t : \mathbb{R}^2 \to L_t$ are diffeomorphisms, and hence $F_t(D^2) = D_t \subset L_t$ are smoothly imbedded disks, $0 < t \leq \eta$. For this, we need the following lemma.

Lemma 9.3.12. *Let $\mathbb{R}^2 = \bigcup_{k=0}^{\infty} \Delta_k$, a union of compact, smoothly imbedded disks where, for every k, $\Delta_k \subset \operatorname{int}(\Delta_{k+1})$. Let $\pi : \mathbb{R}^2 \to L$ be a smooth covering of an orientable surface L such that, for each $k \geq 0$, $\pi(\partial\Delta_k)$ is a loop whose self-intersections are transverse double points. If the sequence $\{m_k\}_{k=0}^{\infty}$ is bounded, where m_k is the number of double points of $\pi(\partial\Delta_k)$, then π is a diffeomorphism.*

Proof. Assume that L is not diffeomorphic to $\mathbb{R}^2$. It is standard that the group $G \subset \operatorname{Diff}(\mathbb{R}^2)$ of covering transformations contains an element of infinite order. Indeed, the fundamental group of a compact, orientable surface $L \neq S^2$ contains elements of infinite order while, in the noncompact case, $\pi_1(L)$ is a nontrivial free group unless $L \cong \mathbb{R}^2$. Choose an element $f \in G$ of infinite order. If $x_0 \in \mathbb{R}^2$, write $x_k = f^k(x_0)$, noting that $k > j$ implies that $x_k \neq x_j$. Otherwise, the covering transformation f^{k-j} would have a fixed point, hence would be the identity.

Choose $x_0 \in \Delta_0$, pass to a subsequence of the Δ_k's, and choose a strictly increasing sequence $\{k_n\}_{n=0}^{\infty}$ such that $\{x_0, x_1, \dots, x_{k_n}\} \subset \Delta_n$, for each $n \geq 0$, but $x_{k_n+1} \notin \Delta_n$. Since $\{x_n\}_{n=0}^{\infty}$ cannot cluster anywhere in $\mathbb{R}^2$, the sequence $\{k_n\}_{n=0}^{\infty}$ is defined and unbounded. For every $n \geq 0$ and $j \geq 1$, the boundaries of the disks Δ_n and $f^j(\Delta_n)$ either coincide or meet (if at all) transversely and in an even number of points. Indeed, the boundaries project to the same loop under π and that loop has only transverse double points as self-intersections, proving our claim. Finally, for $1 \leq j \leq k_n$, the two disks have x_j as a common point; hence they intersect. If $f^j(\Delta_n) \subseteq \Delta_n$, the Brouwer fixed point theorem would imply that f^j has a fixed point in Δ_n, hence that $f^j = \mathrm{id}$. Similarly, $\Delta_n \subseteq f^j(\Delta_n)$ implies that f^{-j} has a fixed point. It follows that their boundaries intersect transversely in an even number of points.

For $1 \leq j \leq k_n$, choose distinct points $p_j, p'_j \in \partial\Delta_n \cap f^j(\partial\Delta_n)$ such that $\pi(p_j) = \pi(p'_j)$. If $1 \leq i < j \leq k_n$, we must have $\{p_i, p'_i\} \cap \{p_j, p'_j\} = \emptyset$. Otherwise, since $\partial\Delta_n$, $f^i(\partial\Delta_n)$, and $f^j(\partial\Delta_n)$ all intersect transversely, $\pi(\partial\Delta_n)$ would have a self-intersection of multiplicity greater than 2, contradicting the hypothesis. It follows that $k_n \leq m_n$, and this contradicts the boundedness of $\{m_n\}_{n=1}^{\infty}$. This contradiction completes the proof. □

If $t \in (0,1]$ has the property that $L_t = L_{t_k}$ for some sequence $t_k \downarrow 0$, we will say that t is recurrent.

Lemma 9.3.13. *There is $\eta \in (0,1]$ such that every $t \in (0,\eta]$ is recurrent.*

Proof. First we remark that the set

$$U = \{x \in D^2 \mid F_0(x) = \lim_{t\to 0} F_t(x) \text{ exists in } M\}$$

is open and contains ∂D^2. Furthermore, each connected component of U is carried by F_0 into a leaf of $\mathcal{F}$. If $U = D^2$, F_0 would define a nullhomotopy of σ_0 in L_0, and so we choose $x \in D^2 \smallsetminus U$. Suppose that there is no η as asserted and choose a sequence $\tau_k \downarrow 0$ of nonrecurrent numbers. Passing to a subsequence, if necessary, and appealing to compactness of M, we assume that $y = \lim_{k\to\infty} F_{\tau_k}(x)$ exists in M and choose a biregular chart W containing y. Since $x \notin U$, the parametrized arc $\tau(t) = F_t(x)$ in a leaf of $\mathcal{L}$ must pass infinitely often through W as $t \downarrow 0$, each time meeting W in an $\mathcal{L}$-plaque. In particular, for each $k \geq 1$, this arc must meet the $\mathcal{F}$-plaque through $F_{\tau_k}(x)$ infinitely often as $t \downarrow 0$, contradicting the fact that τ_k is nonrecurrent. □

Corollary 9.3.14. *Let η be as in Lemma 9.3.13. Then, for $0 < t \leq \eta$, $L_t \cong \mathbb{R}^2$ and $F_t(D^2) = D_t \subset L_t$ is a smoothly imbedded disk.*

Proof. Fix any choice of $t \in (0,\eta]$. Choose $x \in D^2 \smallsetminus U$ as in the proof of Lemma 9.3.13 and, reasoning as in that proof, choose $t_k \downarrow 0$ so that $F_{t_k}(x)$ lies in a fixed $\mathcal{F}$-plaque $P \subset L_t$, $1 \leq k < \infty$, converging to a point $y \in P$. Choose $\widehat{y} \in \mathbb{R}^2$ so that $\pi_t(\widehat{y}) = y$. Fix the choices of lifts $\widehat{F}_{t_k} : D^2 \to \mathbb{R}^2$ so that $\widehat{F}_{t_k}(x)$ converges to $\widehat{y}$ as $k \to \infty$. We claim that the disks

$$\Delta_k = \widehat{F}_{t_k}(D^2) \subset \mathbb{R}^2$$

satisfy the hypotheses of Lemma 9.3.12, at least after passing to a subsequence.

We will show first that, if $K \subset \mathbb{R}^2$ is a compact subset, then K can meet $\partial\Delta_k$ for only finitely many values of k. Indeed, $\pi_t(K)$ is a compact subset of L_t, hence meets only finitely many $\mathcal{F}$-plaques. Since $\sigma_{t_k} \to \sigma_0$ uniformly along the leaves of $\mathcal{L}$, it follows that at most finitely many of the loops σ_{t_k} meet any of these plaques. The assertion follows.

One consequence of the previous paragraph is that $\widehat{y} \in \operatorname{int}\Delta_k$ for all but finitely many values of k. Indeed, since the points $\widehat{F}_{t_k}(x) \in \Delta_k$ converge to $\widehat{y}$ in $\mathbb{R}^2$, the only alternative is that a sequence of points $z_k \in \partial\Delta_k$ will cluster at $\widehat{y}$. That is, given a compact neighborhood C of $\widehat{y}$, $\partial\Delta_k$ would meet C for infinitely many values of k. Another consequence is that, after passing to a subsequence, we can assume that $\Delta_k \subset \operatorname{int}\Delta_{k+1}$, $1 \leq k < \infty$. Indeed, for each k_0, all but finitely many $\partial\Delta_k$ miss Δ_{k_0}. Since all of these disks contain the common point $\widehat{y}$, it must be that $\Delta_{k_0} \subset \operatorname{int}\Delta_{k_1}$, for some $k_1 > k_0$. Finally, given an arbitrary point $z \in \mathbb{R}^2$, let $\tau : [0,1] \to \mathbb{R}^2$ be a continuous path joining $\widehat{y}$ to z. Since $K = \tau([0,1])$ is compact, some $\partial\Delta_k$ misses K. Since τ is continuous and has the endpoint $\widehat{y} \in \operatorname{int}\Delta_k$, it follows that $z \in \Delta_k$ also. This proves that $\mathbb{R}^2$ is the union of this expanding nest of disks.

Finally, the sequence $\{m_k\}_{k=1}^{\infty}$ is bounded by Lemma 9.3.4. All hypotheses of Lemma 9.3.12 have been verified; hence the covering map $\pi_t : \mathbb{R}^2 \to L_t$ is a diffeomorphism. Since $t \in (0, \eta]$ is arbitrary and $F_t = \pi_t \circ \widehat{F}_t$, all assertions are proven. □

At this point, the proof of Proposition 9.3.1 is complete. In fact, taking into account Corollary 9.3.7, we have proven the following apparently stronger statement.

Proposition 9.3.15. *If L_0 contains a vanishing cycle*

$$\sigma_t : S^1 \to L_t, \quad 0 \le t \le 1,$$

then there are $t_0 \ge 0$, as close to 0 as desired, and an exploding disk

$$\varphi : D^2 \times (t_0, 1],$$

parametrized so that the explosion occurs at t_0, such that $\varphi_t(\partial D^2)$ converges uniformly to a loop in L_{t_0} as $t \downarrow t_0$. If L_0 contains a simple vanishing cycle, then one can take $t_0 = 0$.

We will see, in fact, that L_0 must contain a simple vanishing cycle and that the only possible choice is $t_0 = 0$.

9.4. Completion of the Proofs of Novikov's Theorems

At this point, Propositions 9.1.5 and 9.1.6 in Section 9.1 have been proven (see pages 294 and 297). In that section, Theorems 9.1.4 and 9.1.7 were seen to follow from Theorem 9.1.3 and these propositions. The proof of Theorem 9.1.3 is completed by the following, which we will prove using the exploding disk.

Theorem 9.4.1. *If $(M, \mathcal{F})$ is a compact, foliated, transversely orientable 3-manifold and a leaf L_0 of $\mathcal{F}$ contains a vanishing cycle, then L_0 is the boundary torus of a Reeb component of $\mathcal{F}$.*

Proof of Theorem 9.1.3 using Theorem 9.4.1. In Section 9.1, we have already noted that the implication (1) ⇒ (2) in Theorem 9.1.3 is easy, while (2) ⇒ (3) was proven as Proposition 9.2.5. Theorem 9.4.1 is a somewhat stronger version of the implication (3) ⇒ (1). □

The proof of Theorem 9.4.1 will be carried out in three subsections. In the first two, we will use the following temporary hypothesis, removing it in the third.

Temporary Hypothesis. $\partial_{\pitchfork} M = \emptyset$ and L_0 contains a simple vanishing cycle.

9.4.A. The torus. Since L_0 contains a *simple* vanishing cycle σ_0, we obtain an exploding disk as in Proposition 9.3.15 with $t_0 = 0$. Recall from [I, Lemma 10.4.8] (where we assumed that $\partial_{\pitchfork} M = \emptyset$) that the exploding disk gives rise to a nontrivial foliation cycle μ (a holonomy-invariant transverse measure) that bounds. This foliation cycle, as a linear functional on $A^2(M)$, was seen to be the limit of an averaging sequence

$$\mu(\omega) = \lim_{k\to\infty} \frac{\int_{D_k} \omega}{A(D_k)}, \quad \forall\, \omega \in A^2(M),$$

where $A(D_k)$ denotes area and $\{D_k\}_{k=1}^{\infty}$ is (possibly a subsequence of) the sequence of expanding disks in a planar leaf L_* given by the exploding disk of Proposition 9.3.15.

Lemma 9.4.2. *The set* $\operatorname{supp}\mu$ *is contained in* $\overline{L}_*$.

Proof. Indeed, the support of a holonomy-invariant measure μ is the union of leaves $\widetilde{L}$ with the property that any transverse arc J through $\widetilde{L}$ has $\mu(J) > 0$. If $\widetilde{L} \subset \partial_\tau M$, one requires that an endpoint of J is in $\widetilde{L}$. It follows that such an arc meets infinitely many of the disks D_k, hence that $\widetilde{L} \subset \overline{L}_*$. □

Exercise 9.4.3. Let μ be a holonomy-invariant transverse measure, and assume that $\partial M = \emptyset$. Assume minimal regularity for the foliation, namely that it is integral to a continuous plane field. Use Poincaré duality and the theory of foliation cycles to define $\int_\sigma d\mu$, for every closed transversal σ to $\mathcal{F}$. If the foliation cycle μ is a boundary, prove that all such integrals vanish. (In fact, the integral is defined and vanishes for every closed loop, whether or not it is transverse to $\mathcal{F}$.)

Proposition 9.4.4. *If* $\widetilde{L} \subseteq \operatorname{supp}\mu$ *is a leaf of* $\mathcal{F}$, *then* $\widetilde{L} = L_0$ *and is a torus.*

Proof. If $\partial M = \partial_\tau M \neq \emptyset$, double along the boundary. This gives a foliated manifold with the foliation integral to a C^0 plane field. In this foliated manifold, the measure μ survives as a holonomy-invariant measure, and $\widetilde{L}$ meets no closed transversal by Exercise 9.4.3. By a theorem of S. Goodman that we proved in [I, Theorem 6.3.5], it follows that $\widetilde{L}$ is a torus. (We only proved this theorem for C^∞ foliations but, in fact, it holds for foliations integral to a continuous 2-plane field. Details are left for Exercise 9.4.5.) At this point we can return to the original undoubled foliated manifold $(M, \mathcal{F})$.

Fix a transverse arc $J = [0, \varepsilon)$, where 0 is the coordinate of $\{x\} = J \cap \widetilde{L}$ and $L_* \cap (0, \varepsilon)$ clusters at 0. Since $\widetilde{L}$ is a torus, its holonomy on the side defined by J is generated by commuting diffeomorphisms

$$f, g : [0, \eta) \to [0, \varepsilon).$$

Since $L_* \cap J$ accumulates on 0, f and g cannot have common fixed points accumulating on 0. It follows easily that, for $\eta > 0$ sufficiently small, either f or g is a contraction to 0. Let γ be a loop on $\widetilde{L}$ defining this holonomy contraction. Then the path that runs around γ forever in the contracting direction lifts to a path $\widetilde{\gamma}$ on L_* that ultimately escapes every D_k, hence intersects ∂D_k. Thus, choose points $x_k \in \widetilde{\gamma} \cap \partial D_k$, obtaining a sequence clustering only at points in $\widetilde{L}$. Since ∂D_k converges uniformly to $\sigma_0 \subset L_0$ (Proposition 9.3.15), it follows that $\widetilde{L} = L_0$. □

Exercise 9.4.5. Let $\mathcal{F}$ be of class $C^{1,0+}$. That is, $\mathcal{F}$ is integral to a C^0 2-plane field $E \subset T(M)$. By [I, Corollary 5.1.5], $\mathcal{F}$ is actually of class $C^{\infty,0+}$. Show that Goodman's theorem [I, Theorem 6.3.5] holds for such foliations. (Hint. A small homotopy of E smooths this bundle without affecting its Euler class. Of course, $\mathcal{F}$ is not integral to the new plane field, but...)

9.4.B. The Reeb component. Let $\{L_t\}_{0<t\le 1}$ be the family of leaves given by the exploding disk.

Lemma 9.4.6. *The set* $V = \bigcup_{0<t\le 1} L_t$ *is an open, connected,* $\mathcal{F}$*-saturated set and* $L_0 \subseteq \delta V$.

Proof. By Lemma 9.3.13, we can assume that every $t \in (0, 1]$ is recurrent. Taking $t = 1$, we see that the set V can also be defined using the strict inequality $0 < t < 1$, and so V is open. Since $F(D^2 \times (0, 1])$ is connected and meets every leaf L_t, V is connected. Finally, $F(\partial D^2 \times \{0\}) \subset L_0$ implies that L_0 is a border leaf of V ($L_0 \subseteq \delta V$, *cf.* [I, Definition 5.2.4]). □

Corollary 9.4.7. *There is a basis* $\{\sigma, \tau\}$ *of* $\pi_1(L_0) \cong \mathbb{Z} \oplus \mathbb{Z}$ *such that the corresponding holonomy elements on the side of* L_0 *bordering* V *are* $\widehat{h}_\sigma = \mathrm{id}$ *and* $\widehat{h}_\tau$*, the germ of a contraction.*

Proof. The loop σ_0 on L_0 is homotopically essential. Since it projects along the leaves of $\mathcal{L}$ to closed loops σ_t on L_t, $0 < t \le 1$, we see that $\widehat{h}_{\sigma_0} = \mathrm{id}$. If σ_0 is nondivisible in $\pi_1(L_0)$, it can be completed to a basis $\{\sigma_0, \tau\}$. If it *is* divisible, then find a nondivisible loop σ such that $\sigma^k = \sigma_0$ in $\pi_1(L_0)$. Since $\widehat{h}_\sigma^k = \widehat{h}_{\sigma_0} = \mathrm{id}$ and the foliation is transversely orientable, it follows that $\widehat{h}_\sigma = \mathrm{id}$. If a representative h_τ of the germ $\widehat{h}_\tau$ had fixed points arbitrarily near L_0, it would follow that V contained toroidal leaves, contradicting Corollary 9.3.14. Thus, h_τ (or its inverse) is a contraction near L_0. $\square$

Exercise 9.4.8. If $f, g : [0, \varepsilon) \to [0, \varepsilon)$ are contractions to 0, show that there is a homeomorphism $h : [0, \varepsilon) \to [0, \varepsilon)$ such that $hgh^{-1} = f$. If the contractions f and g are C^r diffeomorphisms onto their images, show that h can be taken to be a C^r diffeomorphism on $(0, \varepsilon)$. Explain why h cannot generally be a diffeomorphism on $[0, \varepsilon)$.

Lemma 9.4.9. *There is a one-sided neighborhood N of L_0 on the side bordering V such that $(N, \mathcal{F}|N)$ is homeomorphic to a corresponding foliated neighborhood $(\widetilde{N}, \widetilde{\mathcal{F}})$ of the boundary T_0 of a Reeb-foliated solid torus. This homeomorphism can be taken to be a C^∞ diffeomorphism away from the boundary tori.*

Proof. There is a basis α_1, α_2 of $\pi_1(T_0)$ such that $\widehat{h}_{\alpha_1} = \text{id}$ and $\widehat{h}_{\alpha_2}$ is the germ of a contraction. By Exercise 9.4.8 and the well known fact that there is a diffeomorphism $\varphi : L_0 \to T_0$ carrying σ to α_1 and τ to α_2, we conclude that the germinal holonomy homomorphism of the leaf L_0 of $\mathcal{F}|N$ is topologically conjugate to that of the leaf T_0 of $\widetilde{\mathcal{F}}|\widetilde{N}$. This conjugacy is of class C^∞ away from L_0 and T_0. But the germ of a foliation at a compact leaf F is determined, up to isomorphism, by the conjugacy class of the holonomy homomorphism of F [I, Theorem 2.3.9], and so we only need to choose the neighborhoods N and $\widetilde{N}$ sufficiently small. $\square$

In the Reeb-foliated solid torus, it is obvious that the boundary torus T_0 can be deformed to a torus T' in the interior, uniformly near T_0 and transverse to the foliation, meeting each leaf in a circle. By Lemma 9.4.9, we find a similar torus $T \subset V$ arbitrarily near L_0, such that the component of $V \smallsetminus T$ bordering L_0 is foliated by cylinders winding in on the toroidal leaf L_0 in the Reeb way.

Lemma 9.4.10. *The closure $\overline{V}$ of V in M is a compact, $\mathcal{F}$-saturated manifold with $\partial\overline{V} = L_0$.*

Proof. Let $L_t \subset V$ be a leaf. By the above, there is a circle $S \subseteq L_t \cap T$ such that one component of $L_t \smallsetminus S$ is a cylinder winding in on L_0. Since $L_t \cong \mathbb{R}^2$, the other component is a disk (and, in particular, $S = L_t \cap T$); hence the asymptote of L_t [I, Definition 4.3.1] is $\lim(L_t) = L_0$, $0 < t \leq 1$. But every leaf in δV must lie in $\lim(L_t)$, for some $t \in (0,1]$; hence $\delta V = L_0$. Also, all $L_t \subset V$ approach L_0 from the same side, and hence it is clear that $\overline{V} = \widehat{V}$ is a compact $\mathcal{F}$-saturated manifold with boundary leaf L_0. $\square$

Theorem 9.4.11. *The closure $\overline{V}$ of V in M is foliated-homeomorphic to a Reeb-foliated solid torus.*

Proof. By the above, T separates the compact, foliated manifold $\overline{V}$ into two parts, a 3-manifold P with foliation $\mathcal{F}|P$ transverse to $\partial P = T$ and a manifold Q homeomorphic to $T^2 \times [0, \varepsilon]$, where $T^2 \times \{0\}$ is identified with L_0 and $T^2 \times \{\varepsilon\}$ is identified with T. The foliation $\mathcal{F}|Q$ is by cylinders, asymptotic to L_0 and meeting T in circles. Some of the leaves of $\mathcal{F}|P$ are disks; hence Reeb stability implies that all are disks and that $\mathcal{F}|P$ fibers P over S^1. The only two possibilities are that P is the solid Klein bottle or the solid torus. In the first case, the orientability of M would be contradicted, hence $P \cong D^2 \times S^1$ with $\mathcal{F}|P$ homeomorphic to the foliation by the disks $D^2 \times \{z\}$. The standard Reeb-foliated solid torus is split by the torus T' into similar pieces $P' \cong D^2 \times S^1$ and Q', the latter being foliated-homeomorphic to Q. The piece P' is foliated-homeomorphic to $D^2 \times S^1$, equipped with the product foliation. The foliation-preserving homeomorphism of Q onto Q', restricted to $T' = \partial Q = \partial P$, extends over P to a foliation-preserving homeomorphism of P to P'. That is, $\overline{V}$ is foliated-homeomorphic to a Reeb-foliated solid torus. □

We have proven Theorem 9.4.1, subject to the temporary hypothesis.

9.4.C. Removing the temporary hypothesis. Our first goal is to reduce to the case $\partial M = \partial_\tau M$ by doubling along $\partial_\pitchfork M$.

Let $(M'', \mathcal{F}'')$ denote the result of doubling. Each leaf L of $\mathcal{F}$ either doubles to a leaf L'' of $\mathcal{F}''$ or, if $L \cap \partial M = \emptyset$, perdures as a leaf L'' of $\mathcal{F}''$. As in [I, Example 3.3.1], we can assume that $\mathcal{F}''$ is again of class C^∞. Suppose that the leaf L_0 of $\mathcal{F}$ has a simple vanishing cycle. Then, as no boundary conditions were assumed in proving Proposition 9.3.15, there is an exploding disk $F : D^2 \times (0, 1] \to M$. Viewed in M'', this is still an exploding disk. Indeed, the fact that $A(D_t) \uparrow \infty$ as $t \downarrow 0$ implies that F cannot be extended to a continuous map of $D^2 \times [0, 1] \to M''$. Consequently, Proposition 9.4.4 implies that L_0'' is a torus bounding a Reeb component of $\mathcal{F}''$, hence that L_0 is either an annulus or a torus.

Proposition 9.4.12. *If L_0 is a leaf of $\mathcal{F}$ having a simple vanishing cycle, then L_0 is a torus.*

Proof. We assume that L_0 is an annulus and obtain a contradiction. The two boundary circles of L_0 must lie in ∂M and each generates $\pi_1(L_0)$. Let σ be one of these circles. The vanishing cycle must be homotopic to $k\sigma$, for some nonzero integer k; hence $k\sigma$ is itself a vanishing cycle. The deformation $(k\sigma)_t \subset L_t$, given by the definition of vanishing cycle, must be of the form $k\sigma_t$, where σ_t is a component of ∂L_t, $0 \le t \le 1$. Since the orientable surface L_t has torsion-free fundamental group, $k\sigma_t \sim 0$ in L_t implies that $\sigma_t \sim 0$ in L_t, $0 < t \le 1$. But it is elementary that the only connected surface with a compact, nullhomotopic boundary component is D^2. Since M is connected, it follows [I,Theorem 6.1.5] that every leaf of $\mathcal{F}$ is a disk, contradicting the fact that $L = L_0$ is an annulus. □

At this point, the argument given in the preceding subsection goes through unchanged to prove that the leaf L_0 having a simple vanishing cycle must be the boundary of a Reeb component of $\mathcal{F}$. The following completes the proof of Theorem 9.4.1, hence of all of Novikov's theorems.

Lemma 9.4.13. *If L_0 has a vanishing cycle, it has a simple one.*

Proof. Assume that L_0 has a vanishing cycle, but not a simple one. By Proposition 9.3.15, choose $0 < t_0 < s_0 < 1$ so that L_{t_0} and L_{s_0} both contain simple vanishing cycles. Thus, these leaves are the boundary tori of Reeb components, and the interior of the Reeb component bounded by L_{t_0} contains L_{s_0}. This contradiction implies that L_0 has a simple vanishing cycle. □

9.5. The Roussarie-Thurston Theorems

In this section, we discuss two general position theorems. The first, due to R. Roussarie [**160**], concerns incompressible surfaces imbedded in Reebless foliated manifolds. This will have critical application in the following chapter. A sharper result, due to W. Thurston [**175**], holds in tautly foliated 3-manifolds.

Let $(M, \mathcal{F})$ be a compact, connected, oriented and transversely oriented, foliated 3-manifold without Reeb components. Assume also that no leaf is a disk or 2-sphere. By [I, Theorem 6.1.5], this latter hypothesis eliminates only the uninteresting product foliations with all leaves disks or spheres. As usual, $\partial M = \partial_\tau M \cup \partial_\pitchfork M$. By Corollary 9.1.9, the existence of the Reebless foliation with nonspherical leaves implies that M is irreducible.

Let $\varphi : T \to M$ be a smooth imbedding of a compact, connected, orientable surface T into M. It is allowed that T have boundary and corners. We often omit reference to φ, viewing T as a surface sitting in M. We also require that all imbedded surfaces be *proper*, in the sense that $\partial T = T \cap \partial M$ and $T \pitchfork \partial M$. We require further that T meet the corners of M precisely in the corners of T, as illustrated in Figure 9.5.1. Thus, a component C of ∂T contains no corners if and only if it lies entirely in $\partial_\tau M$ or entirely in $\partial_\pitchfork M$. If $C \subset \partial_\pitchfork M$, we require either that C itself be transverse to $\mathcal{F}$ or lie entirely in a leaf of $\mathcal{F}$. If the component contains corners, the corners divide C into arcs lying alternately in $\partial_\tau M$ and $\partial_\pitchfork M$. In the latter case, we require that the arc be transverse to $\mathcal{F}$. Finally, all of this is compatible with one further requirement, that T be transverse to $\mathcal{F}$ in a neighborhood of ∂M.

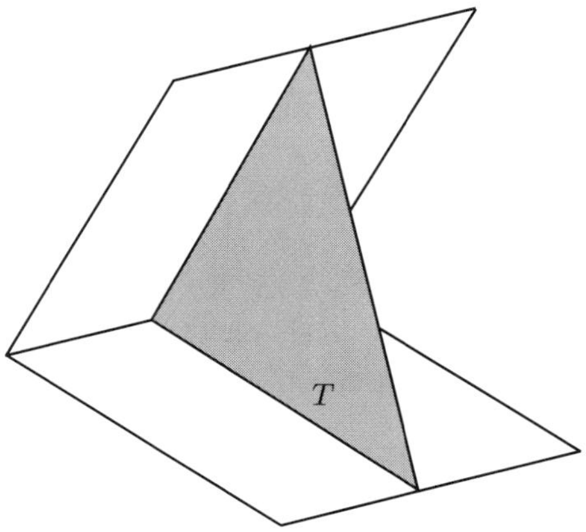

Figure 9.5.1. A properly imbedded surface with corners

Definition 9.5.1. A properly imbedded, connected (smooth) surface T in M is incompressible if one of the following holds:

(1) $T \cong S^2$ and does not bound an imbedded 3-ball in M.

(2) T is not a 2-sphere and, if a loop $\sigma \subset T$ bounds an imbedded disk in M, then σ bounds an imbedded disk in T.

We assume that our imbedding $\varphi : T \hookrightarrow M$ is incompressible. It follows from this definition and Dehn's lemma [**93**, 4.1] that the imbedding φ is injective on fundamental groups. Since M is irreducible, the incompressible surface cannot be a sphere. If it is a disk, we assume that it is $\partial_\tau M$-*incompressible.* That is, there is no imbedded disk $D \subset \partial_\tau M$ such that $\partial T = \partial D$ and $T \cup D$ bounds an imbedded 3-ball.

Definition 9.5.2. An imbedded surface satisfying all of the requirements we have made on $\varphi : T \hookrightarrow M$ will be said to be *admissible* or *admissibly imbedded.*

Lemma 9.5.3. *If the admissibly imbedded surface $T \hookrightarrow M$ is a disk, then ∂T cannot be contained entirely in $\partial_\tau M$ nor in $\partial_\pitchfork M$.*

Proof. Let F be a component of $\partial_\tau M$ such that $\partial T \subset F$. Since the foliation is Reebless, the imbedding of F is π_1-injective (Theorem 9.1.3), implying that ∂T bounds a disk $D \subset F$. By irreducibility, this violates our hypothesis that T is $\partial_\tau M$-incompressible. If $\partial T \subset \partial_\pitchfork M$, this imbedded loop either is transverse to $\mathfrak{F}$ or is a boundary component of a leaf L. The first possibility again violates the hypothesis that $\mathcal{F}$ is Reebless (Theorem 9.1.4). In the second case, ∂T must be nullhomotopic in L (by π_1-injectivity). The only surface having a nullhomotopic boundary component is a disk; hence we have violated the hypothesis that no leaf of $\mathcal{F}$ is a disk. □

Definition 9.5.4. The admissible imbedding $\varphi : T \hookrightarrow M$ is said to be *reduced* or in *Roussarie general position* if all of its tangencies with $\mathcal{F}$ (necessarily in $\text{int}(T)$) are either saddle points (saddle tangencies) or whole circles σ, each of which is an essential loop on a leaf and has trivial germinal holonomy h_σ. It is further required that, in a neighborhood of such a circle σ, T lies entirely on one side of the leaf containing σ and, finally, that no two of these tangencies are to lie on a common leaf. This will be called a circle tangency

Theorem 9.5.5 (Roussarie). *If the foliated manifold $(M, \mathcal{F})$ is Reebless, no leaf being a disk or 2-sphere, and if $\varphi : T \hookrightarrow M$ is admissible, then there is an isotopy of $\varphi = \varphi_0$ through admissible imbeddings φ_t to an imbedding φ_1 in Roussarie general position. Furthermore, $\varphi_t|\partial T = \varphi_0|\partial T$, $0 \le t \le 1$.*

Remark. In [**160**], Roussarie shows that, if T is a torus, then it can further be isotoped either to be transverse to $\mathcal{F}$ or to coincide with a leaf of $\mathcal{F}$. Indeed, once it is in Roussarie general position, the fact that $\chi(T) = 0$ implies that there can be no saddle tangencies. If there are no circle tangencies, the imbedded torus is everywhere transverse to $\mathcal{F}$. Otherwise, an annular "flattening" process along circle tangencies might lead to annuli with holonomy that can then be perturbed to be transverse to $\mathcal{F}$, or it might push T onto a torus leaf. We omit the details, since we will not use this result.

The proof of Theorem 9.5.5 will be divided up into several lemmas. As usual, a preliminary isotopy allows us to assume that φ is in general position with respect to $\mathcal{F}$. In the following discussion, we construct a sequence of isotopies. The reader should note that, in every case, the transversality of T to $\mathcal{F}$ near ∂M guarantees that these constructions can be carried out so as not to perturb T in a neighborhood of ∂M. More precisely, $\varphi_t = \varphi$ in a neighborhood of ∂T. We will not continue to say this explicitly.

Let $\mathcal{F}' = \varphi^{-1}(\mathcal{F})$ and let $p \in T$ be a center of $\mathcal{F}'$. We consider the set $\{R_\alpha\}_{\alpha\in\mathfrak{A}}$ of all the *open* disk neighborhoods R_α of p such that $R_\alpha \setminus \{p\}$ is a union of closed trajectories.

Definition 9.5.6. The compact set $E = \overline{\bigcup_{\alpha\in\mathfrak{A}} R_\alpha}$ is called the adequate neighborhood of the center p.

Exercise 9.5.7. Prove that the set-theoretic boundary ∂E is of one of the following two types:

(a) ∂E is a graph made up of one saddle q and one trajectory.

(b) ∂E is a graph made up of one saddle q and two trajectories.

(In the case that T is a disk, you will need Lemma 9.5.3.)

In case (a), E is an imbedded disk with a singularity at q. In case (b), E is a "pinched" annulus as in Figure 9.2.4. In both cases, the interior of $E \setminus \{p\}$ is foliated by smooth circles. In accordance with the labels of these two cases, we will speak of centers of types (a) and (b).

We smoothly parametrize E by a one-parameter family of maps

$$\gamma_t : S^1 \to T, \quad 0 \le t \le 1,$$

where $\gamma_0(S^1) = p$, $\gamma_1(S^1) = \partial E$, and γ_t is a smooth imbedding with image a closed trajectory of $\mathcal{F}'$, $0 < t < 1$. Since $\mathcal{F}'$ can be parametrized as a flow, each γ_t is chosen to carry the counterclockwise orientation of S^1 to the positive orientation of the trajectory or graph $\gamma_t(S^1)$, $0 < t \le 1$. Also, for each $z \in S^1$, $\gamma_t(z)$ is to describe a trajectory transverse to $\mathcal{F}'$, except at p and q.

Set $C_t = \gamma_t(S^1)$, $0 \le t \le 1$, and let L_t be the leaf of $\mathcal{F}$ that contains C_t. The local picture of a center tangency shows that C_t is nullhomotopic in L_t for suitably small $t > 0$. Since no leaf of $\mathcal{F}$ has a vanishing cycle, C_t is nullhomotopic in L_t, $0 \le t \le 1$. Since no leaf L_t is a sphere, there is a unique disk $D_t \subset L_t$ with $\partial D_t = C_t$, $0 < t < 1$. For notational simplicity, write $C = C_1$ and $L = L_1$.

Consider the curve C. If p is a center of type (a), there is again a unique disk D in L with $C = \partial D$. If p is of type (b), however, there are two possibilities. There is an imbedded circle $C' \subset L$ that coincides with C outside a small neighborhood of the saddle point q. This circle is homotopic to C in L. In particular, C' is nullhomotopic in L, hence bounds a unique disk $D' \subset L$. The two cases depend on whether or not $q \in D'$. If $q \in D'$, the two loops of C will form the boundary of the union of two disks in L, joined exactly at q. If $q \notin D'$, the two loops form the boundary of a pinched annulus, which we will denote by D. This pinched annulus will have a neighborhood $A \cong S^1 \times (-1, 1) \subset L$ that has D as a deformation retract.

Definition 9.5.8. An isotopy φ_t, $0 \le t \le 1$, will be said to respect $\mathcal{F}'$ if $\varphi_t^{-1}(\mathcal{F}) = \mathcal{F}'_t$ is independent of t up to conjugacy.

Lemma 9.5.9. *There is a smooth isotopy of φ through admissible imbeddings to an admissible imbedding, in general position with respect to $\mathcal{F}$, such that all centers are of type* (b).

Proof. Let p be a center of type (a) and let E be its adequate neighborhood. If T does not meet $\operatorname{int}(D)$, then $E \cup D$ will be an imbedded 2-sphere and,

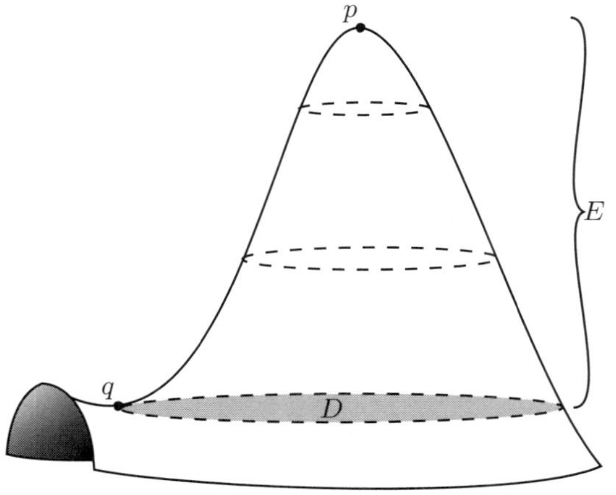

Figure 9.5.2

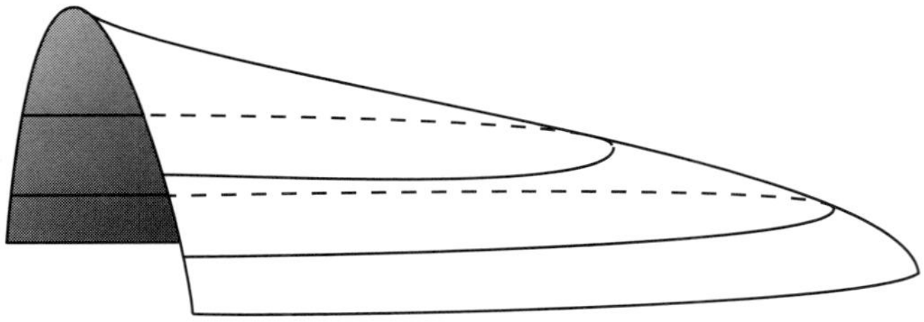

Figure 9.5.3. Cancellation of a saddle and a center

by irreducibility, will bound an imbedded 3-ball with interior disjoint from T (see Figure 9.5.2). The standard technique of cancelling the center p and the saddle $q \in C$, illustrated in Figure 9.5.3, can be applied here to get rid of these two tangencies without affecting the other tangencies.

In general, it is necessary to perform an isotopy of T, respecting $\mathcal{F}'$, so as to make sure that $T \cap \text{int}(D) = \emptyset$. A component σ of $T \cap \text{int}(D)$ is an imbedded circle bounding a disk $\Delta \subset \text{int}(D)$. Choose σ so that $T \cap \text{int}(\Delta)$ is empty. Since T is incompressible in M, σ must bound a disk $\Delta' \subset T$. Then $\Delta' \cup \Delta$ is an imbedded 2-sphere and bounds a 3-ball, and Δ' can be isotoped through D to a position not meeting D. It is not hard to see that the isotopy can be chosen to respect $\mathcal{F}'$. After finitely many such moves, $T \cap \text{int}(D) = \emptyset$.

Finite repetition of this whole process removes all centers of type (a). $\square$

By Lemma 9.5.9, we assume that the imbedding $\varphi : T \to M$ has no centers of type (a).

Let p be a center of type (b), E the adequate neighborhood of p, and let c, c' be the two loops in the figure-eight loop $C = \partial E$. Recall that there are two possibilities:

- c and c' bound imbedded disks in L meeting exactly at a saddle point.
- $c \cup c' = \partial D$, where $D \subset L$ is a pinched annulus.

Lemma 9.5.10. *In the absence of centers of type* (a)*, it is not possible for c and c' to bound disks in L meeting exactly at a saddle point.*

Proof. Otherwise, since T is incompressible in M, it would follow that c and c' also bound disks in T, at least one of which does not contain $c \cup c'$. Let Δ be that disk bounded, say, by c. Then $\text{int}(\Delta)$ contains $k \geq 1$ centers and $k - 1$ saddles. We show that at least one of these centers is of type (a), a contradiction.

Indeed, if $k = 1$, this is evident. If $k > 1$, let $q \in \text{int}(\Delta)$ be a saddle. Consider a graph in $\text{int}(\Delta)$ consisting of q and two trajectories that issue from and return to q. One of these bounds a disk in $\text{int}(\Delta)$ whose interior contains fewer than k centers. By induction on k, we conclude that there is a center of type (a). $\square$

Thus, for each center, $C = c \cup c'$ bounds a pinched annulus D in the leaf L.

Lemma 9.5.11. *After an isotopy of φ that respects $\mathcal{F}'$, we can assume that $T \cap \text{int}(D) = \emptyset$.*

Proof. By the usual argument, using the irreducibility of M and the incompressibility of T, a component σ of $T \cap \text{int}(D)$ can be found that bounds disks in $\text{int}(D)$ and in T, the two of which together bound an imbedded 3-ball with interior disjoint from T. An isotopy, supported in a Euclidean neighborhood of this ball and respecting $\mathcal{F}'$, gets rid of the intersection loop σ. Finite repetition of the process gets rid of all intersections of T with $\text{int}(D)$. $\square$

Thus, for all centers p, we assume that $T \cap \text{int}(D) = \emptyset$. Note that E is also a pinched annulus and that $E \cup D$ is a 2-sphere with two points identified at the saddle q. Since this can be arbitrarily well approximated

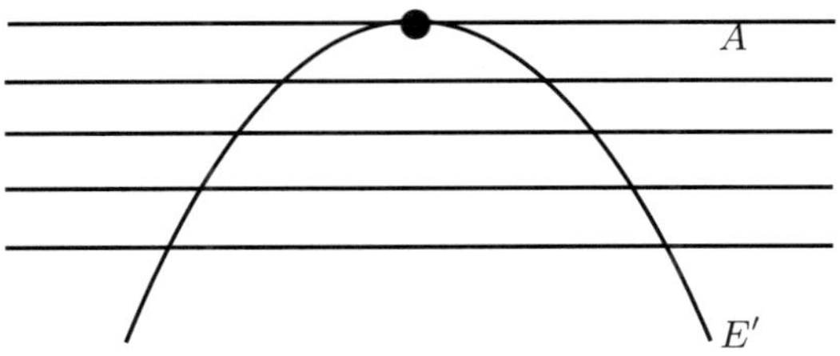

Figure 9.5.4. A cross-section view of a circle tangency

by a true imbedded 2-sphere, it is not hard to conclude the following from the irreducibility of M.

Lemma 9.5.12. *The set $E \cup D$ bounds a solid $B \subset M$, this being a 3-ball with two boundary points identified. Furthermore, $T \cap \text{int}(B) = \emptyset$.*

The last assertion is by Lemma 9.5.11.

Let $D \subset A \subset L$, where D is a deformation retract of the annulus A. There is a normal neighborhood $N \cong A \times (-1,1)$ of $A = A \times \{0\}$ in M such that $\mathcal{F}|N$ is transverse to the interval fibers. We can choose the orientation of the fiber so that E meets $A \times (0, \delta)$ for all small values of $\delta > 0$.

Set $N^+ = A \times [0,1)$, and observe that Lemma 9.5.12 implies that E can be isotoped into N^+ by an isotopy of $\varphi : T \hookrightarrow M$ that is supported in a small annular neighborhood E' of E. One can then perform an isotopy of $\varphi : T \hookrightarrow M$, likewise supported in a small annular neighborhood E'' of E', so that E' is pushed into $N^- = A \times (-1, 0]$. With care, this can be done so that E' and $A = A \times \{0\}$ are tangent along their respective core circles, but so that E'' is elsewhere transverse to $\mathcal{F}$. Thus, we obtain a circle tangency, the cross-section view being as in Figure 9.5.4.

Since $\pi_1(A) = \mathbb{Z}$, the foliation of N near A is determined by a single diffeomorphism of the form $f : (-\delta, \delta) \to (-\delta \pm \eta, \delta \pm \varepsilon)$ with $\delta > 0$ sufficiently small. We summarize these remarks in the following lemma.

Lemma 9.5.13. *There is an isotopy of φ through admissible imbeddings so that the center p and the saddle q are replaced by a circle tangency, but the remaining configuration of tangencies is unchanged. When f has trivial germ at 0, the circle tangency has trivial holonomy.*

After performing all possible such moves, we are left with no centers. There are possibly some saddles and some circle tangencies.

Lemma 9.5.14. *If all tangencies are saddles and/or circles, each circle tangency is essential on its leaf.*

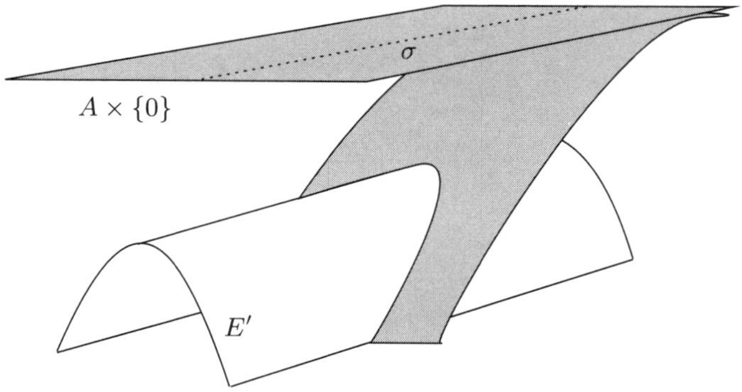

Figure 9.5.5. Removing the circle tangency in Case 1

Proof. Otherwise, since T is incompressible, this circle tangency would also bound a disk Δ in T. Since $\chi(\Delta) = 1$, this would force the existence of a center in Δ. □

Our remaining concern is to eliminate circle tangencies such that f has nontrivial germ at 0, and to do so without introducing new tangencies.

Our hypothesis that the germ of f at 0 is nontrivial means that, for a suitable choice of $\varepsilon > 0$, one of the following holds:

Case 1: f is fixed point free on $(-\varepsilon, 0)$.

Case 2: f is fixed point free on $(0, \varepsilon)$.

Case 3: There is a sequence $\{[a_k, b_k] \subset (-\varepsilon, \varepsilon)\}_{k=1}^{\infty}$ of disjoint intervals , clustering at 0, such that $f|(a_k, b_k)$ is fixed point free, $k \geq 1$.

In Case 1, the leaves of $\mathcal{F}|A \times (-\varepsilon, 0)$ are spiral ramps winding in on $A \times \{0\}$, and a slight displacement of T, supported in E', produces the picture in Figure 9.5.5 in which the tangency is completely removed. Effectively, the center-saddle pair of tangencies that gave rise to the circle tangency have cancelled.

Case 2 is entirely similar, the displacement being "upward" rather than "downward".

In Case 3, $\mathcal{F}|A \times (-\varepsilon, \varepsilon)$ has infinitely many annular leaves $A \times \{b_k\}$ clustering at $A \times \{0\}$, and a small displacement moves the circle tangency to some one of these. But the foliation $\mathcal{F}|A \times (a_k, b_k)$ has all leaves spiraling on $A \times \{b_k\}$ and we are reduced to Case 1. This is summarized in the following lemma, finite iteration of which completes the proof of Theorem 9.5.5.

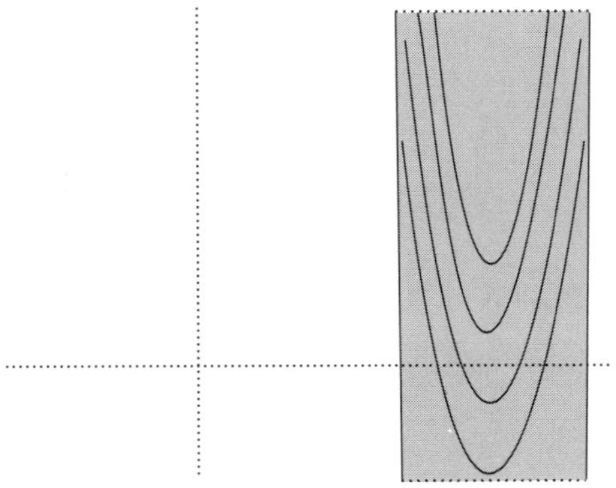

Figure 9.5.6. A Reeb strip in $\mathbb{R}^2$

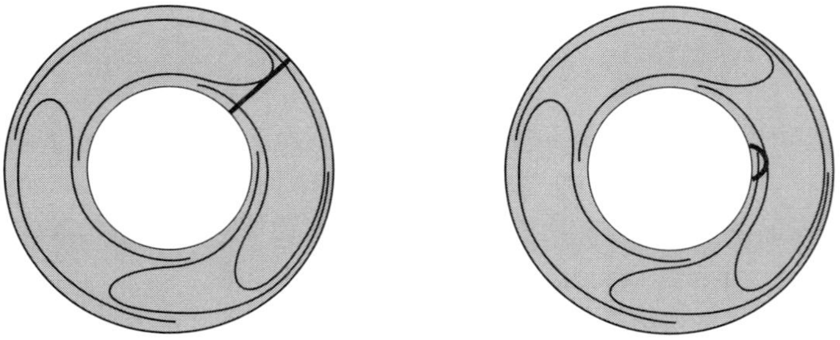

Figure 9.5.7. Two admissible imbeddings with essential circle tangencies

Lemma 9.5.15. *When $f|(-1,0]$ has nontrivial germ at 0, there is an isotopy of φ through admissible imbeddings so that the circle tangency is eliminated, but the remaining configuration of tangencies is unchanged.*

Finally, one observes that circle tangencies with trivial holonomy can be slightly perturbed to nearby leaves so as to make sure that no two such tangencies lie on the same leaf nor on a leaf having a saddle tangency. The saddles were originally on distinct leaves; hence the surviving saddles are on distinct leaves. The proof of Theorem 9.5.5 is complete.

Example 9.5.16. The circles of tangency are generally unavoidable. Consider the Reeb-foliated strip in $\mathbb{R}^2$, situated as in Figure 9.5.6. Rotate about the y-axis, obtaining a thickened cylinder, foliated by cylinders and having only tangential boundary. Quotient out the action of $\mathbb{Z}$ by vertical translations, obtaining a Reebless foliated manifold $(M, \mathcal{F})$, where $M = T^2 \times I$ is a thickened torus having the boundary tori as leaves and foliated in the

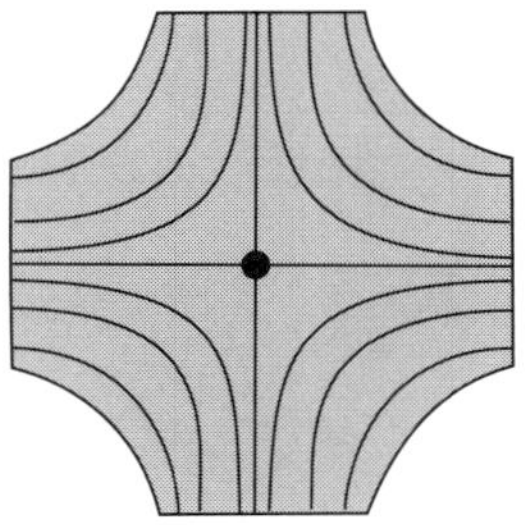

Figure 9.5.8. Induced foliation on an 8-gon in Roussarie general position

interior by cylinders. There are essentially two admissible imbeddings of the annulus A in M, both indicated by thickened arcs in a 2-dimensional cross-section view in Figure 9.5.7. While both are incompressible, the second one is $\partial_\tau M$-compressible (allowed by our definition of an admissible imbedding). There is an obvious circle tangency in each case, and it is intuitively clear that it cannot be isotoped away without introducing centers. It is worthy of note that this foliation contains an "exploding annulus" (see [I, Definition 10.4.4] for the more general concept of an exploding plateau). This is the reason why circles of tangency cannot be eliminated in this example. Exploding plateaus are impossible in taut foliations [I, Theorem 10.4.5], and so we might expect that the circles of tangency can be eliminated in the taut case.

Exercise 9.5.17. If T is in Roussarie general position with no circle tangencies, let s denote the number of saddle tangencies and c the number of arcs (not complete circles) in $\partial T \cap \partial_\tau M$. Note that c is even and prove that

$$\chi(T) = \frac{c}{2} - s.$$

(Hint. This is a corollary of Theorem 4.4.11.)

Example 9.5.18. In Gabai's theory of disk decompositions (Chapter 11), imbedded disks occur in Roussarie general position. Observe that these disks can have no circle tangencies, since such a circle would have to be inessential. These disks appear as $4n$-gons, the edges being alternately arc components of $\partial T \cap \partial_\tau M$ and of $\partial_\pitchfork T$. For the case $n = 2$, the induced foliation has one saddle singularity and is pictured in Figure 9.5.8. The analogous picture with $n > 2$ would have a $2n$-pronged saddle, but an arbitrarily small isotopy of the imbedded disk puts it in Roussarie general position with $n-1$ ordinary (4-pronged) saddles. In all cases, the formula in Exercise 9.5.17 gives 1, the correct Euler characteristic of the disk.

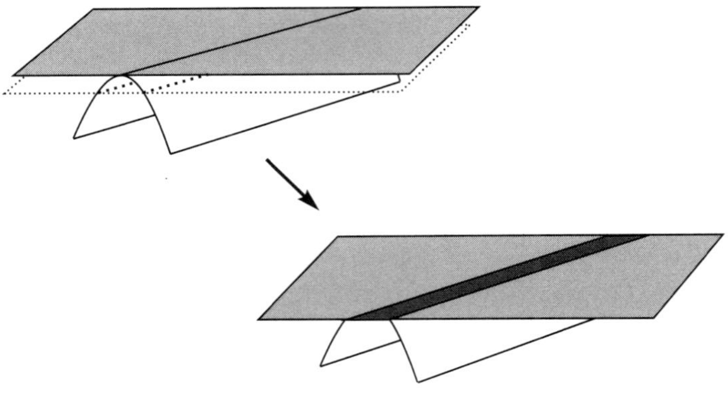

Figure 9.5.9. Flattening a circle of tangency to an annulus

If the hypothesis that $\mathfrak{F}$ is Reebless is strengthened to require that $\mathfrak{F}$ be taut, Thurston has shown (unpublished) that the conclusion can be strengthened to get rid of the circles of tangency. Recall [I, Definition 6.3.3] that $\mathfrak{F}$ is taut if every leaf meets either a closed transversal, or a transverse arc from one component of $\partial_\tau M$ to another.

Definition 9.5.19. An admissible imbedding will be said to be in *Thurston general position* with respect to $\mathfrak{F}$ if all of its tangencies with $\mathfrak{F}$ are saddles.

Theorem 9.5.20 (Thurston). *If $\mathfrak{F}$ is taut and transverse to ∂M and if $\varphi : T \hookrightarrow M$ is admissible, there is an isotopy φ_t, $0 \le t \le 1$, through admissible imbeddings such that $\varphi_0 = \varphi$ and $\varphi_1(T)$ either coincides with a leaf or is in Thurston general position with respect to $\mathfrak{F}$.*

The theorem is true with more general boundary conditions. Thurston has never published a proof.

A "hands on" proof simply takes up where Roussarie's stops, flattening the circular tangencies into annular plateaus (Figure 9.5.9) and continuing until an "event" occurs that obstructs the progress of the moving plateau. By tautness, this event is not an exploding annulus. The plateau may acquire holonomy, in which case we have seen that it can be displaced to be transverse to $\mathfrak{F}$. Other events include limiting on a saddle and one or both closed separatrices, or limiting on another circular tangency. Various cases need to be considered and in all but one, the plateau and limiting obstruction either cancel or coalesce into a saddle. In one case, illustrated in Figure 9.5.10 (which is drawn in a foliated chart about a saddle q), the connectivity of the plateau increases and one presses on. Events continue to arise in which either everything dissipates into saddles or the connectivity of

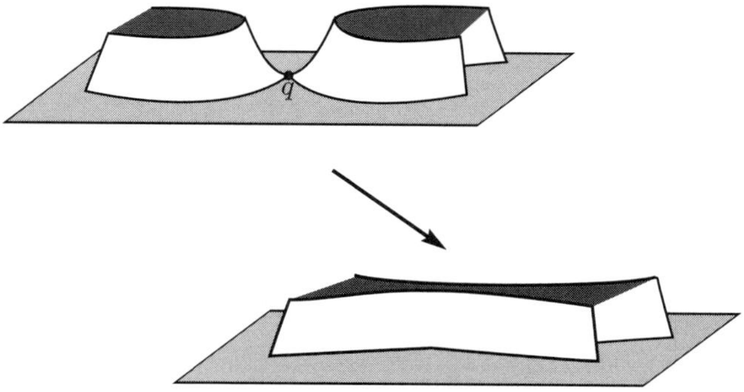

Figure 9.5.10. Increasing the connectivity of the moving plateau

the plateau increases. After finitely many steps, either T has been flattened onto a leaf or T has achieved Thurston general position.

A very different and more satisfactory line of proof, due to J. Hass [**89**], uses the theorem of D. Sullivan that there is a Riemannian metric making all of the leaves of $\mathfrak{F}$ minimal surfaces [I, Corollary 10.5.9]. Results of various authors (*cf.* [**90**]) show that the incompressible surface can also be isotoped to be minimal with respect to this metric, and it follows that all tangencies of this minimal surface with leaves are $2p$-pronged saddles or that the surface coincides with a leaf [**89**, Lemma 2.3]. It is easy, via small isotopies, to convert the $2p$-pronged saddles to a collection of $p-1$ ordinary saddle tangencies. This proof is carried out in closed manifolds, but Theorem 9.5.20 is easily reduced to that case.

Finally, D. Gabai [**75**] has generalized Theorem 9.5.20 to foliations of class $C^{1,0+}$ (we are using the notation of [I, Definition 1.2.24]) and to C^0-immersed, incompressible surfaces T. He does not require M to be orientable nor $\mathfrak{F}$ to be transversely orientable.

We will only need Theorem 9.5.20 in one key step in the proof of Theorem 10.5.3.

Chapter 10

Foliations and the Thurston Norm

We continue to assume that M is a compact, connected, oriented 3-manifold, possibly with boundary. Foliations $\mathcal{F}$ will always be smooth and transversely oriented of codimension one, with $\partial M = \partial_{\pitchfork} M$. Properly imbedded surfaces will also be transversely oriented, hence oriented via the orientation of M.

In this chapter, we introduce a (pseudo)norm on $H_2(M, \partial M; \mathbb{R})$, due to W. Thurston [**175**]. If we restrict our attention to the integer lattice $H_2(M, \partial M; \mathbb{Z}) \subset H_2(M, \partial M; \mathbb{R})$, it is a standard fact (see Lemma 10.3.3) that each class ζ is represented by a compact surface imbedded in M, with $\partial N = N \cap \partial M$. We write

$$\zeta = [N, \partial N] \in H_2(M, \partial M; \mathbb{Z}).$$

When computed for such a class ζ, the Thurston norm measures the "minimal topological complexity" of compact surfaces N representing the class. It will be proven that compact leaves of Reebless foliations realize this minimal topological complexity. This result is due to Thurston and has been used by D. Gabai to determine the genus of many knots and links [**71, 72**].

The norm defines a (possibly noncompact) polyhedral unit ball, each top-dimensional face of which determines a convex, polyhedral cone in the vector space $H_2(M, \partial M; \mathbb{R})$. Following Thurston, we show that, if M can be fibered over S^1 by fibers transverse to ∂M, then the homology classes of fibers are the nondivisible lattice points interior to certain of these cones. In fact, this classifies all fibrations up to isotopy, but we will not give the details of this. Indeed, all rays in the interiors of these cones correspond to the isotopy classes of foliations without holonomy [**25**]. This is a mild extension of a famous result of F. Laudenbach and S. Blank [**117**] that closed, cohomologous 1-forms on compact 3-manifolds are isotopic.

Remark. More generally, the norm can be defined on $H_2(M, A; \mathbb{R})$, where $A \subseteq \partial M$ is a compact 2-dimensional submanifold. In particular, one can allow foliations with boundary and corners and define the Thurston norm on $H_2(M, \partial_\pitchfork M; \mathbb{R})$, but we will not need this and have decided, for simplicity, to discuss only the case $A = \partial M$.

10.1. Compact Leaves of Reebless Foliations

Let $\alpha \in H_2(M, \partial M; \mathbb{Z})$ be a class that is represented by a compact, possibly disconnected, properly imbedded surface $(N, \partial N) \subset (M, \partial M)$. We write $[N, \partial N] = \alpha$.

It will be useful to introduce a sort of "norm" on the surface N and on its relative homology class. This concept is due to Thurston and will be studied carefully in subsequent sections.

Definition 10.1.1. If N is connected, then the Thurston norm of N is defined by

$$|N| = \begin{cases} 0, & \chi(N) > 0, \\ |\chi(N)|, & \text{otherwise.} \end{cases}$$

Generally, if $N_1, N_2, \dots, N_r$ are the connected components of N, then

$$|N| = \sum_{i=1}^{r} |N_i|.$$

Definition 10.1.2. For arbitrary $\alpha \in H_2(M, \partial M; \mathbb{Z})$, the Thurston norm $\xi(\alpha)$ is defined to be the smallest value of $|N|$ for all properly imbedded $(N, \partial N) \subset (M, \partial M)$ representing α.

The following is the main goal of this section.

Theorem 10.1.3 (Thurston). *If L is a compact leaf of a Reebless foliation, let $[L, \partial L] \in H_2(M, \partial M; \mathbb{Z})$. Then $|L| = \xi([L, \partial L])$.*

This says, essentially, that a compact leaf of a Reebless foliation realizes the minimal topology in its homology class. For taut foliations, this is made more precise in the following.

Corollary 10.1.4. *If L is a compact leaf of a taut foliation, then L has the minimal genus of all compact, connected, properly imbedded surfaces N in M such that ∂N and ∂L have the same number of components and $[N, \partial N] = [L, \partial L]$.*

Before proving Theorem 10.1.3, we deduce Corollary 10.1.4.

Lemma 10.1.5. *If $\mathcal{F}$ is taut and L is a compact leaf, then $[L, \partial L] \neq 0$ in $H_2(M, \partial M; \mathbb{Z})$.*

Proof. By tautness, L meets a closed transversal σ to $\mathcal{F}$. The intersection product of $[\sigma]$ with $[L, \partial L]$ is nonzero, proving that the latter class is nonzero. □

Proof of Corollary 10.1.4. The foliation is Reebless and L is a compact leaf. We are given $(N, \partial N) \subset (M, \partial M)$, where N is connected, ∂N and ∂L have the same number c of components, and $[N, \partial N] = [L, \partial L]$. Let

$$g = \text{genus}(L) \quad \text{and} \quad g' = \text{genus}(N).$$

Then $\chi(L) = 2 - 2g - c$ and $\chi(N) = 2 - 2g' - c$.

Case 1. If $g = 0$, then $g \leq g'$, and we are done.

Case 2. Suppose that $g \geq 1$ and $c = 0$. In this case, $\chi(L) = 2 - 2g \leq 0$, and so $|L| = 2g - 2$. The leaf L is not a sphere; hence the Reebless foliation has no spherical leaves. By Corollary 9.1.9, M is irreducible. The only way that $g' = 0$ is that N is a 2-sphere, in which case it bounds a 3-ball. In particular, $[L, \partial L] = [L] = [N] = 0$, contradicting Lemma 10.1.5, and so this case never arises. Thus, $\chi(N) = 2 - 2g' \leq 0$ and $|N| = 2g' - 2$. By Theorem 10.1.3, $2g - 2 \leq 2g' - 2$ and again $g \leq g'$.

Case 3. It remains that $g \geq 1$ and $c > 0$. Then $\chi(L) = 2 - 2g - c < 0$, so $|L| = 2g + c - 2$. By Theorem 10.1.3, $0 < 2g + c - 2 \leq |N|$; hence $|N| = 2g' + c - 2$ and $g \leq g'$. □

We turn to the proof of Theorem 10.1.3. We assume that L is a compact leaf of $\mathcal{F}$ and that $(N, \partial N) \subset (M, \partial M)$ is a properly imbedded 2-manifold, not necessarily connected, such that $[N, \partial N] = [L, \partial L]$ in $H_2(M, \partial M; \mathbb{Z})$. We must show that $|L| \leq |N|$. Without loss of generality, we assume that no leaf of $\mathcal{F}$ is a disk or a sphere. Indeed, since M is connected, Reeb stability would imply that every leaf is a disk or a sphere, so $|L| = 0$ and the desired conclusion is immediate.

Lemma 10.1.6. *There is a properly imbedded surface $(N', \partial N') \subset (M, \partial M)$ such that $|N'| \leq |N|$, $[N', \partial N'] = [N, \partial N]$, and each component of $\partial N'$ is contained in a leaf of $\mathcal{F}$.*

Proof. In this proof, we will write cohomology and homology without explicit reference to the coefficient ring, always understood to be $\mathbb{Z}$.

Near each component V of ∂M that meets N, we will modify N. There are two cases.

Case 1. $L \cap V$ is nullhomologous in $H_1(V)$. Since $H_1(V)$ is a direct summand of $H_1(\partial M)$, we project $p : H_1(\partial M) \to H_1(V)$. If

$$\partial_* : H_2(M, \partial M) \to H_1(\partial M)$$

is the connecting homomorphism for the homology sequence of $(M, \partial M)$, our assumption is that $[L, \partial L]$ is in the kernel of $p \circ \partial_*$. But this class is equal to $[N, \partial N]$, implying that $N \cap V$ is nullhomologous in $H_1(V)$.

As a component of $\partial M = \partial_\pitchfork M$, V is a torus. Thus, if any component σ of $N \cap V$ is itself homologous to zero on V, then it bounds a disk D there. By a standard nesting argument, we can choose σ so that D meets no other component of $N \cap V$. We attach the disk D to N along σ and isotope this part into $\operatorname{int}(M)$. The resulting surface N_0 has the same relative homology class as N and the component containing σ has genus unchanged, but has one less boundary component. It follows that $|N_0| \leq |N|$. Replace N with N_0 and repeat this process until no component of $N \cap V$ is nullhomologous in V.

The previous paragraph allows us to assume that no component of $N \cap V$ is nullhomologous in V. If $N \cap V \neq \emptyset$, it follows that this intersection consists of $2m$ disjoint, essential, simple closed curves on V, half of which are oriented one way and half the opposite way. Choose an oppositely oriented pair σ_1, σ_2 that cobound an annulus A on V meeting no other component of $N \cap V$. Attach this annulus to N and isotope into $\operatorname{int}(M)$. Once again, the resulting surface has the same relative homology class as N and we will show that the norm is not increased. Finite iteration of this process will then produce a new representative surface N', not meeting V and having $|N'| \leq |N|$.

If σ_1 and σ_2 lie on the same component N_0 of N, the above operation does not change the Euler characteristic of N_0, hence leaves $|N|$ unchanged. To see that the Euler characteristic is unaffected, triangulate N_0 and the annulus A so that the induced triangulations on the common boundary agree. Then $\chi(A \cup_{\partial A} N_0) = \chi(A) + \chi(N_0) = \chi(N_0)$.

If σ_1 and σ_2 lie on different components, say N_1 and N_2, then, as above, $N_* = N_1 \cup_{\sigma_1} A \cup_{\sigma_2} N_2$ has $\chi(N_*) = \chi(N_1) + \chi(N_2)$. If $\chi(N_*) \geq 0$, then $|N_*| = 0 \leq |N_1| + |N_2|$. Suppose that $\chi(N_*) < 0$. Then either $\chi(N_i) \leq 0$, $i = 1, 2$, and $|N_*| = |N_1| + |N_2|$, or exactly one of the $\chi(N_i)$'s is positive and $|N_*| < |N_1| + |N_2|$. In all cases, the norm of the surface is not increased by this operation.

Case 2. $L \cap V$ is homologically nontrivial in V, hence is homologous to $k\sigma$, where $k \geq 1$ is an integer and σ is an essential, oriented, simple closed curve on V. Since L is a leaf and $V \subset \partial M$, no component of $L \cap V$ can bound a disk in V. Otherwise, the induced foliation $\mathcal{F}|\partial M$ would have at least one center singularity, contradicting transversality. It follows that each component of $L \cap V$, with orientation induced by the orientations of M, ∂M and L, is isotopic on V to $\pm\sigma$. There will be at least k of these components isotopic to σ.

By the techniques in the previous case, N can be assumed to intersect V only in disjoint, coherently oriented, simple closed curves. Since $p(\partial_*([L, \partial L])) = p(\partial_*([N, \partial N]))$, $N \cap V$ will consist of exactly k simple closed curves, each oriented and isotopic to σ. It is a simple matter, then, to isotope N so that these boundary curves are moved to distinct boundary curves of L in V. □

Lemma 10.1.7. *Whenever N' is as in Lemma* 10.1.6, *then no component of N' is a disk and any spherical component can be discarded.*

Proof. Since M is irreducible (Corollary 9.1.9), any spherical component bounds a ball, hence contributes nothing to the homology class $[N', \partial N']$ and can be suppressed. Suppose there is a disk component D of N'. Then ∂D is a component of $\partial\Lambda$, for some leaf Λ of $\mathcal{F}$. Viewed as a loop on Λ, this boundary component can be homotoped in Λ to a loop missing the boundary. Because D is a disk, this loop is nullhomotopic in $\operatorname{int}(M)$. If it is not nullhomotopic in Λ, then Theorem 9.1.3 would imply that $\mathcal{F}$ is not Reebless. Thus, this boundary component of Λ is nullhomotopic in Λ, so Λ is homeomorphic to a disk, contradicting our ongoing assumption. □

We next modify N' to an incompressible 2-manifold, at worst decreasing the norm, and leaving the other properties unchanged. The construction makes use of the following fundamental fact from 3-dimensional topology.

Theorem 10.1.8 (The Loop Theorem). *Let M be an arbitrary* 3*-manifold and $F \subset \partial M$ a connected* 2*-manifold. If* $\ker(\pi_1(F) \to \pi_1(M))$ *is nontrivial, then there is a proper imbedding $f : D^2 \to M$ such that $f(\partial D^2) \subset F$ and* $0 \neq [f|\partial D^2] \in \pi_1(F)$.

Exercise 10.1.9. Let M be a compact, connected, irreducible 3-manifold and let $T \subseteq \partial M$ be a toroidal boundary component. If T is compressible, prove that M is homeomorphic to a solid torus $S^1 \times D^2$.

A somewhat weaker statement of the loop theorem is called "Dehn's Lemma" because it was first claimed, but inadequately proven, by M. Dehn in 1910. A correct proof was given by C. D. Papakyriakopoulos in 1957, and a more general version of the above theorem was proven by J. Stallings in 1960. A proof of Stallings' version, hence of the above, will be found in [**93**, pp. 39-50].

Corollary 10.1.10. *Let N be a compact, properly imbedded* 2*-manifold in the* 3*-manifold M. If S is a component of N and* $\ker(\pi_1(S) \to \pi_1(M))$ *is* *nontrivial, then there is a smoothly imbedded disk* $D \subset \mathrm{int}(M)$ *such that* $D \cap N = \partial D$ *and* ∂D *is essential in* N.

Proof. Let $f : (D^2, \partial D^2) \to (M, S)$ be a smooth map, transverse to N and such that $f|\partial D^2$ is essential in S. This exists because some essential loop on S is nullhomotopic in M. We can assume that $f^{-1}(\partial M) = \emptyset$. These conditions ensure that $f^{-1}(N)$ consists of a family of disjoint, simple closed curves in D^2 that are carried by f to loops in $N \smallsetminus \partial M$, one of which is ∂D^2. Finally, we choose f so as to minimize the number of components of $f^{-1}(N)$. We will find a disk $\Delta \subset D^2$ such that $f^{-1}(N) \cap \Delta = \partial\Delta$ and $f(\partial\Delta)$ is essential in the component F of N containing it.

If $\partial D^2 = f^{-1}(N)$, we take $\Delta = D^2$. Otherwise, let $\sigma \neq \partial D^2$ be one of the components of $f^{-1}(N)$, chosen so that the disk $\Delta \subset \mathrm{int}(D^2)$ with $\partial\Delta = \sigma$ does not meet any other of the components. The loop $f(\sigma)$ is essential in a component F of N. If not, we could modify f in a neighborhood of Δ so as to decrease the number of components of $f^{-1}(N)$.

Produce M' by cutting M apart along N. One of the two copies of F in $\partial M'$ satisfies the hypotheses of the loop theorem. While M' may have corners, this will not matter because the loop theorem is purely topological. That theorem produces an imbedded disk $D \subset M'$ that meets $\partial M'$ exactly in a loop on F that is essential there. Regluing, we obtain an imbedded disk $D \subset \mathrm{int}(M)$ that meets N exactly in $\partial D \subset F$, an essential loop in F. By a small perturbation, we make this a smooth imbedding. □

Lemma 10.1.11. *Let* N' *be as in Lemma* 10.1.6. *Then there is a properly imbedded, incompressible* 2*-manifold* $(N'', \partial N'') \subset (M, \partial M)$, *coinciding with* N' *in a neighborhood of* ∂M, *such that* $[N'', \partial N''] = [N', \partial N']$ *and* $|N''| \leq |N'|$.

Proof. If N' is already incompressible, let $N'' = N'$. Otherwise, Corollary 10.1.10 provides an imbedded disk $D \subset \mathrm{int}(M)$, meeting N' exactly in ∂D, an essential curve on N'. Thus, we cut N' apart along ∂D, attach two copies of D (and round the corners), obtaining an imbedded surface with the same relative homology class as N'. This surgery affects only one component F of N'.

If the essential loop ∂D is also homologically nontrivial on F (*i.e.*, nonseparating), then F is replaced by a component F' with $\chi(F') = \chi(F) + 2$. By Lemma 10.1.7, F' is not a disk and, if it is a sphere, it is nullhomologous. But F would also be nullhomologous and could have been discarded. We can assume, therefore, that $\chi(F') \leq 0$, hence $\chi(F) < 0$, and it follows that $|F'| < |F|$.

If the loop separates F, there result two components, F' and F'', and $\chi(F') + \chi(F'') = \chi(F) + 2$. By Lemma 10.1.7, neither F' nor F'' is a disk. Also, the separating loop on F is homotopically essential, so neither F' nor F'' can be a sphere. Thus, $\chi(F')$ and $\chi(F'')$ are both nonpositive and $\chi(F) < 0$. It follows that $|F'| + |F''| < |F|$.

This procedure does not change the number of boundary components of N and it strictly reduces $|N|$, implying that the procedure can only be repeated a finite number of times. So we must arrive at an incompressible 2-manifold N'', as desired. □

Thus, each component S of N'' satisfies the hypotheses of Theorem 9.5.5. Applying that theorem, we isotope each component of S to a position in which the boundary components remain unchanged, but the tangencies of the components of S with $\mathcal{F}$ are isolated saddles and/or circles on S.

Observe that isotoping each component into this nice position with respect to $\mathcal{F}$ may result in various components intersecting. But each component S individually defines a class $[S, \partial S] \in H_2(M, \partial M)$, and the sum of these classes is $[N'', \partial N''] = [L, \partial L]$. The class $[S, \partial S]$ is invariant under the isotopy, so we can work with the individual components and be unconcerned with whether or not they intersect after the isotopies. Thus, we assume that all components have only saddle or circle tangencies.

At this point, the Euler class of the tangent bundle $\tau = \tau(\mathcal{F})$ to $\mathcal{F}$ enters the picture. For oriented 2-plane bundles ζ over a compact manifold M (or, more generally, over any compact space with a weakly regular cell structure), this is the class $e(\zeta)$ discussed in Chapter 4. Furthermore, if $A \subset M$ is a "reasonable" subspace (*e.g.*, a submanifold or subcomplex), and if v is a nowhere zero section of $\zeta|A$, one obtains a relative Euler class $e(\zeta, v) \in H^2(M, A; \mathbb{Z})$, this being precisely the obstruction to extending v to a nowhere zero section of ζ. When $\zeta = \tau$, we will take $A = \partial M$ and let v be a nowhere 0, inwardly pointing vector field along ∂M that is tangent to $\mathcal{F}$. Of course, if $\partial M = \emptyset$, v will be an "empty" section and $e(\tau, v) = e(\tau)$. Let $\tau' = \tau|S$ and $v' = v|\partial S$. If $j : S \hookrightarrow M$ is the inclusion, then $\tau' = j^*(\tau)$ and, by naturality of the Euler class, $j^*(e(\tau, v)) = e(\tau', v')$.

Remark that τ is oriented by the transverse orientation of $\mathcal{F}$, together with the orientation of M. Similarly, the orientation of S and that of M determine a transverse orientation for S. In the following definition, the term "positive" for normal vectors to S and $\mathcal{F}$ refers to the transverse orientations.

Definition 10.1.12. A saddle tangency $q \in S$ of S with $\mathcal{F}$ is of positive type if a positive normal vector to $\mathcal{F}$ at q is also a positive normal vector to

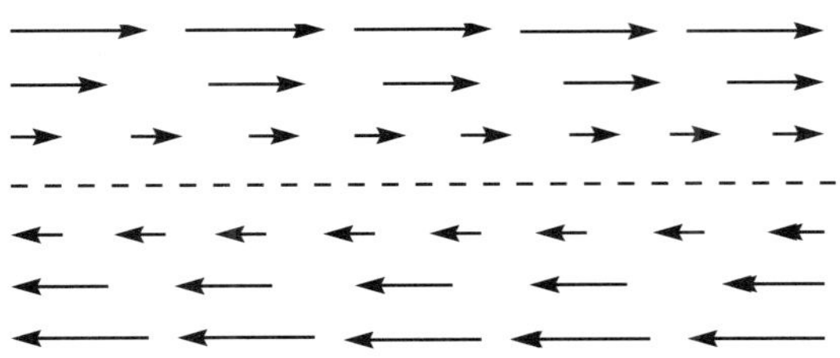

Figure 10.1.1. The field w near a circle tangency

S at q. Otherwise, the saddle is of negative type. The number of positive saddles on S will be denoted by I_+ and the number of negative ones by I_-.

Lemma 10.1.13. *With the above notation, we have*

(1) $-I_- - I_+ = \chi(S)$;

(2) $I_- - I_+ = e(\tau, v)([S, \partial S])$.

Proof. Let $\Sigma \subset S$ be the set of points of tangency. The foliation $\mathcal{F}$ induces a nonsingular foliation $\mathcal{F}'$ on $S \smallsetminus \Sigma$ with a nowhere vanishing tangent field w that can be oriented as follows. At $p \in S \smallsetminus \Sigma$, the tangent space to the induced foliation is $\tau_p(\mathcal{F}) \cap T_p(S)$, and we uniquely specify the direction of w_p in this space by requiring that (ν_p, ν_p', w_p) give the correct orientation to $T_p(M)$, where ν_p is the positively oriented normal to S and ν_p' the positively oriented normal to $\mathcal{F}$. Remark that we can damp w off to 0 on Σ, producing thereby a smooth section w both of $\tau' = \tau|S$ and of $T(S)$.

First assume that there are no circle tangencies. Then, in the language of Example 4.4.12, I_- is the sum of the indices of the negative saddles and $-I_+$ is the sum of the indices of the positive saddles. The second assertion then follows from that example. The first assertion is the Poincaré-Hopf theorem (Theorem 4.4.11).

If there are circle tangencies, the local cross-section in Figure 9.5.4 makes it clear that, near such a tangency, w is as pictured in Figure 10.1.1 (where the dashed line is the tangency). Then w can be modified near the tangent circle, as indicated in Figure 10.1.2, so as to be nowhere zero there. If w is to be a tangent field to S, interpret the modification in Figure 10.1.2 as such. If w is to be a section of τ', the turning arrows in the figure are to be interpreted as tangents to $\mathcal{F}$ with point of origin in S. This reduces us to the argument of the previous paragraph. □

Corollary 10.1.14. *For S as above, $|e(\tau, v)([S, \partial S])| \leq |S|$.*

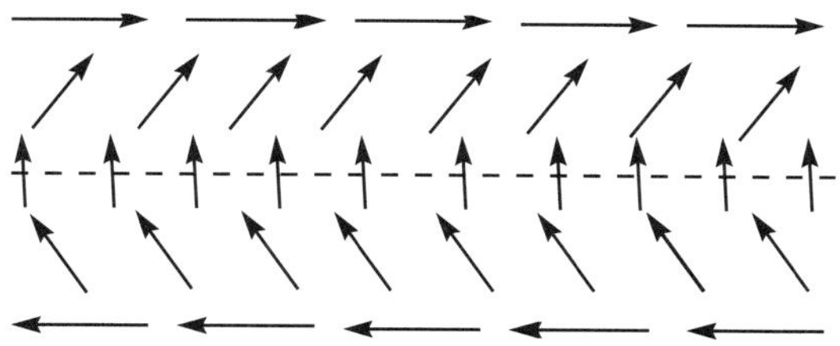

Figure 10.1.2. Modification of the field w near a circle tangency

Proof. Indeed, since S is neither a sphere nor a disk,

$$|S| = |\chi(S)| = |-I_- - I_+| \geq |I_- - I_+| = |e(\tau, v)([S, \partial S])|.$$

□

Proof of Theorem 10.1.3. If the leaf L is a sphere or a disk, the theorem is trivial, so we exclude those cases. By Lemmas 10.1.6 and 10.1.11, we have produced $(N'', \partial N'') \subset (M, \partial M)$ such that $[L, \partial L] = [N'', \partial N'']$, $|N''| \leq |N|$, and each component of N'' is as in the above corollary. Let $S_1, S_2, \ldots, S_k$ be the components of N''. Then

$$\begin{aligned} |L| &= |\chi(L)| \\ &= |e(\tau, v)([L, \partial L])| \\ &= |e(\tau, v)([N'', \partial N''])| \\ &= \left|\sum_{i=1}^{k} e(\tau, v)([S_i, \partial S_i])\right|. \end{aligned}$$

But

$$\left|\sum_{i=1}^{k} e(\tau, v)([S_i, \partial S_i])\right| \leq \sum_{i=1}^{k} |e(\tau, v)([S_i, \partial S_i])| \leq \sum_{i=1}^{k} |S_i| = |N''| \leq |N|.$$

□

10.2. Knots, Links, and Genus

One of the main purposes of Thurston's definition of the homology norm was to study the genus of knots and links.

Definition 10.2.1. A link $\lambda = \lambda_1 \cup \cdots \cup \lambda_r$ is a smooth imbedding of a disjoint union of r copies of S^1 into S^3. If $r = 1$, then the link λ is called a knot. The link λ is oriented by the standard orientation of S^1.

Remark. We often view S^3 as $\mathbb{R}^3 \cup \{\infty\}$. If λ is a link (or a knot), we can assume that $\infty \notin \lambda$ and picture λ as a link in $\mathbb{R}^3$. It is also standard practice to identify two links λ and λ' that are isotopic via an isotopy of S^3 or of $\mathbb{R}^3$.

Definition 10.2.2. Let λ be a link (respectively, a knot) and let $N(\lambda)$ be an open, tubular neighborhood of λ in S^3 such that the complement

$$M(\lambda) = S^3 \setminus N(\lambda)$$

has smooth boundary. Then $M(\lambda)$ is called the link complement (respectively, the knot complement) determined by λ.

Lemma 10.2.3. *Let $\lambda = \lambda_1 \cup \cdots \cup \lambda_r$ be a link. Then there is a canonical identification $H_2(M(\lambda), \partial M(\lambda)) = H_1(\lambda) = \mathbb{Z}^r$.*

Proof. By excision and homotopy invariance,

$$H_2(M(\lambda), \partial M(\lambda)) = H_2(S^3, \lambda).$$

By the exact sequence of the pair, $H_2(S^3, \lambda) = H_1(\lambda)$. Since each component of λ is oriented, the identification $H_1(\lambda) = \mathbb{Z}^r$ is also canonical. □

Remark. It is a standard, nonelementary fact in algebraic topology that cohomology is a *representable functor*. More precisely, let π be an abelian group and let $K(\pi, n)$ be a CW-complex with the property that

$$\pi_i(K(\pi, n)) = \begin{cases} 0 & \text{if } i \neq n, \\ \pi & \text{if } i = n. \end{cases}$$

Such a space always exists [**135**, page 5]. Then, for "reasonable" spaces X (*e.g.*, triangulable spaces, such as compact, smooth manifolds), there is a canonical identification $H^n(X; \pi) = [X, K(\pi, n)]$, the set of homotopy classes of maps of X into $K(\pi, n)$. In order to describe this, first note that

$$\begin{aligned} H^n(K(\pi, n); \pi) &= \operatorname{Hom}(H_n(K(\pi, n), \pi) \quad \text{(universal coefficient theorem)} \\ &= \operatorname{Hom}(\pi_n(K(\pi, n)), \pi) \quad \text{(Hurewicz theorem)} \\ &= \operatorname{Hom}(\pi, \pi). \end{aligned}$$

This latter group contains a fundamental class $\iota = \mathrm{id}$, and the canonical identification of $\alpha \in H^n(X; \pi)$ with $[f] \in [X, K(\pi, n)]$ is given by the requirement that $\alpha = f^*(\iota)$.

Corollary 10.2.4. *Let λ be a link. Then there is a compact, connected, imbedded, oriented surface S in $\mathbb{R}^3$ such that $\partial S = \lambda$. Such a surface S is called a spanning surface for λ.*

Proof. Clearly, $S^1 = K(\mathbb{Z},1)$, and so $H^1(M(\lambda)) = [M(\lambda), S^1]$. Here, the fundamental class $\iota \in H^1(K(\mathbb{Z},1);\mathbb{Z}) = \mathrm{Hom}(\mathbb{Z},\mathbb{Z})$ is identical with the usual orientation class for S^1. By Lefschetz duality and Lemma 10.2.3, $H^1(M(\lambda)) = H_1(\lambda)$ and we take $\alpha \in H^1(M(\lambda))$ to correspond to the class $[\lambda_1]+\cdots+[\lambda_r] \in H_1(\lambda)$. We represent α by a smooth map $f : M(\lambda) \longrightarrow S^1$. By Sard's theorem, there is a regular value $z \in S^1$ both for f and for $f|\partial M(\lambda)$. Then $S' = f^{-1}(z)$ is a smoothly imbedded, compact surface in $M(\lambda)$, oriented via the orientation of $M(\lambda)$ and the transverse orientation determined by the standard orientation of S^1. Furthermore,

$$\partial S' = S' \cap \partial M(\lambda)$$

(by the relative version of the submersion theorem). Since $[z] \in H_0(S^1)$ is the Poincaré dual of ι, it is an exercise in the duality theorems to show that $[S', \partial S'] \in H_2(M(\lambda), \partial M(\lambda))$ is the Lefschetz dual of α. By the choice of α, $\partial S'$ is isotopic to λ. Choose a representative surface $(S, \partial S)$ to have the fewest components of any surface homologous to $(S', \partial S')$.

We must show that S is connected. But α is nondivisible; hence so is its dual $[S, \partial S]$. If S has $k \geq 2$ components, there is an arc τ, issuing from the positive side of some component of S and ending on the positive side of another component. Otherwise, all components are homologous and $[S, \partial S]$ is divisible by k. Remove a small disk from S at each endpoint of τ and attach a cylinder bounding a tubular neighborhood of τ. This produces a surface Σ having $k-1$ components and all the required properties. By the minimality of the number of components of S, it must be that $k = 1$. □

We emphasize that a link λ always comes with an orientation on each component and that the orientation of the spanning surface must induce that of λ. A spanning surface is also called a *Seifert surface* for λ.

Definition 10.2.5. Let λ be a link. The genus of λ is the smallest integer g that occurs as the genus of a spanning surface of λ. A knot of genus 0 (*i.e.*, a spanning surface is a disk) is called the unknot.

Exercise 10.2.6. The simplest nontrivial knot κ is the trefoil knot (Figure 10.2.1). Prove that the genus of this knot is 1.

Example 10.2.7. We construct a more complicated knot. In Figure 10.2.2 we see the unknot λ, imbedded in a solid torus $V = D^2 \times S^1$ so as to be knotted there. Let κ be the trefoil knot and let $i : V \to S^3$ be an imbedding such that $i(\mathrm{int}(V)) = N(\kappa)$ and $i(\{0\} \times S^1) = \kappa$. Then $i(\lambda)$ is a knot called the "Whitehead double" of the trefoil κ. The spanning surface depicted in Figure 10.2.2 is carried by i to a spanning surface for $i(\lambda)$. One finds a cellular subdivision of this surface with six vertices, nine edges, and two

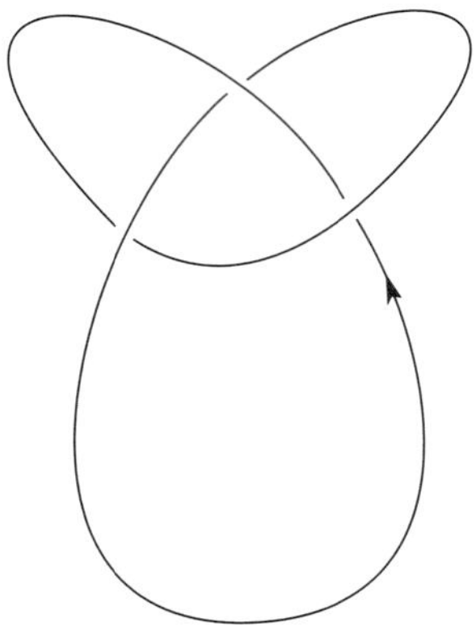

Figure 10.2.1. The trefoil knot

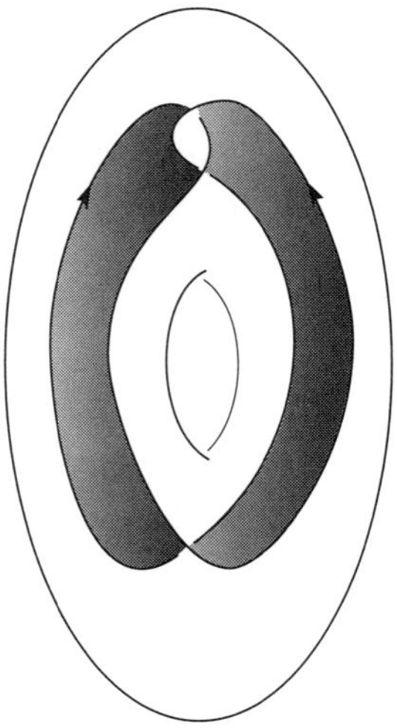

Figure 10.2.2. A knot and spanning surface in the solid torus

faces; hence $\chi(S) = -1$. Since the surface has one boundary component, the formula

$$\chi(S) = 2 - 2g - |\partial S|$$

implies that this spanning surface has genus $g = 1$. In fact, it can be shown that $i(\lambda)$ is not the unknot, so its genus is 1.

Example 10.2.8. The simplest nontrivial link λ that is not a knot is pictured in Figure 10.2.3, along with a spanning surface. This surface is an

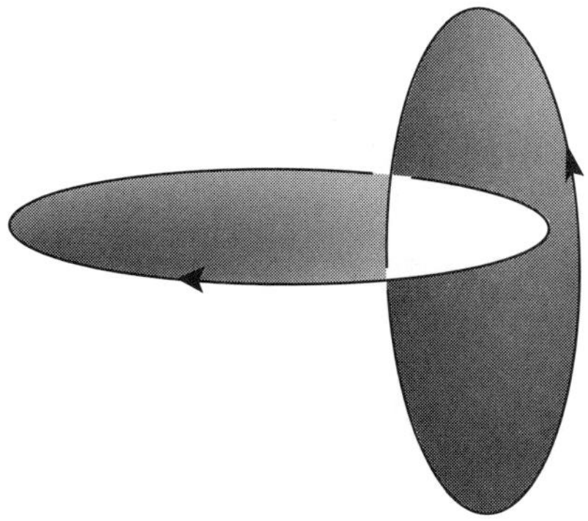

Figure 10.2.3. Simply linked rings and spanning surface

imbedded annulus (it has *two* twists, hence is not a Möbius strip). Since the genus of the annulus is zero, the genus of the link λ is zero.

Example 10.2.9. Let $\lambda = \lambda_1 \cup \lambda_2$ be the Whitehead link. Two isotopic pictures of this link are given in Figure 10.2.4, a spanning surface S being given in the second view. It is easy to find a cellular subdivision of this spanning surface having fourteen vertices, twenty-one edges, and five faces; hence the Euler characteristic is $\chi(S) = -2$. Since $|\partial S|$ (the number of boundary components) is 2, the usual formula for $\chi(S)$ implies that $g = 1$. It is intuitively plausible (and true) that λ does not bound an annulus in $\mathbb{R}^3$, hence that the genus of the Whitehead link is one.

Example 10.2.10. In Figure 10.2.5 we see a link λ' that is analogous to the Whitehead link. (In [**175**, p. 108] this link is called the Whitehead link.) A spanning surface of genus one can be constructed and, again, it is intuitively plausible (and true) that this link does not bound an annulus, hence has genus one.

Example 10.2.11. The danger of this "intuitive plausibility" is seen by considering the link λ'' in Figure 10.2.6. Two views are given, the second with a spanning surface that is clearly an annulus. From the first view it seems implausible that this link bounds an annulus.

Example 10.2.12. Consider the link λ with three components, two views of which are given in Figure 10.2.7. This link is referred to as the "Borromean rings" or, sometimes, as the "Ballantine ale rings". Observe that no one of the three rings can be separated from the others, although any two are unlinked. From the second view of λ in Figure 10.2.7, it can be seen that there is an isotopy of S^3 that interchanges two components, carrying

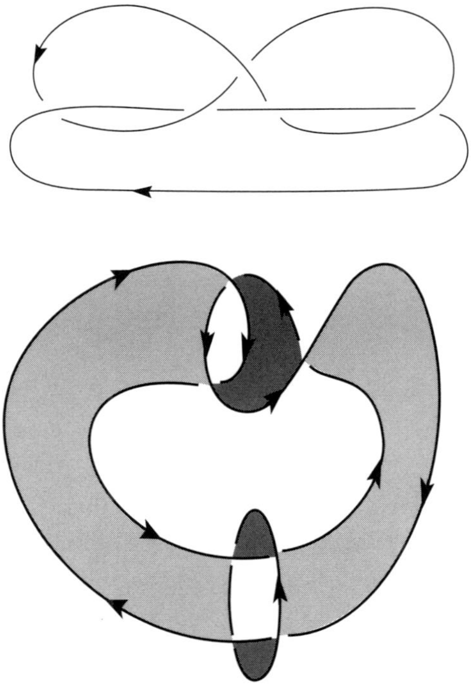

Figure 10.2.4. Two views of the Whitehead link

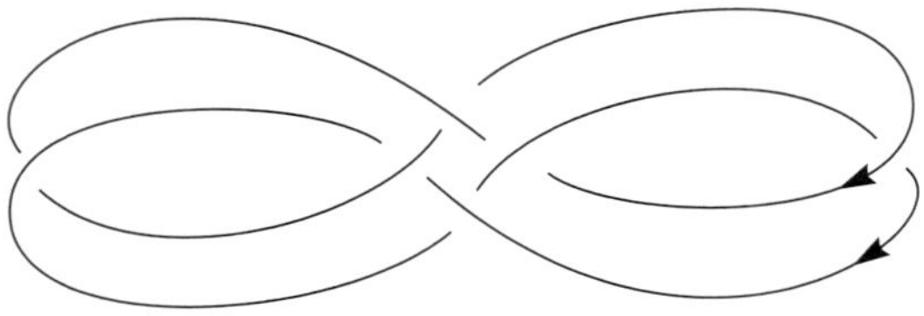

Figure 10.2.5. A Whitehead-like link λ'

the orientation of each to that of the other, and reverses the orientation of the third. Indeed, rotate the figure 180° about a suitable vertical line in the plane of the paper, then drag the ring around to its original position, reversing only its orientation. The first view shows that a suitable isotopy permutes the components cyclically, preserving their orientations. It follows

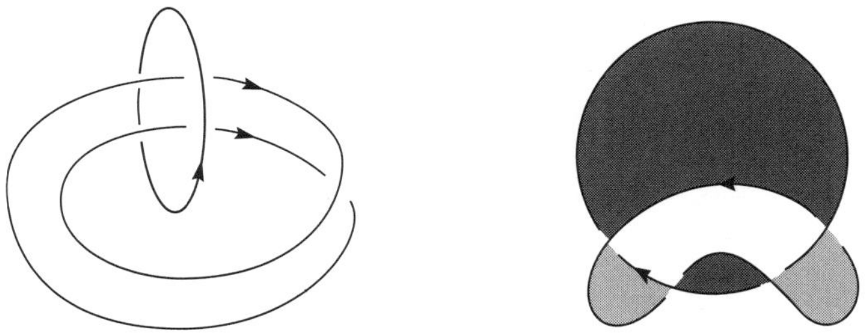

Figure 10.2.6. Two views of λ'', one with spanning surface

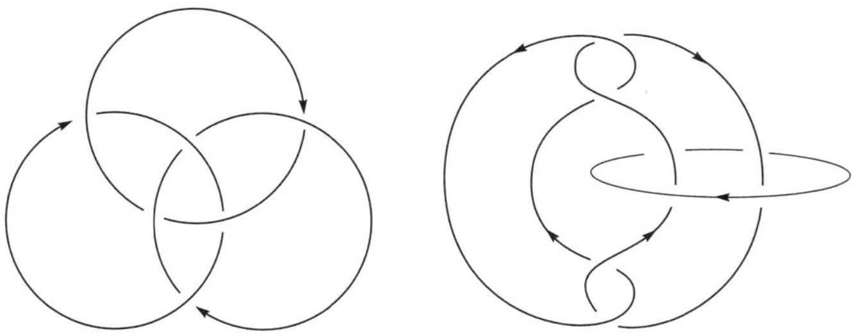

Figure 10.2.7. The Borromean rings

that the genus of this link is not dependent on the orientations of the components. It is clear that a connected spanning surface S, since it has three boundary components, has Euler characteristic $\chi(S) < 0$. In fact, we will see later that the maximum value of $\chi(S)$ realized by a spanning surface is -3, and hence the genus of λ is one.

Example 10.2.13. As a final example, let us consider the three-link chain (Figure 10.2.8). The spanning surface S indicated in Figure 10.2.8 contains the point at infinity. It is orientable (why?) and admits a cellular decomposition having twelve vertices, eighteen edges, and five faces. Since $|\partial S| = 3$, it follows that

$$\chi(S) = -1 = 2 - 2g - 3,$$

hence that the genus of the link λ is zero. What do you think will happen in this example if one of the components of λ is reversed? What if all are reversed?

We will return to many of these examples later.

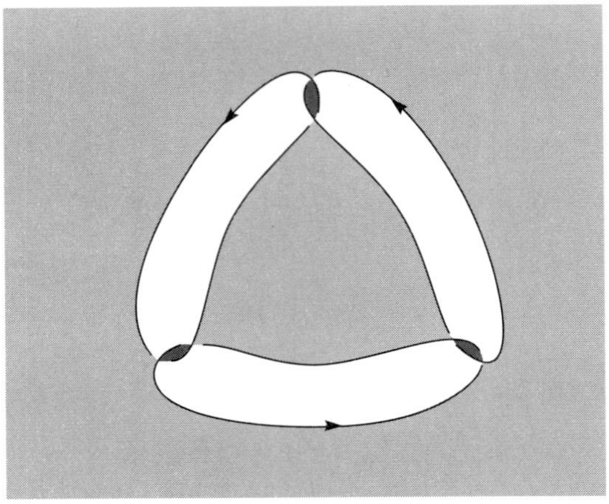

Figure 10.2.8. The 3-link chain with spanning surface

10.3. The Norm on Real Homology

As usual, M will denote a compact, oriented 3-manifold with possibly empty boundary ∂M. We turn to the study of homology and cohomology modules with coefficients in the real number field $\mathbb{R}$.

Recall the definition of the Thurston pseudonorm ξ from Section 10.1, defined on classes $\alpha \in H_2(M, \partial M; \mathbb{Z})$ that can be represented by properly imbedded surfaces. In fact, all of the classes can be so represented (Lemma 10.3.3). By Lefschetz duality, $H_2(M, \partial M; \mathbb{Z})$ is canonically equal to $H^1(M; \mathbb{Z})$. Our main goal in the present section is to prove the following.

Theorem 10.3.1. *The Thurston pseudonorm is defined on every element of* $H_2(M, \partial M; \mathbb{Z})$ *and extends uniquely to a nonnegative, real-valued function* ξ*, defined on* $H_2(M, \partial M; \mathbb{R})$*, that is continuous, convex, and linear on rays through the origin. This extension vanishes exactly on the subspace* $\mathcal{N}$ *spanned by the classes represented by compact, connected, properly imbedded surfaces of nonnegative Euler characteristic.*

We generally abuse terminology and refer to ξ as "the Thurston norm", even though, generally, it is only a pseudonorm. When the subspace $\mathcal{N}$ is trivial, then ξ is an honest norm and has a dual, defined as follows.

Definition 10.3.2. If ξ is a norm on $H^1(M; \mathbb{R}) = H_2(M, \partial M; \mathbb{R})$, then a dual norm ξ^* on the dual vector space $H_1(M; \mathbb{R}) = H^2(M, \partial M; \mathbb{R})$ is defined by the formula

$$\xi^*(\alpha) = \sup_{\xi(a)\leq 1} (\alpha(a)).$$

Even when they are genuine norms, ξ and ξ^* do not come from inner products. Indeed, we will see that their unit balls B_ξ and B_{ξ^*} are polyhedra, not ellipsoids.

We view $H_2(M, \partial M; \mathbb{Z})$ as the integer lattice in the real vector space $H_2(M, \partial M; \mathbb{R})$. In order to streamline notation, we set

$$\Lambda = H_2(M, \partial M; \mathbb{Z}) = H^1(M; \mathbb{Z}),$$
$$\Gamma = H_2(M, \partial M; \mathbb{R}) = H^1(M; \mathbb{R}).$$

The first step in proving Theorem 10.3.1 is the following.

Lemma 10.3.3. *Every element $a \in \Lambda$ is represented by a compact, oriented, properly imbedded surface $(S, \partial S) \subset (M, \partial M)$. If a is divisible by the positive integer k, then S is necessarily a union of k disjoint subsurfaces, each representing a/k.*

Proof. As in the proof of Corollary 10.2.4, the Poincaré dual of a is represented by a smooth classifying map $f_a : M \to S^1$, and for a value $z \in S^1$ that is regular both for f_a and for $f_a|\partial M$ the surface $(S, \partial S)$ is as required. It is left as Exercise 10.3.4 to prove that *every* properly imbedded surface $(S, \partial S) \subset (M, \partial M)$ representing a is so obtained. Suppose, then, that $(S, \partial S) \in a = kb$, $S = f_a^{-1}(z)$. Consider the homotopy commutative diagram

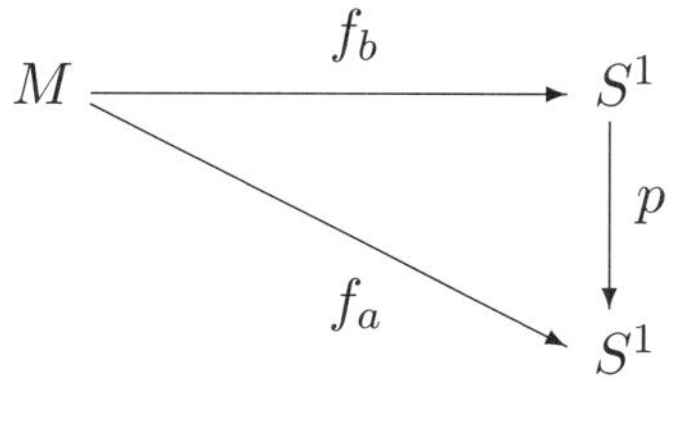

where p is a k-fold covering map. By the covering homotopy theorem, f_b can be replaced by a homotopic map (again called f_b) that makes the diagram truly commutative. Since z is a regular value of f_a, it follows that

$$p^{-1}(z) = \{z_1, \dots, z_k\}$$

consists entirely of regular values of f_b and

$$S = f_b^{-1}(z_1) \cup \cdots \cup f_b^{-1}(z_k)$$

is the union of k disjoint subsurfaces, each representing $b = a/k$. ☐

Exercise 10.3.4. Suppose that $(S, \partial S)$ is a compact, oriented, properly imbedded smooth surface representing a nontrivial class $a \in \Lambda$. Show that there is a smooth classifying map $f_a : M \to S^1$ such that $S = f_a^{-1}(z)$ for a suitable regular value z of f_a and $f_a|\partial M$. (Hint. Begin with a carefully chosen 1-form in a normal neighborhood of S that represents the Poincaré dual of a.)

Corollary 10.3.5. *For $a \in \Lambda$ and k any integer, $\xi(ka) = |k|\xi(a)$.*

Proof. By the definition of ξ on Λ, it is clear that $\xi(-a) = \xi(a)$; hence we can assume that $k \geq 0$. Any surface $(S, \partial S)$ representing ka is the union of k disjoint surfaces, each of which represents a. By taking S such that $|S|$ is minimal we see that $\xi(ka) \geq k\xi(a)$. On the other hand, if Σ represents a and has $|\Sigma|$ minimal, we can take k disjoint copies of Σ, each slightly displaced from the original copy, obtaining thereby a representative of ka. It follows that $\xi(ka) \leq k\xi(a)$. □

Lemma 10.3.6. *If $a, b \in \Lambda$, then $\xi(a + b) \leq \xi(a) + \xi(b)$.*

Proof. Let $(S, \partial S)$ and $(S', \partial S')$ be oriented surfaces representing a and b, respectively, and such that $|S|$ and $|S'|$ are minimal. By standard techniques, we slightly perturb S and S' into general position.

We show first that, without loss of generality, we can assume that no component σ of $S \cap S'$ is a closed loop bounding a disk D on one of the surfaces. Suppose such a component σ exists with, say, $D \subset S$. If D contains any other component τ of $S \cap S'$, then τ bounds a disk $\Delta \subset \operatorname{int}(D)$ by the Jordan-Brouwer separation theorem. Repeating this observation finitely often, we see that no generality is lost in assuming that $\operatorname{int}(D) \cap S' = \emptyset$. Now cut S' along σ and sew in two copies of D, displaced to either side of the original. This leaves S unchanged and replaces $(S', \partial S')$ with a homologous surface $(\Sigma, \partial\Sigma)$, no longer meeting S in the offending loop σ. Triangulate S' so that σ is a union of vertices and edges. The contribution of these vertices and edges to $\chi(S')$ is zero. There is a corresponding cellular subdivision of Σ having two vertices and two edges for each vertex and edge of σ, two new 2-cells, and otherwise having the same vertices, edges, and faces as S'. It is clear, then, that $\chi(\Sigma) = \chi(S') + 2$, hence that $|\Sigma| \leq |S'|$. A strict inequality contradicts the minimality of $|S'|$, while equality allows us to replace S' with Σ.

We must also show that, without loss of generality, it can be assumed that no arc σ of $S \cap S'$ with endpoints in, say, ∂S can be completed by an arc τ in ∂S so as to form a loop γ bounding a disk on S. The minor modifications of the preceding argument that prove this will be left to the reader.

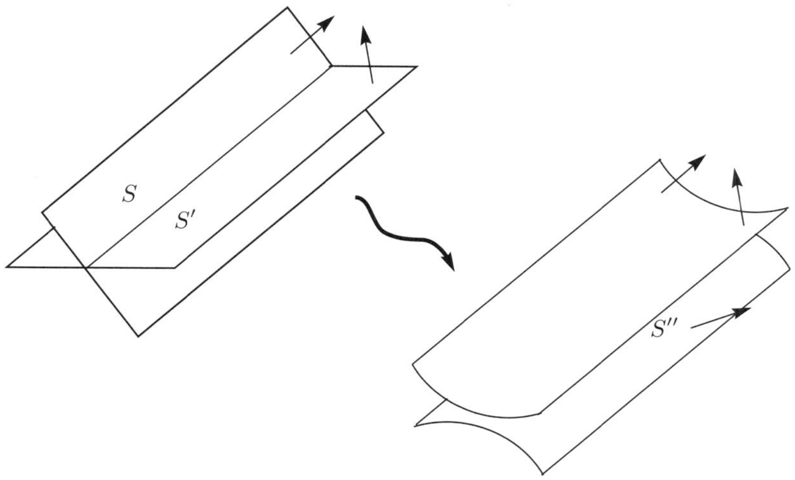

Figure 10.3.1. Cut and paste along $S \cap S'$

The oriented surfaces $(S, \partial S)$ and $(S', \partial S')$ unite to form a cycle representing $a+b$, but this may no longer be an imbedded surface. Since they are in general position, one sees that there is a unique way to cut and paste S and S' along $S \cap S'$ to form a new oriented surface $(S'', \partial S'')$ that carries this same cycle and can be slightly perturbed so as to be smooth and imbedded (see Figure 10.3.1). One has $\chi(S'') = \chi(S) + \chi(S')$ and, by the previous two paragraphs, S'' has no sphere or disk as a component that was not already a component of S or S'. Consequently, $|S''| = |S| + |S'|$. Of course, $|S''|$ may not be minimal for the representatives of $a+b$, so we conclude that $\xi(a+b) \leq \xi(a) + \xi(b)$. □

Corollary 10.3.7. *If $a, b \in \Lambda$ and if n and m are integers, then*

$$\xi(ma + nb) \leq |m|\xi(a) + |n|\xi(b).$$

This corollary is immediate by Corollary 10.3.5 and Lemma 10.3.6.

Corollary 10.3.8. *The function ξ on Λ extends uniquely to a continuous, nonnegative function ξ on Γ that is linear on rays. This extension is convex.*

Proof. Let $\rho \in \Gamma$ be a point with rational coordinates. The ray $\mathbb{R}^+\rho$ through ρ passes through some points of Λ and, by Corollary 10.3.5, ξ can be extended to $\mathbb{R}^+\rho$ by linearity. This extension is unique. Furthermore, ξ is now defined on the rational vector space $\Phi \subset \Gamma$ of rational points, and Corollary 10.3.7 extends to the case in which $a, b \in \Phi$ and $m\,n \in \mathbb{Q}$.

The space Φ is dense in Γ, so there is at most one continuous extension of ξ to Γ. Such an extension will also be linear on rays, since it is linear on a dense family of rays. By the extension of Corollary 10.3.7 to Φ and $\mathbb{Q}$, together with continuity, it follows that $\xi(r\gamma + r'\gamma') \leq |r|\xi(\gamma) + |r'|\xi(\gamma')$, $\forall \gamma, \gamma' \in \Gamma$, $\forall r, r' \in \mathbb{R}$.

We construct the continuous extension in the obvious way via Cauchy sequences of rational points. First, let $\{\rho_i\}_{i=1}^{\infty}$ be a sequence of rational points converging to the origin in Γ. If $\{e_1, \ldots, e_n\}$ is a basis of the integer lattice Λ, we can write

$$\rho_i = \sum_{j=1}^{n} r_i^j e_j,$$

where each $r_i^j \in \mathbb{Q}$ and $r_i^j \longrightarrow 0$ as $i \longrightarrow \infty$, $1 \leq j \leq n$. If E denotes the maximum value of $\xi(e_j)$, $1 \leq j \leq n$, then the above extension of Corollary 10.3.7 implies that

$$\xi(\rho_i) \leq E \sum_{j=1}^{n} |r_i^j|,$$

hence that $\xi(\rho_i) \longrightarrow 0$ as $i \longrightarrow \infty$.

Suppose, now, that $\{\rho_i\}_{i=1}^{\infty}$ and $\{\rho_i'\}_{i=1}^{\infty}$ are Cauchy sequences in Φ converging to the same point $b \in \Gamma$. Arguing as in the previous paragraph, we see that the sequences $\{\xi(\rho_i)\}$ and $\{\xi(\rho_i')\}$ are bounded; hence we can pass to commonly indexed subsequences so as to assume, with no loss of generality, that they converge. But the extension of Corollary 10.3.7 to Φ implies that $|\xi(\rho_i) - \xi(\rho_i')| \leq \xi(\rho_i - \rho_i')$, and hence the previous paragraph implies that $\lim_{i\to\infty} \xi(\rho_i) = \lim_{i\to\infty} \xi(\rho_i')$. □

All that remains for the proof of Theorem 10.3.1 is the following.

Proposition 10.3.9. *The function $\xi : \Gamma \longrightarrow \mathbb{R}^+$ vanishes exactly on the subspace $\mathcal{N} \subseteq \Gamma$ spanned by the elements of Λ that have connected representative surfaces $(S, \partial S)$ with $\chi(S) \geq 0$.*

Proof. Let Γ_0 be the subset of Γ on which ξ vanishes. By convexity and linearity on rays, we see that Γ_0 is a vector subspace. What we need to show is that Γ_0 is exactly the subspace spanned by the set $\mathcal{N} = \{a \in \Lambda \mid \xi(a) = 0\}$.

Let $v \in \Gamma_0$ and let $\{e_1, \ldots, e_n\}$ be a basis of the integer lattice Λ. There is a smallest subspace V of Γ, spanned by some subset of $\{e_1, \ldots, e_n\}$ and containing the ray $\mathbb{R}^+v$. If this ray does not meet $\Lambda \smallsetminus \{0\}$, then it descends to a dense geodesic in the torus $V/(V \cap \Lambda)$. That is, $\mathbb{R}^+v$ passes arbitrarily close to $\Lambda \smallsetminus \{0\}$. But, if tv is sufficiently near a lattice point $a \neq 0$, the integer

$$\xi(a) = \xi(a - tv) < 1.$$

Hence $\xi(a) = 0$. Thus, since t can be taken arbitrarily large, v can be arbitrarily well approximated by real multiples of elements of $\mathcal{N} \cap \Lambda$. It follows that Γ_0 is spanned by $\mathcal{N} \cap \Lambda$. $\square$

Remark. When the pseudonorm ξ fails to be a norm, it passes to a well-defined norm on the quotient space $\Gamma/\mathcal{N}$. Indeed, if $v \in \Gamma$ and $w \in \mathcal{N}$, then

$$\begin{aligned} \xi(v+w) &\leq \xi(v) + \xi(w) \\ &= \xi(v), \\ \xi(v+w) &= \xi(v - (-w)) \\ &\geq \xi(v) - \xi(-w) \\ &= \xi(v). \end{aligned}$$

Thus, if B is the unit ball of the quotient norm, consider the quotient projection $p : \Gamma \longrightarrow \Gamma/\mathcal{N}$ and obtain a noncompact "unit ball" $p^{-1}(B)$ for the pseudonorm ξ on Γ. In this situation, Definition 10.3.2 of the dual norm yields a ξ^* that is sometimes infinite. It will be finite *and* a norm, however, on the subspace of $\mathrm{Hom}(\Gamma, \mathbb{R})$ that annihilates $\mathcal{N}$. We will always assume that ξ^* denotes this finite norm.

10.4. The Unit Ball in the Thurston Norm

In this section we will show that the unit ball

$$B_\xi = \{w \in H_2(M, \partial M; \mathbb{R}) \mid \xi(w) \leq 1\}$$

is a polyhedron. Before giving the proof, we will illustrate what is happening by determining the unit ball for some of the examples in Section 10.2.

Exercise 10.4.1. Let $V = \{\lambda_1, \ldots, \lambda_p\} \subset \partial B_\xi$ be an affinely independent set. Prove that the affine p-simplex Δ_p, spanned by V, is a subset of B_ξ. Generally, $\Delta_p \not\subset \partial B_\xi$, but if an interior point λ of Δ_p has norm $\xi(\lambda) = 1$, prove that $\Delta_p \subset \partial B_\xi$.

In the cases in which $M = M(\kappa)$ is the complement of a *knot*, the unit ball is uninteresting. Indeed, $H_2(M(\kappa), \partial M(\kappa); \mathbb{R}) = \mathbb{R}$ and the unit ball will always be an interval containing 0.

Consider first Example 10.2.8. Each component of $\lambda = \lambda_1 \cup \lambda_2$ bounds a punctured disk in $S^3 \smallsetminus (\lambda_1 \cup \lambda_2)$. That is, under the canonical identification $H_2(M(\lambda), \partial M(\lambda)) = H_1(\lambda)$ of Lemma 10.2.3, there are two annular cycles in $(M(\lambda), \partial M(\lambda))$ corresponding to the generators λ_1 and λ_2 of $H_1(\lambda)$. Thus, $\xi(\lambda_i) = 0$, $i = 1, 2$, so Theorem 10.3.1 implies that $\xi \equiv 0$. The pseudonorm is completely degenerate.

Of considerably more interest is Example 10.2.9, the Whitehead link $\lambda = \lambda_1 \cup \lambda_2$. We take it as evident that neither component λ_i bounds a disk (the Whitehead link is *linked*). It is also true that neither component bounds a singly punctured disk.

Remark. A rigorous proof of this last assertion uses the notion of the *linking number* $\ell(\lambda_1, \lambda_2)$, this being well defined, up to sign, as the intersection number of λ_1 with a compact, immersed surface N in S^3 that is in general position and is bounded by λ_2. It can be shown that $\ell(\lambda_1, \lambda_2) = \ell(\lambda_2, \lambda_1)$. (For more details, see [**158**].) Either view in Figure 10.2.4 shows that this linking number is 0 (!), hence that a punctured disk spanning one component of λ in the complement of λ must be punctured an even number of times.

But each component of λ bounds a doubly punctured disk (*cf.* Figure 10.2.4), determining a cycle in $H_2(M(\lambda), \partial M(\lambda))$ that is a "pair of pants" (a disk from which two disks have been removed). The Thurston norm of a pair of pants is 1. We have proven the following.

Lemma 10.4.2. $\xi(\lambda_i) = 1$, $i = 1, 2$.

Recall that we have found a spanning surface S for the Whitehead link λ with $|S| = 2$ (Figure 10.2.4). This proves that the genus of the Whitehead link is at most 1. We asserted that the genus is exactly 1 and we are now ready to prove this.

Lemma 10.4.3. $\xi(\lambda_1 + \lambda_2) = 2$. *In particular, the genus of the Whitehead link is* 1.

Proof. We show that λ does not bound an imbedded annulus. We assert that there is an isotopy of λ that reverses the orientation of λ_2 while preserving that of λ_1. (Perhaps the easiest way to see this is from the first view in Figure 10.2.4. Just rotate the picture 180 degrees around a vertical axis in the plane of the paper. The figure-eight component λ_1 is carried onto itself in an orientation-preserving manner and the other component λ_2 goes onto itself with a reverse of orientation.) If $\lambda_1 + \lambda_2$ bounds an annulus, it follows that $\lambda_1 - \lambda_2$ does too. But then $\xi(\lambda_1 + \lambda_2) = 0 = \xi(\lambda_1 - \lambda_2)$, and an easy application of convexity implies that $\xi(\lambda_1) = 0 = \xi(\lambda_2)$, a contradiction.

By the above paragraph, $2 \geq \xi(\lambda) > 0$. Since any spanning surface for λ has two boundary components, its Thurston norm must be an even integer, so $\xi(\lambda_1 + \lambda_2) = \xi(\lambda) = 2$. □

Corollary 10.4.4. *Let* $\lambda = \lambda_1 \cup \lambda_2$ *be the Whitehead link and let* $\{\lambda_1, \lambda_2\}$ *be identified with the standard unit basis in* $H_2(M(\lambda), \partial M(\lambda)) = H_1(\lambda) = \mathbb{R}^2$. *Let* $\{\lambda_1^*, \lambda_2^*\}$ *be the dual basis of* $H^1(\lambda)$. *Then* B_ξ *is the diamond in* $\mathbb{R}^2$ *with vertices* $(0, \pm 1)$, $(\pm 1, 0)$ *and* B_{ξ^*} *is the square in* $\mathbb{R}^2$ *with vertices* $(\pm 1, \pm 1)$.

Proof. By 10.4.3 and the symmetry that reverses one component of λ, but not the other, $\xi(\pm\lambda_1 \pm \lambda_2) = 2$. By Exercise 10.4.1, the only compact convex body in $\mathbb{R}^2$ whose boundary contains the points $(0, \pm 1)$, $(\pm 1, 0)$, $(\pm 1/2, \pm 1/2)$ is the diamond with vertices the unit vectors $(\pm 1, 0)$, $(0, \pm 1)$.

Evidently, B_{ξ^*} has boundary containing the eight points $(0, \pm 1)$, $(\pm 1, 0)$, $(\pm 1, \pm 1)$, and this implies that B_{ξ^*} is the square with corners $(\pm 1, \pm 1)$. □

It turns out that the Whitehead-like link in Figure 10.2.5 has the same unit ball B_ξ and the same dual ball B_{ξ^*} as the Whitehead link. Proving this requires that one show that this link does not bound an annulus. One can use the method of disk decomposition, treated in the next chapter, to show this, but we omit the details.

Next, consider Example 10.2.12. The following lemma contains, in particular, the fact, asserted in that example, that the minimal norm of a surface spanning the Borromean rings is 3, hence that this link has genus 1.

Lemma 10.4.5. *Let $\lambda = \lambda_1 \cup \lambda_2 \cup \lambda_3$ be the Borromean rings. The Thurston pseudonorm ξ on $H_2(M(\lambda), \partial M(\lambda))$ is a norm, and it is characterized by*

(1) $\xi(\pm\lambda_i) = 1$, $1 \le i \le 3$,

(2) $\xi(\pm\lambda_i \pm \lambda_j) = 2$, $i \ne j$,

(3) $\xi(\pm\lambda_1 \pm \lambda_2 \pm \lambda_3) = 3$.

Proof. By the first view in Figure 10.2.7, there is an isotopy of S^3 that cyclically permutes the rings. It follows that $\xi(\lambda_i)$ is independent of i. By the second view in Figure 10.2.7, there is a disk with one handle in S^3 having one λ_i as boundary and missing the other two, so $\xi(\lambda_i) \le 1$.

Any surface S spanning λ has three boundary components; hence $\chi(S)$ is odd and negative. In particular, $\xi(\lambda) \ne 0$. Since, by convexity,

$$\xi(\lambda_1 + \lambda_2 + \lambda_3) \le \xi(\lambda_1) + \xi(\lambda_2) + \xi(\lambda_3),$$

it follows that $\xi(\lambda_i) \ne 0$, hence that $\xi(\lambda_i) = 1$, $1 \le i \le 3$. This also implies that $\xi(\lambda) = 1$ or 3.

As observed in Example 10.2.12, there are symmetries of S^3 interchanging two components of λ in an orientation-preserving way and reversing the orientation of the third. But the rings can also be permuted cyclically by an isotopy of S^3 preserving their orientations, and so $\xi(\pm\lambda_1 \pm \lambda_2 \pm \lambda_3)$ is independent of the choices of signs. These symmetries can also be used to show that $\xi(\pm\lambda_i \pm \lambda_j)$ is independent of the choices of signs and of $i \ne j$.

If $\xi(\lambda) \ne 3$, then it is 1 and the inequality

$$2\xi(\lambda_1 + \lambda_2) \le \xi(\lambda_1 + \lambda_2 + \lambda_3) + \xi(\lambda_1 + \lambda_2 - \lambda_3)$$

implies that $\xi(\lambda_1 + \lambda_2) \le 1$.

To complete the proof, we must show that $\xi(\lambda_1+\lambda_2) = 2$. We assume this to be false and derive a contradiction. Because $\xi(\lambda_1+\lambda_2) \leq \xi(\lambda_1)+\xi(\lambda_2)$, the value of the norm cannot be greater than 2. Since $\xi(\lambda_1+\lambda_2) = \xi(\lambda_1-\lambda_2)$, the usual convexity argument shows that this value cannot be 0, since $\xi(\lambda_1) = 1$. We must have $\xi(\lambda_1 + \lambda_2) = 1$. Since any surface (assumed connected with minimal norm) $(\Sigma, \partial\Sigma) \subset (M(\lambda), \partial M(\lambda))$ that represents $\lambda_1 + \lambda_2$ must have at least two boundary components, the formula that relates $\chi(\Sigma)$, the genus g of Σ and the number of components of $\partial\Sigma$ implies that $g = 0$ and that Σ has three boundary components. Therefore, there is an annulus in S^3 that is bounded by $\lambda_1 \cup \lambda_2$ and is pierced once by λ_3. There is also, visibly, a disk in S^3 that is bounded by λ_1, is pierced twice by λ_3 and is disjoint from λ_2. The disk and annulus, together, after being slightly perturbed, form an immersed disk in general position that is bounded by λ_2 and is pierced three times by λ_3. But this implies that the linking number of λ_3 with λ_2 is an odd number, whereas Figure 10.2.7 makes it evident that this linking number is 0. □

Corollary 10.4.6. *Let $\lambda = \lambda_1 \cup \lambda_2 \cup \lambda_3$ be the Borromean rings and identify $\{\lambda_1, \lambda_2, \lambda_3\}$ with the standard unit basis in $H_1(\lambda) = \mathbb{R}^3$, the dual basis $\{\lambda_1^*, \lambda_2^*, \lambda_3^*\}$ being similarly identified as the standard basis of $H^1(\lambda)$. Then B_ξ is the octahedron with vertices the set $\{(\pm1, 0, 0), (0, \pm1, 0), (0, 0, \pm1)\}$. The dual ball B_{ξ^*} is the cube with vertices the set $\{(\pm1, \pm1, \pm1)\}$.*

Proof. Apply Exercise 10.4.1 and Lemma 10.4.5 as in the proof of Corollary 10.4.4. □

Finally, we consider Example 10.2.13, the three-link chain $\lambda = \lambda_1 \cup \lambda_2 \cup \lambda_3$.

Lemma 10.4.7. *For the 3-link chain λ, the Thurston pseudonorm ξ on $H_2(M(\lambda), \partial M(\lambda))$ is a norm, and it is characterized by*

(1) $\xi(\pm\lambda_i) = 1$, $1 \leq i \leq 3$,
(2) $\xi(\pm\lambda_i \pm \lambda_j) = 2$, $i \neq j$,
(3) $\xi(\pm(\lambda_1 + \lambda_2 + \lambda_3)) = 1$,
(4) $\xi(\pm(\lambda_{\sigma(1)} + \lambda_{\sigma(2)} - \lambda_{\sigma(3)})) = 3$, *where σ is any permutation of* $\{1, 2, 3\}$.

Proof. We have observed that the spanning surface S in Figure 10.2.8 has genus 0, hence is a pair of pants. Since any spanning surface has three boundary components, this one of genus zero realizes the minimal Thurston norm; hence $\xi(\lambda_1 + \lambda_2 + \lambda_3) = 1$. This proves (3).

From Figure 10.2.8 it is clear that certain symmetries of S^3 permute the components of λ cyclically while preserving orientations, so $\xi(\lambda_i)$ is independent of i. By convexity and the fact that ξ is not totally degenerate, this common value cannot be 0. Each λ_i is represented, as a class in $H_2(M(\lambda), \partial M(\lambda)) = H_1(\lambda)$, by a pair of pants (again evident from Figure 10.2.8); hence $\xi(\lambda_i) = 1$, $1 \le i \le 3$. This proves (1).

By the symmetries noted in the preceding paragraph, $\xi(\lambda_i + \lambda_j)$ is independent of $i \neq j$. This number cannot be zero, since

$$2 = \xi(2\lambda_1 + 2\lambda_2 + 2\lambda_3) \le \xi(\lambda_1 + \lambda_2) + \xi(\lambda_2 + \lambda_3) + \xi(\lambda_3 + \lambda_1).$$

By (1) and convexity, this number is at most 2, so it remains to exclude the possibility that $\xi(\lambda_1 + \lambda_2) = 1$.

Suppose that $\xi(\lambda_1 + \lambda_2) = 1$ and deduce a contradiction. The only possible spanning surface S of Thurston norm 1 for $\lambda_1 + \lambda_2$ in $(M(\lambda), \partial M(\lambda))$ is a pair of pants. Indeed, $|\partial S| \ge 2$, so we have $2 - 2g = |\partial S| - 1 > 0$ and necessarily $g = 0$; hence $|\partial S| = 3$. It follows that there is an imbedded annulus A in S^3, bounded by $\lambda_1 \cup \lambda_2$ and pierced once by λ_3. Let $\gamma = \lambda_1 \cup \lambda_2$ and note that A represents the class

$$\lambda_1 + \lambda_2 \in H_2(M(\gamma), \partial M(\gamma)) = H_1(\gamma).$$

From Figure 10.2.8, one sees a doubly twisted annulus A', pierced twice by λ_3, that also represents $\lambda_1 + \lambda_2 \in H_2(M(\gamma), \partial M(\gamma))$. Let us view λ_3 as an element of $H_1(M(\gamma))$. Since λ_3 pierces A once, the homological intersection product $[A] \cdot [\lambda_3] = [A'] \cdot \lambda_3 \in H_0(M(\gamma)) = \mathbb{Z}$ must be ± 1. But this implies that λ_3 meets A' in an odd number of points, a contradiction. This establishes the two cases of (2) in which the signs are the same.

To prove (4), it will be enough to prove that $\xi(\lambda_1 + \lambda_2 - \lambda_3) = 3$. By (1), this number is at most 3. Since a spanning surface has exactly three boundary components, the Thurston norm is odd, so the only possible values for the norm are 1 and 3. But

$$\begin{aligned} 4 = 2\xi(\lambda_1 + \lambda_2) &\le \xi(\lambda_1 + \lambda_2 + \lambda_3) + \xi(\lambda_1 + \lambda_2 - \lambda_3) \\ &= 1 + \xi(\lambda_1 + \lambda_2 - \lambda_3), \end{aligned}$$

and this forces the number to be 3.

To complete the proof of (2), it remains to show that $\xi(\lambda_1 - \lambda_3) = 2$, the remaining cases following by the cyclic symmetry of the 3-link chain. But this number is at most 2, and the inequality

$$2\xi(\lambda_1 + \lambda_2 - \lambda_3) \le \xi(\lambda_1 - \lambda_3) + \xi(\lambda_2 - \lambda_3) + \xi(\lambda_1 + \lambda_2),$$

together with what has already been proven, implies that it is also at least equal to 2. □

As in Corollary 10.4.6, one deduces the following from Lemma 10.4.7 via Exercise 10.4.1.

Corollary 10.4.8. *For the* 3*-link chain and the usual identifications of the triples* $\{\lambda_1, \lambda_2, \lambda_3\}$ *and* $\{\lambda_1^*, \lambda_2^*, \lambda_3^*\}$ *as the standard basis in* 3*-space, the ball* B_ξ *is obtained as the union of the regular octahedron with vertices* $\{(\pm 1, 0, 0), (0, \pm 1, 0), (0, 0, \pm 1)\}$ *and the two tetrahedra with vertices* $\{(1, 0, 0), (0, 1, 0), (0, 0, 1), (1, 1, 1)\}$ *and* $\{-(1, 0, 0), (0, -1, 0), (0, 0, -1), (-1, -1, -1)\}$ *respectively. This makes* B_ξ *a parallelepiped. The dual ball* B_{ξ^*} *is an octahedron obtained by removing two opposite corners of a cube.*

In our analysis of the unit ball in each of these examples, the fact that it was a polyhedron was always proven using Exercise 10.4.1. It is not *a priori* clear from the definition that B_ξ will always be a polyhedron. This is so, however, being a delicate but elementary cosequence of the fact that ξ is $\mathbb{Z}$-valued on Λ.

Definition 10.4.9. A linear functional $L : \mathbb{R}^n \longrightarrow \mathbb{R}$ is said to be integral linear if it is $\mathbb{Z}$-valued on the integer lattice $\mathbb{Z}^n$.

Definition 10.4.10. A norm ζ on $\mathbb{R}^n$ is an integral norm if it is $\mathbb{Z}^+$-valued on the integer lattice $\mathbb{Z}^n$.

Proposition 10.4.11. *Let* ζ *be an integral norm on* $\mathbb{R}^n$*. Let* $a \in \mathbb{Z}^n$ *be nondivisible. Then there are an integral linear functional* L *on* $\mathbb{R}^n$ *and a* $\mathbb{Z}$*-basis* $\{a_1, a_2, \ldots, a_n = a\}$ *of* $\mathbb{Z}^n$ *such that* L *coincides with* ζ *on the cone* $\mathcal{C}$ *of* $\mathbb{R}^+$*-linear combinations of* $\{a_1, a_2, \ldots, a_n\}$*. In particular,*

$$\mathcal{C} \cap \partial B_\zeta = \{v \in \mathcal{C} \mid L(v) = 1\}$$

is a nondegenerate affine $(n-1)$*-simplex.*

We will prove this by induction on n, the case $n = 1$ being obvious. Suppose, then, that the assertion holds for some value of $n \geq 1$, let ζ be an integral norm on $\mathbb{R}^{n+1}$, and let $a \in \mathbb{Z}^{n+1}$ be nondivisible. By a suitable change of coordinates in $\mathbb{R}^{n+1}$ preserving $\mathbb{Z}^{n+1}$, we can assume that

$$a \in \mathbb{R}^n \times \{0\} = \mathbb{R}^n.$$

Inductively, there are a basis $\{a_1, a_2, \ldots, a_n = a\}$ of $\mathbb{Z}^n = \mathbb{Z}^n \times \{0\}$ and a linear functional $L : \mathbb{R}^n \longrightarrow \mathbb{R}$ that agrees with $\zeta|\mathbb{R}^n_+$. Again a suitable unimodular coordinate change allows us to assume that $a_i = e_i$, $1 \leq i \leq n$, where $\{e_1, e_2, \ldots, e_{n+1}\}$ is the standard basis of $\mathbb{R}^{n+1}$. Of course, ζ and L are replaced by their conjugates under these coordinate changes, but we continue to denote these conjugate objects by the same symbols.

Set $m_i = \zeta(e_i)$ and $\varepsilon_i = e_i/m_i$, $1 \leq i \leq n$, and let Δ denote the affine $(n-1)$-simplex spanned by $\{\varepsilon_1, \ldots, \varepsilon_n\}$. Note that the cone $\mathcal{C}$ of $\mathbb{R}^+$-linear combinations of the ε_i's is just the "first quadrant" $\mathbb{R}^n_+$ of $\mathbb{R}^n$ and that $\Delta = \{L = 1\} \cap \mathbb{R}^n_+$.

Let $v_0 = e_{n+1}$ and, inductively, let $v_p = e_1 + \cdots + e_n + v_{p-1}\ \forall p \geq 1$. Note that $\{e_1, \dots, e_n, v_p\}$ is a basis of the integer lattice $\mathbb{Z}^{n+1}$, $\forall p \geq 0$.

Let $L_p : \mathbb{R}^{n+1} \longrightarrow \mathbb{R}$ be the unique linear extension of L that satisfies $L_p(v_p) = \zeta(v_p)$. This is an integral linear functional. Set $\beta_p = v_p/\zeta(v_p)$ and let Δ_p be the affine simplex spanned by Δ and β_p, $p \geq 0$. The vertices of Δ_p lie on ∂B_ζ; hence, by convexity, $\Delta_p \subset B_\zeta$. For each $p \geq 0$, the ray $\mathbb{R}^+\beta_{p+1}$ meets $\mathrm{int}(\Delta_p)$ in a point $t_{p+1}\beta_{p+1}$, where $t_{p+1} \leq 1$.

Lemma 10.4.12. *For some integer $p \geq 0$, $t_{p+1} = 1$.*

This lemma will complete the proof of Proposition 10.4.11. Indeed, if $t_{p+1} = 1$, then $\beta_{p+1} \in \mathrm{int}(\Delta_p)$ and Exercise 10.4.1 implies that $\Delta_p \subset \partial B_\zeta$. Then $\{e_1, \dots, e_n, v_p\}$ is the desired basis of $\mathbb{Z}^{n+1}$ and L_p the desired integral linear functional.

In order to prove Lemma 10.4.12, we compute the (x_k, x_{n+1})-slope ν_k^p of the hyperplane $\Pi_p = \{L_p = 1\}$, $1 \leq k \leq n$. More precisely, ν_k^p is the slope dx_k/dx_{n+1} of the line ℓ_k^p in which Π_p meets the (x_k, x_{n+1})-plane.

Lemma 10.4.13. *For $1 \leq k \leq n$, the sequence $\{\nu_k^p\}_{p=0}^\infty$ is a nondecreasing subsequence of $\mathbb{Z} \cdot (1/m_k)$. Furthermore, $\nu_k^{p+1} = \nu_k^p$ if and only if $t_{p+1} = 1$.*

Lemma 10.4.14. *For $1 \leq k \leq n$, the sequence $\{\nu_k^p\}_{p=0}^\infty$ is bounded above.*

The proof of Lemma 10.4.13 will be given shortly, while the proof of Lemma 10.4.14 will be left as Exercise 10.4.15. First we make use of these lemmas in the following proof.

Proof of Proposition 10.4.11. Since $\mathbb{Z} \cdot (1/m_k)$ has no cluster point, the fact that the nondecreasing subsequence $\{\nu_k^p\}_{p=0}^\infty$ is bounded above implies that $\nu_k^{p+1} = \nu_k^p$ for sufficiently large values of p. Then Lemma 10.4.13 implies Lemma 10.4.12. As already remarked, the inductive step for the proof of Proposition 10.4.11 is an immediate consequence of Lemma 10.4.12. □

Proof of Lemma 10.4.13. By the inductive hypothesis,

$$L_p(e_i) = m_i, \qquad 1 \leq i \leq n.$$

Since $e_{n+1} = v_p - \sum_{i=1}^n pe_i$, we see that

$$L_p(e_{n+1}) = \zeta(v_p) - \sum_{i=1}^n pm_i,$$

hence that

$$L_p(x_1, \dots, x_n, x_{n+1}) = \sum_{i=1}^n m_i x_i + \left(\zeta(v_p) - \sum_{i=1}^n pm_i\right) x_{n+1}.$$

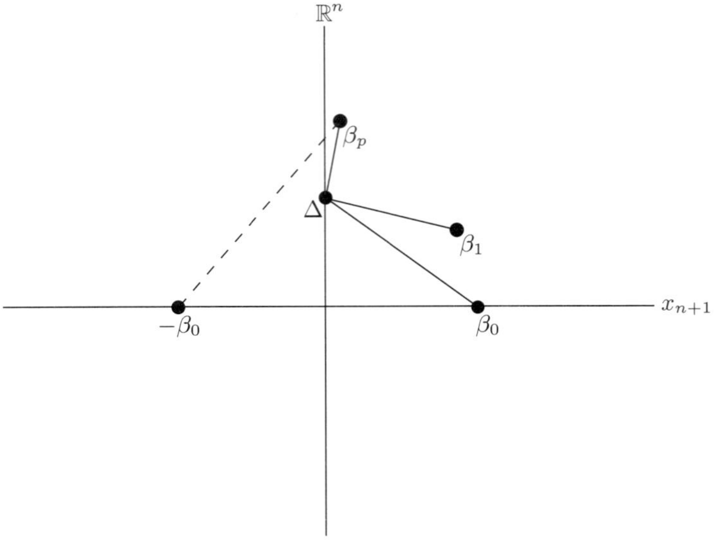

Figure 10.4.1. The $(n+1)$-simplices Δ_p swing around the "hinge" Δ

The equation

$$1 = L_p(0, \ldots, 0, x_k, 0, \ldots, 0, x_{n+1}) = m_k x_k + \left(\zeta(v_p) - \sum_{i=1}^{n} p m_i\right) x_{n+1}$$

describes the line ℓ_k^p, and differentiating with respect to x_{n+1} gives the formula

$$(*) \qquad \nu_k^p = \frac{\sum_{i=1}^n p m_i - \zeta(v_p)}{m_k} \in \frac{1}{m_k} \cdot \mathbb{Z}.$$

By the definition of v_{p+1}, together with the convexity of ζ, we have

$$(**) \qquad \zeta(v_{p+1}) \leq m_1 + \cdots + m_n + \zeta(v_p);$$

hence equation $(*)$ implies that $\nu_k^{p+1} \geq \nu_k^p$, with equality holding if and only if equality holds in $(**)$. By definition,

$$\beta_{p+1} = \frac{m_1 \varepsilon_1 + \cdots + m_n \varepsilon_n + \zeta(v_p)\beta_p}{\zeta(v_{p+1})},$$

and this lies in the interior of Δ_p (equivalently, $t_{p+1} = 1$) if and only if

$$\frac{m_1 + \cdots + m_n + \zeta(v_p)}{\zeta(v_{p+1})} = 1,$$

which is to say, equality holds in $(**)$. □

Remark. An intuitive interpretation of Lemma 10.4.13 is offered in Figure 10.4.1. The n-simplex Δ acts as a sort of "hinge" for the successive simplices Δ_p (represented in the figure by solid lines), $p \geq 0$. This figure, of course, is stylized, the subspace $\mathbb{R}^n$ being represented as 1-dimensional. The figure becomes exact, however, when interpreted as projection into the (x_k, x_{n+1})-plane. In this interpretation, the vertical line represents the x_k-axis and the point labeled "Δ" is the vertex ε_k of Δ. So long as β_{p+1} fails to lie in Δ_p, the slope ν_k^{p+1} is strictly greater than ν_k^p. From this picture, it is also easy to give an intuitive proof of Lemma 10.4.14. The dashed line in the figure represents the (projection of the) line segment I_p that joins $-\beta_0$ to β_p. If there were no finite upper bound to the sequence $\{\nu_k^p\}_{p=0}^{\infty}$, then for large enough values of p, I_p would have (x_k, x_{n+1})-slope strictly greater than ν_k^p. This would imply that (the projection of) I_p would cross the x_k-axis at a point above the vertex ε_k of Δ, hence that I_p would not lie entirely in B_ζ. But the endpoints of I_p do lie in B_ζ, contradicting convexity.

Exercise 10.4.15. Guided by the above remark, write down a careful proof of Lemma 10.4.14, thus completing the proof of Proposition 10.4.11.

Theorem 10.4.16. *Let ζ be an integral norm on $\mathbb{R}^n$, ζ^* the dual norm on $\mathbb{R}^{n*}$. Then the unit balls B_ζ and B_{ζ^*} are compact, convex polyhedra. The ball B_ζ is defined by a set of inequalities $L_i(x) \leq 1$, $1 \leq i \leq q$, where the linear functionals $L_1, \dots, L_q$ are integral linear and constitute the set of vertices of B_{ζ^*}.*

Proof. Let Λ' denote the subset of $\Lambda = \mathbb{Z}^n$ consisting of the nondivisible elements. Note that the union of the rays $\mathbb{R}^+ a$, as a ranges over Λ', is dense in $\mathbb{R}^n$. The integral linear functionals L_a, associated to $a \in \Lambda'$ as in Proposition 10.4.11, are not generally uniquely determined by a, but there is a subset $\Lambda'' \subset \Lambda'$ such that L_a is unique, for each $a \in \Lambda''$, and such that the union of the associated rays $\mathbb{R}^+ a$ is again dense in $\mathbb{R}^n$. Indeed, we take Λ'' to be the set of points $b \in \Lambda'$ such that the ray $\mathbb{R}^+ b$ meets the interior of one of the simplices Δ_a, $a \in \Lambda'$, given by Proposition 10.4.11.

For each $a \in \Lambda''$, the hyperplane $\Pi_a = \{L_a = 1\}$ meets ∂B_ζ in an $(n-1)$-dimensional piece containing $a/\zeta(a)$. Since B_ζ is convex, it lies in the halfspace $\{L_a \leq 1\}$. Since the union of the rays $\mathbb{R}^+ a$, $a \in \Lambda''$, is dense, we see that

$$B_\zeta = \{x \in \mathbb{R}^n \mid L_a(x) \leq 1, a \in \Lambda''\}.$$

But it is also clear that $\zeta^*(L_a) = 1$, so compactness of B_{ζ^*} and the fact that L_a is *integral* linear imply that only finitely many distinct functionals L_a occur, say $L_1, \dots, L_q$. Then B_ζ is a polyhedron with top-dimensional faces in the planes $\Pi_i = \{L_i = 1\}$, $1 \leq i \leq q$.

Let $v \in \partial B_\zeta$ be a vertex and let $\{L_{i_1}, \dots, L_{i_r}\} \subseteq \{L_1, \dots, L_q\}$ be the subset such that $L_{i_j}(v) = 1$, $1 \le j \le r$. Let $\{t_j \ge 0\}_{j=1}^r$ be such that $\sum_{j=1}^r t_j = 1$. Then $L = \sum_{j=1}^r t_j L_{i_j}$ satisfies

$$\begin{aligned} L(v) &= 1, \\ L(w) &\le 1, \quad \forall w \in B_\zeta. \end{aligned}$$

That is, the convex hull $\mathcal{S}_{i_1 \cdots i_r}$ of $\{L_{i_1}, \dots, L_{i_r}\}$ lies in ∂B_{ζ^*}. But v is a vertex, so $\{L_{i_1}, \dots, L_{i_r}\}$ contains a basis of $\mathbb{R}^{n*}$ and $\mathcal{S}_{i_1 \cdots i_r}$ is a top-dimensional face of B_{ζ^*} with vertices $\{L_{i_1}, \dots, L_{i_r}\}$. □

Corollary 10.4.17. *If the Thurston norm ξ is a true norm, then the unit balls B_ξ and B_{ξ^*} are compact, convex polyhedra. If ξ is only a pseudonorm, then B_ξ is a product $P \times \mathbb{R}^k$, where P is a compact, convex polyhedron of codimension k.*

For the last assertion, see the remark on page 345.

Corollary 10.4.18. *If ξ is a true norm, then the image of* $\mathrm{Diff}(M)$ *in the group* $\mathrm{Aut}(H_2(M, \partial M; \mathbb{R}))$ *is a finite subgroup.*

Proof. Indeed, if $\varphi \in \mathrm{Diff}(M)$, the induced automorphism

$$\varphi_* \in \mathrm{Aut}(H_2(M, \partial M; \mathbb{R}))$$

is a linear isometry of the Thurston norm. In particular, φ_* permutes the vertices of B_ξ. Since ξ is a norm, these vertices contain a basis of $H_2(M, \partial M; \mathbb{R})$; hence φ_* is determined by this induced permutation. □

Exercise 10.4.19. Give examples in which M contains a homologically nontrivial, properly imbedded annulus or torus and the image of $\mathrm{Diff}(M)$ in $\mathrm{Aut}(H_2(M, \partial M; \mathbb{R}))$ is infinite.

There are extreme cases in which the norm is totally degenerate ($\xi \equiv 0$). For these, the unit ball B_ξ is the entire space $H_2(M, \partial M; \mathbb{R})$. The following exercise leads the reader through an important class of examples.

Exercise 10.4.20. Suppose that M fibers over S^1 with connected fiber of nonnegative Euler characteristic. Prove that the Thurston norm for M is totally degenerate and that every nontrivial class in $H^1(M; \mathbb{R})$ is represented by a closed, nowhere zero 1-form. Here is a suggested outline (also *cf.* [**25**, Section 4]).

(1) If the fiber is D^2 or S^2, then our ongoing hypothesis that M is orientable forces the fibration to be a product.

(2) If the fiber is the annulus A^2, there are two cases. Either the fibration is a product, or M is the total space of a nonorientable interval bundle over the Klein bottle and $H_2(M, \partial M; \mathbb{R}) = \mathbb{R}$.

(3) If the fiber is the torus T^2, there are subcases according to the type of the monodromy map $\varphi : T^2 \to T^2$ of the bundle. As in [**8**, Introduction], φ is isotopic to a linear automorphism $\varphi \in \mathrm{Sl}(2, \mathbb{Z})$ and, according to whether $|\operatorname{tr}\varphi|$ is (a) greater than 2, (b) equal to 2 or (c) less than 2, this linear automorphism is
   (a) an Anosov diffeomorphism;
   (b) a power of a Dehn twist or the composition of such with an automorphism of period 2;
   (c) a periodic automorphism. (In fact, $\varphi^{12} = \mathrm{id}$.)

   In all cases except that in which φ is a power of a Dehn twist, $H_2(M, \partial M; \mathbb{R}) = \mathbb{R}$ (periodic of period 1 is really the 0th power of a Dehn twist). In the Dehn twist case, M is the total space of an orientable circle bundle over T^2 (see Exercise 4.4.7). If this bundle is trivial, then $M = T^3$ and, otherwise, $H^1(M; \mathbb{R}) = \mathbb{R}^2$. (Hint. The determination of $H^1(M; \mathbb{R})$ can be achieved elegantly using the Serre spectral sequence.)

10.5. Foliations without Holonomy

One of the striking applications of Thurston's norm has been to give a homological description of the set of smooth foliations $\mathcal{F}$ without holonomy on a given compact, orientable 3-manifold M. If $\partial M \neq \emptyset$, we assume that $\mathcal{F} \pitchfork \partial M$. For the definition and basic properties of smooth foliations without holonomy, the reader is referred to [I, Chapter 9].

In particular, we recall that such a foliation $\mathcal{F}$ is topologically isotopic to a foliation $\mathcal{F}_\omega$ defined by a closed, nowhere zero 1-form ω [I, Corollary 9.5.9]. (In the case that $\partial M \neq \emptyset$, ω is transverse to the boundary.) In fact, there is a continuous, holonomy-invariant measure μ for $\mathcal{F}$. finite on compact, transverse arcs and defining a class $[\mu] \in H^1(M; \mathbb{R})$ that coincides with $[\omega]$ under the de Rham isomorphism. This cohomology class $[\mu]$ is evidently invariant under topological isotopies. If $c > 0$, then $c\mu$ is also holonomy invariant, so the topological isotopy class of $\mathcal{F}$ actually determines an open ray $\{c[\mu] \mid c > 0\}$ issuing from the origin in the vector space $H^1(M; \mathbb{R})$. These remarks, combined with the following deep result, establish a one-to-one correspondence between the topological isotopy classes of foliations $\mathcal{F}$ without holonomy and the rays obtained as above.

Theorem 10.5.1 (Laudenbach and Blank [**117**]). *If ω and ω' are cohomologous nonsingular 1-forms, then the foliations $\mathcal{F}_\omega$ and $\mathcal{F}_{\omega'}$ are smoothly isotopic.*

Accordingly, we use the symbol $[\mathcal{F}]$ to denote both the isotopy class of $\mathcal{F}$ and the corresponding ray (called a "foliated ray") in $H^1(M; \mathbb{R})$. In the special case that $\mathcal{F}$ has all leaves compact, hence fibers M over S^1, we call $[\mathcal{F}]$ a "fibered ray". For alternative proofs of Theorem 10.5.1, the reader is referred to [**154**] and [**25**]. Here, we assume this theorem and show how the Thurston norm reduces the task of locating the foliated or fibered rays to a finite problem.

Definition 10.5.2. By a Thurston cone $\mathcal{C}(\Delta)$, we will mean the closure of the union of all open rays $\{c\alpha \mid c > 0\}$ determined by points α in a top-dimensional face Δ of $B_\xi \subset H^1(M;\mathbb{R})$. In the case that $\xi \equiv 0$, we agree that $\Delta = \infty$ is the unique face of $B_\xi = H^1(M;\mathbb{R}) = \mathcal{C}(\Delta)$.

Remark. If the ball B_ξ is compact, taking the closure of the union of the open rays through Δ only adds the vertex 0. Generally, however, it adds the subspace $\mathcal{N}$ on which ξ vanishes.

Our goal in this section is to prove the following.

Theorem 10.5.3 (Thurston [**175**]). *Each foliated ray $[\mathcal{F}]$ lies in the interior of a Thurston cone $\mathcal{C}(\Delta)$. Furthermore, every open ray in the interior of $\mathcal{C}(\Delta)$ is a foliated ray. A foliated ray is fibered if and only if it meets the integer lattice.*

The cones $\mathcal{C}(\Delta)$ given by Theorem 10.5.3 are sometimes called "foliation cones". The corresponding faces Δ of B_ξ are sometimes called "foliated faces" or "fibered faces". There are examples [**175**] in which some, but not all, of the top dimensional faces of B_ξ are foliated.

Example 10.5.4. In the next chapter, we will see (Example 11.3.5) that the spanning surface for the Whitehead link $\lambda = \lambda_1 \cup \lambda_2$, depicted in Figure 10.2.4, is the fiber of a fibration $\pi : M(\lambda) \to S^1$. By Corollary 10.4.4, the unit ball B_ξ can be viewed as the diamond in $\mathbb{R}^2$ with vertices $(\pm 1, 0)$ and $(0, \pm 1)$. By Theorem 10.5.3, the face connecting $(1,0)$ and $(0,1)$ must subtend a foliation cone. Obviously, the negative of this face will also be foliated. The symmetry that reverses the orientation of one component of the link while preserving that of the other can then be used to show that every ray except the four through the vertices of B_ξ is a foliated ray. In the case of the Whitehead-like link (Example 10.2.10), we have claimed that the unit ball B_ξ is the same as for the Whitehead link, but here it can be shown that two of the faces are foliated while two are not.

We turn to the proof of Theorem 10.5.3. The following is an elementary observation. Recall that our ongoing hypotheses include connectivity of M.

Lemma 10.5.5. *If $\pi : M \to S^1$ is a fibration, then the connected components of a fiber are mutually diffeomorphic and are themselves fibers of a fibration $\pi' : M \to S^1$.*

Accordingly, we will always assume that fibrations over the circle have connected fiber. The following is just a rephrasing of Exercise 10.4.20.

Lemma 10.5.6. *If there is a fibration $\pi : M \to S^1$ such that the fiber has nonnegative Euler characteristic, then $\xi \equiv 0$, and the assertions of Theorem* 10.5.3 *hold.*

Accordingly, we assume always that the fiber of a bundle over the circle has negative Euler characteristic. In particular, the norm is not totally degenerate and B_ξ has honest (possibly noncompact) top-dimensional faces.

Lemma 10.5.7. *Let $\mathcal{F}$ and $\mathcal{F}'$ be two foliations of M, both of which are transverse to ∂M, with respective tangent bundles τ and τ'. Relative to some fixed Riemannian metric, let v and v' be unit normal fields to τ and τ', respectively. If a continuous choice of angle θ_x from v'_x to v_x is strictly less than π, $\forall x \in M$, then the 2-plane bundles τ and τ' are isomorphic. In particular, the Euler classes $e(\tau)$ and $e(\tau')$ are equal.*

Proof. The fixed Riemannian metric enables us to consider, at each point $x \in M$, the group I_x of orientation-preserving isometries of $T_x(M)$, a compact group isomorphic to $SO(3)$. If $v_x \neq v'_x$, there is a unique 2-plane in $T_x(M)$ containing v and v', and we obtain a well defined $A_x \in I_x$ that leaves this 2-plane invariant, rotating v'_x into v_x through the angle θ_x. If $v_x = v'_x$, take $A_x = \text{id}$. Then A_x depends continuously on x and carries τ'_x isomorphically onto τ_x. □

Proposition 10.5.8. *Let τ be the tangent bundle to a foliation $\mathcal{F}_\alpha$, defined by a closed, nonsingular* 1*-form α that is transverse to ∂M. Then $e(\tau)$ is a vertex of B_{ξ^*} and the ray through $[\alpha]$ meets the interior of the face Δ of B_ξ that is defined by the linear equation $-e(\tau) = 1$. Furthermore, there is an open neighborhood $U \subset \operatorname{int} \mathcal{C}(\Delta)$ of $[\alpha]$ such that the open cone of rays through U consists entirely of points that, as classes in $H^1(M;\mathbb{R})$, are represented by closed, nonsingular* 1*-forms on M.*

Proof. Since $H^1(M;\mathbb{R})$ is the union of Thurston cones, there is a cone $\mathcal{C}(\Delta)$ such that the ray through $[\alpha]$ lies in that cone. Choose a basis of $H^1(M)$ and let $\omega_1, \dots, \omega_k$ be closed 1-forms representing this basis. For sufficiently small real numbers $\varepsilon_1, \dots, \varepsilon_k$, the closed form $\omega = \alpha + \sum_{i=1}^k \varepsilon_i \omega_i$ is nonsingular and transverse to ∂M. The corresponding classes $[\omega]$ form an open neighborhood U of $[\alpha]$ in $H^1(M;\mathbb{R})$, and the rays through U are foliated and form an open cone $\mathcal{C}(U)$.

By Lemma 10.5.7, we guarantee that the tangent bundles τ of the corresponding foliations are mutually isomorphic by keeping the ε_i's small. Thus $e(\tau)$ is constant for all the foliations $\mathcal{F}_\omega$, $[\omega] \in \mathcal{C}(U)$. Those choices of the

ε_i's for which the period group $P(\omega)$ is rational form a dense set in a neighborhood of $0 \in \mathbb{R}^k$, and so the corresponding set of cohomology classes is dense in U. By [I, Corollary 9.3.8], the leaves of such an $\mathcal{F}_\omega$ are compact and fiber M over S^1. Let $b = [F', \partial F']$, where F' is a leaf of such an $\mathcal{F}_\omega$. This is the Poincaré dual of $[\omega]$, and τ is isomorphic to the tangent bundle to $\mathcal{F}_\omega$. Hence

$$|e(\tau)(b)| = |\chi(F')| = |F'| = \xi(b),$$

where the last equality is by Theorem 10.1.3. It follows that $|e(\tau)(b)| = \xi(b)$ holds on a dense set of $b \in \mathcal{C}(U)$, hence on all of $\mathcal{C}(U)$. The norm being linear on $\mathcal{C}(U)$, it follows that this cone is contained in the interior of a Thurston cone $\mathcal{C}(\Delta)$. Furthermore, the face Δ of B_ξ satisfies the linear equation $-e(\tau) = 1$, and so $e(\tau)$ is a vertex of B_{ξ^*}. □

In order to complete the proof of Theorem 10.5.3, we must show that *every* ray through $\operatorname{int}(\Delta)$ consists of the cohomology classes of closed, nonsingular 1-forms.

Lemma 10.5.9. *Let $(S, \partial S)$ be a compact, connected surface in $(M, \partial M)$, realizing the minimal norm in its homology class and such that the ray $\mathbb{R}^+[S, \partial S]$ meets $\operatorname{int}(\Delta)$. Then, $e(\tau) \cdot [S, \partial S] = \chi(S)$.*

Proof. By assumption, $[S, \partial S]/\xi([S, \partial S]) \in \operatorname{int}(\Delta)$. Since $e(\tau)|\Delta \equiv -1$ and $\xi([S, \partial S]) = -\chi(S)$, the conclusion is immediate. □

Let $U \subset \operatorname{int} \mathcal{C}(\Delta)$ be the open set of Proposition 10.5.8, and let ω be a closed, nonsingular 1-form with rational periods such that $\mathbb{R}^+[\omega]$ meets U. Let $(S, \partial S)$ be as in Lemma 10.5.9, such that $[S, \partial S] \notin \mathbb{R}^+[\omega]$. By Theorem 9.5.20, we assume that S has only saddle tangencies with $\mathcal{F}_\omega$. Lemma 10.5.9 then implies that these tangencies are all of *positive* type (Definition 10.1.12).

Lemma 10.5.10. *There is a closed 1-form η, representing the Poincaré dual of $[S, \partial S]$, such that the half-open line segment $\{t[\omega] + (1-t)[\eta] \mid 0 < t \leq 1\}$ consists entirely of classes represented by closed, nonsingular 1-forms.*

Proof. The dual form η can be chosen so that its support is in a small normal neighborhood $N = S \times (-1, 1)$ of S. Indeed, for arbitrary $\varepsilon \in (0, 1]$, one can rechoose η to be supported in $S \times [-\varepsilon, \varepsilon]$ [**11**, Proposition 6.25].

At each point $x \in M$, choose a small neighborhood U_x of x and a nonvanishing vector field v_x on U_x such that $\omega(v_x) > 0$ everywhere in U_x. If $x \notin S$, choose U_x to be disjoint from S. If $x \in S$, we can choose the vector field to be transverse also to S and point in the positive direction relative to the transverse orientation of S. Indeed, if $x \in S$ is one of the saddle tangencies, the fact that this tangency is of positive type makes this choice easy. But if $x \in S$ is not a tangent point, S is transverse to $\mathcal{F}_\omega$ in a small neighborhood of x and the choice is again easy. Using a partition of unity, these local fields can be assembled to a global field v positively transverse both to S and to $\mathcal{F}_\omega$. Thus, $\omega(v) > 0$ everywhere on M, and, if $\varepsilon > 0$, as in the previous paragraph, is chosen sufficiently small, it is also true that $\eta(v) \geq 0$ on M. The assertion follows. □

Since the rays through the classes $[S, \partial S]$, as in the above lemma, are dense in $\mathcal{C}(\text{int}(\Delta))$ and the rays $\mathbb{R}^+[\omega]$ are dense in $\mathcal{C}(U)$, it follows that the ray through any point of $\text{int}(\Delta)$ meets some half-open line segment $t[\omega] + (1-t)[\eta]$, $0 < t \leq 1$, as in the lemma. The proof of Theorem 10.5.3 is complete.

Corollary 10.5.11. *If ξ is not totally degenerate and* $\dim H_2(M, \partial M; \mathbb{R})$ *is at least* 2*, then there is a compact, properly imbedded, orientable surface $(S, \partial S) \subset (M, \partial M)$ that is not homologous to the fiber of a fibration of M over S^1.*

Proof. If there are no fibrations, we are done. Otherwise, consider first the case in which ξ is a norm. By the proof of Theorem 10.4.16, each vertex of B_ξ is a rational point of $H_2(M, \partial M)$; hence the ray through a vertex must meet the integer lattice, say in $a \in \Lambda$. Since the dimension is at least 2, no vertex lies in the interior of a top-dimensional face. Thus, by Theorem 10.5.3, no representing surface $(S, \partial S) \in a$ can be the fiber of a fibration. If ξ is only a pseudonorm, recall that the space $\mathcal{N}$ on which ξ vanishes identically is spanned by a nontrivial sublattice of the integer lattice. If an element of that sublattice represents the fiber of a fibration, Exercise 10.4.20 implies that $\xi \equiv 0$, contrary to hypothesis. □

Chapter 11

Disk Decomposition and Foliations of Link Complements

We will define an operation, called *disk decomposition*, on a certain class of 3-manifolds, called *sutured manifolds*. To each compact, imbedded surface $R \subset S^3$ with boundary λ, we associate a sutured manifold

$$M = S^3 \smallsetminus N(R)^\circ.$$

Here, $N(R)$ denotes a normal neighborhood of R in S^3 and $N(R)^\circ$ its interior. If the geometric operation can be successfully applied to M, we will be able to construct a taut, depth one foliation $\mathcal{F}$ in the complement $S^3 \smallsetminus N(\lambda)^\circ$ of a tubular neighborhood $N(\lambda)$ of the link $\lambda = \partial R$. This construction is due to D. Gabai [**71**].

Throughout this chapter, we will denote the interior of a manifold N by N°.

11.1. A Basic Example

Before giving technical details, we consider a simple example (see Figure 11.1.1). This example is actually key to the general construction. Let Δ be the octagon

$$\Delta = \{(x, y); |xy| \leq \frac{\pi}{2}, \max\{|x|, |y|\} \leq \pi\}.$$

On $\Delta \times \mathbb{R}$ consider the foliation $\mathcal{H}$ with leaves given by

$$z = \tan(xy) + \text{constant},$$

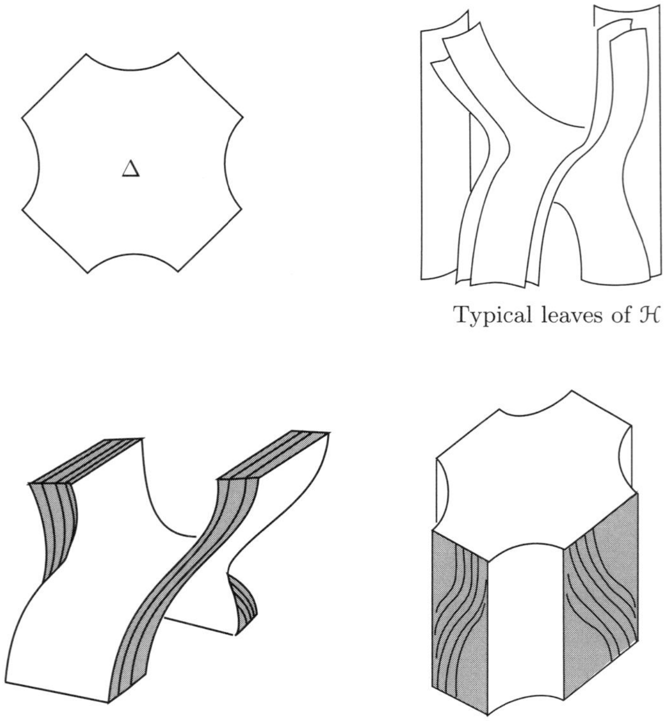

Typical leaves of $\mathcal{H}$

$\mathcal{H}$ in $\operatorname{int}(\Delta \times \mathbb{R})$

$\mathcal{H}|\partial_{\pitchfork}(\Delta \times \mathbb{R})$

Figure 11.1.1

together with the obvious boundary leaves

$$\{(x, y, z) \mid |xy| = \frac{\pi}{2}\}.$$

This foliation is invariant by translations in the z-direction and rotation by π of the Δ-factor. The manifold M, obtained as the quotient of $\Delta \times \mathbb{R}$ by the $\mathbb{Z}$-action generated by

$$T : (x, y, z) \mapsto (-x, -y, z + 1),$$

is a solid torus, and, since $\mathcal{H}$ is invariant by T, this $\mathbb{Z}$-action induces a foliation $\mathcal{F}$ on M. As T maps interior leaves of $\mathcal{H}$ to different ones, the leaves of $\mathcal{F}|M^\circ$ are like the ones pictured above. Furthermore, the leaf space $M^\circ/\mathcal{F}$ of $\mathcal{F}|M^\circ$ is a circle and the quotient map

$$p : M^\circ \to M^\circ/\mathcal{F}$$

defines a fibration, the fibers being the leaves, and the monodromy map is rotation by π. The tangential boundary $\partial_\tau M$ falls into two annuli, denoted

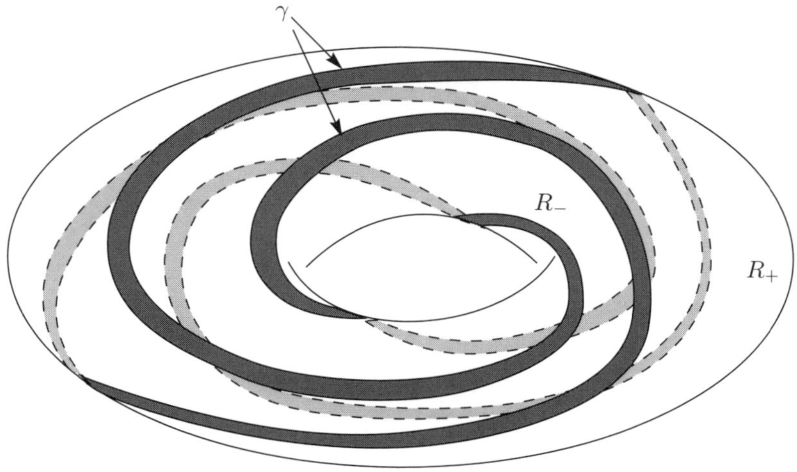

Figure 11.1.2. The sutured solid torus M

by R_+ and R_-, and the transverse boundary also falls into two annuli. Following Gabai [**71**], we set $\gamma = \partial_{\pitchfork} M$. This decomposition of ∂M, depicted in Figure 11.1.2, makes M into a *sutured manifold* in a sense soon to be defined.

Exercise 11.1.1. In the above foliated manifold, show how to glue R_+ to R_- (leaving γ free) to get a foliated manifold $(N, \mathcal{G})$ with toroidal boundary $\partial N = \partial_{\pitchfork} N$ and such that, if $\lambda \subset S^3$ is the $(4,2)$-torus link (Figure 11.1.3), then $(N, \partial N) \cong (S^3 \smallsetminus N(\lambda)^\circ, \partial N(\lambda))$. The manifold N is called the *link complement*. Show that the annuli $R_\pm$ are identified to a Seifert surface R of λ, this being the sole compact leaf of the taut depth one foliation $\mathcal{G}$. It is clear that R has minimal genus as a Seifert surface of λ. The juncture $J \subset R$ for the depth one leaves is a properly imbedded arc, as indicated in Figure 11.1.3.

Observe that N does not fiber over the circle with R as fiber. To see this, split N along R, obtaining M again. In order for N to fiber, the induced map on fundamental groups

$$\pi_1(R) \to \pi_1(M)$$

would have to be an isomorphism. But one easily checks that this map takes the generator of $\pi_1(R)$ to the square of the generator of $\pi_1(M)$.

Finally, note that similar constructions can be made starting with a $4n$-gon Δ, $n \geq 1$, and making the identification to a solid torus with a rotation of the (x, y)-plane through $k\pi/n$ radians, $0 \leq k \leq n$.

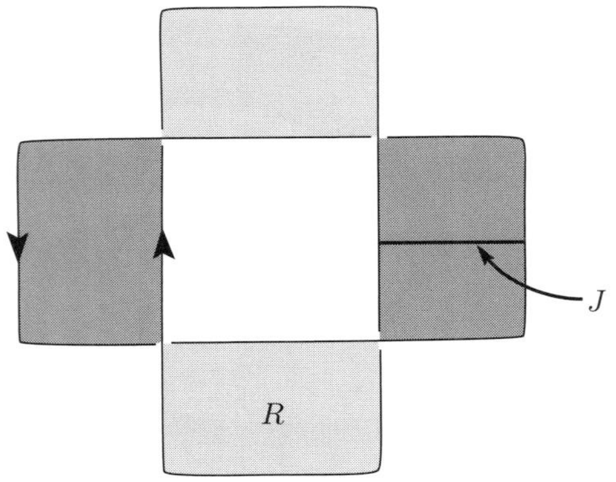

Figure 11.1.3. The $(4, 2)$-torus link with Seifert surface R

11.2. Sutured Manifolds

The following structure was invented by D. Gabai [**70**].

Definition 11.2.1. A sutured manifold is a pair (M, γ) consisting of a compact, oriented 3-manifold M, together with the disjoint union $\gamma = A(\gamma) \cup T(\gamma) \subset \partial M$, where $A(\gamma)$ is the union of pairwise disjoint annuli and $T(\gamma)$ is the union of pairwise disjoint tori. The interior of each component of $A(\gamma)$ contains a *suture*, this being an oriented, simple closed curve that is homologically nontrivial in γ. We denote the union of the sutures by $s(\gamma)$. Every component of $R(\gamma) = \partial M \setminus \gamma^\circ$ is also to be oriented, and we denote by $R_+(\gamma)$ (respectively, $R_-(\gamma)$) the union of those components of $R(\gamma)$ whose normal vectors point out of (respectively, into) M. Finally, the orientations on $R(\gamma)$ must be *coherent* with respect to $s(\gamma)$. That is, if a component δ of $\partial R(\gamma)$ is given the induced orientation, then δ must represent the same homology class in $H_1(\gamma)$ as some suture.

Example 11.2.2. Let M be the 3-ball with $\gamma = A(\gamma)$ a normal neighborhood of the (oriented) equator. The surfaces $R_\pm$ are disks. In picturing this (Figure 11.2.1) and other sutured manifolds, we draw $A(\gamma)$ as a curve, following the standard practice of making little or no visual distinction between $A(\gamma)$ and $s(\gamma)$. We will also make a practice of shading R_+.

Example 11.2.3. Let M be the solid torus. We have already discussed many sutured structures on M in Section 11.1. In Figure 11.2.2 we depict what is essentially the complement of the sutured manifold in Figure 11.1.2.

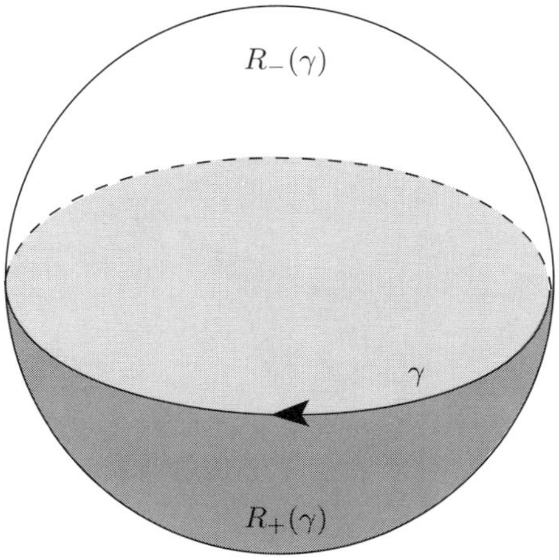

Figure 11.2.1. The sutured 3-ball with equatorial suture

In this case, the two sutures wind once longitudinally around ∂M while winding twice meridianally. This can be thought of as a "fattening" of the Seifert surface R in Figure 11.1.3. More precisely, $M = R \times I$ with

$$A(\gamma) = \partial R \times I,$$
$$T(\gamma) = \emptyset.$$

Note that this example is critically dependent on the orientations of the components of the $(4, 2)$-torus link λ. If the orientation of one component is reversed, the Seifert surface becomes the one depicted in Figure 11.2.3 and the corresponding sutured manifold $(R \times I, \partial R \times I)$ is no longer a sutured solid torus.

Quite generally, if (M, γ) is a sutured manifold imbedded in the closed oriented 3-manifold N, then $(N \setminus M^\circ, \gamma)$ is a sutured manifold in a canonical way. Note that the labeling of $R_-(\gamma)$ and $R_+(\gamma)$ must be interchanged, but we will generally ignore this when no confusion will arise. For the time being, we will consider sutured manifolds (M, γ) imbedded in S^3 with $T(\gamma) = \emptyset$.

Typically, our sutured manifolds will arise, as in Example 11.2.3, from a compact, oriented, imbedded surface $R \subset S^3$, every component of which has nonempty boundary. Such a surface is a Seifert surface of the knot or link $\lambda = \partial R$. Take $M = N(R) = R \times I$ to be a normal neighborhood of R in S^3 and obtain the sutured manifold

$$(M, \gamma) = (M, \partial R \times I).$$

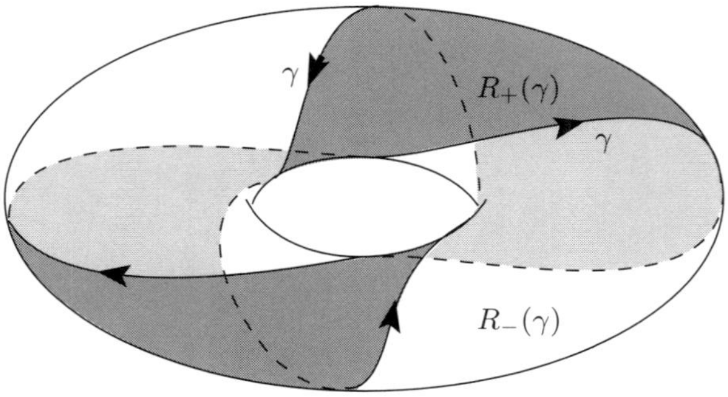

Figure 11.2.2. A sutured solid torus

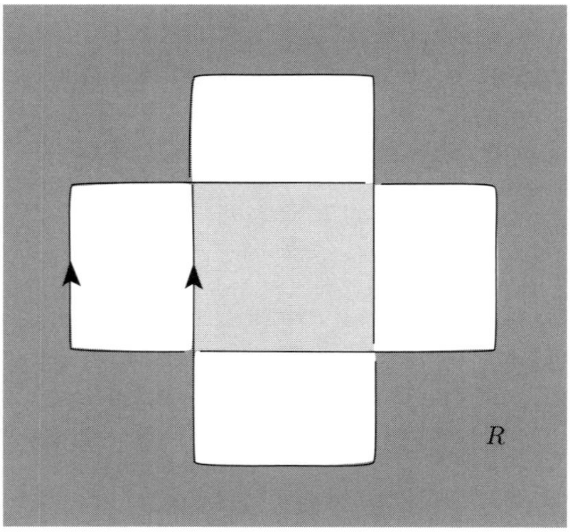

Figure 11.2.3. The result of changing orientation on one component of λ

The orientation of R induces a canonical orientation on $s(\gamma) = \partial R \times \{1/2\}$, hence on the components $R_+(\gamma)$ and $R_-(\gamma)$ of $R \times \partial I$. While this sutured manifold is topologically tame, the complementary one $(S^3 \smallsetminus M^\circ, \gamma)$ can be quite complicated. If this complement $N = S^3 \smallsetminus M^\circ$ can be tautly foliated so that $\partial_\pitchfork N = \gamma$ and $\partial_\tau M = R_\pm(\gamma)$, then gluing this by the identity map along $R_\pm(\gamma)$ to the product foliation of $M = R \times I$ produces a taut foliation of the link complement $S^3 \smallsetminus N(\lambda)^\circ$ that is transverse to the link boundary and has the Seifert surface R as a compact leaf.

Hereafter, to save verbiage, tautness for foliations of sutured manifolds should be understood in the following sense.

Definition 11.2.4. A foliation $\mathcal{F}$ of a sutured manifold (M, γ) is taut if the following properties all hold:

(1) $\mathcal{F}$ is transversely oriented and the components of $R(\gamma)$ are leaves whose transverse orientations agree with the transverse orientations of $R_\pm(\gamma)$ given by the sutured structure.

(2) $\gamma = \partial_\pitchfork M$ and the foliation induced by $\mathcal{F}$ on each component S of $\partial_\pitchfork M$ is transverse to a fibration of S over S^1.

(3) Each leaf of $\mathcal{F}$ meets either a closed transverse circle or a compact, properly imbedded transverse arc with one endpoint in $R_+(\gamma)$ and the other in $R_-(\gamma)$.

Remark. The condition (2) ensures that $\mathcal{F}|\partial_\pitchfork M$ has no 2-dimensional Reeb components. In fact, these properties are equivalent, but the proof is not trivial.

11.3. Operations on Sutured Manifolds

There are many operations that can be performed on a sutured manifold to obtain a new one. For the moment we will consider only one of these, *disk decomposition*. As we will see, many sutured manifolds coming from knots can be reduced by a sequence of disk decompositions to a union of sutured 3-balls $(D^2 \times I, \partial D^2 \times I)$. Then a foliation can be constructed, starting with the product foliation on these 3-balls and putting things back together.

Let (M, γ) be a sutured manifold and let $D \times I$ be a 2-handle H properly imbedded in M. That is, D is a 2-disk, $H \subset M$ and

$$A = \partial D \times I = H \cap \partial M.$$

Suppose that A does not meet $T(\gamma)$, $A \pitchfork A(\gamma)$, $A \pitchfork s(\gamma)$ and each arc of $A(\gamma) \cap (\partial D \times \{t\})$ intersects $s(\gamma)$ exactly once, $0 \le t \le 1$. We are going to use this 2-handle to decompose (M, γ) into a new sutured manifold (M', γ').

Selecting an orientation of ∂D, we give $(D \times I, \partial D \times I)$ a sutured manifold structure and set $D_+ = R_-(\partial D \times I)$, $D_- = R_+(\partial D \times I)$. This seemingly perverse notation comes from the fact that D_+ (respectively, D_-), or rather a translate into $D \times I^\circ$, is going to become part of $R_+(\gamma')$ (respectively, $R_-(\gamma')$). Set

$$D \times I = N(D_+) \cup (D \times J) \cup N(D_-),$$

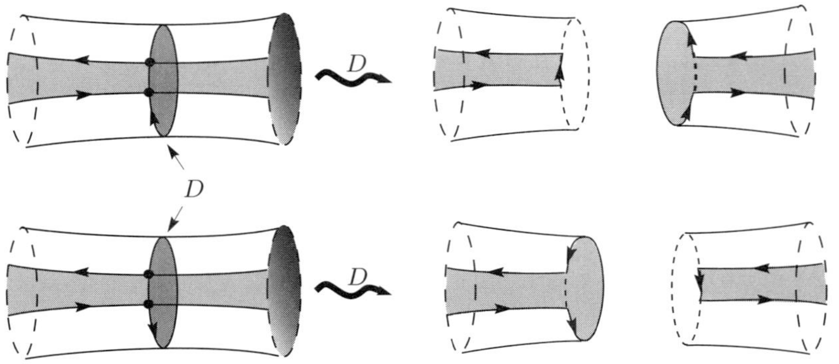

Figure 11.3.1. Local views of disk decomposition

where $J = (1/3, 2/3)$ and $N(D_+)$ is the component of $D \times I \setminus D \times J$ containing D_+, $N(D_-)$ the component containing D_-. Finally set

$$M' = M \setminus (D \times J),$$
$$\gamma' = (\gamma \cap M') \cup (N(D_+) \cap R_-(\gamma)) \cup (N(D_-) \cap R_+(\gamma)).$$

We say that D defines a sutured manifold decomposition and write

$$(M, \gamma) \overset{D}{\rightsquigarrow} (M', \gamma').$$

Less precisely, thinking of $A(\gamma')$ and $s(\gamma')$ as visually indistinguishable, we construct $s(\gamma')$ by connecting the endpoints of $s(\gamma) \cap \partial M'$ along the circles $\partial D \times \{1\}$ and $\partial D \times \{0\}$ using the following rule. If, at an endpoint $(x, 1) \in D \times \{1\}$ of $s(\gamma) \cap \partial M'$, this oriented curve connects to its clockwise (respectively, counterclockwise) neighboring arc (in $\partial D \times \{1\}$) in an orientation-coherent way, then at the endpoint $(x, 0)$, $s(\gamma) \cap \partial M'$ connects to its counterclockwise (respectively, clockwise) neighboring arc in an orientation-coherent way.

In Figure 11.3.1 we give local views of this operation.

Example 11.3.1. Let us return to the sutured solid torus in Figure 11.1.2. As indicated in Example 11.2.3, this arises from the complement of the (4, 2)-torus link by cutting apart along the annular Seifert surface. Let D be the properly imbedded, oriented disk depicted in Figure 11.3.2, thicken D to a 2-handle, and carry out the disk decomposition, obtaining the sutured manifold in Figure 11.3.3. The reader should see that this sutured manifold is homeomorphic to the sutured ball in Figure 11.2.1.

The sutured manifold in the above example is said to be *disk decomposable* because the decomposition produces the ball with equatorial suture.

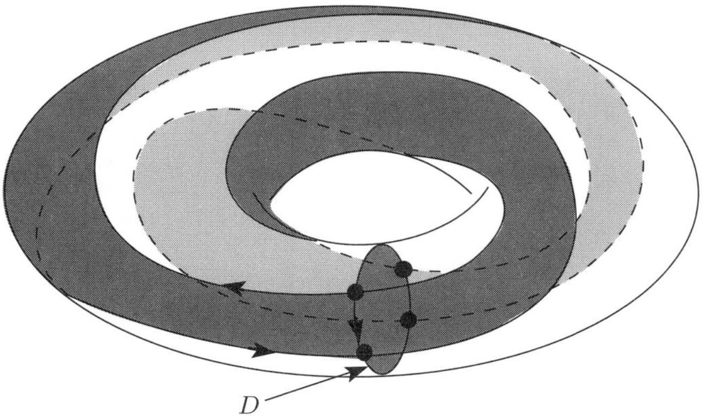

Figure 11.3.2. A decomposing disk D

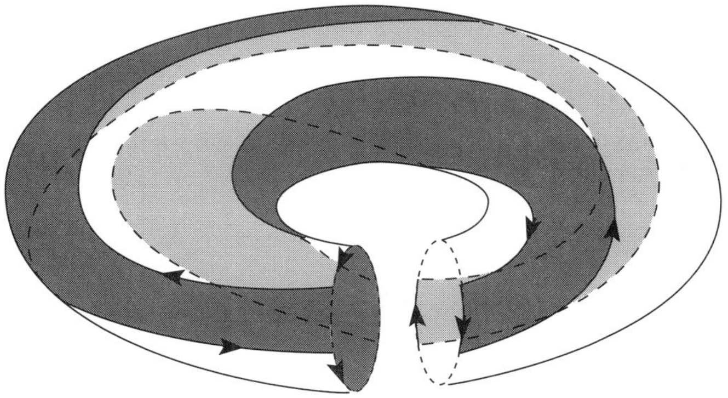

Figure 11.3.3. The result of decomposing along D

More generally, one allows a sequence of disk decompositions resulting in a finite union of such sutured balls.

Definition 11.3.2. A sutured manifold (M, γ) is said to be disk decomposable if there is a sequence of disk decompositions

$$(M, \gamma) \overset{D_1}{\rightsquigarrow} (M_1, \gamma_1) \overset{D_2}{\rightsquigarrow} \dots \overset{D_n}{\rightsquigarrow} (M_n, \gamma_n),$$

where $(M_n, \gamma_n) = (R \times I, \partial R \times I)$, $R_+(\gamma_n) = R \times 1$, and R is a disjoint union of disks.

Remark. Topologically, the components of the manifolds M_k at each stage of a disk decomposition are handlebodies or 3-balls.

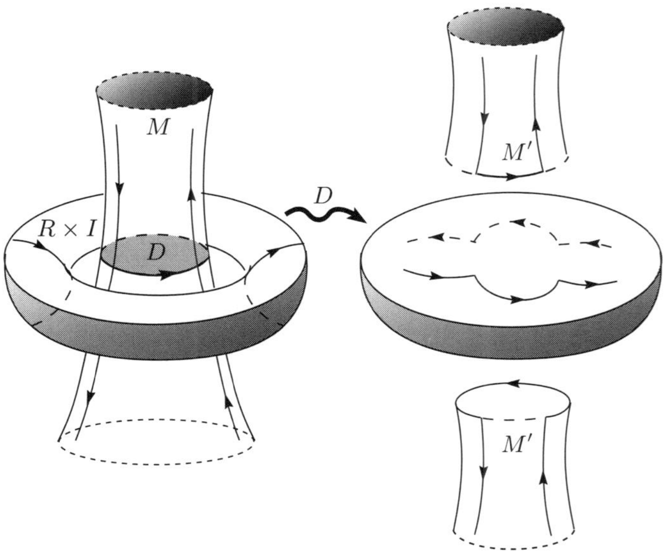

Figure 11.3.4. A disk decomposition adds a 2-handle to $(R \times I, \partial R \times I)$

Definition 11.3.3. A compact, connected surface $R \subset S^3$, with no closed component, is said to be disk decomposable if the sutured manifold

$$(M, \gamma) = (S^3 \setminus (R \times I)^\circ, \partial R \times I)$$

is disk decomposable.

Given a surface R in S^3, it is quite cumbersome to draw the sutured manifold $(S^3 \setminus (R \times I)^\circ, \partial R \times I)$. However, to look for a disk decomposition of R we need not care about its complement. Just note that, at the same time we cut $(M, \gamma) = (S^3 \setminus (R \times I)^\circ, \partial R \times I)$ along a 2-handle, we are adding a 2-handle to $(R \times I, \partial R \times I)$ (see Figure 11.3.4).

Example 11.3.4. In Figure 11.3.5 we picture the disk decomposition of the complement of the $(4, 2)$-torus link (Example 11.3.1) from the point of view of the fattened Seifert surface $(R \times I, \partial R \times I)$. The double-headed arrow indicates the thickening operation, replacing R with $R \times I$. The squiggly arrow, as usual, indicates the disk decomposition, but now from the point of view of $R \times I$, to which the 2-handle is attached. The last arrow is just the homeomorphism of the resulting sutured manifold with the sutured ball. Note that this sutured ball is the complement of the sutured ball in Figure 11.3.3.

Example 11.3.5. In Figure 11.3.6, we picture the Whitehead link Wh with Seifert surface R, together with the sutured manifold $(R \times I, \partial R \times I)$. In

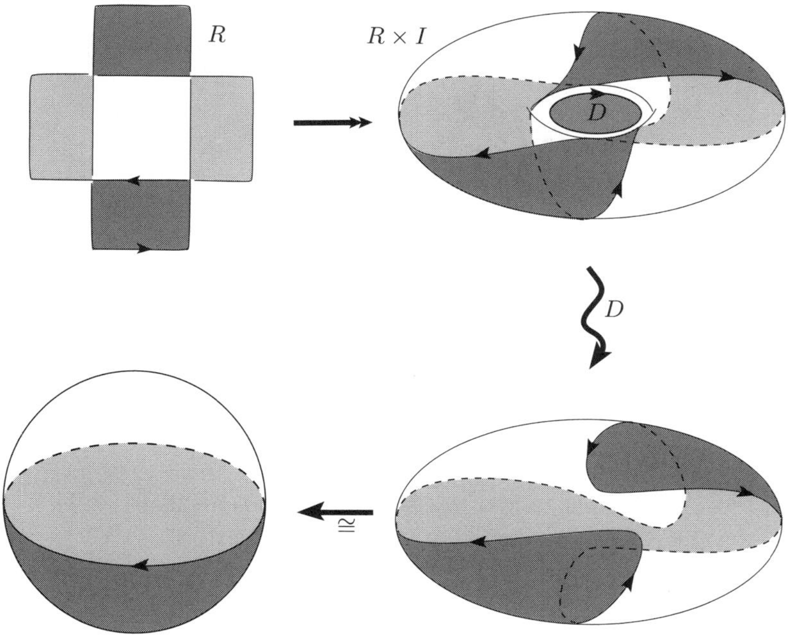

Figure 11.3.5. Complementary view of Example 11.3.1

Figure 11.3.7, we show a sequence of three disk decompositions from the viewpoint of $R \times I$. It should be clear that the final sutured manifold is homeomorphic to the ball with equatorial suture. Notice that, in each disk decomposition in this example, the disk meets $A(\gamma)$ twice. By part (c) of

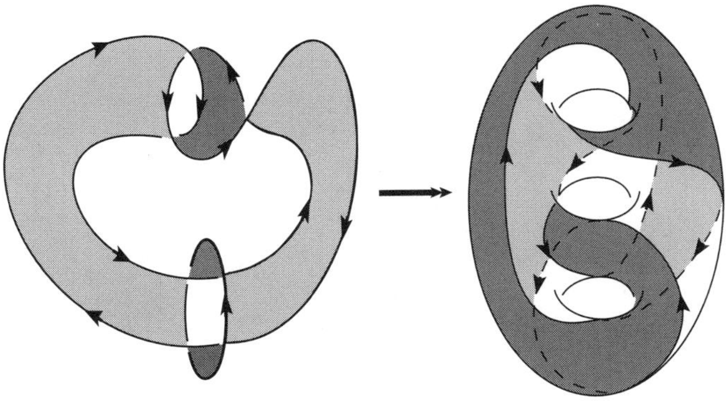

Figure 11.3.6. The link Wh and associated sutured manifold

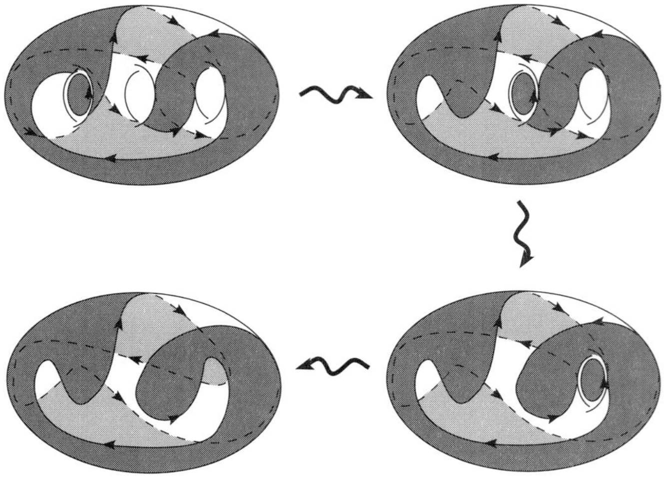

Figure 11.3.7. A disk decomposition for Wh

the main theorem (Theorem 11.4.1) of this chapter. it follows that Wh is a fibered link.

Example 11.3.6. In Figure 11.3.8, we depict a 4-component link with Seifert surface R, associated sutured manifold $(R \times I, \partial R \times I)$ and a decomposing disk D. In Figure 11.3.9, we depict first the result of the decomposition by D, then a second and third disk decomposition, ending with the sutured ball with connected $\gamma = A(\gamma)$. Note that, before the final disk decomposition, we perform an isotopy to tighten the sutured structure. This

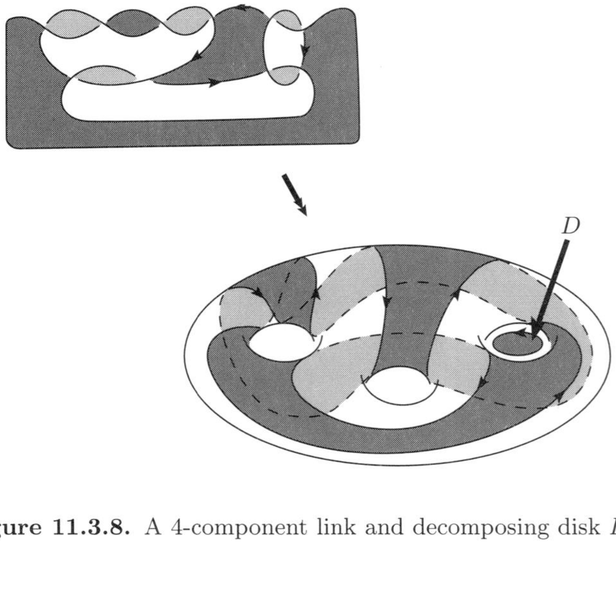

Figure 11.3.8. A 4-component link and decomposing disk D

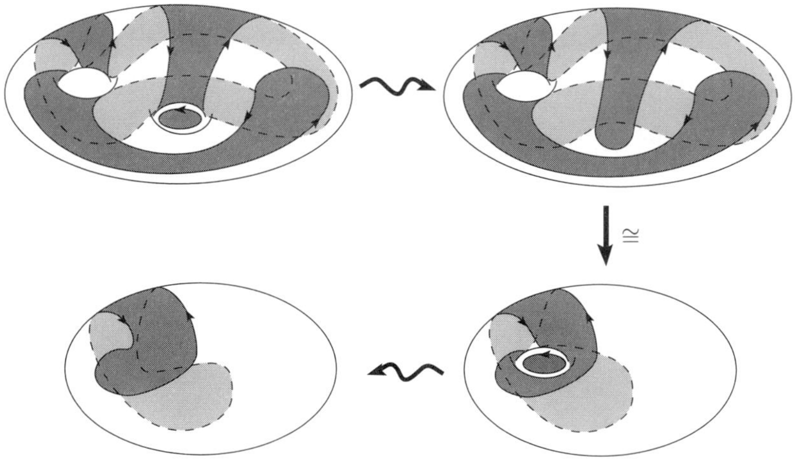

Figure 11.3.9. Tightening sutures before the last decomposition

is not entirely for esthetic reasons. The isotopy pulls an inessential loop of the suture up through the doughnut hole. Without this move we might naively attach the 2-handle so as to produce a sutured ball in which γ is the union of two disjoint annuli. This would fail to verify disk decomposability and would be fatal to the project of constructing a foliation. Once again, each decomposing disk meets the sutures twice, and the resulting foliation will be a fibration.

Example 11.3.7. Consider the knot pictured in Figure 11.3.10, along with the Seifert surface R and the associated sutured manifold $(R \times I, \partial R \times I)$.

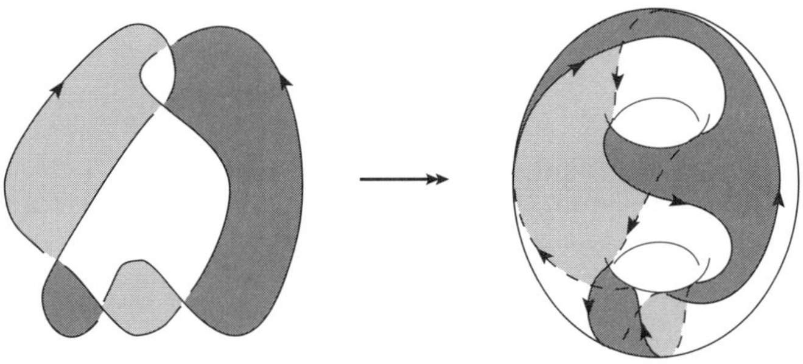

Figure 11.3.10. A nonfibered knot

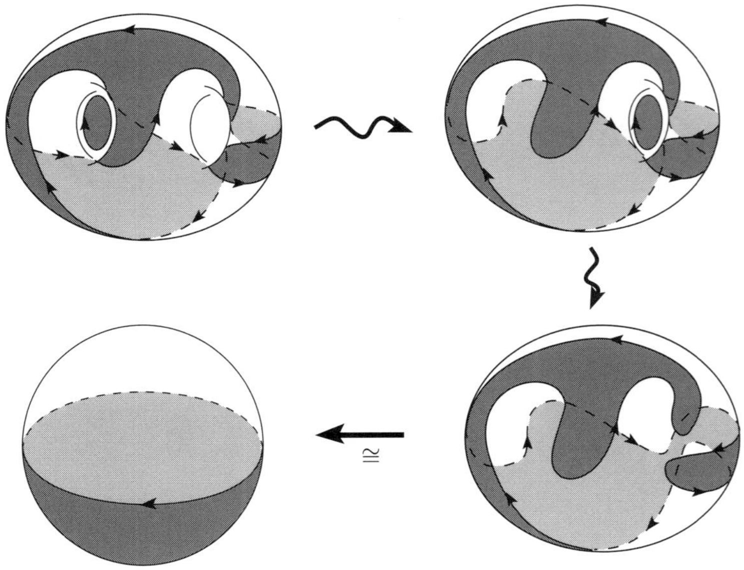

Figure 11.3.11. The second disk meets the sutures four times

This knot is known to be nonfibered. The second disk in the disk decomposition in Figure 11.3.11 meets the sutures four times; hence Theorem 11.4.1 yields a depth one foliation.

When working with the sutured manifold $(R \times I, \partial R \times I)$, the criterion for disk decomposability is dual to that of the one for $(S^3 \smallsetminus (R \times I)^\circ, \partial R \times I)$. We will say that the surface R is disk decomposable if there is a sequence of operations

$$(R \times I, \partial R \times I) = (N_1, \gamma_1) \rightsquigarrow (N_2, \gamma_2) \rightsquigarrow \cdots \rightsquigarrow (N_p, \gamma_p),$$

each operation consisting of adding a 2-handle to (N_k, γ_k). That is, we perform disk decomposition in $(S^3 \smallsetminus N_k^\circ, \gamma_k)$, terminating the process when

a) N_p is connected, and

b) ∂N_p is a union of spheres $S_1, \ldots, S_r$ and $S_i \cap s(\gamma_p)$ is connected. That is, $(S^3 \smallsetminus N_p^\circ, \gamma_p)$ is a disjoint union of sutured 3-balls $(D \times I, \partial D \times I)$.

Before stating the Main Theorem, we give an example to show how the disk decomposition is used to build a foliation in a sutured manifold (M, γ) that is tangent to $R(\gamma)$ and transverse to $A(\gamma)$.

Example 11.3.8. Recall the disk decomposition of Example 11.3.1, the complement of the sutured manifold of Section 11.1. We ended up with a sutured 3-ball. The complement is also a sutured ball that we can foliate as

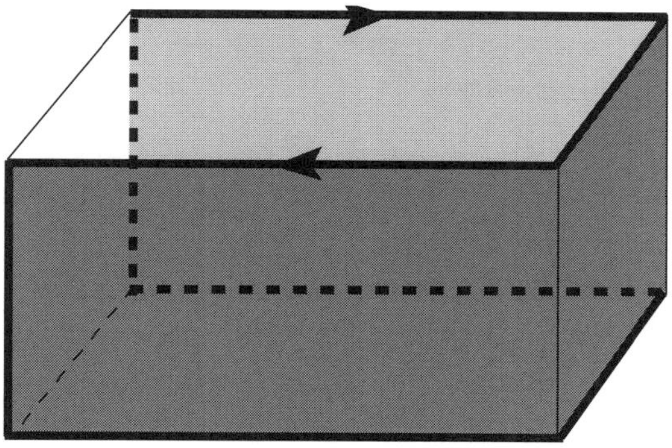

Figure 11.3.12. Deform $(D \times I, \partial D \times I)$ to a sutured box

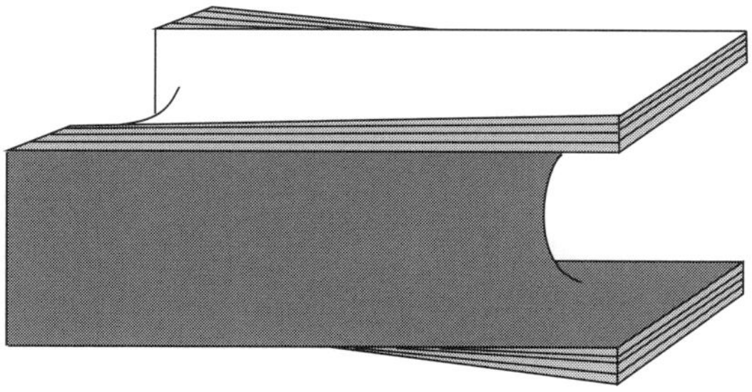

Figure 11.3.13. The foliated sutured box as a stack of chairs

a product $D \times I$. That is, the ball is foliated by disk leaves $D \times \{t\}$, $t \in I$, and the sutured structure has $\gamma = \partial D \times I$. To reglue and reconstruct the original sutured manifold of Section 11.1 we deform it into the shape of a box, sutured as in Figure 11.3.12. The foliation is carried along with this deformation. Now push the faces R_+ (the shaded face) and R_- (unshaded) toward one another to get something like a nested stack of chairs (Figure 11.3.13). Next twist through π radians and glue R_- to R_+ as indicated in Figure 11.3.14. This gives a foliation (with both convex and concave corners) of a solid torus T that can be imbedded in the interior of the foliated solid torus $M = D \times \mathbb{R}/\mathbb{Z}$ of Section 11.1 in such a way as to respect the foliations. The complement $M \setminus T^\circ$ is a pair of thickened annuli, foliated by strips $I \times [0, \infty)$ that spiral toward the annular components of $\partial_\tau M$ and have corners, convex ones fitting into the concave corners of T and concave ones fitting onto the

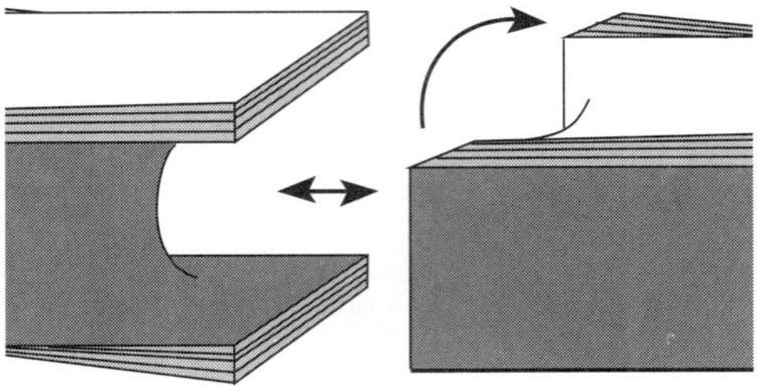

Figure 11.3.14. Glue R_- to R_+ with a half twist

convex corners of T, all so that the foliations match up. Intuitively, this extends the foliation of T by pushing the legs and arms of the chairs so as to spiral towards the respective tangential boundary annuli of M. This foliation is the one of Section 11.1.

11.4. The Main Theorem

We turn to the main result of this chapter. Recall the definition of tautness for foliations of sutured manifolds (M, γ) (Definition 11.2.4). We emphasize that such a foliation $\mathcal{F}$ of M induces no 2-dimensional Reeb components on $\gamma = \partial_\pitchfork M$. Note also that a knot complement is sutured with $\gamma = T(\gamma)$.

Theorem 11.4.1 (Gabai). *Let R be a compact, connected, oriented surface in S^3 that is not a disk, $\partial R \neq \emptyset$, and let λ be the oriented link ∂R. If R is disk decomposable, then:*

(a) *There exists a taut, depth one, C^∞ foliation $\mathcal{F}$ of the link complement $E(\lambda) = S^3 \smallsetminus N(\lambda)^\circ$, transverse to $\partial E(\lambda)$, for which R is the unique compact leaf.*

(b) *The Seifert surface R is norm minimizing for the link λ.*

(c) *If R has a disk decomposition such that, for each term*

$$(M_{k-1}, \gamma_{k-1}) \overset{D_k}{\rightsquigarrow} (M_k, \gamma_k),$$

the disk D_k intersects $s(\gamma_{k-1})$ in exactly two points, then λ is a fibered link with fiber R.

In part (c) of this theorem, the disk decomposition will be called a *product decomposition*. The theorem will be a consequence of the following.

Theorem 11.4.2. *Let (M, γ) be a sutured manifold such that $R(\gamma) \neq \emptyset$. If (M, γ) is disk decomposable, then there is a taut, transversely oriented, C^∞ foliation $\mathcal{F}$ on M, transverse to γ and having the components of $R(\gamma)$, with their given transverse orientation, as sole compact leaves. Furthermore, if N is a component of M with $\pi_1(N)$ nontrivial, then the projection*

$$p : N \smallsetminus R(\gamma) \to \text{ leaf space of } \mathcal{F}|(N \smallsetminus R(\gamma))$$

is a fibration over the circle.

Given a disk decomposable surface $R \subset S^3$ as in Theorem 11.4.1, apply Theorem 11.4.2 to the sutured manifold $(S^3 \smallsetminus (R \times I)^\circ, \partial R \times I)$. The foliation given by Theorem 11.4.2 yields a foliation $\mathcal{F}$ of $S^3 \smallsetminus N(\lambda)^\circ$, obtained by gluing $R_+(\gamma)$ to $R_-(\gamma)$. This foliation has R as unique compact leaf. Since R is connected and not closed, $\mathcal{F}$ has no toroidal leaves. By inspection of the foliation induced in $\partial R \times I$, one concludes that $\mathcal{F}$ is taut. By tautness, the Seifert surface is norm minimizing (Theorem 10.1.3). The fact that $\mathcal{F}$ is proper of depth one follows from the last assertion in Theorem 11.4.2. All of this proves parts (a) and (b) of Theorem 11.4.1.

We state part (c) of Theorem 11.4.1 as a separate theorem.

Theorem 11.4.3. *Let R and $\lambda = \partial R$ be as in* Theorem 11.4.1. *Then λ is a fibered link with fiber R if and only if R has a product decomposition.*

Proof. Observe that λ is a fibered link with fiber R if and only if

$$(S^3 \smallsetminus N(\lambda)^\circ) \smallsetminus (R \times I)^\circ \cong R \times I.$$

Suppose λ is a fibered link. Since R is connected and $\partial R \neq \emptyset$, there is a set $\alpha_1, \dots, \alpha_n$ of pairwise disjoint, properly imbedded arcs in R such that

$$R \smallsetminus \bigcup_{i=1}^{n} N(\alpha_i)^\circ = \text{2-disk.}$$

(This can be seen, for instance, by viewing R as a "disk with ribbons" as in [**125**, pp. 44-45].) Let $D_i = \alpha_i \times I$. Then

$$(R \times I, \partial R \times I) \overset{D_1}{\rightsquigarrow} \cdots \overset{D_n}{\rightsquigarrow} (D^2 \times I, \partial D^2 \times I)$$

is a product disk decomposition as in Theorem 11.4.1, part (c). A simple case of this is illustrated in Figure 11.4.1.

Conversely, suppose $(M_0, \gamma_0) = (S^3 \smallsetminus (R \times I)^\circ, \partial R \times I)$ has a product disk decomposition

$$(M_0, \gamma_0) \overset{D_1 1}{\rightsquigarrow} \cdots \overset{D_n}{\rightsquigarrow} (M_n, \gamma_n)$$

with $(M_n, \gamma_n) = (D \times I, \partial D \times I)$, D a disjoint union of 2-disks. Then (M_n, γ_n) is a product sutured manifold, and, by induction, one observes that each (M_k, γ_k) is a product sutured manifold. Indeed, if (M_{k+1}, γ_{k+1})

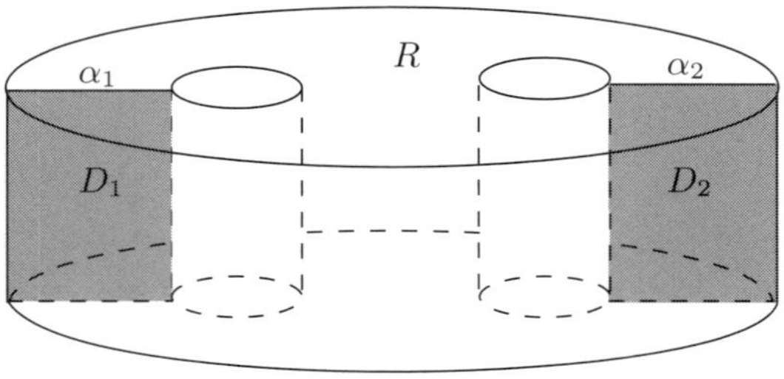

Figure 11.4.1. $R \times I$ admits a product decomposition

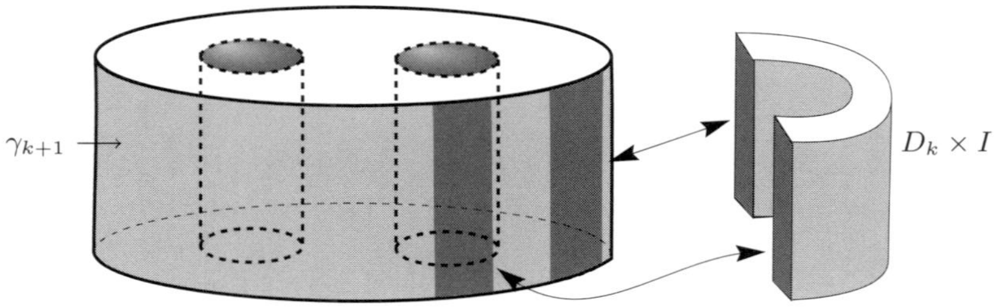

Figure 11.4.2. Reattaching product 2-handles

is such a product, then (M_k, γ_k) is obtained by reattaching the 2-handle $D_k \times I$ as indicated in Figure 11.4.2. This results in an oriented (hence trivial) I-bundle, so finite induction implies that

$$(M_0, \gamma_0) \cong (R \times I, \partial R \times I).$$

□

We turn to the proof of Theorem 11.4.2, beginning with a couple of elementary lemmas.

Lemma 11.4.4. *Let $\mathfrak{F}$ be a transversely oriented, C^∞ foliation of the connected sutured manifold (M, γ), transverse to γ and having the components of $R(\gamma)$ as sole compact leaves. Let $\mathfrak{L}$ be a smooth one-dimensional foliation transverse to $\mathfrak{F}$ and tangent to γ, so that $\mathfrak{L}|A(\gamma)$ is a foliation by compact, properly imbedded arcs. Then the following are equivalent:*

(1) *$\mathfrak{F}$ is a taut, depth one foliation.*

(2) *There is a smoothly imbedded circle* $\Sigma \subset M^\circ$ *that is transverse to* $\mathcal{F}|M^\circ$, *meeting each leaf of that foliation exactly once.*

(3) $\mathcal{L}$ *can be chosen to have a closed leaf in* M° *that meets each leaf of* $\mathcal{F}|M^\circ$ *exactly once.*

(4) $\mathcal{F}|M^\circ$ *fibers* M° *over* S^1.

In this case, there is a C^0 *flow* Φ_t *on* M *having the leaves of* $\mathcal{L}$ *as flow lines, stationary at the points of* $R(\gamma)$, *smooth on* M° *and carrying the leaves of* $\mathcal{F}$ *diffeomorphically onto one another.*

Proof. Indeed, by the well understood structure of depth one foliations near a compact leaf [I, Section 9.4], we find a compact, $\mathcal{F}$-transverse arc $J \subset M^\circ$, near a component of $R(\gamma)$, having endpoints on the same depth one leaf L and meeting each of the remaining leaves of $\mathcal{F}|M^\circ$ exactly once. The "waterfall" construction in [I, Lemma 3.3.7] then provides the imbedded circle Σ, proving that (1) $\Rightarrow$ (2). Given this transverse circle, it is easy to extend a smooth, nowhere zero vector field tangent to Σ to a smooth, nonsingular vector field on M, everywhere transverse to $\mathcal{F}$ and agreeing with the transverse orientation of $\mathcal{F}$, which is tangent to γ and induces a foliation by properly imbedded arcs on $A(\gamma)$. (The tautness of $\mathcal{F}$ is essential here, since it prevents 2-dimensional Reeb components in $A(\gamma)$.) The foliation $\mathcal{L}$ is defined by this field. The converse is evident, so (2) $\Leftrightarrow$ (3). By (2), $\mathcal{F}|M^\circ$ is without holonomy and does not have leaves that are dense in M°. Hence (2) $\Rightarrow$ (4) is given by [I, Theorem 9.1.4]. Note that the proof shows that this fibration is identical with the projection

$$\pi : M^\circ \to \Sigma,$$
$$\pi(L) = L \cap \Sigma, \quad \text{for each leaf } L \text{ of } \mathcal{F}|M^\circ.$$

The implication (4) $\Rightarrow$ (1) is clear. The flow Φ_t is given by lifting the rotation flow on $\Sigma \cong S^1$ to a smooth flow along $\mathcal{L}|M^\circ$. It is evident that this flow, oriented by the transverse orientation of $\mathcal{F}$, preserves $\mathcal{F}|M^\circ$ and "takes forever" to reach $R_+(\gamma)$ in forward time and to reach $R_-(\gamma)$ in backward time. Evidently, this flow extends continuously to a flow on M that is stationary at $R(\gamma)$. □

Lemma 11.4.5. *Let* R *be a compact, oriented, connected but not simply connected surface. If* M *is* (a) $R \times I$ *or* (b) *the total space of a fibration over* S^1 *with* R *as fiber, then* M *has a smooth, depth one, transversely oriented foliation* $\mathcal{F}$ *and, in the respective cases,*

(a) $\mathcal{F}$ *is transverse to* $\partial R \times I$ *and* $R \times \{0,1\}$ *are the only compact leaves;*

(b) $\mathcal{F}$ *is transverse to* ∂M *and a fiber* R *is the only compact leaf.*

This is a standard and elementary construction. Note that (b) reduces to (a) by cutting M along a fiber. To construct $\mathcal{F}$ in case (a), let α be a simple, closed, nonseparating curve or a properly imbedded, nonseparating arc in R. On $(R \setminus N(\alpha)^\circ) \times I$ consider the product foliation. Let $\partial N(\alpha) = \alpha \times \{0, 1\}$. The product foliation extends to a foliation $\mathcal{F}$ on $R \times I$ by identifying $(x, 0, t)$ to $(x, 1, f(t))$, $x \in \alpha$, where $f : I \to I$ is a diffeomorphism C^∞-tangent to the identity at $0, 1 \in I$, with $f(t) < t$ for all $t \in (0, 1)$.

The proof of Theorem 11.4.2 is by induction. Suppose we have obtained (M_1, γ_1) from (M_0, γ_0) by a disk decomposition

$$(M_0, \gamma_0) \overset{D}{\rightsquigarrow} (M_1, \gamma_1),$$

and (M_1, γ_1) has a foliation $\mathcal{F}_1$ such that for each component N of M_1, $\mathcal{F}_1|N$ is either a product foliation or a foliation as in the conclusion of Theorem 11.4.2.

The construction of $\mathcal{F}_0$ when $D \cap s(\gamma_0)$ consists of two points is basically the same as in the proof of Theorem 11.4.3, Figure 11.4.2 being insensitive to whether the foliation $\mathcal{F}_{k+1}$ is a product or has depth 1. Thus, we assume that $D \cap s(\gamma_0) = 2n > 2$ points. For simplicity of exposition, we prove Theorem 11.4.2 for $D \cap s(\gamma_0) = 4$ points, the other cases being analogous.

There are five cases to be considered.

Case 1. The manifold M_1 is connected, $\pi_1(M_0) \neq 1$ and $\pi_1(M_1) = 1$. In this case, $(M_1, \gamma_1) = (D^2 \times I, S^1 \times I)$ is the sutured ball and (M_0, γ_0) is one of the sutured solid tori in Figure 11.4.3. A foliation for (a) was constructed in Section 11.1, and a foliation for (b) can be obtained by the same method, replacing

$$T : (x, y, z) \mapsto (-x, -y, z + 1)$$

with

$$T : (x, y, z) \mapsto (x, y, z + 1).$$

Alternatively, the product foliation of $M_1 = D^2 \times I$ induces a foliation on M_0 via the construction in Example 11.3.8, the half twist in Figure 11.3.14 being omitted for (b). In any case, the resulting foliation evidently has all the required properties.

Case 2. The manifold M_1 is connected and $\pi_1(M_1) \neq 1$. Applying Lemma 11.4.5, if necessary, we assume that the foliation $\mathcal{F}_1$ is as in the conclusion of Theorem 11.4.2. Recall that (M_1, γ_1) is obtained from (M_0, γ_0) by removing a 2-handle $D \times I$ and modifying the sutures in a certain way. In Figure 11.4.4, we recall how this is done, the shading indicating $R_+(\gamma_i)$, $i = 0, 1$. Take $\mathcal{F}_0$ to be $\mathcal{F}_1$ in the complement of the neighborhood of D in M_0 depicted in the figure. To finalize the construction we have to give the description of $\mathcal{F}_0$ in this neighborhood of D.

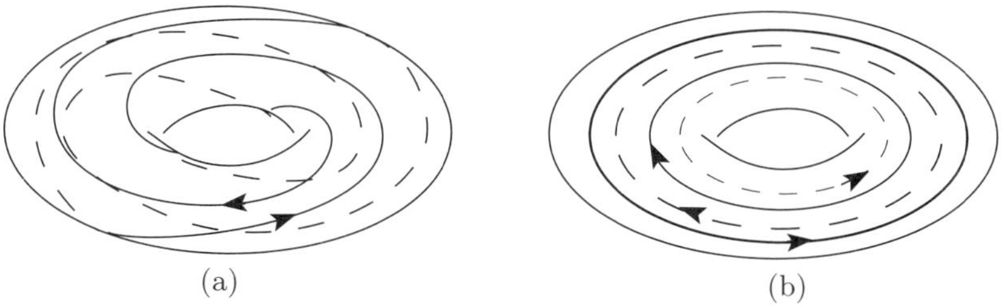

Figure 11.4.3. Two sutured solid tori

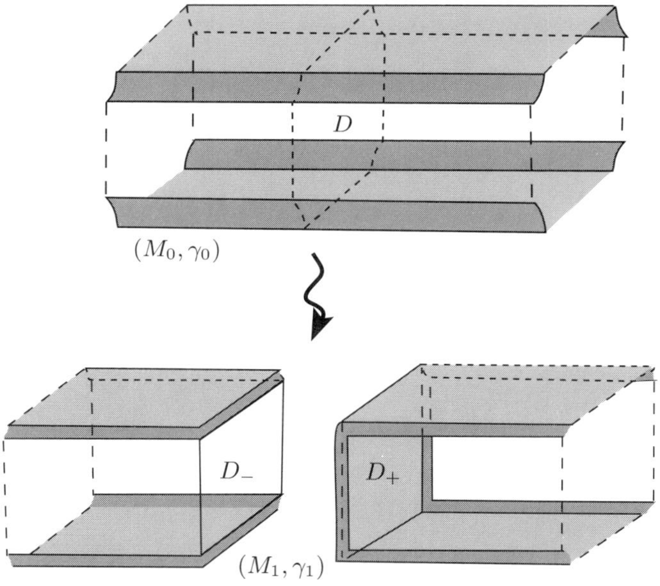

Figure 11.4.4. Remove a 2-handle and modify the sutures

To begin with, look at the parts of (M_1, γ_1) to be joined from a new point of view as in Figure 11.4.5. These views emphasize the role of $A(\gamma_1)$ as $\partial_{\pitchfork} M_1$ and that $\partial A(\gamma_1)$ consists of convex corners for $\mathcal{F}_1$. By Lemma 11.4.4, we fix a transverse, 1-dimensional foliation $\mathcal{L}_1$ with a closed leaf Σ meeting each depth one leaf of $\mathcal{F}_1$ exactly once and such that the leaves of $\mathcal{F}_1|A(\gamma_1)$ are compact arcs, properly imbedded in $A(\gamma_1)$. Put Σ so deep in the interior of M_1 that the following cut and paste moves do not affect it. Fix leaves $J_\pm$ of $\mathcal{L}_1|A(\gamma_1)$, these being compact arcs having one endpoint at a vertex of $D_\pm$. The leaf-preserving flow Φ_t can be used to parametrize $J_\pm$ as $[-\infty, \infty]$, the parametrization being uniquely determined by the choice of $0 \in J_\pm$. We make this choice so that both transverse arcs meet the same leaf L_0 of $\mathcal{F}_1$ at

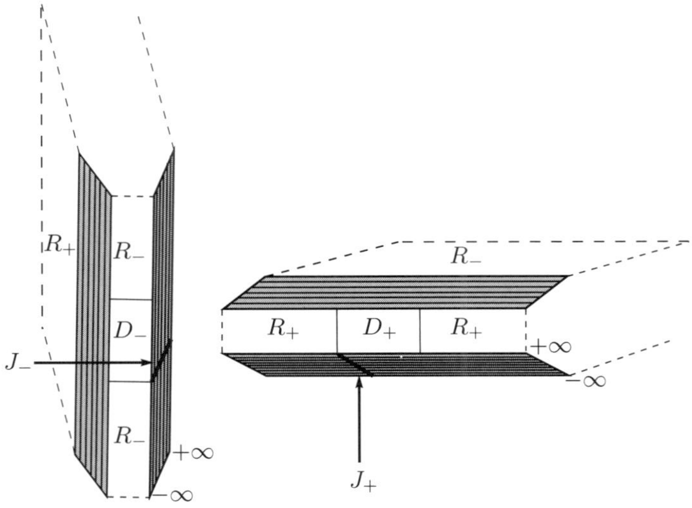

Figure 11.4.5. A new point of view of relevant parts of (M_1, γ_1)

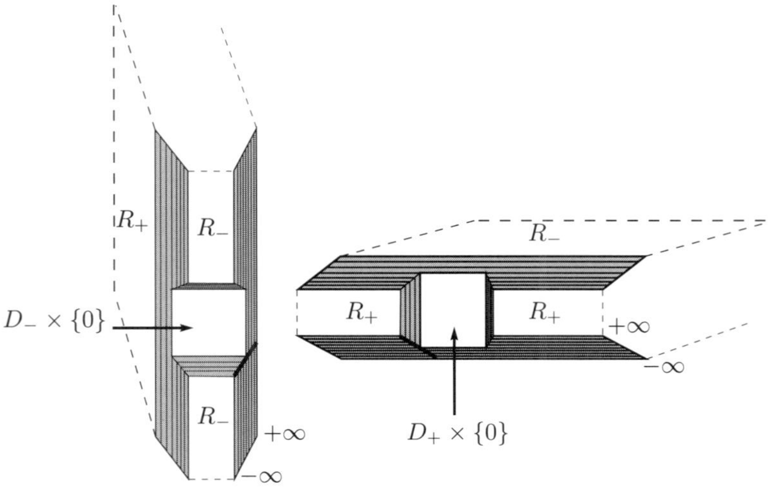

Figure 11.4.6. Cutting out product foliated neighborhoods of $D_\pm$

0. These arcs are also indicated in Figure 11.4.5. Recall that we normalize Φ_t to have minimum period 1 on Σ, so L_0 meets $J_\pm$ at the integer points.

By Reeb stability, the foliation near $D_\pm$ is trivial, so we may use the flow Φ_t to produce respective normal neighborhoods $D_+ \times [0, +\infty]$ and $D_- \times [-\infty, 0]$ in which the restriction of $\mathcal{F}_1$ is just the product foliation.

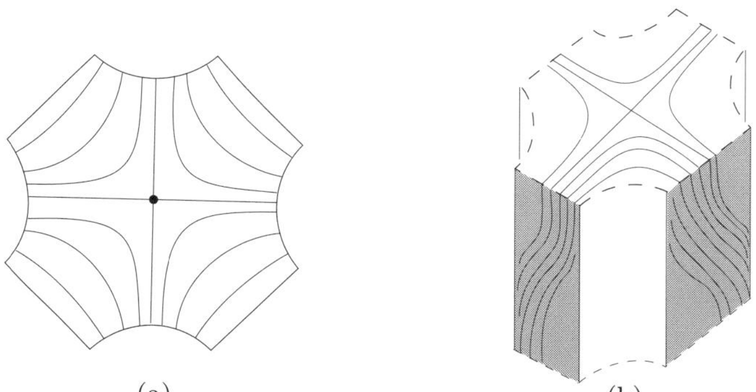

(a) (b)

Figure 11.4.7. (a) The induced foliation $\mathcal{F}|\Delta \times \{0\}$ and (b) the foliated neighborhood $\Delta \times (-\varepsilon, \varepsilon)$

Since the local transverse parametrizations are only determined up to an integer translation, these normal neighborhoods can be kept as close to $D_\pm$ as desired. In Figure 11.4.6, we cut out these blocks, using a diffeomorphism to bevel the edges so that the two parts will fit together like "Lincoln logs", matching the transverse parameters t, the foliations $\mathcal{L}_1$ and $\mathcal{F}_1$, and identifying $D_- \times \{0\}$ to $D_+ \times \{0\}$. This creates a new sutured manifold (M_0, γ_0) and a foliation $\mathcal{F}_0$ tangent to $R(\gamma_0)$, each interior leaf of which meets Σ exactly once. This is because the identifications respect the transverse flow parameter, hence attach only parts of each interior leaf to parts of the same leaf. This construction also creates a transverse, 1-dimensional foliation $\mathcal{L}_0$ such that $\mathcal{L}_0|A(\gamma_0)$ is a foliation by compact, properly imbedded arcs. Our picture introduces corners into the leaves, but these are easily rounded so as to make $\mathcal{F}_0$ smooth. At any rate, Lemma 11.4.4 and the structure of $\mathcal{L}_0|A(\gamma_0)$ guarantees that $\mathcal{F}_0$ is taut and of depth one with $R_\pm(\gamma_0)$ as sole compact leaves. Similar pictures can be drawn for the case that D meets the sutures $2n$ times, $n > 2$.

Alternatively, one can use the foliation in Section 11.1 to describe this construction. This is a smooth foliation $\mathcal{F}$ of $\Delta \times \mathbb{R}$, where Δ is an octagon, and the cross-section $\Delta \times \{0\}$ meets $\mathcal{F}$ in the level curves

$$\tan(xy) = \text{constant}$$

as in Figure 11.4.7(a). A foliated neighborhood $\Delta \times (-\varepsilon, \varepsilon)$ as pictured in Figure 11.4.7(b) will be our model for $\mathcal{F}_0$ in a normal neighborhood of the decomposing disk D. The disks $D_\pm$ in M_1 are isotopic to properly imbedded octagonal disks $\Delta_\pm$ that meet the sutures four times and meet $\mathcal{F}_1$ in level curves, also as in Figure 11.4.7(a). To see this at an intuitive level, concentrate on D_+ (for definiteness), first replacing it with the octagonal

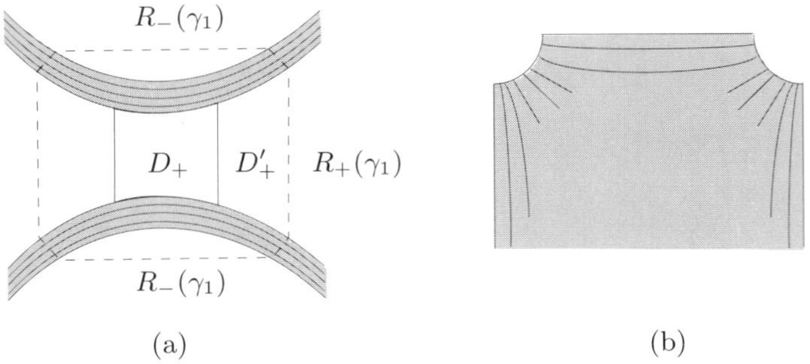

Figure 11.4.8. (a) The octagonal disk $D'_+ \subset \partial M_1$ and (b) a transverse slice of $\mathfrak{F}_1$ near D'_+

disk $D'_+ \subset \partial M_1$ indicated in Figure 11.4.8(a). Using Reeb stability and the fact that $\mathfrak{F}_1$ is transverse to γ_1, it is easy to see the structure of $\mathfrak{F}_1$ in a neighborhood of D'_+ in M_1. A transverse slice of this view is indicated in Figure 11.4.8(b), and the reader is invited to visualize an arbitrarily small deformation of D'_+, fixing $\partial D'_+$ pointwise and producing a properly imbedded, octagonal disk Δ_+ meeting $\mathfrak{F}_1$ with the one saddle tangency. It follows that a normal neighborhood N of $\Delta_\pm$ has foliation $\mathfrak{F}_1|N$ isomorphic to the model in Figure 11.4.7(b).

A more rigorous approach uses the Roussarie-Thurston general position theorem (Theorem 9.5.20). This gives an isotopy of $D'_\pm$ to a properly imbedded disk $\Delta_\pm$ having only saddle tangencies with the taut foliation $\mathfrak{F}_1$. One easily sees that the induced foliation near $\partial\Delta_\pm$ is as desired, and an application of the Poincaré-Hopf Theorem shows that there is only one saddle. At any rate, these observations show that the model foliation in a normal neighborhood of D fits smoothly with the portion of $\mathfrak{F}_1$ that foliates the complement in M_0 of a smaller normal neighborhood of D. Again one uses the transverse parameter t to match each leaf to itself, guaranteeing that the resulting foliation $\mathfrak{F}_0$ of M_0 has interior leaves each meeting Σ exactly once.

If D meets the sutures $2n$ times, then the saddle singularity is $2n$-pronged and the foliation of Section 11.1, based on a $4n$-gon, provides the local model.

Case 3. M_1 has exactly two components N_1 and N_2, and $\pi_1(N_1) = 1$. In this case, $(N_1, \gamma_1 \cap N_1)$ is the sutured box (Figure 11.3.12), there is a diffeomorphism

$$\varphi : (M_0, \gamma_0) \cong (N_2, \gamma_1 \cap N_2)$$

of sutured manifolds and we set $\mathfrak{F}_0 = \varphi^*(\mathfrak{F}_1|N_2)$.

Case 4. M_1 has exactly two components, N_1 and N_2, neither of which is simply connected. If $\mathcal{F}_1|N_i$ is a product for $i = 1$ and/or 2, one can use Lemma 11.4.5 to replace this with a depth one foliation. Let $\Sigma \subset N_1^\circ$ be a transverse circle meeting each interior leaf in exactly one point. One uses the method of Case 2, taking care that the transverse parameter for $\mathcal{F}_1|N_1$ is identified to the transverse parameter of $\mathcal{F}_1|N_2$ by the identity map. This guarantees that each interior leaf of one foliation is attached to only one interior leaf of the other, guaranteeing that the interior leaves of the new foliation $\mathcal{F}_0$ each meet Σ in exactly one point.

The proof of Theorem 11.4.2, hence of Theorem 11.4.1, is complete.

Remark. If, in the above constructions, we do not preserve the transverse parameter when gluing, instead matching the leaves by $t \mapsto t/k$, $k > 1$ an integer, the leaves of the resulting foliation will each meet the transverse circle Σ exactly k times. This will still guarantee trivial germinal holonomy on each interior leaf and these leaves will be proper, so the foliation will be of depth one (and taut). In this way, infinitely many inequivalent foliations may be associated to a given disk decomposition. Another way to think of this is the content of the following exercise.

Exercise 11.4.6. Let $\mathcal{F}$ be a smooth depth one foliation associated to a disk decomposition of the sutured manifold (M, γ). Set $M_0 = M \smallsetminus R(\gamma)$. Show that there is a closed, nonsingular 1-form ω defined on M_0, defining $\mathcal{F}|M_0$ and blowing up at $R(\gamma)$. The fact that the foliation defined by ω extends smoothly to a foliation of M by throwing in the components of $R(\gamma)$ as leaves will be indicated by saying that "ω blows up nicely at $R(\gamma)$". Suppose that the disks of the decomposition all live as properly imbedded, nonseparating disks $D_1, D_2, \dots, D_r$ in the original sutured manifold M. Show that one can choose 1-forms η_i, Poincaré dual to D_i, $1 \leq i \leq r$, such that all linear combinations

$$\omega' = \omega + \sum_{i=1}^{r} a_i \eta_i$$

having nonegative coefficients a_i are everywhere nonsingular 1-forms on M_0 that blow up nicely at $R(\gamma)$. Show that the associated foliation $\mathcal{F}(\omega')$ has depth one if and only if all the coefficients a_i are rational. In any case, the foliation is taut and has holonomy only along the compact boundary leaves. Observe that the corresponding classes $a[\omega'] \in H^1(M; \mathbb{R})$, $a > 0$, form an open convex cone in $H^1(M; \mathbb{R})$.

11.5. Applications

In this section we offer a few applications of the theory of disk decomposition. We will verify Property R for knots that admit a disk decomposable Seifert surface, we will show that the Murasagi sum of depth one knots is a depth one knot, and we will show that Seifert's algorithm produces a minimal genus spanning surface for all alternating links.

If λ is a knot or link, we will use the notation $E(\lambda)$ for the manifold $S^3 \setminus N^\circ(\lambda)$. (Here, $N^\circ(\lambda)$ denotes the interior of a tubular neighborhood $N(\lambda)$ of λ.) One calls $E(\lambda)$ the *link complement* or *link exterior*. A Seifert surface R for λ will often be viewed as a properly imbedded submanifold of the link complement by shaving off a small collar neighborhood $R \cap N(\lambda)$ of $\partial R = \lambda$. The sutured manifold

$$(M(R), \gamma(R)) = (S^3 \setminus N(R), \partial R \times I)$$

will be thought of as the result of cutting the sutured manifold $(E(\lambda), \partial E(\lambda))$ along R.

If κ is a knot, we recall that a simple closed curve m on $\partial E(\kappa)$ that is essential there, but bounds a disk in $N(\kappa)$, is called a *meridian* for the knot. At the other extreme is a *longitude* $\ell = \partial R$ of the knot, where R is a Seifert surface, properly imbedded in $E(\kappa)$ as above. We always tighten meridians and longitudes so that they intersect in a single point. Note that $[m]$ and $[\ell]$ form a free abelian basis of $H_1(\partial E(\kappa))$.

11.5.A. Property R. If the disk decomposable surface R spans a knot κ, we will see that the foliation in Theorem 11.4.1 must meet $\partial E(\kappa)$ in circles. This will imply that performing longitudinal surgery on the knot κ yields an irreducible three-manifold and Property R (Definition 11.5.5) will follow easily.

Exercise 11.5.1. Let κ be a knot, $M = E(\kappa)$ and let $N = N(\kappa)$. Using the Mayer-Vietoris sequence for the decomposition $M \cup N$ of S^3, prove that the connecting homomorphism

$$\partial_* : H_3(S^3) \to H_2(M \cap N)$$

is an isomorphism, concluding that $H_2(M) = 0$. (Hint. Recall the definition of ∂_*.) Also, show that $H_1(M) = \mathbb{Z}$ and is generated by the class $[m]$ of the meridian.

As in the exercise, let $M = E(\kappa)$. We study the homological properties of a connected, incompressible and ∂-incompressible surface S, properly imbedded in M. The boundary ∂S of S is a collection of simple closed curves in ∂M, all of them parallel to an essential, simple closed curve $\gamma \subset \partial M$. Write

$$[\gamma] = a[\ell] + b[m] \in H_1(\partial M),$$

a nondivisible class. By the exercise, $H_2(M) = 0$, so the homology sequence of the pair $(M, \partial M)$ gives an exact sequence

$$0 \to H_2(M, \partial M) \xrightarrow{\partial_*} H_1(\partial M) \xrightarrow{\iota} H_1(M).$$

If S does not separate M, then $[S]$ is nontrivial in $H_2(M, \partial M)$. Since ∂_* is injective, $\partial_*[S] = n[\gamma]$, $n \neq 0$. But $\iota \circ \partial_* = 0$, so $\partial_*[S] \in \ker(\iota)$. By the exercise, $[m]$ generates $H_1(M)$, so $\ker(\iota)$ is a free abelian group of rank one generated by $[\ell]$. It follows that $b = 0$ and $a = \pm 1$, so $[\gamma] = \pm[\ell]$.

Proposition 11.5.2. *Let S be a compact leaf of a taut foliation $\mathcal{F}$ on M that is transverse to ∂M. Then ∂S is connected and parallel to the longitude ℓ.*

Proof. Since S is a leaf, it is connected. Since $\mathcal{F}$ is taut, S is homologically nontrivial, so S meets ∂M. Otherwise, $\partial_*[S] = 0$, implying that $[S] = 0$. This also shows that ∂S is not homologically trivial in $H_1(\partial M)$ and, by the discussion above, $\partial_*[S] = n[\ell]$, $n \neq 0$. Since $\mathcal{F}|\partial M$ has no Reeb components, $|n|$ is exactly the number of components of $S \cap \partial M$. It follows that $[S] = n[R]$, where $[R]$ is a generator of $H_2(M, \partial M) \cong \mathbb{Z}$. But a connected, nonseparating surface must represent a primitive homology class; hence $n = \pm 1$. □

Theorem 11.5.3. *If κ is a nontrivial knot in S^3 and $\mathcal{F}$ is a depth one foliation of $M = E(\kappa)$ with the properties in Proposition* 11.5.2, *then $\mathcal{F}|\partial M$ is a foliation by longitudinal circles.*

Proof. We have seen that each compact leaf of $\mathcal{F}$ meets ∂M in exactly one longitude. Each depth one leaf L has finitely many ends, each of which spirals in on one ar another compact leaf. Let e be an end of L, $N(e) \subset L$ a neighborhood of e spiraling on a compact leaf S. Corresponding to this spiral, there is an element of contracting holonomy g on S. Indeed, since there are no higher depth leaves, g generates the holonomy group of S. This holonomy is produced by a *juncture* $J \subset S$ (see [I, Definition 8.4.2]). This is a compact, properly imbedded 1-manifold $J \subset S$ with the property that the holonomy around a loop σ in S is the power of g given by the homological intersection number $\sigma \cdot J$. In particular, we can choose J so that no component is trivial in $H_1(S, \partial S)$. Since ∂S is connected, it follows that no component of J is a properly imbedded arc in S, and from this it follows that loops on ∂S have trivial holonomy. In particular, $N(e)$ meets ∂M in a sequence of circles parallel to ∂S. Since the depth one leaf is a finite union of a compact core and neighborhoods $N(e)$ of isolated ends [I, Corollary 9.4.7], it follows that $L \cap \partial M$ is a countable union of circles. Since all leaves are either compact or of depth one, the assertion follows. □

This theorem is not true for links. The compact leaf of the foliation of the complement of the (4,1)-torus link, constructed in Section 11.1, has a juncture that is an arc whose boundary components lie in different components of the boundary of the Seifert surface.

Corollary 11.5.4. *If $R \subset S^3$ is a disk decomposable surface and $\partial R = \kappa$ is a nontrivial knot, then longitudinal surgery on κ yields an irreducible three-manifold Q.*

Proof. By Theorem 11.4.1, there is a taut, depth one foliation $\mathcal{F}$ on $E(\kappa)$. By Theorem 11.5.3, $\mathcal{F}|\partial E(\kappa)$ is the foliation by longitudinal circles. The manifold Q is obtained by capping off the circles of $\mathcal{F}|\partial N(\kappa)$ with disks. Hence $\mathcal{F}$ extends to a taut foliation of Q, and so Q either is irreducible or $Q = S^2 \times S^1$ with the product foliation (Corollary 9.1.9). In the latter case, R is a disk and κ is unknotted. □

Definition 11.5.5. A knot κ has property P if nontrivial Dehn surgery along κ always gives a nonsimply connected three-manifold. The knot has property R if longitudinal surgery on κ does not yield $S^2 \times S^1$.

The Poenaru conjecture is that longitudinal surgery on a nontrivial knot never yields a connected sum $S^2 \times S^1 \# N$, where N is a homotopy sphere. It follows that, if longitudinal surgery on κ gives an irreducible three manifold, then κ has property R and cannot be a counterexample to the Poenaru conjecture.

Corollary 11.5.6. *If the nontrivial knot κ has a disk decomposable Seifert surface, then κ has Property* R *and satisfies the Poenaru conjecture.*

11.5.B. Murasugi Sum. The Murasugi sum is a geometric operation that associates to two oriented surfaces R_1 and R_2 in S^3 a new oriented surface $R \subset S^3$. If $\lambda_i = \partial R_i$ is the corresponding oriented link, $i = 1, 2$, then $\lambda = \partial R$ is a sort of generalized connected sum of λ_1 and λ_2.

It will be convenient to identify S^3 with the one-point compactification $\mathbb{R}^3 \cup \{\infty\}$, using coordinates (x, y, z) for the finite points. Since S^3 is a homogeneous space, the point ∞ can be moved to any point of a figure, changing the appearance of the figure radically, but not its topology. The equatorial 2-sphere S will be taken to be the xy-plane, compactified by the additional point ∞. Similarly, the compactifications of $\{z \geq 0\}$ and $\{z \leq 0\}$ are 3-balls denoted by B_1 and B_2, respectively, and called the northern and southern hemispheres. Note that $S = B_1 \cap B_2$.

Definition 11.5.7. The compact, connected, oriented surface $R \subset S^3$ is a Murasugi sum of compact, connected, oriented surfaces R_1 and R_2 in S^3 if the following three conditions are satisfied:

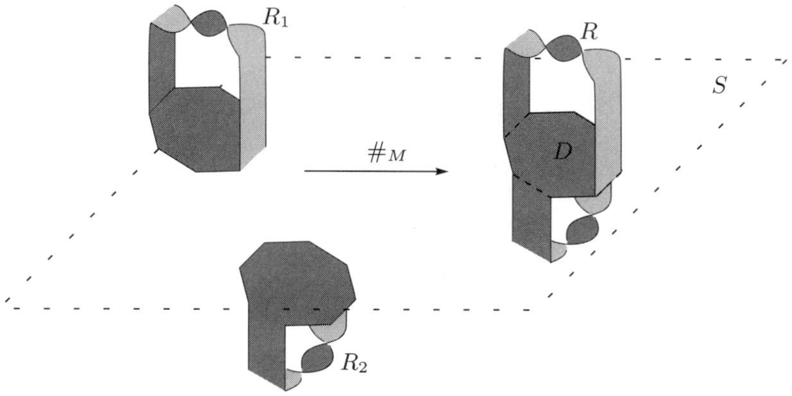

Figure 11.5.1. The Murasugi sum operation

(1) $R = R_1 \cup_D R_2$, where D is a $2n$-gon, oriented compatibly with R_1 and R_2.

(2) The edges of D are numbered consecutively $s_1, s_2, \dots, s_{2n}$ in such a way that

$$\begin{aligned} s_{2k+1} &\subset \partial R, \quad 0 \le k \le n-1, \\ s^\circ_{2k} &\subset R^\circ, \quad 1 \le k \le n. \end{aligned}$$

(3) $R_i \subset B_i$, $i = 1, 2$, and $R_i \cap S = D$, $i = 1, 2$.

In this case, we write $R = R_1 \#_M R_2$. We will also say that the oriented link $\lambda = \partial R$ is the Murasugi sum of the oriented links $\lambda_i = \partial R_i$, $i = 1, 2$, and write $\lambda = \lambda_1 \#_M \lambda_2$.

For the case $n = 4$, this is illustrated in Figure 11.5.1. Observe that, when $n = 2$, $\lambda_1 \#_M \lambda_2$ is just the connected sum operation. When $n = 4$ and s°_2, s°_6 lie in one R°_i, s°_4, s°_8 in the other (as in Figure 11.5.1), Murasugi sum is also called "plumbing".

Consider the disk $E_0 = S \setminus D^\circ$, oriented "oppositely" to D. More precisely, if the orientation of D agrees (respectively, disagrees) with the orientation inherited from S, then the orientation of E_0 disagrees (respectively, agrees) with the orientation inherited from S. Note that E_0 is isotopic, through disks E_t such that $E_t \cap R = \partial E_t \subset D$, $0 \le t \le 1$, to a disk E_1 with boundary a sequence of consecutive arcs (say, $2m$ of them), alternately on opposite sides of D (Figure 11.5.2). By fattening R to $R \times I$, we see that E_1 can be viewed as a properly imbedded decomposing disk in the sutured

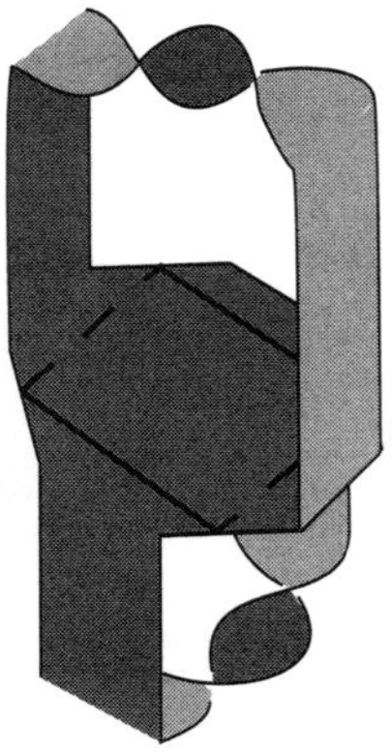

Figure 11.5.2. $\partial E_1 = E_1 \cap R \subset D$ has edges on alternate sides of D

manifold

$$(M(R), \gamma(R)) = (S^3 \setminus (R \times I)^\circ, \partial R \times I),$$

crossing the sutures $2m$ times.

Lemma 11.5.8. *Disk decomposition of* $(M(R), \gamma(R))$ *by the disk* E_1 *gives the disjoint union* $(M(R_1), \gamma(R_1)) \amalg (M(R_2), \gamma(R_2))$.

Proof. First note that, by moving ∞ to a point in $D \times \{0\} \subset R_1 \times I$, the surface $\partial(R_1 \times I)$ can be pictured as in Figure 11.5.3, where the annuli $\partial R_1 \times I$ are seen myopically as linked, simple closed curves. This surface divides S^3 into complementary sutured manifolds, the one containing B_2 being $(R_1 \times I, \partial R_1 \times I)$, the one contained in B_1 being $(M(R_1), \gamma(R_1))$. There is a similar picture of $(M(R_2), \gamma(R_2))$ with the roles of B_1 and B_2 interchanged.

It is clear that the properly imbedded disk E_0 separates M; hence so does the properly imbedded, isotopic disk E_1. One also sees that one component of the decomposed sutured manifold is exactly $(M(R_1), \gamma(R_1))$ as pictured in Figure 11.5.3. Similarly, the other component is $(M(R_2), \gamma(R_2))$. □

Theorem 11.5.9. *Let* $R = R_1 \#_M R_2$, *where* $\lambda = \partial R$ *and* $\lambda_i = \partial R_i$, $i = 1, 2$, *are the associated links. If* $E(\lambda_i)$ *admits a smooth, taut, transversely oriented foliation* $\mathcal{F}_i$ *of depth one, transverse to the boundary and having* R_i *as sole compact leaf,* $i = 1, 2$, *then* $E(\lambda)$ *admits such a foliation* $\mathcal{F}$ *having* R *as sole compact leaf.*

Recall that, if either of the $S^3 \setminus N^\circ(\lambda_i)$ are fibered, they also admit depth one foliations of the required type.

The proof of Theorem 11.5.9 follows from Lemma 11.5.8 via Case 4 in the proof of Theorem 11.4.2 (see page 385).

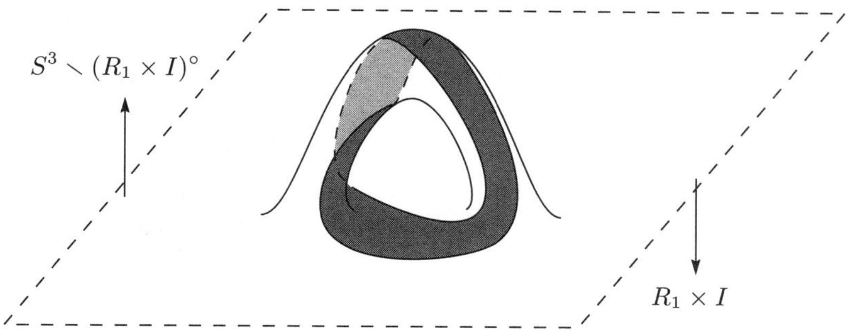

Figure 11.5.3. The surface $\partial(R_1 \times I)$ with sutures

11.5.C. Seifert's Algorithm. For use in the following subsection, we recall Seifert's algorithm for constructing an oriented spanning surface R for an oriented knot or link, illustrating it by two examples. We always assume that links are nonsplit, this being equivalent to the connectivity of R.

Example 11.5.10. Consider the oriented link in Figure 11.5.4. At each crossing point, alter the projection to eliminate the crossing, joining segments so that the arrows give the same local orientation. This yields a family of disjoint, oriented circles as in Figure 11.5.4, called the *Seifert circles*. In addition, as indicated in Figure 11.5.4, the Seifert circles may bound a disjoint family of 2-cells in S^2. Those bounded by Seifert circles with counterclockwise orientation are transversely oriented toward the viewer and are given a darker shading. The others are oriented away from the viewer. One now restores the crossings, connecting the 2-cells by appropriately twisted, oriented bands as in Figure 11.5.5. This yields an oriented spanning surface for the link. It should be remarked that sometimes the 2-cell bounded by a Seifert circle should be chosen to contain ∞ in order to obtain disjoint 2-cells in S^2. (Consider, for example, the usual projection of the trefoil knot and apply Seifert's algorithm.)

Example 11.5.11. For some links λ, it is impossible to choose the 2-cells in Seifert's algorithm to lie in S^2 and also be disjoint. There may be an essential nesting of the Seifert circles, as indicated in Figure 11.5.6. In this case, choose a Seifert circle C such that both components of $S^2 \smallsetminus C$ contain Seifert circles. Those on one side of C should be slightly displaced into the northern hemisphere B_1, and those on the other side of C into the southern hemisphere B_2, all in such a way that the Seifert circles bound disjoint 2-cells in S^3. The 2-cell D bounded by C should lie in S^2. Once again, appropriately twisted bands are added to finish the construction of

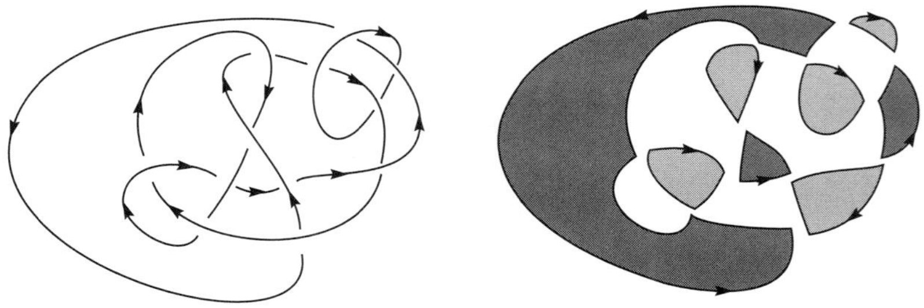

Figure 11.5.4. The Seifert circles associated to an oriented link

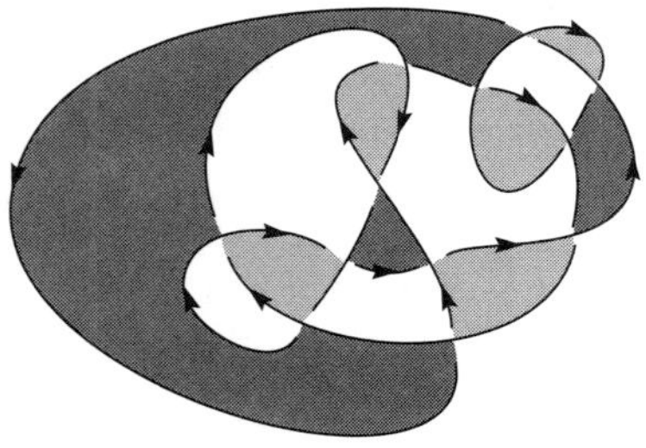

Figure 11.5.5. The spanning surface produced by Seifert's algorithm

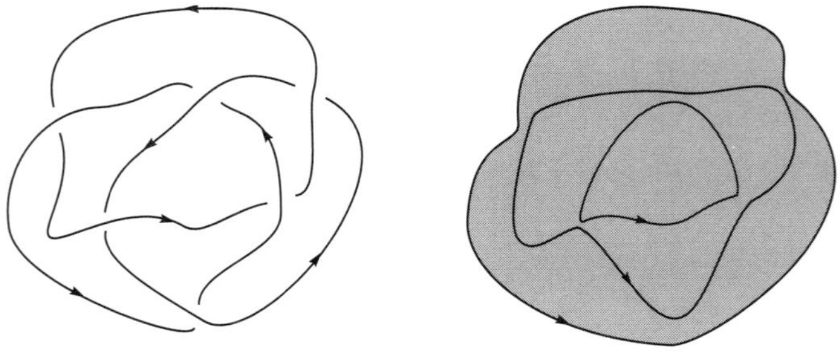

Figure 11.5.6. The Seifert circles are nested

the surface R spanning a link $\lambda' = \partial R$, isotopic to λ. It is evident that $R = R_1 \#_M R_2$ is a Murasugi sum, $R_i \subset B_i$, $i = 1, 2$. For the link in Figure 11.5.6, this is illustrated in Figure 11.5.7.

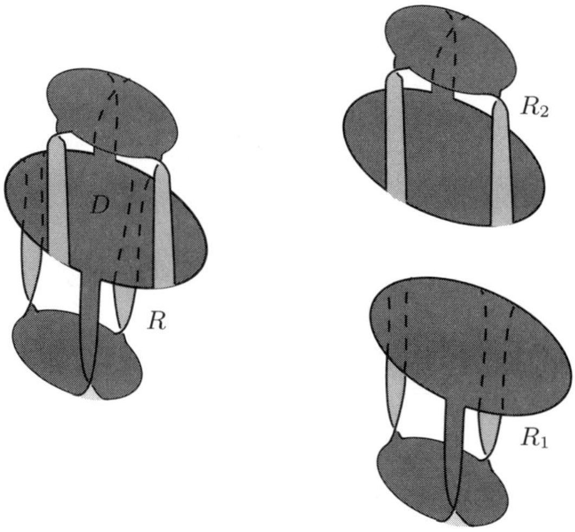

Figure 11.5.7. Seifert's algorithm produces a Murasugi sum when the circles are nested

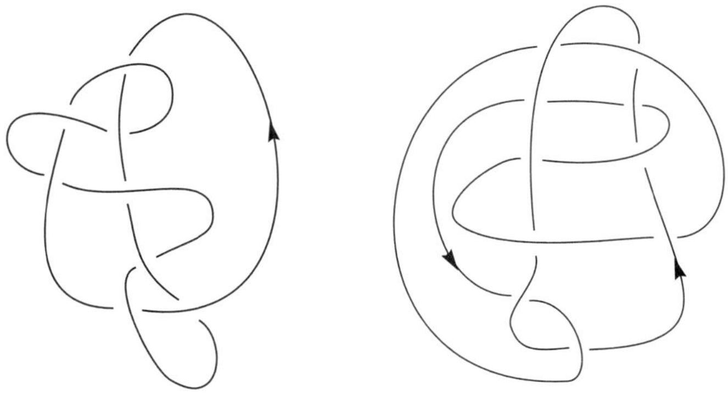

(a) (b)

Figure 11.5.8. (a) An alternating knot and (b) a nonalternating link projection

11.5.D. Alternating Knots and Links. Recall that an oriented knot or link projection (onto $\mathbb{R}^2 \cup \{\infty\} = S^2$) is said to be alternating if upper crossings and lower crossings alternate while running along the knot or link. A knot or link is called alternating if it possesses an alternating projection; otherwise, it is nonalternating. Examples are given in Figure 11.5.8.

Consider the case of an oriented, nonsplit link projection for which the Seifert circles are not nested. Then the surface R produced by Seifert's algorithm gives rise to a finite, connected graph $G \subset S^2$ having one vertex

Figure 11.5.9. The two senses of twist

for each 2-cell bounded by a Seifert circle and one edge connecting two vertices for each twisted band connecting the corresponding 2-cells. There are exactly two types of twisted bands to glue on, those twisted in the positive or negative sense as pictured in Figure 11.5.9. For a nonsplit, alternating link, observe that either all of the bands are twisted in the positive sense, or all in the negative sense. Thus, in this case, the Seifert surface R can be recovered from the graph G by knowing this one added piece of information. Observe that, by first untwisting trivial twists in R before constructing G, we can assume that no interior point of an edge separates G.

Exercise 11.5.12. Prove that $\chi(R) = \chi(G)$.

Exercise 11.5.13. Assume that R, as above, spans an oriented, nonsplit, alternating link λ and is not a disk. Let $D \subset S^2$ be a disk such that $G \cap D = \partial D$. Suppose D intersects G in n vertices $v_1, \ldots, v_n$ and n edges $v_i v_{i+1}$, $i = 1, \ldots, n \pmod n$.

(1) Show that D represents a properly imbedded disk E in the sutured manifold $(M(R), \gamma(R))$ that meets $s(\gamma)$ in n points.

(2) Show that disk decomposition along E yields a new sutured manifold $(M(R_1), \gamma(R_1))$, where R_1 is a connected surface spanning an oriented, nonsplit alternating link $\lambda_1 = \partial R_1$ with unnested Seifert circles.

(3) Show that the graph G_1 of R_1 is obtained as a quotient of G as follows:
   (a) Identify the vertices v_i, $1 \le i \le n$, if i is even (respectively, odd) to a single vertex v.
   (b) Identify each edge $v_{2k}v_{2k+1}$ (respectively, $v_{2k+1}v_{2k+2}$) with the edge $v_{2k-1}v_{2k}$ (respectively, $v_{2k}v_{2k+1}$), where the index is written mod n.

The statements in parenthesis indicate the possibly different graphs we can obtain corresponding to the two different orientations of E (see Figures 11.5.10 and 11.5.11, where the corresponding Seifert surfaces are labeled R_1 and R_1', the graphs G_1 and G_1'). Finally,

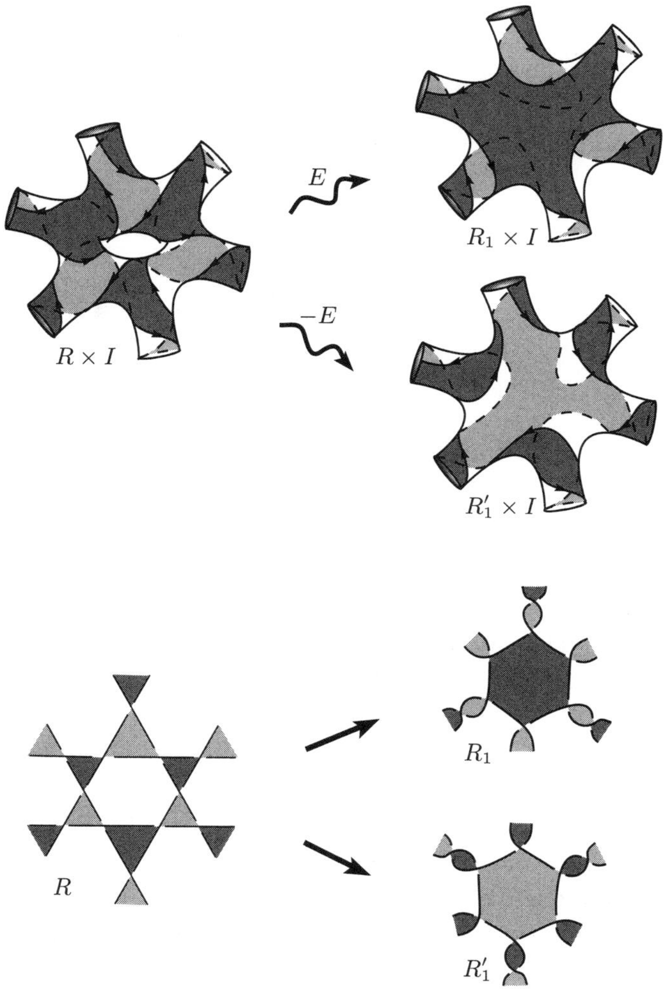

Figure 11.5.10. Respective disk decompositions by E and $-E$ and the resulting Seifert surfaces

(4) using the previous exercise, conclude that $\chi(R_1) = \chi(R) + 1$ and that, when $\chi(R) = 0$,

$$(M(R_1), \gamma(R_1)) = (D^2 \times I, \partial D^2 \times I),$$

the sutured ball.

Theorem 11.5.14. *Let λ be a nonsplit alternating link. If R is a Seifert surface obtained by applying Seifert's algorithm to an alternating projection of λ, then R is disk decomposable.*

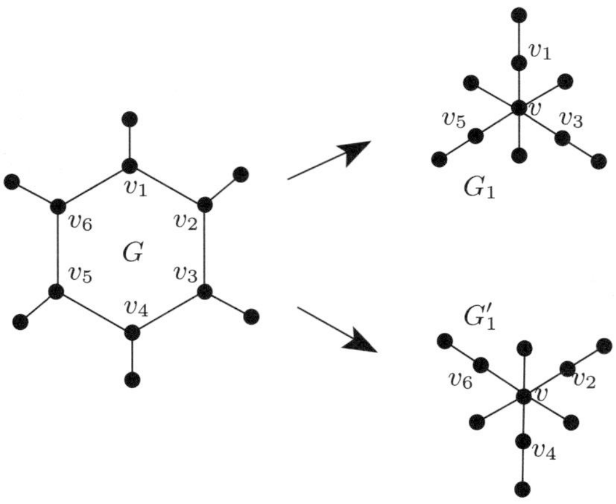

Figure 11.5.11. The quotient graphs of G corresponding to the disk decompositions in Figure 11.5.10

Proof. The proof will be an induction on $|\chi(R)|$. We start the induction with the case $|\chi(R)| = 0$. First, note that the Seifert circles are unnested. Indeed, if $R = R_1 \#_M R_2$, one easily computes

$$(*) \qquad \chi(R) = \chi(R_1) + \chi(R_2) - 1,$$

implying that $\chi(R) < 0$. The disk decomposition of Exercise 11.5.13 then yields the sutured ball $(D_2 \times I, \partial D^2 \times I)$ and we are done.

For the inductive step, consider first the case in which the Seifert circles are unnested. Exercise 11.5.13 then gives a disk decomposition yielding an oriented, nonsplit, alternating link $\lambda_1 = \partial R_1$ such that the Seifert circles are again unnested and $|\chi(R_1)| < |\chi(R)|$. By the inductive hypothesis, we are done.

In case the Seifert circles are nested, Example 11.5.11 shows that

$$R = R_1 \#_M R_2,$$

where R_i also results from an application of Seifert's algorithm to an alternating link projection, $i = 1, 2$. Equation $(*)$ implies that

$$|\chi(R_i)| < |\chi(R)|, \quad i = 1, 2.$$

By the disk decomposition of Lemma 11.5.8 and the inductive hypothesis, we are done. □

By part (b) of Theorem 11.4.1, we get a proof of a result due independently to K. Murasugi [**138**] and R. Crowell [**41**].

Corollary 11.5.15. *Application of Seifert's algorithm to an alternating link produces a minimal genus spanning surface for that link.*

11.6. Higher Depth

Knot and link complements and, more generally, certain sutured manifolds have taut, finite depth foliations. Gabai's constructions use properly imbedded surfaces that are not necessarily disks to decompose these sutured manifolds and construct, inductively, the foliations. These constructions sacrifice smoothness, in general, because their holonomy may violate the Kopell lemma [I, Lemma 8.1.1]. The details of this construction will be found in [**70**] and are too complex to treat here. We will content ourselves with a brief sketch of the facts.

Theorem 11.6.1 (Gabai). *Let M be a compact, connected, irreducible and oriented 3-manifold such that ∂M, if nonempty, is a union of tori. Let $S \subset M$ be any connected, properly imbedded, norm-minimizing surface representing a nontrivial class in $H_2(M, \partial M)$. Then there is a taut, transversely oriented, finite depth foliation $\mathcal{F}$, transverse to ∂M, tangent to a C^0 2-plane field, having S as a compact leaf and such that $\mathcal{F}|\partial M$ has no 2-dimensional Reeb components.*

Recall that, for foliations of codimension 1, the condition of tangency to a C^0 plane field is equivalent to smoothness of class $C^{\infty,0+}$ [I, Corollary 5.1.5].

The idea is to view M as a sutured manifold (M, γ) with $T(\gamma) = \partial M$, then to cut this sutured manifold apart along S, obtaining a new sutured manifold (M_1, γ_1). It is then proven that either this is a product sutured manifold, or there exists another decomposing surface $S_1 \subset M_1$, related to the sutured manifold structure in a way analogous to that in which a disk of a disk decomposition is related to the sutured structure. Cutting apart along S_1 and following rules entirely analogous to those for disk decompositions, we produce a new sutured manifold (M_2, γ_2). It is proven that this procedure can be iterated in finitely many steps to produce a disjoint union of product sutured manifolds. The proof of this delicate theorem follows the general lines of Haken's decomposition theorem [**107**, IV.12]. It is shown that, at each step, the decomposition can be chosen to reduce a certain invariant, called the *complexity* of the sutured manifold. This sequence of decompositions is therefore finite and is called a "sutured manifold hierarchy".

It is shown that the decomposing surfaces in the above hierarchy can be chosen to have very nice properties. It is evident that they can be chosen to be connected. Much less evident is the fact that, at the ith stage of the hierarchy, the decomposing surface (call it S_i) of (M_i, γ_i) can be chosen so that, whenever it meets a component R_0 of $R(\gamma_i)$, it does so either in a family of parallel, properly imbedded arcs or in a family of parallel, essential circles. One assumes inductively that, after decomposition along S_i, the resulting sutured manifold (M_{i+1}, γ_{i+1}) has the desired sort of foliation, transverse to $A(\gamma_{i+1}) \cup T(\gamma_{i+1})$ and tangent to $R(\gamma_{i+1})$, with convex corners. One can then reassemble the original manifold M_i, no longer as a sutured manifold, but so that the foliation has both concave and convex corners in R_0, as in Example 11.3.8. Much as in that example, one then constructs a collar for R_0 that is a sort of "spiral ramp", also foliated to have convex and and concave corners that fit into the complementary ones in R_0. One completes the foliation by introducing pieces of leaves spiraling towards (a new) R_0. One carries this out for each component of $R(\gamma_i)$ that is met by the decomposing surface S_i. It is at this stage that, in the case that S_i meets a component of R in a family of circles, one may produce a violation of the Kopell lemma.

We mention two of the striking corollaries of Theorem 11.6.1.

Corollary 11.6.2. *Let λ be an oriented, nonsplit link in S^3 and let S be a Seifert surface for λ. Then S has minimal genus among all Seifert surfaces if and only if it is a leaf of a taut, finite depth $C^{\infty,0+}$ foliation $\mathfrak{F}$ of the link complement, $\mathfrak{F}$ being transverse to $\partial N(\lambda)$, such that $\mathfrak{F}|\partial N(\lambda)$ has no Reeb components.*

Corollary 11.6.3. *A nontrivial link in S^3 is nonsplit if and only if it is the union of cores of the Reeb components of some foliation of S^3 of class $C^{\infty,0+}$.*

Remark. There is no apparent obstruction to making Gabai's construction smooth of class C^1. Indeed, D. Pixton [**150**] has shown that the Kopell lemma fails for C^1 diffeomorphisms.

Appendix A

C^*-Algebras

This appendix collects some basic concepts from the theory of C^*-algebras that have been used in Chapter 1. For the sake of continuity in the exposition, some proofs have been included, and others have been outlined in exercises.

Several references have been used in composing this appendix, especially J. Dixmier [**47**], A. A. Kirillov [**114**], M. Takesaki [**168**] and M. A. Naimark [**139**]. Two other references that are very useful introductions to C^*-algebras are P. A. Fillmore [**65**] and K. R. Davidson [**43**]. As their titles indicate, they are devoted to examples.

A.1. Bounded Operators

There are numerous references for this section, for instance K. Yosida [**191**] and G. K. Pedersen [**146**], among others.

Definition A.1.1. A Banach space is a complex vector space endowed with a norm so that, as a metric space, it is complete.

The basic example is the space of continuous functions $f : X \to \mathbb{C}$ on a compact space X, endowed with the *supremum norm*

$$\|f\| = \sup_{x\in X}|f(x)|.$$

Definition A.1.2. Let X, Y be Banach spaces. A linear map $T : X \to Y$ is a *bounded operator* if there exists a constant C such that $\|T(x)\| \leq C\|x\|$, for all $x \in Y$. A bounded linear functional on X is a bounded operator $X \to \mathbb{C}$.

It is well known that boundedness is equivalent to continuity.

The space $\mathcal{B}(X, Y)$ of bounded linear operators $T : X \to Y$ is a normed linear space with the *operator norm*

$$\|T\| = \sup_{x \neq 0 \in X} \frac{\|Tx\|}{\|x\|}.$$

Besides the operator topology, there are several other important topologies on $\mathcal{B}(X, Y)$. A sequence T_n *converges strongly* to T if $T_n x \to Tx$ in Y for every $x \in X$, and it *converges weakly* if $\lim_n g(T_n x) = g(Tx)$, for every bounded linear functional g on Y and every $x \in X$. These topologies are all the same in the finite dimensional case.

It is apparent that convergence in norm implies strong convergence, which in turn implies weak convergence. These implications are, in fact, strict, as the following exercise illustrates for weak and strong convergence.

Exercise A.1.3. Let H be the space of absolutely summable sequences $\xi = (\xi_n)_{n=1}^{\infty}$ of complex numbers, with norm $\|\xi\|$ defined by

$$\|\xi\|^2 = \sum_{n=1}^{\infty} |\xi_n|^2.$$

The shift operator $S : H \to H$ is defined by setting

$$S(\xi_1, \xi_2, \dots) = (0, \xi_1, \xi_2, \dots)$$

Show that $S^n \to 0$ weakly but not strongly.

The following result, regarding bounded linear operators, is used in Section 1.5.

Theorem A.1.4. *Let $\{T_\alpha\}$ be a net of bounded linear operators in $\mathcal{B}(X, Y)$. Suppose that, for each $x \in X$, the net $\{T_\alpha x\}$ is bounded in Y and that it is convergent for a dense set of elements $x \in X$. Then there exists a bounded linear operator $T : X \to Y$ such that $T_\alpha x \to Tx$ for every $x \in X$.*

This is one of the category results in Banach space theory. The proof is a combination of the results in [**146**, Section 2.2].

A.2. Measures on Hausdorff Spaces

This section states the main definitions and results from integration theory that have been used in the main part of the book. Standard references are N. Bourbaki [**12**], G. B. Folland [**67**], G. K. Pedersen [**146**], and W. Rudin [**161**].

Let X be a locally compact Hausdorff space, and let $C_c(X)$ denote the space of continuous, compactly supported functions on X. The natural topology on $C_c(X)$ is the *inductive limit topology*, defined as follows. A net $\{f_\alpha\}$ in $C_c(X)$ converges to f if there is a compact set $K \subset X$ containing the supports of all f_α and f, and such that f_α converges uniformly to f on K.

A linear functional I on $C_c(X)$ is positive if $I(f) \geq 0$ for every non-negative function $f \in C_c(X)$. The following basic result was explained in Section 1.6.5.

Proposition A.2.1. *A positive linear functional I on $C_c(X)$ is continuous; that is, for every compact set $K \subset X$ there is a constant C_K such that*

$$|I(f)| \leq C_K \|f\|,$$

for every f with compact support in K.

If μ is a positive Borel measure on X (*i.e.*, a measure on the Borel σ-algebra of X) that is finite on compact sets, then every compactly supported continuous function on X is μ-integrable, and the integral $\int f \cdot \mu$ defines a positive linear functional on $C_c(X)$. Conversely, the Riesz representation theorem, to be stated shortly, associates a Borel measure to each positive linear functional, and this measure will be unique if a regularity condition is required.

Definition A.2.2. A *Radon measure* on a Hausdorff space X is a positive measure μ defined on all Borel subsets of X, and that satisfies

(1) $\mu(K) < \infty$ for every compact subset K of X, and
(2) $\mu(B) = \sup\{\mu(K) \mid K \subset B, K \text{ compact}\}$, for every Borel subset B of X.

Condition (2) above is called *inner regularity.* A related condition, called outer regularity, is $\mu(B) = \inf\{\mu(U) \mid U \supset B, U \text{ open}\}$, for every Borel set B.

The following facts have been used in the main text. Proofs can be found in the references cited at the beginning of this section.

Exercise A.2.3. Let X be a locally compact Hausdorff space. Let μ be a σ-finite Radon measure on X and let R be a Borel subset of X. Show that the measure μ_R defined by

$$\mu_R(E) = \mu(E \cap R)$$

is a Radon measure on X.

Exercise A.2.4. Let $f : X \to Y$ be a continuous mapping between locally compact Hausdorff spaces and let μ be a finite Radon measure on X. Show that the push-forward measure $f_*\mu$ on Y defined by

$$f_*\mu(B) = \mu(f^{-1}B)$$

is a Radon measure on Y.

Exercise A.2.5. Let X be locally compact Hausdorff and second countable. Show that every Borel measure on X that is finite on compact sets is a Radon measure.

The following theorem, the Riesz representation theorem, is one of the fundamental results of the theory.

Theorem A.2.6. *Let X be a locally compact Hausdorff space. Let I be a positive linear functional on $C_c(X)$. Then there is a unique Radon measure μ on X such that*

$$I(f) = \int_X f(x) \cdot \mu(x),$$

for every $f \in C_c(X)$.

A signed Radon measure is a Borel measure, the positive and negative parts of which are Radon measures. A complex Radon measure is a Borel measure, the real and imaginary parts of which are signed Radon measures.

The space $C_0(X)$ of continuous functions on X that vanish at infinity is a Banach space with the supremum norm. The following theorem describes its dual.

Theorem A.2.7. *Let I be a continuous linear functional on $C_0(X)$. Then there is a complex Radon measure μ on X such that*

$$I(f) = \int_X f(x) \cdot \mu(x),$$

for every $f \in C_0(X)$.

Some results regarding measures on metric spaces are collected next.

A polish space is a topological space that is homeomorphic to a separable complete metric space.

Theorem A.2.8. *Let μ be a finite measure on a polish space. Then μ is a Radon measure.*

An important result about measures in metric spaces is the disintegration theorem, which is now reviewed.

Theorem A.2.9. *Let $f : X \to Y$ be a Borel map between polish spaces, let μ be a probability measure on X, and let $\nu = f_*\mu$. Then there exists a Borel mapping $y \mapsto \mu_y$ from Y into the space of probability measures on X such that*

(1) *$\mu_y(f^{-1}\{y\}) = 1$, for ν-almost all $y \in Y$, and*

(2) *for every bounded Borel function* φ *on* X,

$$\int \varphi(x) \cdot \mu(x) = \int \left(\int \varphi(x) \cdot \mu_y(x) \right) \nu(y).$$

Furthermore, if $y \mapsto {\mu'}_y$ *is another mapping with these properties, then* $\mu_y = {\mu'}_y$ *for* ν*-almost all* $y \in Y$.

A.3. Hilbert Spaces

This section summarizes the basic terminology pertaining to Hilbert spaces.

Definition A.3.1. A *Hilbert space* is a complex vector space H with an inner product $\langle \cdot | \cdot \rangle$, linear in the first variable and conjugate linear in the second, such that H is complete for the norm $\|\cdot\|$ defined by $\langle \cdot | \cdot \rangle$.

Thus, if I is an index set, the set of elements (ξ_i) of the product $\mathbb{C}^I$ such that $\sum_i |\xi_i|^2 < \infty$ is a Hilbert space with inner product

$$\langle (\xi_i) | (\zeta_i) \rangle = \sum_i \xi_i \overline{\zeta_i}.$$

Another familiar example is the space of (classes of) square integrable functions on a measure space (X, μ), with inner product

$$\langle f_1 | f_2 \rangle = \int_X f_1(x) \overline{f_2(x)} \cdot \mu(x).$$

Definition A.3.2. A linear map $T : H_1 \to H_2$ between Hilbert spaces that satisfies

$$\|T(\xi)\| \leq C \|\xi\|,$$

for some constant C and for all $\xi \in H_1$ is called a *bounded operator.*

Definition A.3.3. If the linear operator $T : H_1 \to H_2$ preserves inner products, then it is called an *isometry.* If furthermore T is surjective, then it is called a *unitary operator* or an isomorphism.

Definition A.3.4. If H is a Hilbert space with inner product $\langle \cdot | \cdot \rangle$, a collection of elements $\{\xi_i\}$ such that

(1) $\langle \xi_i | \xi_i \rangle = 1$ and $\langle \xi_i | \xi_j \rangle = 0$, if $i \neq j$,

(2) the linear span of $\{\xi_i\}$ is dense in H,

is called an *orthonormal system.* If $\{\xi_i\}$ satisfies property (2), then it is said to be *total* in H.

The dimension of a Hilbert space is the cardinality of an orthonormal system. It is well known that two given Hilbert spaces are isomorphic if and only if they have the same dimension. A Hilbert space is *separable* if it admits a countable orthonormal system.

Let H_1, H_2 be two Hilbert spaces. The algebraic tensor product $H_1 \otimes_{\mathbb{C}} H_2$ is a vector space with inner product given by

$$\langle \xi_1 \otimes \zeta_1 | \xi_2 \otimes \zeta_2 \rangle = \langle \xi_1 | \xi_2 \rangle \langle \zeta_1 | \zeta_2 \rangle.$$

Definition A.3.5. The completion of $H_1 \otimes_{\mathbb{C}} H_2$ with respect to the norm defined by this inner product is a Hilbert space, called the *tensor product* of the Hilbert spaces H_1 and H_2 and denoted by $H_1 \otimes H_2$.

Exercise A.3.6. If $T_i : H_i \to H_i$, $(i = 1, 2)$, are bounded operators, show that there is bounded operator $S = S_1 \otimes S_2$ on $H_1 \otimes H_2$ that satisfies $S(\xi_1 \otimes \xi_2) = S_1(\xi_1) \otimes S_2(\xi_2)$. Furthermore, $\|S\| = \|S_1\| \, \|S_2\|$.

Definition A.3.7. Let $\{H_i\}_{i \in I}$ be a family of Hilbert spaces. The *Hilbert sum* $H = \bigoplus_{i \in I} H_i$ is the Hilbert space obtained by completing the algebraic direct sum $\sum_{i \in I} H_i$ for the norm given by the inner product $\langle (\xi_i) | (\zeta_i) \rangle = \sum_{i \in I} \langle \xi_i | \zeta_i \rangle$.

The concept of Hilbert sum can be further generalized. Let X be a locally compact Hausdorff space.

Definition A.3.8. A field of Hilbert spaces over X consists of a family of Hilbert spaces $\{H_x\}_{x \in X}$, indexed by the points of X. A vector field ξ is an element of the vector space $\prod_{x \in X} H_x$.

Definition A.3.9. The field $\{H_x\}_{x \in X}$ is a *measurable field* of Hilbert spaces if there is given a vector subspace F of the product $\prod_{x \in X} H_x$ satisfying the following conditions:

(1) For every $\xi \in F$, the map $x \in X \mapsto \|\xi_x\|$ is Borel measurable.

(2) If $\zeta = \{\zeta_x\}$ is a vector field such that, for every $\xi \in F$, the map $x \in X \mapsto \langle \xi_x | \zeta_x \rangle$ is Borel measurable, then $\zeta \in F$.

(3) There exists a sequence $\xi_n \in F$ such that, for each $x \in X$, the family $\xi_{n,x}$ is total in H_x.

Elements of F are called measurable vector fields. The sequence $\{\xi_n\}$ in (3) is called a *fundamental family of measurable vector fields.*

Property (3) implies that the Hilbert spaces H_x are separable.

Definition A.3.10. Let $\{H_x\}$ be a measurable field of Hilbert spaces on X, and let μ be a Radon measure on X. A measurable field $\xi = \{\xi_x\}$ is square-integrable if $\xi \in F$ and

$$\int_X \|\xi_x\|^2 \cdot \mu(x) < \infty.$$

The square-integrable sections form a subspace of F endowed with the inner product

$$\langle \xi | \zeta \rangle = \int_X \langle \xi_x | \zeta_x \rangle \cdot \mu(x),$$

and they form a Hilbert space called the *Hilbert integral*, or direct integral, of the field $\{H_x\}$, and denoted by

$$\int^{\oplus} H_x \cdot \mu(x).$$

Example A.3.11. The concept of a Hilbert integral generalizes that of a Hilbert sum of Hilbert spaces. For example, given H_i indexed by a countable set I, give I the measurable structure in which every subset is measurable and take μ to be the counting measure on I. (It is evident that the concept of Hilbert integral can be further generalized so as to include all Hilbert sums.)

In the above example, every Hilbert space H_i is a subspace of the direct sum, but that is not the general case.

Example A.3.12. Let X be a locally compact, Hausdorff, second countable space, μ a Radon measure on X. If all the Hilbert spaces $H_x = \mathbb{C}$ are one-dimensional and the measurable vector fields are the measurable functions $f : X \to \mathbb{C}$, then the Hilbert integral is isomorphic to $L^2(X, \mu)$.

Exercise A.3.13. Let X be as in the above example. Let H be a fixed Hilbert space and take the constant field $H_x = H$ for every $x \in X$. Take as fundamental family of vector fields all the mappings $f : X \to H$ such that $x \in X \mapsto \langle f(x)\xi | \zeta \rangle$ are measurable for every $\xi, \zeta \in H$. Show that the Hilbert integral is isomorphic to the Hilbert tensor product $L^2(X, \mu) \otimes H$.

Let $\{H_{1,x}\}$, $\{H_{2,x}\}$ be fields of Hilbert spaces over X, and let

$$H_1 = \int^{\oplus} H_{1,x} \cdot \mu(x),$$

$$H_2 = \int^{\oplus} H_{2x} \cdot \mu(x)$$

be their respective Hilbert integrals. A field of operators consists of giving, for each $x \in X$, an operator $T_x : H_{1,x} \to H_{2,x}$. Such a field is called measurable if for every measurable field ξ, the field is also measurable. This is equivalent to requiring that the functions $x \in X \mapsto \langle T_x \xi_{i,x} | \zeta_{j,x} \rangle$ are measurable, where ξ_1 and ζ_j are fundamental sequences of vector fields for H_1 and H_2, respectively.

If $\{T_x\}$ is a measurable field, the mapping $x \in X \mapsto \|T_x\|$ is measurable. Suppose that it is also essentially bounded, and let λ be its essential upper bound. Then T defines a bounded linear operator $T : H_1 \to H_2$ with norm $\|T\| = \lambda$. Such operators are called *decomposable*.

Example A.3.14. Let $f \in L^\infty(X, \mu)$ be an essentially bounded function. Then f induces a decomposable bounded operator $M_f : H_1 \to H_2$ given by multiplication: $M_f(\xi_x) = f(x)\xi_x$. Such an operator is called *diagonalizable*.

Exercise A.3.15. Show that the operator norm $\|M_f\| = \|f\|_\infty$.

The question arises as to what conditions are necessary, and sufficient, for a bounded linear operator $T : H_1 \to H_2$ to be decomposable.

Theorem A.3.16. *A bounded linear operator $T : H_1 \to H_2$ is decomposable if and only if it commutes with all the diagonalizable operators, that is, if*

$$TM_f = M_fT$$

for every $f \in L^\infty(X, \mu)$.

The theory of fields of Hilbert spaces is deeper than this brief survey can convey. A complete account can be found in [**48**].

A.4. Topological Spaces and Algebras

An *algebra* is a vector space over the complex numbers endowed with a multiplication that is associative, and that is distributive with respect to the vector space structure. If the underlying vector space is a topological vector space, then it is typically assumed (or evident from the context) that the multiplication is separately continuous in each of the two variables involved.

A left (respectively, right) ideal in an algebra A is a subalgebra $I \subset A$ such that $aI \subset I$ (respectively, $Ia \subset I$) for all $a \in A$. A left and right ideal is called a *two-sided ideal.*

Example A.4.1. Let X be a locally compact Hausdorff space. The space $C_c(X)$ of compactly supported continuous functions $f : X \to \mathbb{C}$ is an algebra with the obvious pointwise algebraic operations.

Endowed with the inductive limit topology, the space $C_c(X)$ is a topological algebra.

If $x_0 \in X$, then $\{f \in C_c(X) \mid f(x_0) = 0\}$ is a two-sided ideal in $C_c(X)$.

Definition A.4.2. An algebra A whose underlying vector space is a Banach space and whose multiplication and norm satisfy

$$\|xy\| \leq \|x\|\|y\| \qquad \text{for all } x, y \in A$$

is called a *Banach algebra.*

An element $u \in A$ such that $ux = xu = x$, for all $x \in A$, is called a *unit*. A Banach algebra A need not have a unit element. If it does contain a unit $\mathbf{1}$, then it is assumed that $\|\mathbf{1}\| = 1$. There is an artifice that adds a unit to a given unitless Banach algebra. Indeed, let A be such, and let $A_1 = A \oplus \mathbb{C}$. Addition in A_1 is coordinatewise, multiplication is defined via the distributive law, and the norm is given by $\|(a, \lambda)\| = \|a\| + |\lambda|$.

Examples of Banach algebras abound. Familiar ones are the matrix algebras $\mathfrak{gl}(n, \mathbb{C})$. A generalization of these examples is the following infinite-dimensional version.

Example A.4.3. Let H be a Hilbert space and let $\mathcal{B}(H)$ be the algebra of all bounded linear operators on H. Then $\mathcal{B}(H)$, endowed with the operator norm

$$\|T\| = \sup_{\xi \neq 0} \frac{\|T\xi\|}{\|\xi\|},$$

is a Banach algebra with unit.

Example A.4.4. Let X be a topological space and let $C(X)$ be the space of bounded continuous functions $f : X \to \mathbb{C}$ endowed with the usual algebraic operations and supremum norm

$$\|f\| = \sup_{x \in X} |f(x)|.$$

Then $C(X)$ is a Banach algebra with unit.

Two fundamental examples of Banach algebras, related to the C^*-algebra of a foliated space, are the following.

Example A.4.5. Let X be a locally compact Hausdorff topological space. The space $C_0(X)$ of continuous functions on X that vanish at infinity, endowed with the supremum norm, is a Banach algebra. It has a unit if and only if X is compact.

Exercise A.4.6. Let G be a locally compact, metrizable topological group, and let μ be a left-invariant Haar measure on G. Let $L^1(G)$ be the space of (equivalence classes of) integrable functions on G. Show that the product

$$f * g(x) = \int_G f(y) g(xy^{-1}) \cdot \mu(y)$$

is well defined and makes $L^1(G)$ into a Banach algebra.

The following example is related to the construction of the diffusion semigroup associated to the Laplace operator on a foliated space.

Exercise A.4.7. Let $C_0(0,\infty)$ be the space of continuous, absolutely integrable functions on $(0,\infty)$. Define the convolution of two such functions f and g to be the function

$$f * g(t) = \int_0^t f(t-s)g(s) \cdot ds.$$

Show that the L^1-norm

$$\|f\| = \int_0^\infty |f(t)| \cdot dt$$

satisfies $\|f * g\| = \|f\|\|g\|$. Its completion is a Banach algebra, denoted by $L^1(0,\infty)$.

One carries out a similar construction by considering the space of continuous functions $f : [0,\infty) \to X$, where X is a Banach space. There is no natural norm in this situation, but there is a natural topology, which is uniform convergence on compact subsets of $[0,\infty)$.

A.5. C^*-Algebras

Definition A.5.1. An *involution* of an algebra A is a conjugate linear map $* : A \to A$ such that $(ab)^* = b^*a^*$, for all $a, b \in A$.

A Banach algebra with involution is called an *involutive Banach algebra.*

Definition A.5.2. A Banach algebra A, together with an involution

$$* : A \to A$$

that satisfies

$$\|a^*a\| = \|a\|^2,$$

is called a *C^*-algebra.*

Definition A.5.3. If A is an algebra with involution, a norm $\|\cdot\|$ on A such that $\|a^*a\| = \|a\|^2$, for all $a \in A$, is called a C^*-norm.

Exercise A.5.4. If A is a C^*-algebra, show that the involution $* : A \to A$ is an isometry.

Most of the examples of Banach algebras previously described admit a natural structure of C^*-algebras.

Example A.5.5. If $A = C_0(X)$ is the Banach algebra of continuous functions on a locally compact Hausdorff space X that vanish at infinity, then the involution $f \mapsto f^*$, given by $f^*(x) = \overline{f(x)}$, makes A into a C^*-algebra.

Example A.5.6. Let H be a Hilbert space, with inner product $\langle\cdot|\cdot\rangle$. The Banach algebra $\mathcal{B}(H)$ of bounded linear endomorphisms of H has a natural involution, assigning to an operator T its adjoint operator T^*, which is given by

$$\langle T^*(\xi)|\zeta\rangle = \langle\xi|T(\zeta)\rangle.$$

This involution makes $\mathcal{B}(H)$ into a C^*-algebra.

The example in the next exercise is relevant to foliated spaces.

Exercise A.5.7. Let X be a locally compact space and let A be a C^*-algebra. Let $C_0(X, A)$ denote the space of continuous maps $f : X \to A$ that vanish at infinity on X.

(1) Show that, with the pointwise operations, $C_0(X, A)$ is an algebra with involution.

(2) Show that

$$\|f\| = \sup_{x\in X}\|f(x)\|$$

makes $C^*(X, A)$ a C^*-algebra. (The only property that requires some consideration is completeness.)

If A is a C^*-algebra without unit, then $A_1 = A \oplus \mathbb{C}$ is an involutive Banach algebra with unit, but its norm needs to be modified for it to be a C^*-algebra. Define the norm of an element $(a, \lambda) \in A_1$ by

$$\|(a, \lambda)\| = \sup_{x\in A, x\neq 0} \frac{\|ax + \lambda x\|}{\|x\|}.$$

It is now an interesting exercise in the definitions to verify that A_1 is a C^*-algebra with unit $(0, 1)$, and that the inclusion $a \in A \mapsto (a, 0)$ is a norm-preserving embedding of C^*-algebras.

In any case, a C^*-algebra always admits an *approximate unit*, that is, a net $\{u_\alpha \in A\}$ (a sequence if A is separable) such that:

(1) $u_\alpha \leq u_\beta$ if $\alpha \leq \beta$,

(2) $\|u_\alpha\| \leq 1$, and

(3) $\lim_\alpha u_\alpha x = \lim_\alpha x u_\alpha = x$, for all $x \in A$.

Example A.5.8. Let X be a locally compact separable space. Let

$$K_1 \subset K_2 \subset \cdots \subset K_n \subset \cdots \subset X$$

be an increasing sequence of compact subsets that exhaust X, each contained in the interior of the next one, and, for each n, let $f_n : X \to [0, 1]$ be a function that is equal to 1 on K_n and equal to 0 outside K_{n+1}. Then $\{f_n\}$ is an approximate unit for the C^*-algebra $C_0(X)$.

Another important concept in C^*-algebras, especially with regard to representation theory, is that of a positive linear functional.

Definition A.5.9. A *positive linear functional* ρ on an algebra with involution A is a linear map $\rho : A \to \mathbb{C}$ such that $\rho(a^*a) \geq 0$ for every $a \in A$.

If A is a $*$-algebra, a positive linear functional ρ is called a *state* if $\|\rho\| = 1$.

The basic properties of positive linear functionals on C^*-algebras are summarized in the following proposition.

Proposition A.5.10. *Let A be a C^*-algebra and let ρ be a positive linear functional on A. Then ρ has the following properties.*

(1) Cauchy-Schwarz inequality: $|\rho(a^*b)|^2 \leq \rho(a^*a)\rho(b^*b)$, *for all* $a, b \in A$.

(2) Boundedness: *ρ is bounded, and if A has a unit $\mathbf{1}$, then* $\rho(\mathbf{1}) = \|\rho\|$.

One of the basic examples of positive linear functionals is the following.

Example A.5.11. Positive linear functionals on $C_0(X)$, where X is a locally compact Hausdorff space, correspond to (finite) positive Radon measures on X, via the Riesz representation theorem.

A.6. Representations of Algebras

Definition A.6.1. A *representation* of an algebra A on a Hilbert space H is a homomorphism $\pi : A \to \mathcal{B}(H)$. If A has an involution and π satisfies $\pi(a^*) = \pi(a)^*$, then π is called an *involutive representation.*

Representations of C^*-algebras are always assumed to be involutive.

If the algebra A has a topology, then some continuity requirements may be imposed on π. However, there is no need for such requirements in the case of representations of C^*-algebras.

Proposition A.6.2. *Let A be an involutive Banach algebra, let B be a C^*-algebra, and let $\pi : A \to B$ be an involutive homomorphism of algebras. Then $\|\pi(a)\| \leq \|a\|$ for all $a \in A$.*

An algebra always has a representation, called the zero representation, which sends each element into the zero operator.

Definition A.6.3. A representation π of an algebra A on a Hilbert space H is called *nondegenerate* if the collection of vectors

$$\pi(A)H = \{\pi(a)\xi \mid a \in A, \xi \in H\}$$

spans H. The representation π is *faithful* if the homomorphism

$$\pi : A \to \mathcal{B}(H)$$

is injective.

Definition A.6.4. Let π_1, π_2 be two representations of an algebra A on Hilbert spaces H_1 and H_2. They are called unitarily equivalent if there is a conjugating isomorphism $U : H_1 \to H_2$. That is, $\pi_2(a)U = U\pi_1(a)$ for all $a \in A$.

The representation π_1 is contained in the representation π_2 if there is an isometry $T : H_1 \to H_2$ such that $\pi_2(a)T = T\pi_1(a)$ for all $a \in A$.

An illustration of unitary equivalence and containment is as follows. Let $\pi : A \to B(H)$ be a representation. The set $V \subset H$ consisting of those $\xi \in H$ such that $\pi(a)\xi = 0$, for all $a \in A$, is a closed subspace of H, invariant under π. Its orthogonal complement, $V^\perp$, is also invariant. Let π_1 be the representation on the direct sum $V \oplus V^\perp = H$ given by $\pi_1(a)(\xi, \zeta) = (0, \zeta)$. Then π_1 is equivalent to π, and π_1 contains the zero representation and a nondegenerate representation.

The representation π in this example is then equivalent to the direct sum of two representations. The concept of the sum of representations can be generalized.

If π_i is a family of representations of an algebra A on Hilbert spaces H_i, and for each $a \in A$ there is a constant C_a such that $\|\pi_i(a)\| \le C_a$, then there is a representation $\pi = \bigoplus_i \pi_i$ of A on the Hilbert sum $H = \bigoplus_i H_i$, defined coordinatewise in the obvious way, which is called the *direct sum of the representations* π_i.

Exercise A.6.5. Let $\{H_x\}_{x \in X}$ be a field of Hilbert spaces over (X, μ) (as in Section A.3) and let $\{\pi_x\}$ be a family of representations of a C^*-algebra A on H_x. If $\{\pi_x\}$ preserves the measurable fields of $\{H_x\}$, show that there is a representation of A on the Hilbert space $\int^\oplus H_x \cdot \mu(x)$.

This representation is called the *direct integral of the representations* π_x.

Definition A.6.6. Let $\pi : A \to \mathcal{B}(H)$ be a representation of an algebra A. A vector ξ is called a *cyclic vector* if $\pi(A)\xi$ is dense in H. The representation π is said to be a *cyclic representation* if it has a cyclic vector.

It is apparent that a representation that is the sum of nondegenerate representations is nondegenerate, and that a cyclic representation is nondegenerate. These two facts are complemented by the following result.

Proposition A.6.7. *Every involutive representation π of an involutive algebra A is equivalent to a direct sum of cyclic representations and the zero representation.*

Proof. By the discussion prior to the statement, it may be assumed that the Hilbert space H of the representation decomposes as a direct sum of invariant subspaces $H = H_0 \oplus H_1$, and that π is the zero representation on H_0 and nondegenerate on H_1. Assume that $H_1 \neq 0$ and let $\xi_0 \in H_1$ be a nonzero vector. Let V be the closure of the set of all vectors $\pi(a)\xi_0$, $a \in A$. Then V is a closed subspace of H_1, hence a Hilbert space, that is invariant under $\pi(A)$. Its orthogonal complement $V^\perp$ in H_1 is also invariant. Indeed, if $\xi \in V^\perp$ and $a, b \in A$, then

$$\langle \pi(a)\xi | \pi(b)\xi_0 \rangle = \langle \xi | \pi(a^*b)\xi_0 \rangle = 0,$$

so that $\pi(a)\xi$ is orthogonal to V.

Write $\xi_0 = \xi_1 + \xi_2$ with $\xi_1 \in V$ and $\xi_2 \in V^\perp$. Then $\pi(a)\xi_0 = \pi(a)\xi_1 + \pi(a)\xi_2$, and hence $\pi(a)\xi_1 \in V$ because of the invariance of V and $V^\perp$ under π. It follows that $\pi(A)\xi_1$ is dense in V.

A standard maximality argument gives the result. Consider the collection of families of closed invariant subspaces $\{V_i\}$ of H_1 such that $V_i \perp V_j$ for $i \neq j$, and such that the restriction of π to V_i is cyclic. By the first paragraph, this collection is nonempty. It this collection is orderd by inclusion, then it is apparent that every chain has a maximal element, namely the union. By Zorn's lemma, there is then a maximal collection $\{V_i\}_{i \in I}$. Let V be the closure of the subspace generated by V_i. If $\xi \in V^\perp$, then $\pi(a)\xi = 0$ for all $a \in A$, for otherwise the collection $\{V_i\}$ would not be maximal.

Thus $H_1 = \bigoplus_{i \in I} V_i$, and consequently π decomposes into a direct sum of cyclic representations and the zero representation. □

Another important concept of representation theory is that of irreducibility.

Definition A.6.8. A representation π of an algebra A on a Hilbert space H is called *irreducible* if π leaves invariant no proper subspaces. The representation π is *topologically irreducible* if $\pi(A)$ has no proper, *closed*, invariant subspaces.

Theorem A.6.9. *For C^*-algebras the two concepts of irreducible representations are equivalent.*

Exercise A.6.10. Show that an irreducible representation of a C^*-algebra is cyclic, and that every nonzero vector is a cyclic vector.

A fundamental theorem on the representations of C^*-algebras is the following.

Theorem A.6.11. *Let A be a C^*-algebra. Then there is a family of representations π_i such that $\|x\| = \sup_i \|\pi_i(x)\|$ for any $x \in A$.*

Definition A.6.12. A C^*-algebra A is called *simple* if it has no closed two-sided ideals other that $\{0\}$ and A.

If π is a representation of a C^*-algebra A, then its kernel

$$\ker \pi = \{a \in A \mid \pi(a) = 0\}$$

is a closed two-sided ideal of A. Thus every nonzero representation of a simple C^*-algebra is faithful. The following exercise completes this observation.

Exercise A.6.13. Let A be a C^*-algebra and let I be a closed two-sided ideal.

(1) Show that the quotient space A/I admits the structure of a C^*-algebra such that the projection $A \to A/I$ is a homomorphism of C^*-algebras.
(2) Use Theorem A.6.11 to show that A is simple if an only if every nonzero representation of A is faithful.

Definition A.6.14. Let π be a representation of A and S a collection of representations of A. Then π is said to be *weakly contained* in S if the kernel of π contains the intersection of the kernels of the representations in S.

Exercise A.6.15. Let $C_0(X)$ be the C^*-algebra of Example A.5.5. For $x \in X$ and $f \in C_0(X)$, $\pi_x(f) = f(x)$ defines a representation. If $x_n \to x$ in X, show that the reprsentation π_x is weakly contained in the family $\{\pi_{x_n}\}$.

Another elementary application of representation theory is to give a procedure for constructing a C^*-algebra out of a given algebra (with certain properties). This technique is used in the construction of the C^*-algebra of a foliated space.

Exercise A.6.16. Let A be an algebra with involution. Let $\mathcal{R}$ be a set of involutive representations of A such that $\|a\|' = \sup_{\pi \in \mathcal{R}} \|\pi(a)\| < \infty$, for all $a \in A$.

(1) Show the set $I = \{a \in A \mid \pi(a) = 0 \text{ for all } \pi \in \mathcal{R}\}$ is a two-sided ideal in A.
(2) Show that $\|\cdot\|'$ induces a C^*-norm on the quotient A/I.

Let S be a locally compact second countable group with left-invariant Haar measure $\mu = ds$. The inverse map $i(s) = s^{-1}$ transforms the measure μ into a right-invariant Haar measure $i_*\mu = ds^{-1}$. These measures are mutually absolutely continuous, and the Radon-Nikodym derivative $\delta = ds^{-1}/ds$ is a homomorphism $S \to \mathbb{R}_+^*$, called the modular function of ds.

The space $C_c(S)$ of compactly supported functions on S is endowed with a product

$$f_1 * f_2(t) = \int_S f_1(s) f_2(s^{-1}t) \cdot ds$$

and an involution

$$f^*(s) = \overline{f(s^{-1})}$$

making it into an algebra with involution. There are now two possible ways of completing this algebra in order to obtain a C^*-algebra.

The first uses the involutive representation $R : C_c(S) \to \mathcal{B}(L^2(S, ds))$ given by

$$R(f)(\xi)(t) = \int_G f(s)\xi(ts) \cdot ds.$$

Exercise A.6.17. Show that R is indeed a representation of the involutive algebra $C_c(S)$ on $L^2(S, ds)$, that is, R satisfies

(1) $R(f_1 * f_2) = R(f_1)R(f_2)$, and
(2) $R(f^*) = R(f)^*$.

Furthermore, show that R is faithful, continuous for the inductive limit topology on $C_c(S)$, and $\|R(f)\| < \infty$ for all $f \in C_c(S)$.

This representation R is called the (right) *regular representation* of S. It is the integrated form of the group representation of S on $L^2(S, ds)$ induced by the action of S on itself given by right multiplication.

Definition A.6.18. The reduced C^*-algebra $C_r^*(S)$ of S is the completion of $C_c(S)$ with respect to the norm $\|f\| = \|R(f)\|$.

The other way of associating a C^*-algebra to S is similar, but uses all the representations of the involutive algebra $C_c(S)$. That is, the full C^*-algebra $C^*(S)$ of S is the completion of $C_c(S)$ with respect to the norm

$$\|f\| = \sup_{\pi} \|\pi(f)\|,$$

where the supremum is taken over all the involutive representations π of the algebra $C_c(S)$ on a separable Hilbert space H that are continuous for the inductive limit topology on $C_c(S)$ and the weak operator topology on $\mathcal{B}(H)$. (The Hilbert space H is fixed so that the supremum is actually taken over a set of real numbers.)

Theorem A.6.19. *The homomorphism $C^*(S) \to C_r^*(S)$ is an isomorphism if and only if S is amenable.*

For a proof, see [**47**, Chapter 3].

There is a very important method of manufacturing representations out of positive linear functionals on a C^*-algebra.

Example A.6.20. Let π be a representation of A on the Hilbert space H and let $\xi \in H$. Then $\rho(a) = \langle \pi(a)\xi|\xi\rangle$ is a positive linear functional on A.

The following theorem is the converse to this example.

Theorem A.6.21. *Let ρ be a positive linear functional on the C^*-algebra A. Then there are a Hilbert space H_ρ, a vector $\xi_\rho \in H_\rho$ and a representation $\pi_\rho : A \to \mathcal{B}(H_\rho)$ such that*

(1) *ξ_ρ is a cyclic vector for π_ρ, and*
(2) *$\rho(a) = \langle \pi_\rho(a)\xi|a\rangle$, for all $a \in A$.*

The triple $(H_\rho, \xi_\rho, \pi_\rho)$ is called the *Gelfand-Naĭmark-Segal representation* associated to the positive linear functional ρ. The following exercise outlines the proof of this theorem.

Exercise A.6.22. Let A be a C^*-algebra and let ρ be a positive linear functional on A. For simplicity, assume that A has a unit (the general case can be obtained by means of an approximate unit).

(1) Show that the map $(a, b) \in A \times A \to \mathbb{C}$, given by $\tau(a, b) = \rho(b^*a)$, is linear in the first variable and conjugate linear in the second.
(2) Let $N = \{a \in A \mid \rho(b^*a) = 0, \text{ for all } b \in A\}$. Show that N is a left ideal of A.
(3) Show that the map $(a, b) \mapsto \rho(a^*b)$ induces a nondegenerate inner product on the quotient vector space A/N.
(4) Let H_ρ be the completion of A/N with respect to the inner product defined in (3). Show that left multiplication of A on itself induces a representation of A on H_ρ.
(5) Let $\xi_\rho = [\mathbf{1}]$ be the element of H_ρ represented by the unit $\mathbf{1}$ of A. Show that ξ_ρ is a cyclic vector for π_ρ.
(6) Assume that ρ is obtained from a cyclic representation of A on H with cyclic vector ξ. Show that the operator $U : H_\rho \to H$ given by $U([a]) = \pi(a)\xi$ is unitary and conjugates the representations π and π_ρ.

Exercise A.6.23. Let π be a representation of the C^*-algebra A on the Hilbert space H. Let $\xi \in H$ be a unit vector and $\rho(a) = \langle \pi(a)\xi|\xi\rangle$. Show that the Gelfand-Naĭmark-Segal representation of A associated to ρ is contained in π.

A.7. The Algebra of Compact Operators

There is, besides the dimension, a fundamental difference between the finite dimensional algebras $\mathfrak{gl}(n, \mathbb{C})$ and the algebra $\mathcal{B}(H)$ of Example A.4.3. The difference is that, while the former are simple, the latter is not. In order to exhibit an important ideal in $\mathcal{B}(H)$, a definition is required.

Definition A.7.1. An operator $T : H \to H$ is said to be a *compact operator* if it maps bounded subsets of H to relatively compact ones.

The set of compact operators on H is denoted by $\mathcal{K}(H)$, or simply by $\mathcal{K}$. A compact operator $T : H \to H$ is automatically bounded because the image of the unit ball in H is a bounded set.

Proposition A.7.2. *The compact operators on H form a two-sided ideal of $\mathcal{B}(H)$.*

Proof. To show that $\mathcal{K}$ is a linear subspace it suffices to notice that, if B_r is the ball of radius r in H, then T is compact if and only if $T(B_1)$ is relatively compact, and then that

$$(S + T)(B_1) \subset S(B_1) + T(B_1) \quad \text{and} \quad (\lambda T)(B_1) = |\lambda|(T(B_1)).$$

That $\mathcal{K}$ is a two-sided ideal means that, if $S \in \mathcal{B}(H)$ and T is compact, then ST and TS are also compact. But if S is bounded, then $S(B_1) \subset B_{\|S\|}$; hence $TS(B_1)$ is relatively compact. Also, since S is continuous, it preserves compactness; hence $ST(B_1)$ is relatively compact. □

Proposition A.7.3. *If T_n is a sequence of compact operators and $T_n \to T$, then T is compact. That is, $\mathcal{K}$ is (sequentially) closed in $\mathcal{B}(H)$.*

Proof. Let ξ_k be a sequence of vectors in B_1. Since each T_n is compact, a diagonal argument provides a subsequence ξ_{k_i} such that $\lim_i T_n(\xi_{k_i})$ exists for each n. Then the standard $\varepsilon/3$-argument gives

$$\lim_{i,j\to\infty} \|T(\xi_{k_i}) - T(\xi_{k_j})\| \leq 2\|T_n - T\|$$

for each n. Thus $T(\xi_{k_i})$ is a Cauchy sequence, hence convergent. □

The parenthetical word will be removed in a moment, after it is shown that $\mathcal{K}(H)$ is separable if H is separable.

Exercise A.7.4. An operator $T : H \to H$ is of *finite rank* if it has finite dimensional image. Show that operators of finite rank are compact. Find an example of a compact operator that is not of finite rank.

Let H be a Hilbert space with inner product $\langle\cdot|\cdot\rangle$, and let H' be its dual. There is a conjugate linear isomorphism $H \to H'$ that assigns to a vector $\xi \in H$ the linear functional $v_\xi : \zeta \mapsto \langle\zeta|\xi\rangle$. Dirac's "bra-ket" notation is particularly appealing: denote the element $\xi \in H$ by $\langle\xi|$, and for $\zeta \in H$, let $|\zeta\rangle$ denote the linear map $H \to \mathbb{C}$ given by $\langle\xi| \mapsto \langle\xi|\zeta\rangle$. Then the elements $\langle\xi| \otimes |\zeta\rangle$ generate a dense subspace of the Hilbert tensor product $H \otimes H'$.

Define an involution on $H \otimes_{\mathbb{C}} H'$ by $(\langle\xi| \otimes |\zeta\rangle)^* = \langle\zeta| \otimes |\xi\rangle$. The algebraic isomorphism $H \otimes_{\mathbb{C}} H' \to \operatorname{Hom}(H, H)$, given by $\langle\xi| \otimes |\zeta\rangle\langle\eta| = \langle\eta|\zeta\rangle\langle\xi|$, is a homomorphism of algebras that commutes with the involution. Since the operator norm of $\langle\xi|\otimes|\zeta\rangle$ is $\|\xi\|\|\zeta\|$, its image belongs to $\mathcal{B}(H)$. The bounded operators on H lying in the image of this homomorphism are the finite rank operators.

Exercise A.7.5. Show that the algebra of finite rank operators on H is weakly dense in $\mathcal{B}(H)$.

Theorem A.7.6. *The closed subspace of $\mathcal{B}(H)$ generated by the image of $H \otimes H'$ is the algebra $\mathcal{K}$ of compact operators.*

It is useful to have a more palpable description of this algebra. Let (X, μ) be a measure space and let $L^2(X, \mu)$ denote the space of (equivalence classes of) square integrable functions on X. Let $k(x, y)$ be a square integrable function on $X \times X$. Then k defines an operator

$$K : L^2(X, \mu) \to L^2(X, \mu)$$

by

$$Kf(x) = \int_X k(x, y)f(y) \cdot \mu(y).$$

This operator has the property that it takes bounded sets in $L^2(X, \mu)$ to relatively compact sets; that is, it is compact.

Exercise A.7.7. Suppose that X is a compact metric space and μ a Borel measure. Show that, if $k(x, y)$ is a continuous function on $X \times X$, then the operator K described in the previous paragraph is a compact operator on $L^2(X, \mu)$. The proof uses the Ascoli-Arzela theorem.

Let X be a manifold and $\Gamma_c(X, \mathcal{D}^{1/2})$ the space of compactly supported half-densities on X (see Section 1.1), endowed with the inner product

$$\langle\varphi|\phi\rangle = \int_X \varphi \cdot \overline{\phi}.$$

Its completion with respect to the associated norm is the Hilbert space $L^2(X)$ of square integrable half-densities on M.

An element $\varphi \in \Gamma_c(X \times X, \mathcal{D}^{1/2})$ defines an operator on $L^2(X)$ by convolution. There is a canonical linear map

$$\Gamma_c(X, \mathcal{D}^{1/2}) \otimes \Gamma_c(X, \mathcal{D}^{1/2}) \to \Gamma_c(X \times X, \mathcal{D}^{1/2}),$$

uniquly determined by the condition that it sends $\phi \otimes \psi$ to the half-density $\pi_1^*\phi \cdot \pi_2^*\overline{\psi}$. The image of this map is dense in $\Gamma_c(X \times X, \mathcal{D}^{1/2})$, and also in the space of finite rank operators on $L^2(X)$.

Proposition A.7.8. *The space $\Gamma_c(X \times X, \mathcal{D}^{1/2})$ is dense in the space of compact operators on $L^2(X)$.*

The representation theory of the algebra $\mathcal{K} = \mathcal{K}(H)$ of compact operators on a (separable) Hilbert space H is quite simple. In fact, if

$$\pi : \mathcal{K} \to \mathcal{B}(H)$$

is a nontrivial representation, then it is faithful, for its kernel is a two-sided ideal of $\mathcal{K}$, hence trivial.

Theorem A.7.9. *Every representation of $\mathcal{K}$ is equivalent to a direct sum of the trivial representation and copies of the inclusion representation*

$$\mathcal{K} \hookrightarrow \mathcal{B}(H).$$

There are several proofs available. One is essentially that given in Proposition 1.5.2.

A.8. Representations of $C_0(X)$

This section describes the representation theory of the C^*-algebra of continuous functions vanishing at infinity on a locally compact Hausdorff space X. From the material discussed in Section A.2 (especially the Riesz representation theorem), it is apparent that this theory should be an extension of the measure theory on such spaces.

The basic example of a representation of $C_0(X)$ is the following.

Example A.8.1. If μ is a Radon measure on X, then $\pi(f)(\xi) = f\xi$ defines a representation of $C_0(X)$ on the Hilbert space $L^2(X, \mu)$, called the *multiplication representation* induced by μ.

As the next theorem shows, every cyclic representation of $C_c(X)$ is of the form given by this example.

Theorem A.8.2. *Every cyclic representation π of $C_c(X)$ is equivalent to the multiplication representation on $L^2(X, \mu)$ induced by a Radon measure μ on X.*

Proof. Let H_π be the representation space of π, and let $\xi_0 \in H_\pi$ be a cyclic vector. The map $\tau(f) = \langle \pi(f)\xi_0 | \xi_0 \rangle$ is a positive linear functional on $C_c(X)$. Therefore, because of the Riesz representation theorem, there is a Radon measure μ on X such that

$$\tau(f) = \int_X f(x) \cdot \mu(x).$$

Let π_μ denote the representation of $C_c(X)$ on $L^2(X,\mu)$ given by

$$\pi_\mu(f)(g) = fg.$$

Note that, in fact, π_μ is a representation of $C(X^\bullet)$, where $X^\bullet$ is the one-point compactification of X. Let $U : f \in C_c(X) \mapsto U(f) = \pi(f)(\xi_0) \in H_\pi$. Since

$$\|\pi(f)(\xi_0)\|^2 = \int_X |f(x)|^2 \cdot \mu(x),$$

U extends to a unitary operator of $L^2(X,\mu)$ into H_π. It is apparent that $U \circ \pi_\mu(f) = \pi(f) \circ U$. Finally, U has dense image because the vectors $\pi(f)\xi_0$, $f \in C_c(X)$, span H_π. □

Corollary A.8.3. *If π is a representation of $C_c(X)$ on a Hilbert space H, then $\|\pi(f)\| \leq \|f\|$. In particular, π extends to a representation of $C_0(X)$.*

Proof. The representation π decomposes as a direct sum $\pi = \pi_0 + \bigoplus_\alpha \pi_\alpha$ on $H = H_0 + \bigoplus_\alpha H_\alpha$, where π_0 is the zero representation and each π_α is cyclic. On the one hand, $\|\pi(f)\| = \sup_\alpha \|\pi_\alpha(f)\|$ and, on the other hand, $\|\pi_\alpha(f)\| \leq \|f\|$ by Theorem A.8.2. □

Corollary A.8.4. *A cyclic representation of $C_0(X)$ is equivalent to the multiplication representation on $L^2(X,\mu)$, for some finite Radon measure on X.*

Proof. It only remains to add that the positive linear functional τ in the proof of Theorem A.8.2 is bounded. If this were not the case, there would exist a sequence of functions $f_n \in C_0(X)$ such that $0 \leq f_n \leq 1$, while $\tau(f_n) \geq 2^n$. Then the function $f = \sum_n 2^{-n} f_n \in C_0(X)$ would satisfy $\tau(f) = \infty$, a contradiction. □

Exercise A.8.5. Let π_μ and $\pi_{\mu'}$ be two representations of $C_0(X)$ associated to finite measures μ and μ' on X. Show that the representations are unitarily equivalent if and only if the measures are mutually absolutely continuous.

It follows that a representation π of $C_0(X)$ is unitary equivalent to a direct sum of representations given by finite Radon measures on X and the zero representation. That is, if π is nondegenerate, then there is a countable family of measures μ_n on X such that π is equivalent to the multiplication representation on the Hilbert sum $\bigoplus_{n=1}^\infty L^2(X,\mu_n)$. This collection of measures is not unique, but they can be chosen so that μ_{n+1} is absolutely continuous with respect to μ_n, for each n. The measure μ_1 is called the *basis measure* of the representation π.

Exercise A.8.6. Show the following.

(1) The representation π is faithful if and only if the support $\operatorname{supp} \mu_1 = X$.

(2) There exists an open subset $Y \subset X$ such that $f \in \ker \pi$ if and only if $\operatorname{supp} f \subset Y$.

Hence, combining this with Exercise A.8.5, we see that there are a measure μ on X and a measurable function $n : X \to \{0, 1, 2, \cdots, \infty\}$ such that the representation π is equivalent to the representation $\pi_{\mu,n}$ constructed as follows. Let $H_1 = L^2(X, \mu)$ and let H_k be the subspace of H_1 consisting of all functions that vanish outside $\{n(x) \geq k\}$. Construct the Hilbert sum $H = \bigoplus_{k=1}^{\infty} H_k$, and construct a representation $\pi_{\mu,n}$ of $C_0(X)$ on H by defining $\pi_{\mu,n}(f)$ to be the operator of multiplication by the function f.

Theorem A.8.7. *Let X be a locally compact, second countable Hausdorff space and let π be a representation of $C_0(X)$. Then there exist a Radon measure μ on X and a measurable function $n : X \to \{0, 1, \cdots, \infty\}$ such that π is equivalent to the representation $\pi_{\mu,n}$. Furthermore, two representations $\pi_{\mu,n}$ and $\pi_{\nu,m}$ are equivalent if and only if the measures μ and ν are mutually absolutely continuous and the functions n and m coincide almost everywhere.*

The details can be found in [**114**, Chapter 4]. A related version is given in [**47**, Chapter 8].

Corollary A.8.8. *Let π be a representation of $C_0(X)$. Then there exist a field of Hilbert spaces $\{H_x\}_{x \in X}$ over X and a Radon measure μ such that π is equivalent to the multiplication representation on the Hilbert integral $H = \int^{\oplus} H_x \cdot \mu(x)$.*

Proof. Let μ be the basis measure and let $n : X \to \{0, 1, \cdots, \infty\}$ be the measurable function associated to π as above. Let H be a Hilbert space with orthonormal system $\{\xi_i\}_{i=1}^{\infty}$ and let $\{H_x\}$ be the field so that H_x is the subspace of H generated by the first $n(x)$ vectors of $\{\xi_i\}$. Finally, let the measurable fields be those Borel functions $f : X \to H$ such that $x \to \langle f(x) | \xi_i \rangle$ is measurable for every i. □

A.9. Tensor Products

The theory of tensor products of C^*-algebras is somewhat technical. A full account can be found in Takesaki [**168**]. Since tensor products appear as building blocks of the C^*-algebras associated to foliated spaces, a very basic discussion needs to be included.

The algebraic tensor product $A \otimes_{\mathbb{C}} B$ of two C^*-algebras is an algebra with involution, but it is not complete with the product norm. Moreover, there is, in general, more than one C^*-norm on $A \otimes_{\mathbb{C}} B$ such that its completion is a C^*-algebra. A C^*-algebra A with the property that, for any C^*-algebra B, the algebraic tensor product $A \otimes_{\mathbb{C}} B$ has a unique C^*-norm making it a C^*-algebra upon completion is called *nuclear*. When A is nuclear, the completion of $A \otimes_{\mathbb{C}} B$ for this unique C^*-norm is denoted by $A \otimes B$.

Theorem A.9.1. *The C^*-algebra $C_0(X)$, X a locally compact Hausdorff space, is nuclear.*

Let A be a C^*-algebra and let $C_0(X, A)$ be the C^*-algebra of continuous functions $X \to A$ that vanish at infinity, as described in Exercise A.5.7. The bilinear map $C_0(X) \times A \to C_0(X, A)$ that assigns to (f, a) the map $x \in X \mapsto f(x)a$ induces an involutive homomorphism

$$\pi : C_0(X) \times A \to C_0(X, A).$$

This homomorphism is, in fact, injective and thus can be used to define a C^*-norm on $C_0(X) \otimes_{\mathbb{C}} A$. Since $C_0(X)$ is nuclear, this must be the unique C^*-norm on $C_0(X) \otimes_{\mathbb{C}} A$, and thus it extends to the completion. The image $\pi(C_0(X) \otimes A)$ is then closed in $C_0(X, A)$. It is also dense; hence it is all of $C_0(X, A)$.

The next item on our agenda is to describe the basic representation theory of the C^*-algebras $C_0(X) \otimes A$.

Theorem A.9.2. *Let π be a nondegenerate representation of $C_0(X, A)$ on H. Then there are unique nondegenerate representations π_X and π_A of $C_0(X)$ and A, respectively, on H such that*

$$\pi(f \otimes a) = \pi_X(f)\pi_A(a) = \pi_A(a)\pi_X(f).$$

Conversely, if π_X and π_A are commuting nondegenerate representations of $C_0(X)$ and A on H, then

$$\pi(f \otimes a) = \pi_X(f)\pi_A(a)$$

is a nondegenerate representation of $C_0(X, A)$.

Sketch of the proof. The case in which $C_0(X)$ and A have a unit (hence X is compact) is particularly simple.

Given π, use the isomorphism $C_0(X) \otimes A \cong C_0(X, A)$ to define

$$\pi_X(f) = \pi(f \otimes \mathbf{1}) \quad \text{and} \quad \pi_A(a) = \pi(\mathbf{1} \otimes a).$$

Then

$$\pi_X(f)\pi_A(a) = \pi(f \otimes a) = \pi_A(a)\pi_X(f).$$

If there is a vector $\xi \in H$ such that $\pi_X(f)\xi = 0$ for all $f \in C_0(X)$, then $\pi(f \otimes a)\xi = 0$ for all $a \in A$ and all f. Since the elements $f \otimes a$ generate a dense subspace of $C_0(X, A)$, this says that π is degenerate.

In the general case, if u_i and k_j are approximate units, then

$$\pi_X(f) = \lim_j \pi(f \otimes k_j) \quad \text{and} \quad \pi_A(k) = \lim_i \pi(u_i \otimes k).$$

Complete details of more general results can be found in [**168**]. □

Exercise A.9.3. Let A_1, A_2 be C^*-algebras with representations π_1, π_2 on H_1, H_2, respectively.

(1) Use Exercise A.3.6 to construct an involutive representation π of the algebraic tensor product $A_1 \otimes_{\mathbb{C}} A_2$ on $H_1 \otimes H_2$.

(2) Show that $\pi_1 \otimes I_2$ and $I_1 \otimes \pi_2$ are commuting representations of A_1 and A_2 on $H_1 \otimes H_2$ that decompose π as in Theorem A.9.2.

A.10. Von Neumann Algebras

A von Neumann algebra (also called a ring of operators) is a particular kind of C^*-algebra. Standard references for von Neumann algebras are Dixmier [**48**] and Takesaki [**168**]. An excellent overview is Connes [**38**].

Definition A.10.1. A *von Neumann algebra* is an involutive subalgebra of bounded linear operators on a Hilbert space, that contains the identity operator and that is closed for the strong operator topology.

One of the first surprising facts is that, while the above definition is topological, there is a purely algebraic characterization of von Neuman algebras. This requires a definition.

Definition A.10.2. If R is a ring and $S \subset R$, the *commutant* of S is the set

$$S' = \{a \in R \mid as = sa \text{ for all } s \in S\}.$$

The double commutant is $S'' = (S')'$.

Exercise A.10.3. Let $\mathcal{K}$ be the algebra of compact operators on a Hilbert space H. This is contained in $R = \mathcal{B}(H)$. If $T \in \mathcal{K}'$, show that T commutes with all finite rank operators and so it must be a multiple of the identity. Thus $\mathcal{K}' = \mathbb{C}I$ and $\mathcal{K}'' = \mathcal{B}(H)$.

The following is called the *von Neumann double commutant theorem*, and gives an algebraic characterization of von Neumann algebras.

Theorem A.10.4. *Let A be an involutive subalgebra of $\mathcal{B}(H)$ that contains the identity operator. Then the following are equivalent:*

(1) *The double commutant $A'' = A$.*

(2) *A is closed in the weak operator topology.*

(3) *A is closed in the strong operator topology.*

One way in which von Neuman algebras arise from C^*-algebras is via representation theory; which is, in fact, the only reason for the existence of this section.

Exercise A.10.5. Let π be an involutive representation of an involutive algebra A on a Hilbert space. Show that the commutant $\pi(A)'$ is a von Neumann algebra.

Although von Neumann algebras form a special class of C^*-algebras, they have a markedly different structure, which makes their study largely independent of C^*-algebras. Broadly speaking, if C^*-algebras are viewed as noncommutative topological spaces, then von Neumann algebras are to be viewed as noncommutative measure spaces. One of the basic results exhibiting their peculiarity is the following.

Theorem A.10.6. *Let $W \subset \mathcal{B}(H)$ be a von Neumann algebra. Then the weak closure of the algebra generated by the projections in W is W itself.*

A *projection* on a Hilbert space H is a bounded linear operator $P \in \mathcal{B}(H)$ such that $P \circ P = P$. Geometrically speaking, a projection in a C^*-algebra corresponds to an open subset of the topological space whose complement is also open, and so there may not be many of them. But in a measure-theoretic setting, the complement of a measurable set is measurable.

The following result was explicitely used in the proof of Theorems 1.4.11 and 1.8.1.

Corollary A.10.7. *Let π be a representation of a C^*-algebra A on a Hilbert space H. Then π is irreducible if and only if its commutant consist of multiples of the identity.*

Proof. Suppose that π is irreducible but $\pi(A)' \neq \mathbb{C}I$. Then $\pi(A)'$ is a von Neumann algebra (Exercise A.10.5) that contains, because of Theorem A.10.6, a nontrivial projection P of H onto a proper closed subspace V of H. If $\xi \in V$ and $a \in A$, then $\pi(a)\xi = \pi(a)P(\xi) = P(\pi(a)\xi) \in V$, so that V is invariant under $\pi(A)$.

Conversely, suppose that $V \subset H$ is a closed, proper, invariant subspace for π, and let P be the projection of H onto V. Then $\pi(A)$ preserves the orthogonal decomposition $H = V \oplus V^{\perp}$, from which it follows immediately that $\pi(a)P = P\pi(a)$, for every $a \in A$. ∎

Appendix B

Riemannian Geometry and Heat Diffusion

The purpose of this appendix and of the following one is to introduce and develop some concepts pertaining to function theory on Riemannian manifolds that have been used in Chapter 2. In this appendix, some aspects of the Laplacian and diffusion on Riemannian manifolds will be discussed, often without proofs. In Appendix C, a few concepts from the theory of stochastic processes will be introduced in order to discuss the probabilistic interpretation of some familiar functional equations.

B.1. Geometric Concepts and Formulas

In what follows, X will denote a smooth, connected manifold. The case $\dim X = 1$ will usually be ignored, particularly if its consideration causes some theorem or formula to need a separate statement. A *domain* in X is a connected open subset of X. A *bounded domain* is a domain D such that $\overline{D}$ is compact, and a *regular domain* is a domain whose boundary in X, if nonempty, is a smooth submanifold of codimension one. Thus a bounded regular domain is the interior of a compact manifold-with-boundary. Sometimes the closure of a regular domain may be called a regular domain.

The following notation for certain function spaces associated to a domain D in X will be used. As usual, $C^0(D) = C(D)$ denotes the space of continuous functions on D, and $C_0(D)$ the space of continuous functions that vanish at ∞. The space of functions on D that are k-times differentiable with continuous kth derivatives is denoted by $C^k(D)$, and $C^k(\overline{D})$ denotes the space of functions on $\overline{D}$ that extend to C^k functions on a neighborhood of $\overline{D}$ in X. Finally, $\overline{C}^k(D)$ is the space of functions that are in $C^k(D)$ and whose derivatives of order $\leq k$ extend continuously to $\overline{D}$.

Let X be given a metric tensor g (a section of the bundle of bilinear 2-forms which is symmetric and positive definite) and let vol denote the associated volume density. The expression of g in local coordinates $x_1, \dots, x_n$ on X, where $n = \dim X$, is of the form

$$\sum_{i,j=1}^{n} g_{ij}(x_1, \dots, x_n) dx_i \otimes dx_j,$$

where each $g_{ij} = g_{ji}$ is a smooth function of the coordinates $x_1, \dots, x_n$, and the matrix $g = (g_{ij})$ is positive definite. The entries of the inverse matrix are denoted by g^{ij}. The volume density vol has the local expression

$$\mathrm{vol} = \sqrt{|g|}\, dx_1 \wedge \cdots \wedge dx_n,$$

where $|g| = \det(g)$. The volume density vol defines a measure on X, typically denoted by something like dx in integrals (sometimes omitted if clear from the context).

The metric tensor g defines an inner product by assigning to two vectors v and w (tangent to X at the same point) the number $g(v, w)$, a quantity that is often abbreviated by $\langle v, w\rangle$. If v and w are smooth vector fields on X, then $\langle v, w\rangle$ is a smooth function on X.

The *gradient* of a smooth function f on X is the vector field $\operatorname{grad} f$ that is dual to the 1-form df with respect to the metric tensor g. That is,

$$\langle \operatorname{grad} f, v\rangle = v(f) = df(v),$$

for every vector field v on X. In local coordinates,

$$\operatorname{grad} f = \sum_{ij} g^{ij} \frac{\partial}{\partial x_i} f \frac{\partial}{\partial x_j}.$$

The *divergence* of a vector field v on X is the function $\operatorname{div} v$ on X defined by

$$\operatorname{div} v(x) = \operatorname{tr}(w \to \nabla_w v),$$

where $w \in T_x X$, and ∇ is the Levi-Civita connection of g. If the local expression for v is $v = \sum_i h_i \partial/\partial x_i$, then

$$\operatorname{div} v = \frac{1}{\sqrt{|g|}} \sum_i \frac{\partial}{\partial x_i} \left(h_i \sqrt{|g|} \right).$$

Definition B.1.1. The *Laplacian* of a C^2 function f on X is the function $\triangle f$ defined by

$$\triangle f = \operatorname{div} \operatorname{grad} f.$$

The local expression for the Laplacian of a function f is

$$\triangle f = \frac{1}{\sqrt{|g|}} \sum_{j=1}^{n} \frac{\partial}{\partial x_j} \left(\sum_{i=1}^{n} g^{ij} \sqrt{|g|} \frac{\partial}{\partial x_i} f \right).$$

The Laplacian is a second order, linear differential operator on the space $C^\infty(X)$. Besides linearity, it satisfies two other useful identities, namely

$$\triangle(f_1 f_2) = f_1 \triangle f_2 + f_2 \triangle f_1 + 2\langle \operatorname{grad} f_1, \operatorname{grad} f_2 \rangle \tag{B.1.1}$$

and

$$\operatorname{div}(f_1(\operatorname{grad} f_2)) = f_1 \triangle f_2 + \langle \operatorname{grad} f_1, \operatorname{grad} f_2 \rangle. \tag{B.1.2}$$

Several integration formulas will now be recalled. They usually have the names of the divergence theorem or the Green-Gauss-Stokes-Ostrogradski formula attached.

Theorem B.1.2. *Let v be a C^1 vector field with compact support on the manifold X. Then*

$$\int_X \operatorname{div} v(y) \cdot dy = 0.$$

In particular, if $f \in C^2(X)$ has compact support, then

$$\int_X \triangle f(y) \cdot dy = 0.$$

By equation (B.1.2), this has the following corollary.

Corollary B.1.3. *Let f_1 and f_2 be smooth functions on X such that the function $f_1 \operatorname{grad} f_2$ has compact support. Then*

$$\int_X f_1 \triangle f_2 = - \int_X \langle \operatorname{grad} f_1, \operatorname{grad} f_2 \rangle.$$

Exercise B.1.4. Show that if either f_1 or f_2 has compact support, then

$$\int_X f_1 \triangle f_2 = \int_X f_2 \triangle f_1.$$

Relative versions of these theorems will also be needed. For this, let D be a regular domain in X and let $\mathbf{n}$ denote the outward unit normal to ∂D.

Theorem B.1.5. *If v is a C^1 vector field on $\overline{D}$ that has compact support on $\overline{D}$, then*

$$\int_D \operatorname{div} v = \int_{\partial D} \langle v, \mathbf{n} \rangle.$$

Corollary B.1.6. *Let f_1, f_2 be functions on $\overline{D}$ such that f_1 is C^1 on $\overline{D}$, f_2 is C^2 on $\overline{D}$, and $f_1 \operatorname{grad} f_2$ has compact support on $\overline{D}$. Then*

$$\int_D f_1 \triangle f_2 = -\int_D \langle \operatorname{grad} f_1, \operatorname{grad} f_2 \rangle + \int_{\partial D} f_1 \mathbf{n}(f_2).$$

If f_1 is also of class C^2 and both f_1, f_2 have compact support on $\overline{D}$, then

$$\int_D (f_2 \triangle f_1 - f_1 \triangle f_2) = \int_{\partial D} \Big(f_2 \mathbf{n}(f_1) - f_1 \mathbf{n}(f_2) \Big).$$

If $f \in \overline{C}^k(D)$, then, by definition, f extends to a continuous function on $\overline{D}$ and $\operatorname{grad} f$ extends to a continuous vector field on $\overline{D}$. If $\mathbf{n}$ is the outward normal to ∂D as above and $z \in \partial D$, let γ_z be the geodesic through z with initial vector $\mathbf{n}(z)$, and define the normal derivative of f at z by

$$\mathbf{n}(f)(z) = \lim_{t \to 0-} \langle \operatorname{grad} f(\gamma_z(t)), \gamma_z'(t) \rangle.$$

The Gauss-Green formulas admit the following extension.

Proposition B.1.7. *Let D be a bounded regular domain in X. If*

$$f_1 \in C^0(\overline{D}) \cap C^1(D) \quad \text{and} \quad f_2 \in \overline{C}^1(D) \cap C^2(D),$$

then

$$\int_D f_1 \triangle f_2 = -\int_D \langle \operatorname{grad} f_1, \operatorname{grad} f_2 \rangle + \int_{\partial D} f_1 \mathbf{n}(f_2).$$

If, furthermore, $f_1 \in \overline{C}^1(D) \cap C^2(D)$, then

$$\int_D (f_1 \triangle f_2 - f_2 \triangle f_1) = \int_{\partial D} (f_2 \mathbf{n}(f_1) - f_1 \mathbf{n}(f_2)).$$

B.2. Estimates of Geometric Quantities

The *injectivity radius* $\iota(x)$ at a point x of a Riemannian manifold X is the supremum of those numbers $r > 0$ for which

$$\exp_x : B(0_x, r) \subset T_x X \to B(x, r) \subset X$$

is a diffeomorphism. Here, of course, $B(\xi, r)$ denotes the metric ball of radius r in the indicated space, centered at ξ.

A Riemannian manifold X is said to have *bounded geometry* if it is complete, and there are numbers $r > 0$ and a, b such that the injectivity radius of every point of X is $\geq r$ and all sectional curvatures belong to the interval $[a, b]$. As a matter of fact, only a lower bound for the Ricci curvature is needed for much of what follows. However, our main concern being manifolds that are leaves of a compact foliated space, there is no need for this greater generality.

Proposition B.2.1. *Let $(M, \mathcal{F})$ be a compact foliated space with a smooth metric tensor. Every leaf of M is a Riemannian manifold of bounded geometry. In fact, the injectivity radius and sectional curvatures of all leaves are uniformly bounded.*

Proof. It suffices that the metric tensor be of class C^3, as the curvature involves only derivatives of order ≤ 2.

The leaves are complete manifolds by [I, Exercise 12.1.3]. The sectional curvature is a function on the Grassmann bundle of two-dimensional subspaces of the tangent bundle to the leaves. This is a compact space, and the smoothness conditions on the foliated space and on the metric tensor imply that the sectional curvature is a continuous function on it, hence bounded.

Similarly, the injectivity radius of every point is bounded below, since a lower bound can be found by examining the critical points of the exponential map along the leaves. Due to the standing smoothness hypothesis, the exponential map $\exp : TM \to M$ is a smooth map of foliated spaces. □

Standard comparison theorems for Riemannian manifolds will be reviewed in this section. It will be convenient to recall the description of the metric tensor of (X, g) in polar coordinates. Let $x \in X$ and let $B = B(\varepsilon)$ be a ball about x so that the exponential map $\exp_x : B(0_x, \varepsilon) \to B$ is a diffeomorphism, that is, $\varepsilon \leq \iota(x)$.

Using polar coordinates on T_xX, a diffeomorphism

$$\phi : (0, \varepsilon) \times S^{d-1} \to B \setminus \{x\}$$

defined by $\phi(r, v) = \exp_x(rv)$ is obtained. Gauss's lemma says that radial curves are geodesics; hence the distance function $d(x, \cdot)$ is the coordinate function r.

In these coordinates, the metric tensor g can be written as

$$g = dr \otimes dr + h_{(r,v)},$$

where $h_{(r,v)}$ is the metric induced by g at (r, v) on the sphere $\phi(r \times S^{n-1})$ $(n = \dim X)$.

A straightforward calculation shows that, if f is a function which on $B(\varepsilon)$ depends on the distance $r(y) = d(x, y)$ only, then the Laplacian has the formula

$$\triangle f = \frac{d^2}{dr^2} f + \frac{\nu'}{\nu} \frac{d}{dr} f,$$

where $\nu(r, v)$ is the volume density function in polar coordinates, and ν' denotes $(d/dr)\nu$.

For a real number κ, let g_κ denote the metric tensor (in dimension n) given by

$$g_\kappa = \begin{cases} dr \otimes dr + \dfrac{1}{\kappa}\sin^2\!\left(r\sqrt{\kappa}\right) \displaystyle\sum_{i=2}^{n} dv_i \otimes dv_i, & \kappa > 0 \text{ and } r \in [0, \pi/\sqrt{\kappa}], \\ dr \otimes dr + r^2 \displaystyle\sum_{i=2}^{n} dv_i \otimes dv_i, & \kappa = 0, \\ dr \otimes dr - \dfrac{1}{\kappa}\sinh^2\!\left(r\sqrt{-\kappa}\right) \displaystyle\sum_{i=2}^{n} dv_i \otimes dv_i, & \kappa < 0, \end{cases}$$

Let X_κ denote the Riemannian manifold of dimension n with metric tensor g_κ. This is the simply connected manifold with constant curvature κ. The volume form vol_κ corresponding to g_κ is

$$\mathrm{vol}_\kappa = \begin{cases} \left((1/\sqrt{\kappa})\sin(r\sqrt{\kappa})\right)^{n-1} dr \wedge dv, & \text{if } \kappa > 0, \\ r^{n-1}\, dr \wedge dv, & \text{if } \kappa = 0, \\ \left((1/\sqrt{\kappa})\sinh^2(r\sqrt{-\kappa})\right)^{n-1} dr \wedge dv, & \text{if } \kappa < 0, \end{cases}$$

where $dv = dv_2 \wedge \cdots \wedge dv_n$. Let $\nu_\kappa(r)$ denote the corresponding volume density.

The following results center around Bishop's comparison theorems. Detailed proofs can be found in any standard book on differential geometry, for example in Petersen [**148**]. One version of Bishop's comparison theorem states the following.

Theorem B.2.2. *Let X have sectional curvatures in the interval $[a, b]$. Then*

$$\frac{\nu_b'}{\nu_b} \le \frac{\nu'}{\nu} \le \frac{\nu_a'}{\nu_a},$$

within $B(x, \iota(x))$, for every $x \in X$.

A more elementary result is given in the lemma below, which provides an upper bound for the volume density in terms of a curvature upper bound.

If $x \in X$, the distance function $r(y) = d(x, y)$ is generally not smooth, and its obstruction to being smooth is due to the possibility of conjugate points. These appear as either critical values for the exponential map $\exp_x : T_x(X) \to X$, or points where $\exp_x$ fails to be one-to-one. Because X is complete, there is an open subset of $T_x(X)$, star-shaped about 0_x (the segment domain, see Petersen [**148**], for instance), so that the exponential map $\exp_x$ is a diffeomorphism onto an open dense subset U_x of X.

Lemma B.2.3. *Let X have sectional curvature bounded below by κ. Then*

$$\nu(r, v) \le \nu_\kappa(r).$$

Proof. By the above, the function $r(y) = d(x,y)$ is smooth on the open and dense subset $U_x \setminus \{x\}$ of X, and polar coordinates are also valid on this set. Thus Δr is defined on this set, and

$$\Delta r = \frac{\nu'}{\nu}.$$

Furthermore,

$$(\Delta r)' + \frac{(\Delta r)^2}{(n-1)} \leq (\Delta r)' + |\nabla^2 r|^2 = -\mathrm{Ric}\left(\frac{\partial}{\partial r}, \frac{\partial}{\partial r}\right).$$

Because of the curvature hypothesis, $\mathrm{Ric} \geq (n-1)\kappa$, and so the two displayed equations give the differential inequality

$$\frac{d^2}{dr^2}\nu(r,v)^{\frac{1}{n-1}} \leq \kappa\, \nu(r,v)^{\frac{1}{n-1}}.$$

Near $r = 0$, the function $\nu(r,v)$ behaves like

$$\nu(r,v)^{\frac{1}{n-1}} = r + O(r^2),$$

and

$$\frac{d}{dr}\nu(r,v)^{\frac{1}{n-1}} = 1 + O(r).$$

Therefore $\nu(r,v) \leq \nu_\kappa(r)$ by standard comparison theorems for differential equations. Complete details are given in any of the references mentioned above. □

The final comparison result concerns the asymptotics of the volume function $V(x,r) = \mathrm{vol}\, B(x,r)$.

Lemma B.2.4. *If the sectional curvatures of X are bounded below by κ, then*

$$V(x,r) \leq V_\kappa(r),$$

where $V_\kappa(r)$ is the volume of the ball of radius r in the Riemannian manifold X_κ.

Proof. The exponential map $\exp : T_xX \to X$ is smooth; hence its set of critical values has measure 0. It is also surjective and sends the ball $B(0_x, r) \subset T_xX$ onto $B(x,r) \subset X$. Therefore,

$$\begin{aligned} \mathrm{vol}\, B(x,r) &\leq \int_{B(0_x,r)} \nu(r,v)\, dr \wedge dv \\ &\leq \int_{B(0_x,r)} \nu_\kappa(r)\, dr \wedge dv \\ &= V_\kappa(r), \end{aligned}$$

where the second inequality is by Lemma B.2.3. □

Corollary B.2.5. *If X is a complete manifold of bounded geometry, then X has at most exponential volume growth, in a uniform sense. That is, there are constants $k, C > 0$ such that $V(x,r) \leq ke^{Cr}$, for all $x \in X$.*

B.3. Basic Function Theory

This section reviews some basic properties of functions on a Riemannian manifold X with Laplacian $\triangle$.

Definition B.3.1. A function f on X is said to be *harmonic* if it is of class C^2 and $\triangle f \equiv 0$. (*A fortiori*, f is of class C^∞.)

The following fact is usually referred to as Hopf's theorem.

Lemma B.3.2. *Let X be a compact manifold. If f is a function on X such that $\triangle f \geq 0$, then f is constant. In particular, the only harmonic functions on X are the constants.*

Proof. By Theorem B.1.2,

$$\int_X \triangle f = 0 = \int_X \triangle f^2.$$

Since $\triangle f \geq 0$, the first equality implies that $\triangle f \equiv 0$. Therefore, by the second equality and equation (B.1.1),

$$\begin{aligned} 0 &= \int_X \triangle f^2 \\ &= \int_X (2f\triangle f + 2|\operatorname{grad} f|^2) \\ &= \int_X 2|\operatorname{grad} f|^2, \end{aligned}$$

so $\operatorname{grad} f \equiv 0$ and f is constant (X has been assumed to be connected). □

The following result is called the *maximum principle*. A proof can be found in [**153**, p. 61].

Theorem B.3.3. *Let D be a bounded regular domain. Let u be a continuous function on $\overline{D}$ that is C^2 on D and and has $\triangle u \geq 0$. If u reaches its maximum value K at a point in D, then $u \equiv K$ on $\overline{D}$.*

Corollary B.3.4. *Let D be a bounded regular domain and let u be a continuous function on $\overline{D}$ that is C^2 on D and has $\triangle u \equiv 0$ on D. Then*

$$|u(x)| \leq \sup_{y \in \partial D} |u(y)|,$$

for all $x \in D$.

Proof. Since $\triangle u \equiv 0$, Theorem B.3.3 applies to both u and $-u$. □

A boundary maximum principle will also be needed. The proof can be found in [**153**, p. 65]

Theorem B.3.5. *Let D be a bounded regular domain. Let u be a function in $\overline{C}^1(D) \cap C^2(D)$ such that $\triangle u \geq 0$ on D. If $u \leq K$ and $u(x) = K$ at a point $x \in \partial D$, then $\mathbf{n}(u)(x) > 0$.*

Exercise B.3.6. Let g, f, f' be continuous on $\overline{D}$, f, f' of class C^2 on D, and $\lambda f - \triangle f = \lambda f' - \triangle f' = g$ on D for some $\lambda > 0$. Then

(1) if $g \geq \lambda K$ on D and $f \geq K$ on ∂D, for some constant K, then $f \geq K$ on D, and

(2) if $f \geq f'$ on ∂D, then $f \geq f'$ on D.

Exercise B.3.7. Let X be a compact manifold and $f \in C^2(X)$. If $\lambda > 0$, then

$$\min\{\lambda f - \triangle f\} \leq \lambda f(x) \leq \max\{\lambda f - \triangle f\},$$

for every $x \in X$.

B.4. Regularity Theorems

A typical problem in geometry or analysis is that of finding solutions to partial differential equations of the form

$$\Phi f = \varphi,$$

where Φ is a differential operator and φ is a given function on a domain D. If φ is a smooth function, it is then reasonable to expect that the equation has a solution f that is also smooth. However, for this sort of problem, the space of smooth functions is somewhat bad from a functional analytic point of view. Hence the equation is translated into one on a nicer space, such as the space of distributions or a Sobolev space, where it can be solved. Thus the solutions obtained are not functions in the usual sense and the regularity theorems to be stated below are required to guarantee that such generalized solutions are represented by smooth functions.

A *distribution* on D is a linear functional T on the space $C_c^\infty(D)$ of smooth, compactly supported functions on D, which is continuous in the sense that, whenever a sequence $\{\varphi_k\}_{k=1}^\infty \subset C_c^\infty(D)$ consists of functions with supports in a fixed compact set and converges to zero in the C^r-norm, for each $r \geq 0$, then $T(\varphi_k) \to 0$ as $k \to \infty$. The standard notation for the action of a distribution T on a function φ is $\langle T, \varphi \rangle$.

Example B.4.1. If ν is a Radon measure on D, then integration against ν is a distribution on D (*cf.* the Riesz Representation Theorem A.2.6).

Example B.4.2. If f is a locally integrable function on D (that is, f is integrable on each compact subset of D), then

$$\varphi \mapsto \int_D f(x)\varphi(x) \cdot dx$$

is a distribution. Such a distribution is said to be equal to the function f.

The elliptic regularity theorem, stated below, will assert that a distribution solution to a certain partial differential equation is necessarily equal to a smooth function. This regularity theorem will not be proven here, but the reader is invited to try out one of the first such ones, known as the DuBois-Raymond lemma.

Exercise B.4.3. Let f be an integrable function on the interval $[0,1]$. If

$$\int_0^1 f(x)g(x) \cdot dx = 0$$

for every smooth function g with compact support on $(0,1)$, show that $f = 0$ almost everywhere. If such a condition holds for those functions g such that $\int_0^1 g(x) \cdot dx = 0$, show that $f \equiv$ const. almost everywhere.

The meaning of a distribution solution to the equation

$$\triangle f = \varphi$$

is now explained. By Exercise B.1.4, if $h \in C_c^\infty(D)$, then

$$\int_D f(x)\triangle h(x) \cdot dx = \int_D h(x)\triangle f(x) \cdot dx.$$

Let $\langle \cdot | \cdot \rangle$ denote the inner product

$$\langle f | h \rangle = \int_D fh.$$

If f is a solution to $\triangle f = \varphi$, then, for arbitrary $h \in C_c^\infty(D)$,

$$\langle f | \triangle h \rangle = \langle \varphi | h \rangle.$$

A distribution T is a *solution* of $\triangle f = \varphi$ if $\langle T, \triangle h \rangle = \langle \varphi | h \rangle$, for all $h \in C_c^\infty(D)$.

Because of the duality between functions and distributions, it makes sense to define the action of a differential operator Φ on a distribution T as the distribution ΦT given by $\langle \Phi T, \varphi \rangle = \langle T, \Phi\varphi \rangle$, for every $\varphi \in C_c^\infty(D)$.

Example B.4.4. If $x \in \mathbb{R}$, then the Heaviside function

$$H_x(y) = \begin{cases} 1 & \text{if } y > x, \\ 0 & \text{if } y \leq x, \end{cases}$$

defines a distribution whose derivative is $H'_x = -\delta_x$. Here, δ_x is the Dirac delta function, which represents the distribution equal to the point mass measure at x.

The following statement is the elliptic regularity theorem. A proof is given in [**66**].

Theorem B.4.5. *Let D be a domain in X and let T be a distribution solution to the equation $\triangle T = \varphi$, where φ is a smooth function on D. Then T is equal to a smooth function.*

This regularity theorem will be applied to two particular situations, which are as follows.

Corollary B.4.6. *Let D be a domain in X and let ν be a Radon measure on D such that*

$$\int_D \triangle\varphi(x) \cdot \nu(x) = 0$$

for every compactly supported smooth function φ on D. Then

$$\nu(x) = h(x) \cdot dx,$$

where h is a harmonic function on D.

Proof. A Radon measure is a distribution. □

Corollary B.4.7. *Let f be a locally integrable function on D such that*

$$\int_D f(x)\triangle h(x) \cdot dx = 0,$$

for every $h \in C_c^\infty(D)$. Then f equals a harmonic function, except perhaps on a subset of D of measure 0.

Here is another application of the regularity theorem. A function K on $X \times X$ is called a proper kernel if the projections of $\operatorname{supp} K \subset X \times X$ onto each factor are proper maps, that is, for each compact set $C \subset X$, both $\operatorname{supp} K \cap (X \times C)$ and $\operatorname{supp} K \cap (C \times X)$ are compact.

Lemma B.4.8. *Let X be a manifold, let $K(x, y)$ be a smooth proper kernel on $X \times X$ and let $f(x)$ be a smooth function on X. Then the function F defined by*

$$F(x) = \int_X K(x, y) f(y) \cdot dy$$

is smooth, and

$$\triangle F(x) = \int_X \triangle_x K(x, y) f(y) \cdot dy.$$

Proof. It is clear that $F(x)$ is a smooth function. To show that the Laplace operator commutes with the integral, it suffices to show that for every compactly supported function φ the following identity holds true:

$$\left\langle \int K(x,y)f(y)\cdot dy \,\middle|\, \Delta\varphi(x)\right\rangle = \left\langle \int \Delta_x K(x,y)f(y)\cdot dy \,\middle|\, \varphi(x)\right\rangle,$$

for then this implies that F is a solution to the differential equation

$$\Delta F(x) = \int_X \Delta_x K(x,y)f(y)\cdot dy.$$

Starting from the left-hand side, we get

$$\begin{aligned}\left\langle \int K(x,y)f(y)\cdot dy \,\middle|\, \Delta\varphi(x)\right\rangle &= \int \left(\int K(x,y)f(y)\cdot dy\right)\Delta\varphi(x)\cdot dx\\ &= \int f(y)\left(\int K(x,y)\Delta\varphi(x)\cdot dx\right)\cdot dy\\ &= \int f(y)\left(\int \Delta K(x,y)\varphi(x)\cdot dx\right)\cdot dy\\ &= \int\left(\int \Delta K(x,y)f(y)\cdot dy\right)\varphi(x)\cdot dx\\ &= \left\langle \int \Delta K(x,y)f(y)\cdot dy \,\middle|\, \varphi(x)\right\rangle,\end{aligned}$$

which is the right-hand side. The first equality is by the definition of the inner product $\langle\cdot|\cdot\rangle$, the second and fourth by Fubini's theorem, and the third by the property of the Laplacian, as the functions involved have compact support. Fubini's theorem applies because the function $K(x,y)f(y)\varphi(x)$ has compact support on $X\times X$ if φ has compact support on X and K is a proper kernel. □

Exercise B.4.9. Show that the lemma continues to hold true if K is a smooth function on $X \times X$ and f has compact support.

B.5. The Heat Equation

Let X denote a complete, connected Riemannian manifold and let f be a continuous function on X. A continuous function $u(x,t)$ on $X \times (0,\infty)$, which is C^2 in x and C^1 in t, satisfies the *heat equation* with initial condition f if

$$\frac{\partial}{\partial t}u(x,t) = \Delta_x u(x,t),$$

and

$$u(x,0) = f(x).$$

The second condition means that $u(x,t) \to f(x)$ uniformly on compact subsets of X as $t \to 0$. The function u is interpreted as describing the temporal evolution of the temperature on X prescribed by f at time $t = 0$.

Of critical importance is the fundamental solution to the heat equation, or *heat kernel* of (X, g).

Definition B.5.1. A fundamental solution to the heat equation on X is a continuous function $p(x, y; t)$ on $X \times X \times (0, \infty)$, which is C^2 with respect to x, C^1 with respect to t, and which satisfies

$$\triangle_x p(x, y; t) = \frac{\partial}{\partial t} p(x, y; t),$$

and

$$\lim_{t \to 0} \int_X f(y) p(x, y; t) \cdot dy = f(x),$$

for all continuous, bounded functions f on X, the convergence being uniform on compact subsets.

The function $p(x, y; t)$ represents the temperature at the point x at time t resulting from an initial unit distribution of temperature completely concentrated at the point y. The qualification of fundamental solution is that, if at time $t = 0$ an initial temperature distribution modeled by the function $f(x)$ is given, then the evolution of f should be obtained as a sum of punctual evolutions; that is, the temperature $u(x, t)$ at the point x at time t should be given by

$$u(x, t) = \int_X f(y) p(x, y; t) \cdot dy.$$

Example B.5.2. For Euclidean space $\mathbb{R}^n$ with the standard metric $g = dx_1 \otimes dx_1 + \cdots + dx_n \otimes dx_n$, the Laplacian is

$$\triangle = \sum_{i=1}^{n} \frac{\partial^2}{\partial x_i^2}.$$

The heat kernel is the function

$$p_e(x, y; t) = \frac{1}{\sqrt{4\pi t^n}} e^{-|x-y|^2/4t}.$$

Explicit formulas for the heat kernels on the other constant curvature, simply connected manifolds can be found in [**27**].

Exercise B.5.3. Find an expression for the heat kernel of the Riemannian manifold S^1 with metric tensor $g = d\theta \otimes d\theta$. Do the same for the manifold $X = \mathbb{R} \times S^1$ with the product metric.

Exercise B.5.4. Let X be a manifold with heat kernel $p(x, y; t)$. Let Γ be a discrete group of isometries of X. Determine the heat kernel of X/Γ.

A description of the construction of the heat kernel on a Riemannian manifold is contained in the next section. The method of construction for noncompact manifolds described follows [**50**], and it is also described in detail in [**27**].

The construction starts with the case of compact manifolds. A fundamental solution is found by an iterative method starting with an appropriate candidate function. The next step is to study the heat equation with boundary conditions on a bounded regular domain D. A solution to the heat equation in D with initial datum $f \in C(D)$ is a function u on $D \times (0, \infty)$ that satisfies the following conditions:

(1) It is C^2 in the first variable, C^1 in the second, and

$$\frac{\partial}{\partial t} u(x,t) = \Delta u(x,t).$$

(2) It extends continuously to $\overline{D} \times (0, \infty)$ with $u(x,t) = 0$ if $x \in \partial D$.

(3) $u(x,0) = f(x)$ for $x \in D$. That is to say,

$$\lim_{t \to 0} u(x,t) = f(x)$$

uniformly on compact subsets of D.

The fundamental solution q_D to the heat equation on D is a smooth function on $\overline{D} \times \overline{D} \times (0, \infty)$ that is 0 if x or $y \in \partial D$, and such that, if f is a function on D, then

$$u(x,t) = \int_D f(y) q_D(x,y;t) \cdot dy$$

is the solution to the heat equation on D with initial datum f.

Finally, the heat kernel on a noncompact manifold X is shown to be $p = \sup q_D$, where D runs thru all bounded regular domains of X. The statement of this existence result, as well as other properties of p, will be found in the next section.

B.6. Construction of the Heat Kernel

The construction of the heat kernel of a noncompact manifold of bounded geometry is described in this section. The construction is first carried out for compact manifolds, then for compact manifolds with boundary, and finally, by a process of exhaustion, for noncompact manifolds. What follows is essentially a summary of what is done in [**7**, pp. 204–215] and [**27**, Chapter VI].

The construction of the heat kernel of a compact manifold is by a method of successive approximations, the key being to start with a good initial approximation. Paraphrasing M. Kac, particles of heat diffusing in X enter

X at a point via the tangent space, where they have been nurturing the belief that life is Euclidean. Only after the initial moment do they realize that life has curvature, and external effects will affect their future.

This good initial approximation is called a *parametrix* for the operator $\partial/\partial t - \triangle$, and is a function H on $X \times X \times (0, \infty)$ such that

(1) the function $(\partial/\partial t - \triangle)H$ extends to a continuous function on the space $X \times X \times [0, \infty)$, and
(2) $\int_X H(x, y; t) f(y) \cdot dy \to f(x)$ and $\int_X H(x, y; t) f(x) \cdot dx \to f(y)$ as $t \to 0$, uniformly on compact subsets.

The construction of the parametrix given below originates with S. Minakshisundaram [**131**], and is fully described in the two references mentioned at the begining of this section. An appropriate function to start the approximation scheme should be one that resembles the Euclidean heat kernel, for instance the function

$$E(x, y; t) = \frac{1}{\sqrt{4\pi t^n}} e^{-d(x,y)^2/4t},$$

which is differentiable (in the x, y variables) in a neighborhood

$$U = \{(x, y) \mid d_X(x, y) < \iota(X)\}$$

of the diagonal.

Working on U, and in polar coordinates, with r denoting the distance function, it is verified that

$$\left(\frac{\partial}{\partial t} - \triangle\right) E = \frac{1}{2t}\left(1 - n + r\frac{\nu'}{\nu}\right) E,$$

where $n = \dim X$. Here and below, the Laplacian $\triangle$ acts on the second variable when functions of two variables are involved: $\triangle f(x, y) = \triangle_y f(x, y)$.

As should be expected, E is not a solution. Moreover, its excess from being a solution has a singularity at $t = 0$ of higher order than that of E. To correct this, the function E is replaced with Eu_0, where u_0 is a function on U which is to be found in order to eliminate the factor $1/t$. A calculation shows that

$$\left(\frac{\partial}{\partial t} - \triangle\right)(Eu_0) = \left(\frac{r}{2t}\left(\frac{(\nu r^{n-1})'}{2\nu r^{n-1}} u_0 - u_0'\right) + \triangle u_0\right) E,$$

so that u_0 must satisfy

$$\frac{(\nu r^{n-1})'}{2\nu r^{n-1}} u_0 = u_0',$$

and hence u_0 is given by $u_0 = (\nu r^{n-1})^{1/2}$.

Iteratively, functions u_k are searched for so that the functions

$$F_j = E\big(u_0 + tu_1 + \cdots + t^j u_j\big)$$

satisfy a recursive relation

$$\left(\frac{\partial}{\partial t} - \triangle\right) F_j = t^j \, E \triangle u_j.$$

Due to the fairly explicit nature of the metric and Laplacian on a geodesic ball, the equations that these u_j must satisfy can be explicitly described and the solutions shown to exist. For $j > n/2$ the function $t^j E \triangle u_j$ extends continuously to $t = 0$, and to a C^s function if $j > n/2 + s$.

To pass from U to $X \times X$ a smooth function

$$H_j(x, y; t) = \phi\big(d(x, y)\big) F_j(x, y; t)$$

is defined, where $\phi : \mathbb{R} \to [0, 1]$ (which is 0 if $|s| > r_1 = \iota/3$ and is 1 if $|s| < r_2 = \iota/4$) is a suitable bump function.

The proof of the following lemma gives a stronger version of property (2) of the parametrix.

Lemma B.6.1. *If Y is a topological space and f is continuous on $X \times Y$, then, for $j > n/2$,*

$$\lim_{t \to 0} \int_X H_j(x, y; t) f(z, x) \cdot dx = f(z, y),$$

$$\lim_{t \to 0} \int_X H_j(x, y; t) f(z, y) \cdot dy = f(z, x),$$

uniformly on compact sets.

A proof is given in [**7**], Lemme E.III.3 and Remarque E.III.4.

For $j > n/2 + \ell$ the function

$$K_j = (\partial/\partial t - \triangle) H_j$$

extends to a function of class C^ℓ on $X \times X \times [0, \infty)$. A more precise description of the structure of these functions H_j and K_j is as follows.

Lemma B.6.2. *Let $j > n/2 + 2$ and $K_j = (\partial/\partial t - \triangle) H_j$. Then*

$$H_j = \frac{1}{(4\pi t)^{n/2}} e^{-d^2/4t} \sum_{i=0}^{j} t^i U_i \quad \textit{and} \quad K_j = \frac{1}{(4\pi t)^{n/2}} e^{-d^2/4t} \sum_{i=-1}^{j} t^i V_i,$$

where the U_i's and V_i's are smooth functions on $X \times X$ such that

(1) *the functions U_i, $0 \le i \le j$, and V_k are supported in an r_1-neighborhood of the diagonal, and*

(2) *the functions V_i, $-1 \le i \le j - 1$, are supported in an r_1-neighborhood of the diagonal, and vanish on the smaller r_2-neighborhood.*

To improve the H_j's to the actual heat kernel, an iterative method based on convolution is used. Convolution of two functions A, B that are continuous on $X \times X \times [0, \infty)$ is defined by

$$(A \times B)(x, y; t) = \int_0^t \left(\int_X A(x, z; s) B(z, y; t-s) \cdot dz \right) \cdot ds.$$

The result is that the parametrix H_j satisfies

$$\left(\frac{\partial}{\partial t} - \triangle \right) (A \times H_j) = A + (A \times K_j),$$

as is easily verified. The heat kernel $p(x, y; t)$ is then sought in the form

$$p = H_j - F \times H_j.$$

If p is going to satisfy the heat equation $(\partial/\partial t - \triangle)p = 0$, then the unknown function F must satisfy $F = K_j - F \times K_j$, from which it follows that F admits the formal series expansion

$$F = \sum_{i=1}^{\infty} (-1)^i K_j^{\times i},$$

where $K_j^{\times i}$ denotes the i-fold convolution $K_j^{\times i} = K_j \times \overset{i}{\cdots} \times K_j$.

The uniform convergence of this series on compact subsets of the space $X \times X \times [0, \infty)$ is a consequence of the careful initial choice of the H_j. Indeed, note that the function K_j is of the form

$$K_j(x, y; t) = t^{j-n/2} e^{-d(x,y)^2/4t} k_j(x, y; t),$$

where k_j is a smooth function on $X \times X \times [0, \infty)$ that is supported in a neighborhood of the diagonal, and the convolution is defined in such a way that, if A is an upper bound for k_j on $X \times X \times [0, T]$ and $B = AT^{j-n/2}$, then

$$|K_j{}^{\times 2}| \leq \frac{B^2 T^{j-n/2}}{j + 1 - n/2} \mathrm{vol}(X) e^{-d(x,y)^2/4t}.$$

More generally, an inductive argument (*cf.* the calculations in Section 2.2) shows that

$$|K_j^{\times i}| \leq \frac{AB^{i-1} \mathrm{vol}(X)^{i-1}}{(k - n/2 + 1) \cdots (k - n/2 + i - 1)} e^{-d(x,y)^2/4t}.$$

It follows from these estimates that the series defining F above is dominated by a convergent series, and so it converges uniformly on $X \times X \times [0, T]$, for every $T \geq 0$. Also, $\lim_{t \to 0} F_t = 0$.

The smoothness requirements on p are obtained by taking j strictly greater than $n/2 + 2$ because, by basically the same techniques, it is possible to control the convergence of the derivatives (see [**7**, Lemme E.III.7]). As the heat kernel is unique, it does not matter which j is chosen (as long as it is large); hence smoothness can be improved as far as possible, meaning that p will be smooth if the metric tensor of X is smooth. The end result is an expression for p of the form

$$p(x,y;t) = \frac{1}{(4\pi t)^{n/2}} e^{-d(x,y)^2/4t} \left[\phi(d(x,y)) \sum_{i=0}^{j} t^i u_i(x,y) + O(t^{j+1}) \right].$$

The terms u_i contain a lot of geometric information about X.

This concludes the description of the proof of the following theorem. Further details and references can be found in [**7**] and [**27**].

Theorem B.6.3. *Let X be compact Riemannian manifold. Then there exists a unique fundamental solution $p(x,y;t)$ to the heat equation on X. The function p is symmetric in the first two variables: $p(x,y;t) = p(y,x,t)$. Furthermore, if f is a bounded continuous function on X, then*

$$u(x,t) = \int_X f(y)p(x,y;t) \cdot dy$$

is the solution to the heat equation on X with f as initial condition.

The construction of the heat kernel q_D on a bounded regular domain D (*i.e.*, on a compact manifold with boundary) is reduced to a boundary value problem on a compact manifold M. The compact manifold M is obtained by doubling a neighborhood of $\overline{D}$ in X along its boundary, so that the metric is extended across without changing the geometry of a neighborhood of $\overline{D}$. The heat equation on D describes the temporal evolution of the heat distribution on D of a unit amount of heat deposited at a point $y \in D$; alternatively, it describes the heat distribution on M subject to isolation of the boundary of D. The kernel q_D is then obtained by subtracting from the heat kernel of M the solution to a boundary value problem on $(D, \partial D)$.

More precisely, let $p(x,y;t)$ denote the heat kernel of M, and for each $y \in D$, let $f_y(z,t)$ be the function $-p(z,y;t)$ on $\partial D \times (0,\infty)$. If a solution u_y to the heat equation on $D \times (0,\infty)$ can be found, subject to the conditions $u_y(x,0) = 0$ for $x \in D$ and $u_y(x,t) = f_y(x,t)$ on $\partial D \times (0,\infty)$, then upon setting

$$q_D(x,y;t) = p(x,y;t) + u_y(x,t),$$

it follows that

$$\frac{\partial}{\partial t} q_D(x,y;t) = \Delta_x q_D(x,y;t)$$

and

$$q_D(z,y;t) = 0 \qquad \text{if } z \in \partial D.$$

The function $u = u_y$ is found in the form

$$u(x;t) = -\int_0^t \left(\int_{\partial D} F(x,z;t-s) f(z,s) \cdot dz \right) \cdot ds,$$

where F is some formal series whose terms are iteratively constructed, and it can be shown to converge. As it would be inappropriate to include any further detail here, the reader is referred to Chavel [**27**].

The following proposition summarizes the information obtained from the method of construction and other analysis.

Proposition B.6.4. *Let D be a bounded regular domain in a manifold X of dimension n. The fundamental solution to the heat equation on D is a smooth function q_D on $D \times D \times (0, \infty)$ that extends continuously to the space $\overline{D} \times \overline{D} \times (0, \infty)$. Furthermore:*

(1) $q_D(x, y; t) > 0$ *on* $D \times D$, $q_D(x, y; t) = 0$ *on* $\partial D \times D$.

(2) q_D *is symmetric in the* x, y *variables.*

(3) q_D *satisfies the convolution formula*

$$q_D(x, y; s + t) = \int_D q_D(x, z; s) q_D(z, y; t) \cdot dz,$$

for $s, t > 0$ *and* $x, y \in \overline{D}$.

(4) *The asymptotic behavior of* q_D *is*

$$q_D(x, y; t) = \frac{1}{\sqrt{4\pi t^n}} e^{-d(x,y)^2/4t} H(x, y) + O(t^{1-n/2} e^{-d(x,y)^2/4t}),$$

where $H(x, y)$ *is a smooth function on* $D \times D$ *with* $H(x, x) = 1$.

(5) *If* D_1 *and* D_2 *are domains with heat kernels* q_1 *and* q_2, *respectively, then the difference*

$$q_1(x, y, t) - q_2(x, y, t) = O(t^k),$$

for all $k > 0$ *and* t *small, uniformly on compact subsets of* $D_1 \cap D_2$.

The following maximum principle for parabolic equations is a key element in the proof of these facts.

Lemma B.6.5. *Let D be a bounded regular domain in X and let f be a bounded continuous function on $D \times [0, T]$ that is C^2 on $D \times (0, T)$, and that satisfies*

$$\triangle f \geq \frac{\partial}{\partial t} f$$

on $D \times (0, T)$. If there exists a point (x_0, t_0) in $D \times (0, T]$ such that

$$f(x_0, t_0) = \sup_{D \times [0,T]} f(x, t),$$

then

$$f(x, t) = f(x_0, t_0)$$

for all $x \in D$ and $t \leq t_0$.

This maximum principle has the following intuitively evident proposition as a consequence.

Proposition B.6.6. *Let $D \subset D'$ be bounded regular domains in X with respective heat kernels q_D and $q_{D'}$. Then*

$$\int_D q_D(x,y,t) \cdot dy < 1 \tag{1}$$

for all x in D. Moreover, the heat kernels satisfy

$$q_D \leq q_{D'} \tag{2}$$

on D.

Proof. To prove (2), fix $y \in D$ and let $f(x,t) = q_{D'}(x,y;t) - q_D(x,y;t)$. This function has a continuous extension to $\overline{D} \times [0,\infty)$ that vanishes on the subset $\overline{D} \times \{0\}$ and is nonnegative on $\partial D \times [0,\infty)$. Hence $f(x,t) \geq 0$ on $D \times [0,\infty)$. □

It is therefore permissible to define $p = \sup_D q_D$, where D ranges over all bounded regular domains D of X. This p has the following properties, quoted from Chavel [**27**].

Theorem B.6.7. *The function $p : X \times X \times (0,\infty) \to (0,\infty)$ is smooth, strictly positive, and satisfies*

$$\frac{\partial}{\partial t} p(x,y;t) = \triangle_x p(x,y;t)$$

on $X \times X \times (0,\infty)$. For all $s,t > 0$, it satisfies

$$p(x,y;t+s) = \int_X p(x,z;t)p(z,y;s) \cdot dz$$

on $X \times X$. Moreover, p is the minimal positive fundamental solution to the heat equation on X.

For complete manifolds of bounded geometry the previous theorem is complemented by the following one.

Theorem B.6.8. *If X is a complete manifold of bounded geometry, then p is the unique heat kernel on X. In this case, p is the unique continuous function on $X \times X \times (0,\infty)$ for which*

$$f(x,t) = \int_X f(y)p(x,y;t) \cdot dy$$

always gives a solution to the heat equation on X with initial data f (an arbitrary bounded continuous function on X). Furthermore,

$$\int_X p(x,y;t) \cdot dy = 1.$$

B.7. Estimates for the Heat Kernel

This section contains a brief summary of upper estimates for the heat kernel that will later be used to justify some calculations. In general, for noncompact manifolds of bounded geometry, it is expected that the heat kernel behaves like the Euclidean heat kernel. In fact, the method of construction of the heat kernel implies that it behaves in this way for small times.

Estimates of the heat kernel for large times is a very active area of research. The philosophy is that, for large times, the heat kernel should decay like the Euclidean kernel, with a modification that takes into account the curvature of the manifold. For the purposes of this book, the following theorem of S. Y. Cheng, P. Li and S.-T. Yau [**29**] will suffice. (More general results were obtained by Li and Yau [**120**]. The reader will find a survey in [**27**].)

Theorem B.7.1. *Let X be a complete Riemannian manifold with sectional curvatures between $a < 0$ and $b > 0$. Given $T > 0$, there exists a constant A, depending on $n = \dim X$, on the lower bound for the curvature and on T, such that, for $0 \le t \le T$, the heat kernel has the bound*

$$p(x, y; t) \le \frac{A}{t^{n/2}} \delta(x)^{-\alpha(n)/2} \exp\bigl(-d(x,y)^2/16t\bigr),$$

for some universal constant $\alpha(n) > 0$, where $\delta(x) = \min\bigl\{\iota(x), 1, \pi/12\sqrt{b}\bigr\}$.

The form in which this theorem will be applied is the following.

Corollary B.7.2. *Let X be a complete noncompact manifold of bounded geometry and $\dim X = n$. Given $T > 0$, there exist constants A (depending on T and on the geometric constants of X) and B such that*

$$p(x, y; t) \le \frac{A}{t^{n/2}} \exp\left(-d(x,y)^2/Bt\right).$$

Proof. There is not much to say about this estimate. In fact, $B = 16$ (the actual value will be irrelevant), and the quantity $\delta(x)$ is uniformly controlled over X, producing A. □

The following proposition plays a fundamental role in several places in this appendix and in the next. Roughly speaking, it gives an estimate of the heat at the ends of the manifold

Proposition B.7.3. *Let X be a complete Riemannian manifold of bounded geometry. Then, for $\varepsilon > 0$, there exists $t_\varepsilon > 0$ such that $0 < t < t_\varepsilon$ implies*

$$\sup_{x \in X} \int_{X \smallsetminus B(x,\varepsilon)} p(x, y; t) \cdot dy \le \frac{A}{t^{(n+2)/2}} e^{-\varepsilon^2/Bt},$$

where A and B are constants that depend only on the geometry of X.

Proof. Let $V(x,s)$ denote the volume of the ball $B(x,s)$ of radius s about the point x. The fact that X has bounded geometry implies that $V(x,s)$ grows at most exponentially (*cf.* Corollary B.2.5), so that there exist constants C and k, depending only on the geometry of X, such that

$$V(x,s) \leq ke^{Cs},$$

for $s \geq 0$ and all $x \in X$. Also recall that the $(n-1)$-dimensional volume $A(x,s)$ of the boundary $\partial B(x,s)$ is the derivative $V'(x,s)$ with respect to s, for almost all s.

Fix an upper bound $T = 1$ for time and apply Theorem B.7.1 to obtain a (positive) constant A so that the heat kernel

$$p(x,y;t) \leq \frac{A}{t^{n/2}} e^{-d(x,y)^2/16t},$$

for $0 < t \leq 1$. Then

$$\begin{aligned}
\int_{X \setminus B(x,\varepsilon)} p(x,y;t) \cdot dy &= \int_\varepsilon^\infty \left(\int_{\partial B(x,r)} p(x,z;t) \cdot dz \right) \cdot dr \\
&\leq \int_\varepsilon^\infty \frac{A}{t^{n/2}} e^{-r^2/16t} A(x,r) \cdot dr \\
&= \frac{A}{t^{n/2}} \Bigg[-e^{-\varepsilon^2/16t} V(x,\varepsilon) \\
&\qquad + \frac{1}{8t} \int_\varepsilon^\infty e^{-r^2/16t} r V(x,r) \cdot dr \Bigg] \\
&\leq \frac{A}{8t^{(n+2)/2}} \int_\varepsilon^\infty r e^{-r^2/16t} V(x,r) \cdot dr.
\end{aligned}$$

The first equality is the co-area formula and is standard. The first inequality follows from the estimate for the heat kernel just stated. The equality that follows uses integration by parts (using $V'(x,r) = A(x,r)$) and the fact that $\lim_{r\to\infty} e^{Cr-(r^2/16t)} = 0$. Using the fact that $rV(x,r) \leq kre^{rC} \leq ke^{r(C+1)}$, replacing $C+1$ by C, and absorbing the constant k and the number 8 into A, this inequality can be rewritten as

$$\text{(B.7.1)} \qquad \int_{X \setminus B(x,\varepsilon)} p(x,y;t) \cdot dy \leq \frac{A}{t^{(n+2)/2}} \int_\varepsilon^\infty e^{Cr-(r^2/16t)} \cdot dr.$$

The integral

$$\int_\varepsilon^\infty e^{Cr-(r^2/16t)} \cdot dr.$$

will now be analyzed. The change of variables $u = r - 8Ct$ transforms it into the integral

$$e^{8C^2t} \int_{\varepsilon - 8Ct}^\infty e^{-u^2/16t} \cdot du.$$

Since $0 < t \leq 1$, the factor e^{8C^2t} can be replaced by a constant that can then be absorbed into the constant A in equation (B.7.1).

If $t_\varepsilon > 0$ is sufficiently small, so that $16Ct_\varepsilon \leq \varepsilon$, then $\varepsilon - 8Ct \geq \varepsilon/2$ for all $t < t_\varepsilon$. Hence for $t \leq t_\varepsilon$,

$$\int_{\varepsilon-8Ct}^{\infty} e^{-u^2/16t} \cdot du \leq \int_{\varepsilon/2}^{\infty} e^{-u^2/16t} \cdot du. \tag{B.7.2}$$

The integral $\int_a^\infty e^{-s^2} ds$ is known as the error integral, and the following estimate is available. For $a \geq 0$,

$$\int_a^\infty e^{-s^2} \cdot ds \leq \frac{2/\sqrt{2}}{a + \sqrt{a^2+1}} e^{-a^2}.$$

To apply this to (B.7.2), take $s = u/\sqrt{16t}$, use the fact that $t \leq t_\varepsilon \leq \varepsilon/16C$ and suitably fashion the constant A (and $B = 16$) in the statement out of A and C. □

Exercise B.7.4. If X is a complete manifold of bounded geometry, and r is the distance function from a point $x \in X$, show that the function $\exp(-r(x)^2)$ is integrable.

Upper estimates for the heat kernel that work for all times will also be needed. The following theorem [**28**] will be sufficient.

Theorem B.7.5. *Let X be a noncompact manifold of bounded geometry. For every time $T > 0$ there is a constant $C(T)$ such that the heat kernel $p(x, y; t)$ of X satisfies*

$$p(x, y; t) \leq C(T)t^{-1/2},$$

for all $x, y \in X$ and $t \geq T$.

It is not possible, in general, to have an estimate of the form

$$p(x, y; t) \leq Kt^{-\dim X/2},$$

as in Euclidean space. The reason is that the long time behavior of the heat kernel is tied to the geometry of the manifold via the isoperimetric inequality.

B.8. The Green Function

The purpose of this section is to introduce the Green function for a bounded regular domain in a manifold, which is required for discussing the Dirichlet problem. Informally speaking, the Green function for a domain D arises as follows. Consider the differential equation

$$\triangle G_x = -\delta_x,$$

where δ_x is the distribution given by the point mass measure at $x \in D$. If G_x is a (distribution) solution to this equation, then it becomes a sort of fundamental solution to the differential equation $\triangle f = \varphi$. For if φ is given, then $f(x) = -\langle G_x, \varphi\rangle$ is, formally, a solution.

Let X be an n-dimensional manifold with Laplacian $\triangle$, let D be a bounded regular domain in X, and let $\Sigma = \{(x,x) \mid x \in D\}$ be the diagonal.

Definition B.8.1. A continuous function G on $(\overline{D} \times \overline{D}) \smallsetminus \Sigma$ is a *Green function* for the Laplace operator $\triangle$ on D if

(1) G is of class C^2 on $(D \times D) \smallsetminus \Sigma$;

(2) $G(x,y) = 0$ for $x \in D$, $y \in \partial D$;

(3) $\triangle_x G(x,y) = -\delta_x(y)$ (in the distribution sense);

(4) for each $x \in D$, $\operatorname{grad} G(x, \bullet)$ extends to a continuous vector field on $\overline{D} \smallsetminus \{x\}$; and

(5) near the diagonal Σ, the function G is given by

$$G(x,y) = \begin{cases} g(x,y) + \dfrac{1}{(n-2)\sigma(n)} d(x,y)^{2-n}, & \text{if } \dim D = n > 2, \\ g(x,y) - \dfrac{1}{2\pi} \log d(x,y), & \text{if } \dim D = 2, \end{cases}$$

where g is a continuous function on $\overline{D} \times \overline{D}$, of class C^2 on $D \times D$. (The number $\sigma(n)$ denotes the area of the surface of the unit sphere in $\mathbb{R}^n$.)

This concept, when $n = 2$, should be well known to the reader from basic complex analysis. It is a theorem that a domain D as above has a Green function. Proofs can be found in [**44**] for the case in which D is a compact manifold, and in [**6**] for the general case of a compact manifold with boundary.

Besides existence and uniqueness, the Green function has several other properties. It is symmetric: $G(x,y) = G(y,x)$. For each $x \in D$, the function $G_x(y) = G(x,y)$ is harmonic and nonnegative on $D \smallsetminus \{x\}$. It extends continuously to ∂D, taking the value 0 there, and tends to ∞ as $y \to x$. The maximum principle implies that G is in fact strictly positive on $D \times D \smallsetminus \Sigma$, and that it is unique.

The Green function has the integral representation

$$G(x,y) = \int_0^\infty q_D(x,y;t) \cdot dt,$$

where q_D is the heat kernel of the domain D. It is not difficult to verify that this integral is well defined, and, at least from a formal perspective, that it has the properties required of G in Definition B.8.1 above.

Exercise B.8.2. What is the Green function of the interval $(a, b) \subset \mathbb{R}$, with respect to the standard metric whose Laplacian is $\triangle f = f''$? (Compare Example B.4.4.)

Let G_x denote the function $G_x(y) = G(x, y)$. For each $x \in D$, the vector field $\operatorname{grad}_y G_x$ extends continuously to ∂D, and $\mathbf{n}(G_x)$ is defined on ∂D. The following integral representation will be useful in the next section.

Proposition B.8.3. *If u is a function in $\overline{C}^1(D) \cap C^2(D)$, then*

$$-u(x) = \int_D G(x, y) \triangle u(y) \cdot dy + \int_{\partial D} u(z) \mathbf{n}(G_x)(z) \cdot dz.$$

In particular, if $u = 0$ on ∂D, then

$$-u(x) = \int_D G(x, y)(\triangle u)(y) \cdot dy;$$

and if u is harmonic on D, then

$$u(x) = -\int_{\partial D} u(z) \mathbf{n}(G_x)(z) \cdot dz.$$

This is proved by applying the Green formula to the functions u and G_x. The singularity of G_x at x is integrable, but some care is needed for the proper evaluation. More details are given in [**27**, Chapter VII].

B.9. Dirichlet Problem and Harmonic Measure

This section presents the solution to the Dirichlet problem and integral representation of harmonic functions on bounded regular domains of a Riemannian manifold X. The *Dirichlet problem* is the following statement.

Theorem B.9.1. *Let D be a bounded regular domain in X. Let f be a continuous function on ∂D. Then there is a unique function u continuous on $\overline{D}$ such that $\triangle u \equiv 0$ on D and $u = f$ on ∂D.*

Let $G(x, y)$ denote the Green function of the domain D. When the boundary datum f is smooth on ∂D, the Dirichlet problem on D can be solved as follows. It may be assumed that f is the restriction to ∂D of a smooth function (also called f) on $\overline{D}$. Then the function

$$h(x) = \int_D G(x, y) \triangle f(y) \cdot dy$$

is continuous on $\overline{D}$, vanishes on ∂D, and satisfies $\triangle h = -\triangle f$ on D (by (2) of Definition B.8.1). Hence $u = h + f$ solves the Dirichlet problem for smooth initial conditions f.

In fact, the integral representation for a harmonic function u on D that belongs to $C^2(\overline{D})$ is given by

$$u(x) = -\int_{\partial D} u(z)\mathbf{n}(G_x)(z) \cdot dz.$$

It will now be shown that the solution to the Dirichlet problem with continuous initial datum f on ∂D is given by the same integral representation.

If f is a continuous function on ∂D, then f is a uniform limit of smooth functions f_n on ∂D for which the corresponding solutions u_n to the Dirichlet problem admit an integral representation

$$u_n(x) = \int_{\partial D} f_n(y)\mathbf{n}(G_x)(z) \cdot dz.$$

Since the function $u_n - u_m$ is harmonic on D with continuous boundary values $f_n - f_m$, the maximum principle implies that

$$|u_n(x) - u_m(x)| \leq \sup_{y \in \partial D} |f_m(y) - f_m(y)|,$$

for all $x \in D$. Therefore the sequence u_n is equicontinuous and converges uniformly in $\overline{D}$ to a continuous function u. The regularity theorems (Proposition B.4.7) imply that u equals a harmonic function almost everywhere, hence everywhere because it is continuous.

Theorem B.9.2. *Let D be a bounded regular domain in X with Green function G. If f is a continuous function on ∂D, then*

$$H_D f(x) = -\int_{\partial D} f(z)\mathbf{n}(G_x)(z) \cdot dz$$

is the harmonic extension of f to D.

If $x \in D$, then the assignment

$$f \in C(\partial D) \to H_D f(x)$$

is a positive linear functional on $C(\partial D)$ that takes the constant function $\mathbf{1}$ to the number 1. Such a linear functional can be represented by a measure η_x^D on ∂D, which is called the *harmonic measure* of $x \in D$ on ∂D. The previous discussion shows that the harmonic measure of x is given by

$$\eta_x^D = -\mathbf{n}(G_x)(z)dz,$$

where dz denotes the Riemannian measure on ∂D. The density function is continuous and ≥ 0. This is so because the function G_x is positive on $D \setminus \{x\}$ and equals 0 when $y \in \partial D$. In general, there is no guarantee that the density function is > 0 on ∂D, but the fact that the solution to the Dirichlet problem is unique implies that the support of η_x^D is ∂D. (In general the function G_x may develop critical points, even if the domain D is homotopically trivial; otherwise the Poincaré conjecture would follow easily.)

At any rate, if $x \in D$ then the fact that the function $G_x(y) = G(x,y)$ is harmonic on $D \setminus \{x\}$ and increases to ∞ as $y \to x$ implies that there is a regular value a for G_x such that the inward normal derivative of G_x on the hypersurface $\{G_x(y) = a\}$ is strictly positive (by Theorem B.3.5). Let $U_a \subset D$ be the domain $\{G_x(y) > a\}$, and let G^a denote its Green function. Then $G^a(x,y) = G(x,y) - a$ and $-\mathbf{n}_a(G^a_x)(y) > 0$ on the boundary of U_a (where $\mathbf{n}_a$ is the outward normal to ∂U_a). Since G^a is a smooth function, it readily follows that x has a neighborhood V_x such that the density function of $\eta_y^{D_a}$ on ∂U_a is strictly positive, for every $y \in V_x$. Therefore, it has been shown that, for each $x \in D$, there are neighborhoods $V_x \subset U_x$ of x and a constant C (which depends on the neighborhood, on the point x and its distance to ∂D, and on the geometry of D) such that the harmonic measure η_y^a of y on ∂D_a is "proportional" to the measure η_x^a, in the sense that their Radon-Nikodym derivative is bounded above and below away from zero as follows:

$$\frac{1}{C} \leq \frac{d\eta_y^a}{d\eta_x^a} \leq C.$$

One important consequence of this analysis is the *Harnack principle*, which states the following.

Theorem B.9.3. *Let D be a bounded regular domain in X. Given a point $x \in X$, there are a neighborhood U_x of x and a constant $C > 0$ (depending on the geometry of the domain and on the distance from x to the boundary ∂D) such that, if h is a positive harmonic function on a neighborhood of U_x, then*

$$\frac{1}{C^2} \leq h(y)/h(y') \leq C^2$$

for any two points $y, y' \in U_x$.

Proof. It has been shown that, for each $x \in D$, there exist a pair of neighborhoods $V \subset U$ of x and a constant $C > 0$ such that the densities of the harmonic measure η_y^U of a point $y \in V$ on ∂U satisfy

$$C\mathbf{n}(G_x^U) \leq \mathbf{n}(G_y^U) \leq \frac{1}{C}\mathbf{n}(G_x^U).$$

If h is a positive harmonic function on a neighborhood of U, then it has an integral representation

$$h(y) = \int_{\partial U} h(z) \cdot \eta_y^U(z).$$

If $y \in V$, the above displayed inequality implies that

$$(1/C)h(x) \leq h(y) \leq Ch(x),$$

and a repeated application of this argument concludes the proof. □

Corollary B.9.4. *Let W be a domain in X. If $K \subset W$ is compact, then there exists a constant $C > 0$ such that every positive harmonic function h on W satisfies*

$$\frac{1}{C} \leq \frac{h(x)}{h(y)} \leq C,$$

for every pair of points $x, y \in K$.

Proof. Let D be a bounded regular domain in W that contains K. By the discussion above, for every point x in D there exist a neighborhood V_x and a constant C_x such that every positive harmonic function h on W satisfies $(1/C_x) \leq h(y)/h(y') \leq C_x$, for every $y, y' \in V_x$. The conclusion follows by covering the compact set K with finitely many such neighborhoods V_x. □

Another important consequence of this analysis is *Harnack's theorem.*

Theorem B.9.5. *Let u_n be a sequence of harmonic functions on a domain D such that $u_{n+1}(x) \geq u_n(x)$ for every $x \in D$. Then $u = \sup_n u_n$ either is equal to $+\infty$ in all of D, or else it is a (finite) harmonic function on D.*

Proof. By replacing u_n with $u_n - u_1 + 1$, it may be assumed that $u_n > 0$ on D. If $u(x) < \infty$ at a point $x \in D$, then the previous proposition implies that $u < \infty$ in a neighborhood of x. Therefore the set $\{x \in D \mid u(x) < \infty\}$ is open in D.

This set is also closed. Indeed, let $x_k \to x$ with $u(x_k) < \infty$, let V be the neighborhood of x provided by the Harnack principle, and let C be the corresponding constant. For k sufficiently large we have $x_k \in V$, which implies that $u_n(x) \leq C u_n(x_k) \leq C u(x_k)$. Thus $u(x) < \infty$ because $u_n(x)$ is bounded above. As D is connected, either the set $\{u < \infty\}$ is empty or it equals D.

The case $u < \infty$ is now considered. Let K be a compact subset of D and let C be the constant provided by Harnack's principle. If $m \geq n$, then the function $u_m - u_n$ is harmonic and nonnegative on D, and

$$0 \leq u_m(y) - u_n(y) \leq C\big(u_m(x) - u_n(x)\big),$$

for every pair of points $x, y \in K$. It follows that, if $u(x) < \infty$, then the sequence u_n converges uniformly to u in K. Thus u is continuous on D.

Let B be a bounded domain contained in D. Each function u_n has, on B, an integral representation of the form

$$u_n(x) = \int_{\partial B} u_n(y) \cdot \eta_x^B(y).$$

By passing to the limit, the integral representation

$$u(x) = \int_{\partial B} u(y) \cdot \eta_x^B(y)$$

is obtained. As u is continuous, it must be harmonic. □

The Poisson problem for a domain D asks for a solution u to the differential equation $\triangle u = f$, where f is a given continuous function on D, satisfting the boundary condition $u \equiv 0$ on ∂D. The Green function gives

$$u(x) = -\int_D G(x,y)f(y) \cdot dy$$

as a solution to this problem on a bounded regular domain.

B.10. Diffusion and Resolvent

In this section the heat equation and heat kernel on a manifold X of bounded geometry are reexamined from a slightly different point of view, namely that of the theory of semigroups of operators. General references for semigroups of operators are E. Dynkin [**57**] and K. Yosida [**191**].

The process that, to a bounded function f on X, assigns the solution to the heat equation with initial datum f may be interpreted as the action of a family of operators D_t on the space of bounded functions on X, where

$$D_t f(x) = \int_X f(y)p(x,y;t) \cdot dy.$$

These operators are called the *diffusion operators* of the Riemannian manifold X.

The properties of the heat kernel of a Riemannian manifold of bounded geometry that have been previously set forth translate into the following properties of these operators.

Proposition B.10.1. *Let X be a complete Riemannian manifold of bounded geometry. Let $\mathcal{B}(X)$ denote the Banach space of bounded, measurable functions on X with the uniform norm. Associated to the Laplace operator $\triangle$ and heat kernel p of X there is a family of bounded linear operators $D_t : \mathcal{B}(X) \to \mathcal{B}(X)$ $(t > 0)$ such that*

(1) *D_t is positive: if $f \in \mathcal{B}(X)$ is a nonnegative function, then $D_t f$ is nonnegative;*
(2) *the operators D_t form a semigroup: $D_t D_s = D_{t+s}$;*
(3) *$D_t \mathbf{1} = \mathbf{1}$, where $\mathbf{1}$ denotes the function everywhere equal to 1; and*
(4) *$\|D_t f\| \leq \|f\|$, for all $f \in \mathcal{B}(X)$.*

This is a rephrasing of the properties of p and the fact that D_t is an integral operator with smooth kernel $p(x,y;t)$. Properties (1), (3) and (4) follow from the positivity of the heat kernel and the property

$$\int_X p(x,y;t) \cdot dy = 1$$

(Theorem B.6.8). Property (2) follows from the second assertion in Theorem B.6.7. For convenience, D_0 will denote the identity operator.

Remark. In the discussion below, it will be assumed that X is noncompact. The compact case can be treated similarly, and in fact some of the proofs below work in both settings; but treating both cases at the same time will unnecessarily complicate the exposition.

Definition B.10.2. The *intrinsic domain* of the semigroup D_t acting on $\mathcal{B}(X)$ is the subspace of $\mathcal{B}(X)$ consisting of all f for which

$$\lim_{t\to 0+} \|D_t f - f\| = 0.$$

Exercise B.10.3. Show that the intrinsic domain of D_t is a closed subspace of $\mathcal{B}(X)$ that is invariant under D_t.

Proposition B.10.4. *If $f \in \mathcal{B}(X)$ belongs to the intrinsic domain of D_t, then the function*

$$t \in [0,\infty) \to D_t f \in \mathcal{B}(X)$$

is continuous.

Proof. Continuity at 0 is immediate by the definition of the intrinsic domain. Let $t_n \to t > 0$. Then, by (2) and (4), it follows that

$$\|D_{t_n} f - D_t f\| \le \|D_{|t_n - t|} f - f\|,$$

which, as just observed, converges to 0 as $|t_n - t| \to 0$. □

Proposition B.10.5. *The intrinsic domain of the semigroup D_t contains every bounded, uniformly continuous function of X. In fact, if f is bounded and uniformly continuous, then* $\lim_{t\to 0} \sup_{0\le s\le t} \|D_s f - f\| = 0$.

Proof. Let f be bounded and uniformly continuous on X. Let $\varepsilon > 0$ and choose $\delta = \delta(\varepsilon) \le 1$ so that $|f(x) - f(y)| < \varepsilon$ whenever $d(x,y) < \delta$.

By splitting the integral expression for $D_t f(x)$ over $B(x,\delta)$ and its complement, we obtain that, for $0 < t < t_\delta$ (where t_δ is given by Proposition B.7.3),

$$\begin{aligned}
|D_t f(x) - f(x)| &\le \int_{B(x,\delta)} |f(y) - f(x)| p(x,y;t) \cdot dy \\
&\quad + \int_{X \setminus B(x,\delta)} |f(y) - f(x)| p(x,y;t) \cdot dy \\
&\le \varepsilon + 2\|f\| \int_{X \setminus B(x,\delta)} p(x,y;t) \cdot dy \\
&\le \varepsilon + \frac{2\|f\| A}{t^{(n+2)/2}} \exp(-\delta^2/Bt),
\end{aligned}$$

where the first part of the second inequality is because $|f(x) - f(y)| < \varepsilon$ on $B(x,\delta)$ and $\int_X p = 1$, and where the constants A, B appearing in the last inequality come from Proposition B.7.3. Therefore, for

$$t \le \min\{t_\delta, 2\delta^2/(n+2)B\},$$

we have

$$\sup_{0\le s\le t} \|D_s f - f\| \le \varepsilon + \frac{2\|f\|A}{t^{(n+2)/2}} \exp(-\delta^2/Bt),$$

and hence

$$\lim_{t\to 0} \sup_{0\le s\le t} \|D_s f - f\| \le \varepsilon,$$

for every $\varepsilon > 0$. □

The intrinsic domain of D_t in $\mathcal{B}(X)$ will not be identified. Instead, the action of D_t on the space $C_0(X)$ of continuous functions on X that vanish at infinity will be examined. Endowed with the supremum norm, $C_0(X)$ is a closed subspace of $\mathcal{B}(X)$. It will be shown that $C_0(X)$ is invariant under D_t, and that it is a subspace of the intrinsic domain of D_t acting on $\mathcal{B}(X)$.

Corollary B.10.6. *If $f \in C_0(X)$, then $D_t f \to f$ uniformly as $t \to 0+$.*

Proof. Every function $f \in C_0(X)$ is bounded and uniformly continuous on X. □

Proposition B.10.7. *The semigroup D_t preserves $C_0(X)$. That is, if f is a continuous function on X that vanishes at infinity, then $D_t f \in C_0(X)$, for all $t \ge 0$.*

Proof. Since D_0 is the identity operator, only the case of $t > 0$ needs to be considered. It follows from the estimates for the heat kernel stated in Theorem B.7.1 that, given $T > 0$, there are constants A, B such that

$$p(x,y;t) \le \frac{A}{t^{n/2}} e^{-d(x,y)^2/Bt},$$

for all $x, y \in X$ and all $t \le T$, where $n = \dim X$.

Let $f \in C_0(X)$. Then, given $\varepsilon > 0$, there exists a bounded domain K in X such that $|f(x)| \le \varepsilon$ for x in $X \setminus K$. Let $D = \{x \in X \mid d(x,K) < \delta\}$ (where δ is to be chosen). If $x \in X \setminus D$, then

$$\begin{aligned} D_t f(x) &= \int_X f(y)p(x,y;t)\cdot dy \\ &= \int_K f(y)p(x,y;t)\cdot dy + \int_{X\setminus K} f(y)p(x,y;t)\cdot dy. \end{aligned}$$

The absolute value of the first integral is bounded above by

$$\|f\|\frac{A}{t^{n/2}} e^{-\delta^2/Bt}\mathrm{vol}(K),$$

because $d(x, K) \geq \delta$, and this can be made smaller than the given ε by taking δ (*i.e.*, the domain D) sufficiently large. The absolute value of the second integral is bounded above by ε, because $\|f\| \leq \varepsilon$ on $X \smallsetminus K$ and $\int_X p = 1$. This, in fact, shows that $\|D_t f\| \leq 2\varepsilon$ on $X \smallsetminus D$, for all $t \leq T$. □

Corollary B.10.8. *The family of operators D_t, acting on the Banach space $C_0(X)$, is a diffusion semigroup. That is to say, it has the following three properties:*

(1) $\|D_t\| \leq 1$;

(2) $\lim_{t\to 0+} \|D_t f - f\| = 0$;

(3) $D_{s+t} = D_s D_t$, *for all* $s, t > 0$.

Associated to a semigroup of operators there is the *infinitesimal generator* which, in the case of $C_0(X)$ and D_t, is defined on the subspace of $C_0(X)$ consisting of all functions f for which the limit

$$\lim_{t\to 0+} \frac{D_t f - f}{t}$$

exists (in the Banach space topology of $C_0(X)$). If $\mathcal{D}$ denotes this subspace, then the infinitesimal generator $\triangle_{\mathrm{i}}$ is defined on $\mathcal{D}$ by the above limit. That is to say, if $f \in \mathcal{D}$, then $\triangle_{\mathrm{i}} f$ is the function in $C_0(X)$ such that

$$\lim_{t\to 0+} \left\| \frac{D_t f - f}{t} - \triangle_{\mathrm{i}} f \right\| = 0.$$

Exercise B.10.9. Show that the set $\mathcal{D}$ just defined is a linear subspace of $C_0(X)$.

In general, the domain of the infinitesimal generator of a diffusion semigroup is not easy to describe explicitly, but it often happens that its structure on a subspace of its domain can be described, as will be shown next. To this end, some basic properties of semigroups need elucidation, and for these it is convenient to recall the following notation. A continuous map $\phi : [a, b] \to B$ into a Banach space is differentiable at $s \in (a, b)$ if there exists an $f \in B$ such that $\|(\phi(s + t) - \phi(s))/t - f\| \to 0$ as $t \to 0$. In this case, write $(d/dt)\phi(s) = f$. The integral of ϕ over the interval $[a, b]$ is denoted by $\int_a^b \phi(s) \cdot ds$, and is the element of B constructed in a manner analogous to the way in which the Riemann integral of a continuous function is constructed.

Proposition B.10.10. *The subspace $\mathcal{D}$ is dense in $C_0(X)$. Moreover, if $f \in \mathcal{D}$, then $D_t f \in \mathcal{D}$ and*

$$\frac{d}{dt} D_t f = \triangle_{\mathrm{i}} D_t f = D_t \triangle_{\mathrm{i}} f.$$

Proof. Let $h \in C_0(X)$. If $h_r = \int_0^r D_s h \cdot ds$, then

$$\left\| \frac{1}{r} h_r - h \right\| \leq \frac{1}{r} \int_0^r \|D_s h - h\| \cdot ds,$$

which converges to 0 as $r \to 0$ by property (2) of the semigroup (Corollary B.10.8). To conclude the proof of the first assertion, it must be shown that $h_r \in \mathfrak{D}$. Since D_t is a continuous linear operator,

$$\begin{aligned} D_t h_r - h_r &= \int_0^r D_{t+s} h \cdot ds - \int_0^r D_s h \cdot ds \\ &= \int_r^{r+t} D_s h \cdot ds - \int_0^t D_s h \cdot ds. \end{aligned}$$

Therefore

$$\begin{aligned} \left\| \frac{1}{t} (D_t h_r - h_r) - (D_r h - h) \right\| &\leq \frac{1}{t} \int_r^{r+t} \|D_s h - D_r h\| \cdot ds \\ &\quad + \frac{1}{t} \int_0^t \|D_s h - h\| \cdot ds, \end{aligned}$$

and each of the two terms on the right converges to 0 as $t \to 0$. Hence

$$\triangle_{\mathrm{i}} h_r = D_r h - h.$$

It will now be shown that $D_s \mathfrak{D} \subset \mathfrak{D}$. Let $f \in \mathfrak{D}$. Since D_s is linear and $\|D_s\| \leq 1$,

$$\left\| \frac{1}{t}(D_t D_s f - D_s f) - D_s \triangle_{\mathrm{i}} f \right\| \leq \left\| \frac{1}{t}(D_t f - f) - \triangle_{\mathrm{i}} f \right\|,$$

which converges to 0 as $t \to 0+$. This shows that $\triangle_{\mathrm{i}}$ and D_s commute, and that the right derivative of $D_s f$ is $D_s \triangle_{\mathrm{i}} f = \triangle_{\mathrm{i}} D_s f$. That the left derivative exists and equals $\triangle_{\mathrm{i}} D_s f$ is left as an exercise. □

A part of the domain of the infinitesimal generator $\triangle_{\mathrm{i}}$ will now be identified, and its structure there will be described. The Laplace operator $\triangle$ is defined in a dense subspace of $C_0(X)$, namely on $C_c^2(X)$, and it leaves the (also dense) subspace $C_c^\infty(X)$ invariant because it is a differential operator. However, it should be noted that, in general, $C_c^2(X)$ is not all of the domain of $\triangle_{\mathrm{i}}$.

Exercise B.10.11. Let f be a compactly supported function on X of class C^2. Show that

$$\triangle D_t f(x) = D_t \triangle f(x),$$

for every $x \in X$ and $t \geq 0$.

Proposition B.10.12. *The subspace $C_c^2(X)$ of compactly supported functions of class C^2 on X is contained in $\mathfrak{D}$. Moreover, $\triangle = \triangle_{\mathrm{i}}$ on $C_c^2(X)$.*

Proof. If f is of class C^2 with compact support, then $\triangle f$ is continuous with compact support. By Exercise B.10.11, $\triangle D_t f = D_t \triangle f$ on X, so the identity $\triangle D_t f(x) = (\partial/\partial t) D_t f(x)$, for every $x \in X$, implies that

$$D_t f(x) - f(x) = \int_0^t \triangle D_s f(x) \cdot ds = \int_0^t D_s \triangle f(x) \cdot ds,$$

for each $x \in X$, hence that

$$\begin{aligned}\left| \frac{D_t f(x) - f(x)}{t} - \triangle f(x) \right| &= \frac{1}{t} \left| \int_0^t \big(D_s \triangle f(x) - \triangle f(x) \big) \cdot ds \right| \\ &\leq \frac{1}{t} \int_0^t \| D_s \triangle f - \triangle f \| \cdot ds \\ &\leq \sup_{0 \leq s \leq t} \| D_s \triangle f - \triangle f \|.\end{aligned}$$

By Proposition B.10.5 and Corollary B.10.6, this converges to 0 as $t \to 0$. □

Theorem B.10.13. *Let f be a bounded continuous function of class C^2 on X. Then $(D_t f - f)/t$ converges to $\triangle f$ as $t \to 0$, uniformly on compact sets.*

Proof. Let K be a compact subset of X, and let U_1 and U_2 be the neighborhoods of radius 1 and 2, respectively, around K in X. Then f can be written as a sum $f = f_1 + f_2$ of two bounded, continuous functions of class C^2 such that f_1 has compact support in U_2 and f_2 is zero on U_1.

By linearity of the Laplace operator, $\triangle f = \triangle f_1 + \triangle f_2$, and $\triangle f_2$ also vanishes on U_1. By linearity of the operators D_t,

$$\frac{D_t f - f}{t} = \frac{D_t f_1 - f_1}{t} + \frac{D_t f_2 - f_2}{t},$$

and the first term on the right converges uniformly to $\triangle f_1$ on account of Proposition B.10.12.

It then remains to show that $(D_t f_2 - f_2)/t \to 0$ as $t \to 0$, uniformly on K. If $x \in K$, then $B(x,1) \subset U_1$. Also, $f_2 \equiv 0$ on U_1, and so

$$\begin{aligned}\left| \frac{D_t f_2(x) - f_2(x)}{t} \right| &\leq \frac{1}{t} \int_{X \setminus B(x,1)} |f_2(y)| p(x,y;t) \cdot dy \\ &\leq \frac{\|f\|}{t} \int_{X \setminus B(x,1)} p(x,y;t) \cdot dy \\ &\leq \frac{A\|f\|}{t^{(n+2)/2}} \exp(-1/Bt),\end{aligned}$$

for $t \leq t_1$ and some constants A, B, where $n = \dim X$, as guaranteed by Proposition B.7.3. □

There is another family of operators associated to a diffusion semigroup. They are called the *resolvent operators* (or Green operators), are denoted by R_λ, $\lambda > 0$, and are defined by

$$R_\lambda f(x) = \int_0^\infty e^{-\lambda t} D_t f(x)\, dt.$$

Theorem B.10.14. *The domain $\mathcal{D}$ is exactly $R_\lambda(C_0(X))$, and the following identity holds true:*

$$\triangle_{\mathrm{i}} R_\lambda f - \lambda R_\lambda f = -f.$$

Proof. The proof consists of a formal calculation which is then fully justified on account of the facts that the operators R_λ and D_t commute and that R_λ is a continuous operator with norm $\|R_\lambda\| \le 1/\lambda$.

The formalities are as follows. If $f = R_\lambda h$, then, by a change of variables,

$$\begin{aligned} D_t f &= \int_0^\infty e^{-\lambda s} D_{s+t} h \cdot ds \\ &= e^{\lambda t} f - e^{\lambda t} \int_0^t e^{-\lambda s} D_s h \cdot ds. \end{aligned}$$

Hence

$$\frac{D_t f - f}{t} = \frac{e^{\lambda t} - 1}{t} f - \frac{e^{\lambda t}}{t} \int_0^t e^{-\lambda s} D_s h \cdot ds,$$

which converges to $\lambda f - h$ as $t \to 0$. Therefore $f \in \mathcal{D}$.

This also shows that, for a function $f \in C_0(X)$, $R_\lambda f$ is in $\mathcal{D}$, and that $\triangle_{\mathrm{i}} R_\lambda f = \lambda R_\lambda f - f$. If $f \in \mathcal{D}$, then $R_\lambda \triangle_{\mathrm{i}} f = \triangle_{\mathrm{i}} R_\lambda f$, because R_λ and D_t commute. Hence $R_\lambda \triangle_{\mathrm{i}} f = \lambda R_\lambda f - f$, and so $-f = \triangle_{\mathrm{i}} R_\lambda f - \lambda R_\lambda f$. □

Exercise B.10.15. Complete the details left out in the previous proof.

Exercise B.10.16. If $f \in \mathcal{D}$ and $\triangle_{\mathrm{i}} f - \lambda f = 0$ for some $\lambda > 0$, show that $f \equiv 0$.

Brownian Motion

The heat kernel and diffusion lead to a family of probability measures (the Wiener measures) on the space $\Omega(X)$ of continuous paths in X. For a particular point $x \in X$, the Wiener measure P_x assigns a probability to each Borel set $B \subset \Omega(X)$, interpreted as the probability that a Brownian particle, starting at x, will follow a path in B. This appendix, which builds on the previous one, contains an outline of this theory, adapted to complete Riemannian manifolds of bounded geometry.

C.1. Probabilistic Concepts

Let X be a nonempty set. A *ring of subsets* of X is a nonempty collection $\mathfrak{A}$ of subsets of X that is closed under finite unions and complements. If also $X \in \mathfrak{A}$, then $\mathfrak{A}$ is called a *field.* A *σ-field* is a field that is closed under countable unions.

The family of σ-fields on a set X is a lattice. Thus, given a nonempty collection S of subsets of X, there is always a smallest σ-field that contains all the elements of S (namely, the intersection of all σ-fields that contain S). It is called the σ-field generated by S and is usually denoted by $\sigma(S)$.

Example C.1.1. Let X be a topological space, and let $\mathfrak{B}$ denote the σ-field generated by the open subsets of X. This σ-field is called the Borel σ-field of X.

A pair $(X, \mathfrak{B})$, consisting of a set and a σ-field, is called a *measurable space*. By abuse of notation, the measurable space is often denoted by X alone. A mapping $f : X \to Y$ of measurable spaces, with respective σ-fields $\mathfrak{B}_X$ and $\mathfrak{B}_Y$, is called measurable if $f^{-1}B \in \mathfrak{B}_X$, for every $B \in \mathfrak{B}_Y$.

Exercise C.1.2. Let X be a locally compact Hausdorff space and let $\mathfrak{B}_0$ be the smallest σ-field for which all the compactly supported continuous functions are measurable. Elements of this σ-field are called Baire sets.

(1) Show that every Baire set is a Borel set, but not conversely.

(2) Show that, if X is second countable, then every Borel set is a Baire set.

A *probability space* is a triple $(X, \mathfrak{A}, P)$ consisting of a space X, a σ-field $\mathfrak{A}$ of subsets of X and a measure P defined on $\mathfrak{A}$ with $P(X) = 1$ (called a probability measure). Elements of $\mathfrak{A}$ are called measurable sets or events. If $\mathcal{P}$ is a property that elements of X may or may not have, then the set $\{x \in X \mid x \text{ satisfies } \mathcal{P}\}$ will usually be abbreviated as $\{\mathcal{P}\}$. This has the effect of shortening some expressions.

The term "function" between measurable spaces will mean a measurable function (unless otherwise stated or clear from context). If

$$f : X \to [-\infty, +\infty]$$

is a function (where $[-\infty, +\infty]$ has the Borel σ-field), then its *expected value*, or *expectation*, is

$$E[f] = \int_X f(x) \cdot P(x).$$

This is well defined if f is integrable with respect to P. It also makes sense when f is not integrable but is semi-bounded (that is, either bounded above or bounded below), in which case it is $+\infty$ if f is bounded below and $-\infty$ if f is bounded above.

If $A \subset X$ is measurable and f is a function as above (integrable or nonnegative), then the notation $E[f; A]$ will mean

$$\int_A f(x) \cdot P(x) = E[f \cdot \chi_A],$$

χ_A being the characteristic function of A.

In the next section, a measure space X, together with a family of probability measures parametrized by the points of another space Z, will often be considered. That is, there will be a probability measure P_z on X, defined for each $z \in Z$, with associated expectation E_z. If f is a function on X, then $z \mapsto E_z[f]$ is a function on Z, which will be denoted by $E_\bullet[f]$. When $f = \chi_A$ is the characteristic function of a measurable set A, the notation $P_\bullet[A]$ for $E_\bullet[\chi_A]$ will be used.

One of the fundamental concepts of probability is that of conditional expectation. This is a generalization of a familiar concept from elementary probability theory, that of conditional probability. If A and B are events in

$\mathfrak{A}$ with $P[B] > 0$, then the (conditional) probability of A given B is defined to be

$$P\left[A \mid B\right] = P[A \cap B]/P[B].$$

This concept is generalized as follows. Let $\mathfrak{B}$ be a σ-subfield of $\mathfrak{A}$, and let f be a function on $(X, \mathfrak{A})$ that is either integrable or nonnegative.

Definition C.1.3. The *conditional expectation* of f with respect to the σ-field $\mathfrak{B}$ is a function $E\left[f \mid \mathfrak{B}\right]$ on X that is measurable with respect to $\mathfrak{B}$ and satisfies

$$\int_B E\left[f \mid \mathfrak{B}\right](x) \cdot P(x) = \int_B f(x) \cdot P(x),$$

for all $B \in \mathfrak{B}$.

Again, if $f = \chi_A$ for some $A \in \mathfrak{A}$, then $P\left[A \mid \mathfrak{B}\right]$ has the same meaning as $E\left[\chi_A \mid \mathfrak{B}\right]$.

It is clear that the conditional expectation function $E[f \mid \mathfrak{B}]$ is unique in the "almost everywhere" sense. It is also clear that, if f is $\mathfrak{B}$-measurable, then $E[f \mid \mathfrak{B}] = f$. The interesting fact, however, is that a conditional expectation can always be constructed using the Radon-Nikodym theorem.

Theorem C.1.4. *Let $(Y, \mathfrak{M})$ be a measure space. Let μ be a σ-finite positive measure and let ν be a finite signed measure that is absolutely continuous with respect to μ. Then there exists a μ-integrable function h on Y, unique μ-almost everywhere, such that, for all $B \in \mathfrak{M}$,*

$$\nu(B) \int_B h(x) \cdot \mu(x).$$

Remark. The proof of this theorem actually gives more than is stated. If ν is positive and σ-finite, the proof provides a function h that satisfies the displayed identity. This function h is also unique in the same sense. It is not generally integrable, but it is σ-integrable, meaning that there is a countable partition $Y = \bigcup_n Y_n$ of Y such that h is integrable on each Y_n.

Theorem C.1.5. *Let $(X, \mathfrak{A}, P)$ be a probability space, $\mathfrak{B} \subset \mathfrak{A}$ a σ-subfield, and let $f : X \to \mathbb{R}$ be a function on X that is P-integrable or nonnegative. Then the conditional expectation $E\left[f \mid \mathfrak{B}\right]$ exists.*

Proof. For $B \in \mathfrak{B}$, set

$$\nu(B) = \int_B f(x) \cdot P(x).$$

Then ν is a finite signed measure (a positive σ-finite measure if f is only nonnegative) on the measure space $(X, \mathfrak{B})$, which is absolutely continuous with respect to $P|\mathfrak{B}$. The Radon-Nikodym theorem guarantees the existence of a function h (that belongs to $L^1(X, \mathfrak{B}, P)$ in the finite case) such that

$$\mu(B) = \int_B h(w) \cdot P(w).$$

If f is only nonnegative, then h is nonnegative also, but only σ-integrable. □

For easy reference, two basic properties of conditional expectation are included in the next theorem, and the exercises following it contain other properties.

Theorem C.1.6. *Let $(X, \mathfrak{A}, P)$ be a probability space, let $\mathfrak{B}$, $\mathfrak{B}_1$ and $\mathfrak{B}_2$ be σ-subfields of $\mathfrak{A}$, and let f, h be integrable functions.*

(1) *If f is $\mathfrak{B}$-measurable, then*
$$E\left[fh \mid \mathfrak{B}\right] = fE\left[h \mid \mathfrak{B}\right];$$
(2) *If $\mathfrak{B}_1 \subset \mathfrak{B}_2$, then*
$$E\left[E\left[f \mid \mathfrak{B}_2\right] \mid \mathfrak{B}_1\right] = E\left[f \mid \mathfrak{B}_1\right].$$

Exercise C.1.7. If $B_1 \subset B_2$ and f are as above, show that
$$E\left[E\left[f \mid \mathfrak{B}_1\right] \mid \mathfrak{B}_2\right] = E\left[f \mid \mathfrak{B}_1\right].$$
Also, if $E\left[f \mid \mathfrak{B}_2\right]$ is $\mathfrak{B}_1$-measurable, show that
$$E\left[f \mid \mathfrak{B}_2\right] = E\left[f \mid \mathfrak{B}_1\right].$$

Exercise C.1.8. Let $(X, \mathfrak{A}, P)$ be a probability space, f a function on X. If $\mathfrak{A} = \{\emptyset, X\}$ is the trivial σ-field, what is $E\left[f \mid \mathfrak{A}\right]$?

The following theorem, known as the *monotone class theorem*, is a very useful tool. A proof can be found in [**31**], amongst other places.

Theorem C.1.9. *Let S be a set and let $\mathfrak{A}$ be a family of subsets of S that is closed under finite intersections. Let $\mathfrak{B}$ be a family of subsets of S such that $S \in \mathfrak{B}$ and $\mathfrak{A} \subset \mathfrak{B}$, and such that*

(1) *if $B_n \in \mathfrak{B}$ and $B_n \subset B_{n+1}$, for all $n \geq 0$, then $\bigcup_{n=0}^{\infty} B_n \in \mathfrak{B}$;*
(2) *if $A \subset B$ and $A, B \in \mathfrak{B}$, then $B \setminus A \in \mathfrak{B}$.*

Then the σ-field $\sigma(\mathfrak{A})$ generated by $\mathfrak{A}$ is contained in $\mathfrak{B}$.

The following theorem has the same flavor.

Theorem C.1.10. *Let S be a set and let $\mathfrak{A}$ be a family of subsets of S that is closed under finite intersections. Let $\mathcal{H}$ be a vector space of real-valued functions on S such that*

(1) $\mathbf{1} \in \mathcal{H}$, *and* $\chi_A \in \mathcal{H}$ *when* $A \in \mathfrak{A}$;

(2) *if* $f_n \in \mathcal{H}$ *are nonnegative functions that increase to a bounded function* f, *then* $f \in \mathcal{H}$.

Then $\mathcal{H}$ *contains all the bounded functions on* S *that are* $\sigma(\mathfrak{A})$*-measurable.*

C.2. Construction of Brownian Motion

Let X be a complete, connected Riemannian manifold of bounded geometry. The space $\Omega(X)$ of continuous paths $\omega : [0, \infty) \to X$ can be thought of as the set of possible paths that a Brownian particle, located at $\omega(0)$ at time $t = 0$, might follow as time progresses. The heat kernel will be used to construct a family $\{P_x\}_{x \in X}$ of probability measures on $\Omega(X)$, $P_x[B]$ being the probability that a Brownian particle that starts at the point $x \in X$ at time $t = 0$ will follow a path belonging to the Borel set $B \subset \Omega(X)$. Thus, construction of Brownian motion on a manifold X amounts to giving this family of probability measures possessing certain properties that will be reviewed presently.

In order to construct such measures, we view the space $\Omega(X)$ as a subspace of $X^{[0,\infty)}$, the space of all maps $[0, \infty) \to X$ (not necessarily continuous). While the latter space carries the product topology, the Borel σ-field associated to that topology is too large for most purposes. Instead, the σ-field $\mathfrak{G}$ generated by the cylinder sets is considered, these being sets of the form

$$C = \{\omega \in X^{[0,\infty)} \mid \omega(t_1) \in B_1, \ldots, \omega(t_n) \in B_n\},$$

where $B_1, \ldots, B_n$ are Borel subsets of X and $0 \leq t_1 < t_2 < \cdots < t_n$. That is, C consists of those elements ω in $X^{[0,\infty)}$ that visit B_i at time t_i.

To understand the structure of $(X^{[0,\infty)}, \mathfrak{G})$, it is convenient to view this measure space as the inverse limit of an inverse system of much nicer measure spaces. The collection of finite subsets of $[0, \infty)$ is partially ordered by inclusion. To each finite set $F \subset [0, \infty)$ there is associated the measure space $(X^F, \mathfrak{B}^F)$, where $\mathfrak{B}^F$ is the Borel σ-field associated to the product topology. This σ-field is generated by the cylinder sets $C^F = B_1 \times \cdots \times B_n \subset X^F$, where $F = \{0 < t_1 < \cdots < t_n\}$ is a set of times t_i at which the elements of C^F visit B_i, each B_i being a Borel subset of X. Each inclusion of finite sets $E \subset F$ canonically defines a projection $\pi_{EF} : X^F \to X^E$ that drops the finitely many coordinates in $F \setminus E$. These projections are continuous, hence Borel measurable, and they satisfy

$$\pi_{EF} \circ \pi_{FG} = \pi_{EG},$$

whenever $E \subset F \subset G$. In conclusion, an inverse (or *projective*) system of measure spaces has been constructed, and its inverse limit will now be identified.

Denote by $\pi_t : X^{[0,\infty)} \to X$ the time t evaluation map, $\pi_t(\omega) = \omega(t)$. For nonempty subsets A of $[0, \infty)$, let

$$\pi_A : X^{[0,\infty)} \to X^A$$

be the evaluation map defined by $\pi_A(\omega)(\omega(t))_{t \in A}$. The cylinder subsets C of $X^{[0,\infty)}$ are exactly the sets of the form $C = \pi_F^{-1}(C^F)$, for some finite subset $F \subset [0, \infty)$ and some cylinder set $C^F \subset X^F$.

The inverse system $\{(X^F, \mathfrak{B}^F), \pi_{EF} \mid E, F \subset [0, \infty) \text{ finite}\}$ of measure spaces has $(X^{[0,\infty)}, \mathfrak{G})$ as its inverse limit, with canonical projections

$$\pi_F : X^{[0,\infty)} \to X^F$$

satisfying

$$\pi_{EF} \circ \pi_F = \pi_E.$$

The σ-field $\mathfrak{G}$ generated by the cylinder sets is the smallest for which all the projections π_F are measurable.

The Borel σ-field of a finite product X^F is well understood. If Q is a countable (*i.e.*, at most countably infinite) subset of $[0, \infty)$, then the Borel σ-field $\mathfrak{B}^Q$ of the product space X^Q is what is expected, namely, the smallest σ-field making all the projections $\pi_q : X^Q \to X$ $(q \in Q)$ measurable. It is generated by its cylinder sets, as is readily verified.

Definition C.2.1. A subset A of $X^{[0,\infty)}$ is called a σ-cylinder if it is of the form $A = \pi_Q^{-1} B$, where $Q \subset [0, \infty)$ is countable and B is a Borel subset of X^Q.

Proposition C.2.2. *The collection of σ-cylinders in $X^{[0,\infty)}$ coincides with the σ-field $\mathfrak{G}$.*

Proof. Evidently, $\mathfrak{G}$ contains all σ-cylinders. Since all cylinder sets are σ-cylinders, it only needs to be shown that the σ-cylinders form a σ-field. The complement of a σ-cylinder is a σ-cylinder, while closure under countable unions is shown as follows. If $\{A_n = \pi_{Q_n}^{-1} B_n\}_{n=1}^{\infty}$ is a countable collection of σ-cylinders, then

$$\bigcup_{n=1}^{\infty} A_n \pi_Q^{-1} \bigcup_n \pi_{Q_n Q}^{-1} B_n,$$

where $Q = \bigcup_n Q_n$ is countable and $\pi_{Q_n Q}^{-1} B_n \subset X^Q$ is Borel. □

Corollary C.2.3. *If $A \in \mathfrak{G}$, there is a countable subset $Q \subset [0, \infty)$ with the property that, whenever $\omega \in X^{[0,\infty)}$ agrees with some $\eta \in A$ on the set*

Q, then $\omega \in A$. Without loss of generality, Q can be taken to be dense in $[0,\infty)$.

For each $x \in X$, a probability P_x on the measure space $(X^{[0,\infty)}, \mathfrak{G})$ will now be defined. To begin with, if $F = \{0 \leq t_1 < t_2 < \cdots < t_n\}$ and $B_1 \times B_2 \times \cdots \times B_n \subset X^F$ is a cylinder set C^F, define

$$(*) \quad P_x^F(C^F) \int_{B_1}\int_{B_2}\cdots\int_{B_n} p(x,y_1;t_1)\,p(y_1,y_2;t_2-t_1) \cdots p(y_{n-1},y_n;t_n-t_{n-1})\cdot dy_n\cdots dy_2\cdot dy_1,$$

where $p(z,y;t)$ is the heat kernel. (For the case $t_1 = 0$, see the remark below.)

Proposition C.2.4. *If $E \subset F$ are finite subsets of $[0,\infty)$, C^E is a cylinder set in X^E and $C^F = \pi_{EF}^{-1}(C^E)$, then $P_x^F(C^F) = P_x^E(C^E)$.*

This is an easy consequence of the semigroup property of the heat kernel. Therefore, P_x can be well-defined on any cylinder set $C = \pi_F^{-1}(C^F)$ by the formula $(*)$.

Proposition C.2.5. *The above definition of P_x on cylinder sets extends uniquely to a probability measure on $(X^{[0,\infty)}, \mathfrak{G})$.*

This is a corollary of the Kolmogoroff existence theorem of measures on product spaces [**51**, Chapter 12]. A proof of this proposition uses the Carathéodory-Hahn extension theorem, following the procedure set forth in [**183**, Chapter 11, Section 5]. Briefly, let $\mathcal{S}$ denote the (non-σ-)field generated by the cylinder sets in $X^{[0,\infty)}$. Then P_x extends to a "measure" (as defined in [**183**, p. 205]) on $\mathcal{S}$, hence to an outer measure on the family of all subsets of $X^{[0,\infty)}$. The σ-field of sets that are measurable with respect to this outer measure contains the cylinders, hence contains the σ-field $\mathfrak{G}$. The Carathéodory-Hahn extension theorem [**183**, (11.20)] then guarantees that the restriction of this outer measure to $\mathfrak{G}$ is an honest measure and is the unique measure agreeing with P_x on the cylinder sets. This measure will still be denoted by P_x. Since X itself is a cylinder set and $X^{[0,\infty)} = \pi_t^{-1}(X)$, for any $t \in [0,\infty)$, the fact that $\int_X p(x,y;t)\cdot dy = 1$ implies that P_x is a probability measure.

Remark. The expression $p(x,y;0)$ is not defined as a function but as a measure: $p(x,y;0) = \delta_x(y)$. This is consistent with

$$\int_X p(x,y,0)f(x)\cdot dy = D_0(f)(x) = f(x).$$

Therefore, if $t_1 = 0$ in the above cylinder set C, then

$$P_x(C) = \chi_{B_1}(x) \int_{B_2} \cdots \int_{B_n} p(x, y_2; t_2) \cdots p(y_{n-1}, y_n; t_n - t_{n-1}) \cdot dy_n \cdots dy_2.$$

It follows that the measure P_x is supported on the set of $\omega \in X^{[0,\infty)}$ with $\omega(0) = x$.

Corollary C.2.6. *Let F and $B = \pi_F^{-1}(B_1 \times \cdots \times B_n)$ be as above and let $f : X^F \to \mathbb{R}$ be integrable on $B_1 \times \cdots \times B_n$. Then $\pi_F^*(f) = f \circ \pi_F$ is P_x-integrable on B and*

$$\int_B \pi_F^*(f)(\omega) \cdot P_x(\omega) = \int_{B_n} \cdots \int_{B_1} f(y_1, \ldots, y_n) p(x, y_1; t_1) \cdots p(y_{n-1}, y_n; t_n - t_{n-1}) \cdot dy_1 \cdots dy_n.$$

Proof. This is immediate if $f \equiv 1$, from which it follows that

$$\pi_{F*}(P_x) = p(x, y_1; t_1) \cdots p(y_{n-1}, y_n; t_n - t_{n-1}) \cdot dy_1 \cdots dy_n.$$

Since

$$\int_B \pi_F^*(f) \cdot P_x = \int_{B_1 \times \cdots \times B_n} f \cdot (\pi_{F*} P_x),$$

the assertion is proven. □

Exercise C.2.7. Let $f : X^n \to \mathbb{R}$ be a bounded function. Show that

$$E_x \left[f(\omega(t_1), \cdots, \omega(t_n)) \right]$$

is a continuous function of $x \in X$ and $(t_1, \cdots, t_n) \in \mathbb{R}^n$, $0 < t_1 < \cdots < t_n$.

Going from a probability measure on $X^{[0,\infty)}$ to one on $\Omega(X)$ is a more difficult task, since it is rather obvious that $\Omega(X)$ is not an element of $\mathfrak{G}$ (continuity imposes uncountably many conditions). The space $\Omega(X)$ of continuous paths in X is a metrizable space with the compact-open topology; hence it has its own Borel field generated by open subsets. Cylinder sets can also be used to define a σ-field on $\Omega(X)$ via the inclusion into $X^{[0,\infty)}$, but it is not immediately clear that this induced field shares any kind of relation with the Borel field of $\Omega(X)$.

Exercise C.2.8. Prove that the field $\mathfrak{F}$, generated by the cylinder sets of $\Omega(X)$, coincides with the Borel field of $\Omega(X)$ relative to the compact-open topology.

Corollary C.2.9. *Every probability measure on $(\Omega(X), \mathfrak{F})$ is uniquely determined by its values on the cylinder sets.*

Exercise C.2.10. Is there a nonempty Borel subset of $\Omega(X)$ that is also an element of $\mathfrak{G}$?

Exercise C.2.11. By Proposition C.2.2, membership in an event $A \subset X^{[0,\infty)}$ is detected by a countable number of coordinates. More generally, if f is a measurable function on $(X^{[0,\infty)}, \mathfrak{G})$, show that there are a countable set Q and a measurable function h on X^Q such that $f(\omega) = h(\pi_Q(\omega))$.

Corollary C.2.9 indicates the way to construct probability measures on $\Omega(X)$. It starts with the above construction of the measures P_x on $X^{[0,\infty)}$, after which $\Omega(X)$ is added to the Borel field of $X^{[0,\infty)}$, and then it is shown that P_x induces a probability measure (again denoted by P_x) on $\Omega(X)$. Here is the technical tool needed to guarantee the viability of this procedure.

Proposition C.2.12. *Let $(X, \mathfrak{H}, P)$ be a probability space and let Y be a subset of X with outer measure 1 relative to P. Let $\mathfrak{H}_0$ be the σ-field on Y consisting of the sets of the form $B \cap Y$, with $B \in \mathfrak{H}$. Then defining $P_0(B \cap Y) = P(B)$ produces a probability measure P_0 on $(Y, \mathfrak{H}_0)$.*

Indeed, any measurable subset of $X \setminus Y$ has measure zero. If $A, B \in \mathfrak{H}$ and $A \cap Y = B \cap Y$, then the symmetric difference $A \triangle B \subset X \setminus Y$, so $P(A) = P(B)$ and P_0 is well defined.

Thus, the construction of Brownian motion will be completed by proving the following theorem.

Theorem C.2.13. *The subset $\Omega(X) \subset X^{[0,\infty)}$ of continuous paths has outer measure 1 with respect to P_x.*

In plain language this says that "sample paths are continuous." By an abuse of notation, the induced probability measures on $\Omega(X)$ will again be denoted by P_x. The proof of this theorem is postponed to Section C.4, after first examining some consequences of the semigroup property of the heat kernel that will be needed.

C.3. The Markov Process

In the proof of Theorem C.2.13, the Markov process associated to the probability spaces $(X^{[0,\infty)}, \mathfrak{G}, P_x)$ will play a small role. Once the legitimacy of the probability spaces $(\Omega(X), \mathfrak{F}, P_x)$ is established, the completely analogous Markov process for those is obtained, and this plays a major role. In this section, the process will be defined for both cases, letting $(\Theta(X), \mathfrak{H}, P_x)$ stand either for $(X^{[0,\infty)}, \mathfrak{G}, P_x)$ or, once Theorem C.2.13 has been established, for $(\Omega(X), \mathfrak{F}, P_x)$. In either case, the elements of $\Theta(X)$ will often be referred to as paths.

To describe the Brownian motion process associated to the complete Riemannian manifold (X, g), introduce a filtration of σ-fields in the path space $\Theta(X)$ as follows. Associated to the projection π_t there is a subfield of $\mathfrak{H}$, denoted by $\mathfrak{H}_t$, which is characterized as the one generated by all the $\pi_F^{-1}(\mathfrak{B}^F)$ as F runs over all finite subsets of $[0, \infty)$ with $\max F \leq t$. That is, $\mathfrak{H}_t$ is generated by all cylinder sets in $\Theta(X)$ whose associated sequence $t_1, \ldots, t_n$ is bounded above by t. Hence $\mathfrak{H}_t$ keeps track of happenings up to time t. It is clear that, if $s \leq t$, then $\mathfrak{H}_s \subset \mathfrak{H}_t$. The whole family $\{\mathfrak{H}_t\}_{t\in[0,\infty)}$ is called a filtration of fields.

The interesting property of the maps π_t is that they are adapted to the filtration $\mathfrak{H}_t$, meaning that the map π_t is measurable with respect to $\mathfrak{H}_t$ for each $t \geq 0$.

Exercise C.3.1. It is also convenient to introduce the σ-field $\mathfrak{H}_{t+}$, defined as

$$\mathfrak{H}_{t+} = \bigcap_{s>t} \mathfrak{H}_s.$$

This field keeps track of events that happen, not only up to time t, but also immediately after t.

(1) Show that $\mathfrak{H}_{t+}$ is a filtration of σ-fields that is continuous in the sense that $(\mathfrak{H}_{t+})_+ = \mathfrak{H}_{t+}$.

(2) It is clear that $\mathfrak{H}_t \subset \mathfrak{H}_{t+}$. Show that the inclusion is proper by taking a close look at the event comprising all continuous paths $\omega : [0, \infty) \to \mathbb{R}$ for which $\inf\{s > t \mid \pi_s(\omega) > 0\} = t$.

At this point, we have constructed a measure space $(\Theta(X), \mathfrak{H})$ and a family of projections $\pi_t : (\Theta(X), \mathfrak{H}) \to (X, \mathfrak{B})$ (where $\mathfrak{B}$ is the Borel field of the manifold X) that are adapted to the filtration of σ-fields $\mathfrak{H}_t$ of $\mathfrak{H}$. Such a system is called a *stochastic process* and, by itself, is not much of a structure.

This stochastic process will now be enhanced to a *Markov process.* In order to define this notion, some notation is required. If $A \in \mathfrak{H}$, then $P_\bullet(A)$ is a $\mathfrak{B}$-measurable function on X, and the composition

$$P_{\pi_s}(A) = P_\bullet(A) \circ \pi_s$$

is then $\mathfrak{H}_s$-measurable.

The stochastic process $\{\Theta(X), \mathfrak{H}, \pi_t, \mathfrak{H}_t\}_{0\leq t<\infty}$, together with the family $P = \{P_x\}_{x\in X}$ of probability measures on $\Theta(X)$, satisfies the following defining conditions for a Markov process:

(1) P_x is a probability measure on $\Theta(X)$ that is supported on the fiber $\pi_0^{-1}(x) = \Theta(X)_x$. That is, $\Theta(X)_x$ is the space of paths ω issuing from $x = \omega(0)$ at time 0 and $P_x(\Theta(X)_x) = 1$.

(2) For each $B \in \mathfrak{H}$, the map $x \mapsto P_x(B)$ is measurable.

(3) For every $x \in X$, $t \geq s \geq 0$, $A \subset X$ a Borel set,

$$P_x\left[\{\omega(t) \in A\} \mid \mathfrak{H}_s\right] = P_{\pi_s}\left[\{\omega(t-s) \in A\}\right],$$

almost everywhere with respect to P_x.

For $\Theta(X) = \Omega(X)$, this Markov process is called the *Brownian motion process* of (X, g).

Item (3) above is called the *Markov property.* It is a version of the semigroup property of the heat kernel. Details are left to the following exercise.

Exercise C.3.2. Let $A \subset X$ be measurable, and let

$$B = \{\omega \in \Theta(X) \mid \omega(t_i) \in B_i, 0 \leq t_1 < \cdots < t_n \leq s\} \in \mathfrak{H}_s$$

be a cylinder set. Using Corollary C.2.6 and the semigroup property of the heat kernel, prove that

$$\begin{aligned}
\int_B & P_{\eta(s)}[\{\omega(t-s) \in A\}] \cdot P_x(\eta) \\
&= \int_{B_1} \cdots \int_{B_n} p(x, y_1; t_1) \cdots p(y_{n-1}, y_n; t_n - t_{n-1}) \\
&\qquad \cdot \left[\int_X D_{t-s}(\chi_A)(y) p(y_n, y; s - t_n) \cdot dy\right] \cdot dy_n \cdots dy_1 \\
&= P_x\left[\{\omega(t) \in A\} \cap B\right].
\end{aligned}$$

Deduce the Markov property (3).

There is also a Markov property for functions. For this, a dynamical system on the space $\Theta(X)$ is first introduced. For $t \geq 0$, let

$$\theta_t : \Theta(X) \to \Theta(X)$$

denote the map that takes a path ω to the path $\theta_t\omega$ determined by

$$\theta_t\omega(s) = \omega(s+t).$$

The family $\theta = \{\theta_t\}$ is a semigroup of measurable transformations of the space$(\Theta(X), \mathfrak{H})$.

Exercise C.3.3. Prove that θ_t is a measurable transformation of $(\Theta(X), \mathfrak{H}_s)$ onto $(\Theta(X), \mathfrak{H}_{s-t})$, $s \geq t$.

The following theorem is the version of the Markov property for functions.

Theorem C.3.4. *The Markov property (3) is equivalent to the following version. For every $s \geq 0$ and every bounded measurable function f,*

$$E_x [f \circ \theta_s \mid \mathfrak{H}_{s+}] = E_{\pi_s} [f],$$

almost everywhere with respect to P_x.

Here, $E_{\pi_s} [f] = E_\bullet [f] \circ \pi_s$ is the function $\omega \mapsto E_{\omega(s)} [f]$.

Proof. By the definition of conditional expectation, it needs to be shown that

$$E_x [f \circ \theta_s; A] = E_x [E_{\pi_s} [f]; A],$$

for all $A \in \mathfrak{H}_{s+}$.

It will now be shown that the statement holds true for functions f of the form

$$f(\omega) = f_1(\omega(t_1)) \cdot f_2(\omega(t_2)) \cdots f_n(\omega(t_n)),$$

where f_i are bounded functions on X and $0 < t_1 < \cdots < t_n$, and for sets A of the form

$$A = \{\omega \mid \omega(s_i) \in B_i, i = 1, \cdots, k\},$$

with $h \in (0, t_1)$ and $0 < s_1 < \cdots < s_k \leq s+h$. The monotone class theorems C.1.9 and C.1.10 will do the rest.

The definition of the measure P_x implies that, for f and A as just described,

$$\begin{aligned} E_x [f \circ \pi_s; A] &= E_x \left[\prod_{i=1}^{k} (\chi_{B_i} \circ \pi_{s_i}) (\chi_X \circ \pi_{s+h}) \prod_{j=1}^{n} f_j \circ \pi_{t_j+s} \right] \\ &= \int_{B_1} p(x, x_1; s_1) \cdot dx_1 \cdots \int_{B_k} p(x_{k-1}, x_k; s_k - s_{k-1}) \cdot dx_k \\ &\quad \cdot \int_X F(y,h) p(x_k, y; s+h-s_k) \cdot dy \end{aligned}$$

where $F(y, h)$ is the function

$$\int_X f_1(y_1) p(y, y_1; t_1 - h) \cdot dy_1 \cdots \int_X f_n(y_n) p(y_{n-1}, y_n; t_n - t_{n-1}) \cdot dy_n.$$

A similar calculation then shows that, for such f and A,

$$E_x [F(\pi_{s+h}, h); A] = E_x [f \circ \pi_s; A].$$

Application of C.1.9 implies that this identity holds for all sets $A \in \mathfrak{H}_{s+h}$.

By Exercise C.2.7, the function $h, \omega \mapsto F(\pi_h(\omega), h)$ is continuous and $F(\pi_h(\omega), h) \to F(\pi_0(\omega), 0)$ as $h \to 0$. Since $\mathfrak{H}_{s+h} \supset \mathfrak{H}_{s+}$ for all $h > 0$, letting $h \to 0$ gives

$$E_x [F(\pi_s, 0); A] = E_x [f \circ \pi_s; A],$$

for all $A \in \mathfrak{H}_{s+}$.

The proof is concluded by a straightforward application of the monotone class theorem for functions (Theorem C.1.10). □

The following result can be proven along similar lines.

Proposition C.3.5. *If f is a bounded measurable function on $\Theta(X)$, then*

$$E_x [f \mid \mathfrak{H}_{s+}] = E_x [f \mid \mathfrak{H}_s],$$

for all $s \geq 0$.

That is, if $A \in \mathfrak{H}_{s+}$, then $\chi_A = E_x [\chi_A \mid \mathfrak{H}_s] \in \mathfrak{H}_s$; which is to say that the set A differs from a set in $\mathfrak{H}_s$ by a P_x-null set. This has the following fundamental consequence, known as *Blumenthal's zero-one law.*

Theorem C.3.6. *Let $x \in X$. If $A \in \mathfrak{H}_{0+}$, then $P_x [A]$ is either 0 or 1.*

Proof. The previous proposition says that there is a set $A_0 \in \mathfrak{H}_0$ such that $P_x [A] = P_x [A_0]$. But each $A_0 \in \mathfrak{H}_0$ is of the form $A_0 = \{\omega \mid \omega(0) \in B\}$, for some Borel set $B \subset X$. Hence $P_x [A_0] = \chi_B(x) \in \{0, 1\}$. □

The following exercise outlines a slight extension of the Markov property that will be required in Section C.4.

Exercise C.3.7. Let $\tau : \Theta(X) \to [0, \infty]$ be a measurable function that takes only a finite number of values and satisfies the following property: For every $t \geq 0$, the set $\{\omega \mid \tau(\omega) \leq t\}$ is an element of $\mathfrak{H}_{t+}$. Let $\mathfrak{H}_{\tau+}$ denote the σ-field formed by the measurable sets A for which $A \cap \{\tau(\omega) \leq t\} \in \mathfrak{H}_{t+}$ for all $t \geq 0$. Let $f(s, \omega)$ be a bounded measurable function on $[0, \infty) \times \Theta(X)$. Then

$$E_x [f \circ (\tau, \theta_\tau) \mid \mathfrak{H}_{\tau+}] (\omega) = E_{\pi_\tau(\omega)} [f(\tau(\omega), \bullet)],$$

for P_x-almost all ω in $\{\tau < \infty\}$, where $\theta_\tau(\omega)(t) = \omega(t + \tau(\omega))$ and $\pi_\tau(\omega) = \omega(\tau(\omega))$ on $\{\tau < \infty\}$. The right side of this identity is the value at ω of the function on $\{\tau < \infty\}$ given by

$$\omega \mapsto E_{\pi_\tau(\omega)} [f(\tau(\omega), \bullet)] \int_{\Theta(X)} f(\tau(\omega), \omega') \cdot P_{\pi_\tau(\omega)}(\omega').$$

The following statement contains the interpretation of diffusion of functions in the context that has been developed.

Proposition C.3.8. *If $f : X \to \mathbb{R}$, its diffusion is*

$$D_t f(x) = \int_{\Theta(X)} f \circ \pi_t(\omega) \cdot P_x(\omega) = E_x [f \circ \pi_t].$$

Proof. Indeed,

$$\begin{aligned} E_x [f \circ \pi_t] & \int_{\Theta(X)} (f \circ \pi_t)(\omega) \cdot P_x(\omega) \\ &= \int_{\Theta(X)} \pi_t^*(f)(\omega) \cdot P_x(\omega) \\ &= \int_X f(y) \cdot (\pi_{t*} P_x)(y). \end{aligned}$$

Thus, by Corollary C.2.6,

$$E_x [f \circ \pi_t] \int_X f(y) p(x, y; t) \cdot dy = D_t f(x).$$

□

C.4. Continuity of Brownian Paths

As noted in Section C.2, the final step in the construction of Brownian motion is to prove Theorem C.2.13. Actually, this is one of the most interesting facts of the construction, and it requires some calculations. It is also one of the oldest interesting properties of Brownian motion, appearing in the original work of N. Wiener and of P. Lévy. See the survey by K. Itô and H. P. McKean [**106**] for these and other elegant proofs of continuity of sample paths for Brownian motion on the real line.

The approach taken is also fairly well-known and is based on one of the standard proofs of continuity for Brownian motion on Euclidean space using some estimates of the heat kernel. Essentially, the idea for continuity is that, if we think of Brownian particles as entering the manifold from the tangent space, where they live in a Euclidean environment, the change of ambience should not be so abrupt as to cause them instantly to jump. This idea is made precise with an inequality for the heat kernel given in Proposition B.7.3. The method originates with J. R. Kinney [**113**] and E. Dynkin [**56**, Chapter 6]. This reference further elaborates on this matter and is the basis of our presentation. A variation on this theme, originating with A. N. Kolmogoroff, is described in [**31**].

The function $H(\varepsilon, t)$, defined by

$$H(\varepsilon, t) = \sup_{\substack{x \in X \\ s \leq t}} \int_{X \setminus B(x, \varepsilon)} p(x, y; s) \cdot dy,$$

is nondecreasing in $t > 0$. The property of the heat kernel $p(x, y; t)$ of X that was proven in Proposition B.7.3 has the following consequence.

Corollary C.4.1. *For* $\varepsilon > 0$,

$$\lim_{t\to 0} \frac{H(\varepsilon, t)}{t} = 0.$$

The actual proof of continuity of sample paths requires some preliminary calculations. Let $\varepsilon > 0$, let $a < b$ be points in $[0, \infty)$, and let A denote the set of paths

$$A = \{\omega \in X^{[0,\infty)} \mid d_X(\omega(b), \omega(a)) \geq \varepsilon\},$$

where d_X denotes the distance function of X. The probability of the event A with respect to the measure P_x is computed as follows. Let

$$\pi_{ab}(\pi_b, \pi_a) : X^{[0,\infty)} \to X \times X,$$

and let χ_ε be the characteristic function of $\{(x, y) \in X \times X \mid d_X(x, y) \geq \varepsilon\}$, so that $\chi_A(\omega) = \chi_\varepsilon \circ \pi_{ab}(\omega)$ and $\pi_{0(b-a)} \circ \theta_a = \pi_{ab}$. Then

$$\begin{aligned}
P_x[A] &= E_x\left[\chi_\varepsilon \circ \pi_{0(b-a)} \circ \theta_a\right] \\
&E_x\left[E_x\left[\chi_\varepsilon \circ \pi_{0(b-a)} \circ \theta_a \mid \mathfrak{G}_a\right]\right] \\
&= E_x\left[E_\bullet\left[\chi_\varepsilon \circ \pi_{0(b-a)}\right] \circ \pi_a\right] \\
&= \int_X p(x, y; a) \left(\int_{X\smallsetminus B(y,\varepsilon)} p(y, z; b-a) \cdot dz\right) \cdot dy \\
&\leq \sup_{y\in X} \int_{X\smallsetminus B(y,\varepsilon)} p(y, z; b-a) \cdot dz \\
&\leq H(\varepsilon, b-a),
\end{aligned}$$

where the third equality is by the Markov property for functions (Proposition C.3.4).

Lemma C.4.2. *Let F be a countable subset of an interval $[a, b]$ in $[0, \infty)$. Let $\varepsilon > 0$ and let A be the set of paths $\omega \in X^{[0,\infty)}$ with $d_X(\omega(s), \omega(t)) \geq 4\varepsilon$, for some s, t in F. Then*

$$P_x[A] \leq 2H(\varepsilon, b-a).$$

Proof. It suffices to show that the claim holds true when F is finite, the right-hand side of the inequality to be proven being independent of the number of points in F.

Let

$$\tau : X^{[0,\infty)} \to [a, b] \cup \{\infty\}$$

be defined by

$$\tau(\omega) = \min\left\{t \in F \mid d_X(\omega(t), \omega(a)) \geq 2\varepsilon\right\},$$

with the convention that the infimum of the empty set is $+\infty$. Because of the triangle inequality for the distance d_X, this function τ is finite on A.

If A_1 and A_2 are the subsets of $X^{[0,\infty)}$ defined by

$$A_1 = \{\omega \mid d_X(\omega(a), \omega(b)) \geq \varepsilon\},$$
$$A_2 = \{\omega \mid \tau(\omega) < \infty; d_X(\omega(\tau(\omega)), \omega(b)) \geq \varepsilon\},$$

then $A \subset A_1 \cup A_2$, and so $P_x[A] \leq P_x[A_1] + P_x[A_2]$.

The calculation before the statement of the lemma implies that

$$P_x[A_1] \leq H(\varepsilon, b-a).$$

The probability of A_2 is estimated in much the same way as that of A_1. First note that if $\tau(\omega) < \infty$, then $\tau(\omega) \in [a,b]$, and thus obtain

$$E_{\pi_\tau(\omega)}[\chi_\varepsilon \circ \pi_{0(b-\tau(\omega))}] \leq H(\varepsilon, b-\tau(\omega)) \leq H(\varepsilon, b-a)$$

on $\{\tau < \infty\}$. The first inequality is obtained by a calculation similar to the one prior to the statement of this lemma, and the second inequality is due to the fact that $H(\varepsilon, s)$ is a nondecreasing function of s. Next apply the version of the Markov property given in Exercise C.3.7 to the function $f(s,\omega) = \chi_\varepsilon \circ \pi_{sb}(\omega)$ to get

$$E_x[\chi_\varepsilon \circ \pi_{\tau b} \mid \mathfrak{G}_{\tau+}](\omega) E_{\pi_\tau(\omega)}[\chi_\varepsilon \circ \pi_{0(b-\tau(\omega))}],$$

for P_x-almost all ω in $\{\tau < \infty\}$. Finally, all of this implies that

$$\begin{aligned} P_x[A_2] &= P_x[P_x[A_2|\mathfrak{G}_{\tau+}]] \\ &= E_x[E_x[\chi_\varepsilon \circ \pi_{\tau b} \mid \mathfrak{G}_{\tau+}]; \tau < \infty] \\ &\leq H(\varepsilon, b-a). \end{aligned}$$

□

The next calculation is similar.

Lemma C.4.3. *Let F be a countable subset of $[a,b]$ and let B be the set*

$$B = \bigcup_{\substack{s,t \in F \\ |t-s| < \delta}} \{\omega \mid d_X(\omega(s), \omega(t)) \geq \varepsilon\}.$$

Then

$$P_x[B] \leq 2(b-a)\frac{H(\varepsilon, 2\delta)}{\delta}.$$

Proof. The interval $[a,b]$ can be covered by (overlapping) intervals $I_1, \ldots, I_k$ of length $\leq 2\delta$, where we can take $k \leq (b-a)/\delta$. If A_i denotes the event of the previous lemma associated to the interval I_i, then $B \subset \bigcup_{i=1}^k A_i$. Hence

$$P_x[B] \leq \sum_{i=1}^k P_x[A_i] \leq 2kH(\varepsilon, 2\delta) \leq \frac{2(b-a)}{\delta} H(\varepsilon, 2\delta).$$

□

The final step, which consists in showing that $\Omega(X)$ has outer measure 1 in $X^{[0,\infty)}$, will now be handled. A contradiction will be reached under the assumption that there exists a measurable subset A of $X^{[0,\infty)}$ that is disjoint from $\Omega(X)$ and with $P_x[A] > 0$. By Corollary C.2.3, given such a set A, there is a countable dense subset Q of $[0,\infty)$ such that, if $\omega \in X^{[0,\infty)}$ and there exists $\omega' \in A$ for which $\omega(t) = \omega'(t)$, for all $t \in Q$, then $\omega \in A$.

Let $Q_n = Q \cap [n-1, n+1]$, $n = 1, 2, \ldots$, and, for each $\varepsilon > 0$ and $\delta > 0$, let

$$A(n,\varepsilon,\delta) = \bigcup_{\substack{s,t\in Q_n \\ |t-s|<\delta}} \{\omega \mid \varepsilon \leq d(\omega(s),\omega(t))\}.$$

The set

$$Z = \bigcup_{n=1}^{\infty} \bigcup_{\varepsilon>0} \bigcap_{\delta>0} A(n,\varepsilon,\delta)$$

consists of exactly the paths $\omega \in X^{[0,\infty)}$ that fail to be continuous at some point of Q, so it is disjoint from $\Omega(X)$. It may not be measurable, but, if $P_x[A] > a$, a Borel set $B \subset Z$ with $P_x[B] \leq a$ can be constructed as follows. By Lemma C.4.3, given integers $n, m > 0$, an integer $r > 0$ can be found such that, whenever $\delta < 1/r$,

$$P_x[A(n, 1/m, \delta)] \leq a/2^{m+n}.$$

Thus, if

$$B_{mn} \bigcap_{r>0} A(n, 1/m, 1/r),$$

then $P_x(B_{mn}) \leq a/2^{m+n}$, and the set

$$B = \bigcup_{n>0} \bigcup_{m>0} B_{mn} \subset Z$$

is measurable with

$$P_x(B) \leq \sum_n \sum_m a/2^{m+n} = a.$$

Let $C = X^{[0,\infty)} \setminus B$. Then $\Omega(X) \subset C$ because elements of C have the property of being uniformly continuous when restricted to each Q_n. On the other hand, the intersection $A \cap C$ is not empty because $P_x[C] \geq 1 - a$. If $\omega \in A \cap C$, then ω is uniformly continuous when restricted to each interval Q_n, so there is $\omega' \in \Omega(X)$ that agrees with ω on Q. The description of A then implies that $\omega' \in A$, contradicting the hypothesis that $A \cap \Omega(X) = \emptyset$.

The proof of Theorem C.2.13 is complete.

C.5. Stopping Times

The semigroup property of the heat kernel described in Appendix B is interpreted in terms of Brownian expectation by saying that

$$E_x\left[f\circ\pi_{s+t}\mid\mathfrak{F}_t\right]=E_\bullet\left[f\circ\pi_s\right]\circ\pi_t$$

for $s,t>0$ and f a (reasonable) function on the manifold X.

In this section a more general version of the semigroup property will be presented. This requires several pieces of notation. For $t\geq 0$, the σ-field $\mathfrak{F}_{t+}=\bigcap_{s>t}\mathfrak{F}_s$ was introduced in C.3.1 (under a different name). Its presence allows for the following fact, called the *Markov property.*

Theorem C.5.1. *Let F be a measurable semibounded function on $\Omega(X)$. Let $x\in X$. Then*

$$E_x\left[F\circ\theta_t\mid\mathfrak{F}_{t+}\right]=E_\bullet\left[F\right]\circ\pi_t,$$

almost everywhere with respect to P_x.

This property will now be extended to more general time parameters. To do so, the concept of *stopping time* needs to be introduced.

Definition C.5.2. A measurable function $T:\Omega(X)\to[0,\infty]$ is a *stopping time* if, for each $t\geq 0$, the set of paths $\{\omega\mid T(\omega)\leq t\}$ is $\mathfrak{F}_{t+}$-measurable.

Exercise C.5.3. Prove that a measurable $T:\Omega(X)\to[0,\infty]$ is a stopping time if and only if

$$\{\omega\mid T(\omega)<t\}\in\mathfrak{F}_t,$$

for all $t>0$.

Example C.5.4. The simplest example of stopping time is given by a constant time t_0, *i.e.*, $\omega\mapsto t_0$.

Exercise C.5.5. Determine whether the following functions on $\Omega(\mathbb{R})$ are stopping times.

(1) $T_0(\omega)=|\omega(t_0)|$, where $t_0>0$ is a fixed real number.
(2) $T_1(\omega)=\inf\{t\leq 1\mid\omega(t)=0\}$.

A very important example of stopping time is the first exit time from a subset of X.

Definition C.5.6. The *first exit time* from a subset B of X is the function T_B defined by

$$T_B(\omega)=\inf\{t>0\mid\omega(t)\in X\smallsetminus B\},$$

with the convention that the infimum of the empty set is $+\infty$. That is, if ω never exits B, then $T_B(\omega)=+\infty$.

Proposition C.5.7. *Let D be an open set in the manifold X. Then the first exit time from D is a stopping time. In fact, $\{T_D \leq t\} \in \mathfrak{F}_t$.*

Proof. Let $B = X \setminus D$ and let $B_n = \{x \in X \mid d_X(x, B) < 1/n\}$. Then B_n is a sequence of open sets that decreases to B. Let r denote a positive rational number, and let $T_r(\omega) = \inf\{t \geq r \mid \omega(t) \in B\}$. If $t > 0$, it is clear that

$$\{T_D \leq t\} = \bigcup_{n \geq 1} \{T_{1/n} \leq t\}.$$

On the other hand, if $r < t$, then

$$\begin{aligned}\{T_r \leq t\} &= \bigcup_{s \in [r,t]} \{\omega \mid \omega(s) \in B\} \\ &= \bigcap_n \bigcup_{s \in [r,t]} \{\omega \mid \omega(s) \in B_n\} \\ &= \bigcap_n \bigcup_{s \in [r,t] \cap \mathbb{Q}^+} \{\omega \mid \omega(s) \in B_n\},\end{aligned}$$

and each set $\{\omega \mid \omega(s) \in B_n\}$ belongs to $\mathfrak{F}_t$, proving the second assertion of the proposition. Since $\mathfrak{F}_t \subset \mathfrak{F}_{t+}$, Exercise C.5.3 implies that T_D is a stopping time. □

Exercise C.5.8. If B is a subset of X, the function τ_B, defined by

$$\tau_B(\omega) = \inf\{t > 0 \mid \omega(t) \in B\},$$

is called the *hitting time* of B. If B is closed and, more generally, if B is an F_σ subset of X, prove that τ_B is a stopping time. See [**31**] or [**152**] for more on these matters. As $T_D = \tau_{X \setminus D}$, this remark applies to T_D as well.

By the continuity of Brownian paths, if $\omega(0) \in X \setminus \overline{D}$, then $T_D(\omega) = 0$ and, if $\omega(0) \in D$, then $T_D(\omega) > 0$. Paths for which the behavior of T_D is more delicate are those that start at ∂D. For these paths it is possible that $T_D(\omega) > 0$, even though $\omega(0) \notin D$. The situation is now briefly described. First of all, Blumenthal's zero-one law (Theorem C.3.6) has the following consequence.

Theorem C.5.9. *If $D \subset X$ and $x \in X$, then $P_x[T_D = 0]$ equals zero or one.*

Proof. The set of paths $\{T_D(\omega) = 0\}$ belongs to $\mathfrak{F}_{0+}$. □

Points $x \in X$ for which $P_x[T_D = 0] = 0$ are called *irregular* for D. As observed in the last paragraph, the irregular points for D are contained in $\overline{D}$. Their presence in the boundary ∂D of D poses obstructions to the solvability of the Dirichlet problem. A well-known example of an open set in $\mathbb{R}^3$ whose boundary contains irregular points is the so-called Lebesgue thorn, see [**106**, p. 261]. For the domains considered here, things are always nice.

Theorem C.5.10. *Let D be a regular domain in X. Then ∂D has no irregular points for D.*

Exercise C.5.11. Let $D \subset \mathbb{R}^2$ be $D = \{x \mid 0 < |x| < 1\}$. Does ∂D have irregular points for D?

Two pieces of notation associated to a stopping time T are in order. Let π_T denote the map from $\Omega(X)$ to X defined by $\pi_T(\omega) = \omega(T(\omega))$ on the set of paths $\{T(\omega) < \infty\}$. Let $\mathfrak{F}_{T+}$ denote the collection of measurable sets A such that $A \cap \{\omega \mid T(\omega) \leq t\} \in \mathfrak{F}_{t+}$ for all $t \geq 0$. It is easily verified that $\mathfrak{F}_{T+}$ is a σ-field.

Exercise C.5.12. If T is a stopping time, show that T is $\mathfrak{F}_{T+}$-measurable.

If T is a stopping time, the shift θ_T is defined in the obvious way on $\{T < \infty\}$.

Exercise C.5.13. Show that, if T and S are stopping times, then $T + S \circ \theta_T$ is a stopping time. Here $S \circ \theta_T = \infty$ on $\{T = \infty\}$.

The Markov property of the heat kernel can be extended to stopping times other than the constant times.

Theorem C.5.14. *Let T be a stopping time for the process $(\Omega(X), P)$. If F is a bounded measurable function on $\Omega(X)$, then*

$$E_x \left[F \circ \theta_T \mid \mathfrak{F}_{T+}\right] = E_{\pi_T}\left[F\right],$$

almost everywhere with respect to P_x on $\{T < \infty\}$.

More generally, if F is a bounded measurable function on $[0, \infty) \times \Omega(X)$, then

$$E_x \left[F \circ (T, \theta_T) \mid \mathfrak{F}_{T+}\right](\omega) E_{\pi_T(\omega)}\left[F(T(\omega), \bullet)\right],$$

for P_x-almost all ω in $\{T < \infty\}$.

The meaning of the right side of the last identity is

$$E_{\pi_T(\omega)}\left[F(T(\omega), \bullet)\right] \int_{\Omega(X)} F(T(\omega), \omega') \cdot P_{\pi_T(\omega)}(\omega').$$

A proof of the theorem, as well as illustrations of its use, can be found in [**31**] and [**152**], for instance. Note that the theorem is usually stated for bounded functions F, but applies equally well to semi-bounded functions. Indeed, it can be deduced from the case of bounded functions by approximating an arbitrary nonnegative function by an increasing sequence of bounded ones.

C.6. Some Consequences of the Markov Property

Some properties of exit times for the Brownian motion on a Riemannian manifold will be described in this section. The purpose is not only to use them later, but also to illustrate how to use the Markov property. The following fact is immediate from Theorem B.7.5.

Lemma C.6.1. *Let X be a noncompact, complete Riemannian manifold of bounded geometry, with associated heat kernel $p(x, y; t)$. Let D be a bounded regular domain in X. Then there is a constant $A > 0$ such that, for $t \geq 1$,*

$$\sup_{x \in D} \int_D p(x, y; t) \cdot dy \leq \frac{A}{t^{1/2}} \mathrm{vol}(D).$$

Using this fact, the following result will be established. In the proof, the notation is simplified by using (for the first time) the convention, mentioned in Section C.1, that $E_x[f; B] = E_x[f\chi_B]$ for measurable sets B and functions f. Similarly, write $P_x[B; C] = E_x[\chi_B; C]$.

Proposition C.6.2. *Let X be a noncompact, complete Riemannian manifold of bounded geometry. Let $D \subset X$ be a bounded regular domain. Then*

$$P_x[\{\omega \mid T_D(\omega) < \infty\}] = 1,$$

for all $x \in X$.

Proof. If $x \notin D$, then $P_x[T_D = 0] = 1$ (points in ∂D are taken care of by Theorem C.5.10), so it suffices to consider points x in D. Hence, if $t \geq 1$,

$$\begin{aligned} P_x[T_D > t] &\leq P_x[\omega(t) \in D] \\ &= \int_D p(x, y; t) \cdot dy \\ &\leq \frac{A}{t^{1/2}} \mathrm{vol}(D), \end{aligned}$$

where the last inequality is by Lemma C.6.1 (which needs $t \geq 1$).

It will now be shown, by induction on $n \geq 1$, that

$$(*) \qquad P_x[T_D > nt] \leq \left(\frac{A}{t^{1/2}} \mathrm{vol}(D)\right)^n.$$

The case $n = 1$ was established above; hence we assume that the assertion is true for n. By Proposition C.5.7, the event $\{T_D > nt\}$ is $\mathfrak{F}_{nt}$-measurable, hence by the Markov property (Proposition C.3.4) it follows that

$$\begin{aligned}P_x\left[\{T_D > (n+1)t\}\right] &\le P_x\left[\{T_D > nt\}; \{T_D \circ \theta_{nt} > t\}\right] \\ &= E_x\left[P_x\left[\{T_D > nt\}; \{T_D \circ \theta_{nt} > t\} \mid \mathfrak{F}_{nt}\right]\right] \\ &= E_x\left[P_x\left[\{T_D \circ \theta_{nt} > t\} \mid \mathfrak{F}_{nt}\right]; \{T_D > nt\}\right] \\ &= E_x\left[P_{\pi_{nt}}\left[\{T_D > t\}\right]; \{T_D > nt\}\right].\end{aligned}$$

If $\omega \in \{T_D > nt\}$, then $\omega(nt) \in D$. By the estimate obtained at the beginning of this proof it follows that

$$P_{\pi_{nt}}\left[\{T_D > t\}\right] \le \frac{A}{t^{1/2}}\mathrm{vol}(D)$$

on $\{T_D > nt\}$, which, by the inductive hypothesis and the above string of inequalities and equalities, yields $(*)$ with n replaced by $n+1$.

Let B_n denote the set of paths ω such that

$$nt < T_D(\omega) \le (n+1)t.$$

Then

$$\begin{aligned}E_x\left[T_D\right] &= \sum_{n=0}^{\infty} \int_{B_n} T_D(\omega) \cdot P_x(\omega) \\ &\le \sum_{n=0}^{\infty} (n+1)tP_x\left[\{nt < T_D \le (n+1)t\}\right] \\ &\le \sum_{n=0}^{\infty} (n+1)t\left(\frac{A\mathrm{vol}(D)}{t^{1/2}}\right)^n,\end{aligned}$$

which converges if t is chosen sufficiently large. In particular, T_D is finite almost everywhere with respect to P_x. □

Exercise C.6.3. Under the hypothesis of the previous proposition, show that $E_x\left[T_D^k\right] < \infty$ for all $k \ge 1$.

If X is compact and D is a regular domain in X, then almost all paths that start in D are expected to exit D in finite time (if $\overline{D} \ne X$). But the interior of the complement of D is a finite collection of domains; hence almost all the paths starting there must also exit. By continuity, they must reenter D. This observation is made precise in the following exercise.

Exercise C.6.4. Let X be a compact connected manifold, and let D be a regular domain in X. The aim of this exercise is to show that almost all Brownian paths return to D infinitely often.

(1) Show that it is enough to prove the statement when D is a coordinate chart in X.

(2) Let $T : \Omega(X) \to [0, \infty]$ be the function

$$T(\omega) = \inf\{t > 0 \mid \omega(t) \notin X \setminus D\}.$$

Let $f(x) = P_x[T < \infty]$. It can be proven that f is continuous and, in fact, later on it will be seen that it is harmonic on $X \setminus D$ with $f = 1$ on ∂D. In any case, using the Markov property, show that

$$D_t f(x) \leq f(x),$$

for every $x \in X$.

(3) As a consequence of the previous item, the function

$$D_\infty f = \lim_{t\to\infty} D_t f$$

is continuous and nonnegative on X. Show that it is constant.

(4) The above facts imply that $D_\infty f \equiv a \geq 0$. Show that $f \equiv a$ and that $a = 1$.

(5) Conclude that sample paths return to D infinitely often; *i.e.*, for each $x \in D$, $P_x[\{\omega \mid \omega(t_n) \in D,$ for some sequence $t_n \uparrow \infty\}] = 1$.

This exercise may lead the reader to think that completely the opposite phenomenon takes place on a noncompact manifold. That is, it might suggest that almost all paths wander off to infinity. This is certainly not true in general, and the following exercise, a classical one in fact, examines this point.

Exercise C.6.5. Let D denote the unit ball in $\mathbb{R}^d$. Show that, if $d \leq 2$, then almost all paths return to D infinitely often while, if $d > 2$, almost all paths never return to D.

The results of this section lead to a close relationship between the first exit time and the isoperimetric constant of a domain. The idea is that, for domains with the same volume but varying boundary, the bigger the boundary, the smaller the mean exit time. This is important, because the isoperimetric constant is in turn related to the first eigenvalue for the Dirichlet problem of the domain. That this is so can be verified experimentally as follows. Collect several bottles of equal capacity but with different neck sizes. A bee inside one of those bottles will perform a flight that resembles Brownian motion, and the assertion is rather obvious.

C.7. The Discrete Dirichlet and Poisson Problems

The Dirichlet problem for a bounded regular domain D is the following. Given a continuous function φ on ∂D, find a continuous extension $f : \overline{D} \to \mathbb{R}$ that is harmonic on D. An important case of the Poisson problem is to find a continuous function f on $\overline{D}$ that vanishes on ∂D, is smooth on D and satisfies $\triangle f \equiv -1$ there. This section, which is purely motivational, investigates a discrete analogue of these problems, together with a probabilistic solution that suggests how Brownian motion might be used to solve the continuous versions.

In terms of Euclidean geometry, the condition that f be harmonic in the domain $D \subset \mathbb{R}^2$ is that

$$\triangle f(x,y) = \left(\frac{\partial^2}{\partial x^2} + \frac{\partial^2}{\partial y^2}\right) f(x,y) = 0, \quad \forall\,(x,y) \in D.$$

The idea is to approximate D by a discrete net, say by a set of points of the form $D \cap \Lambda$, where $\Lambda = \{(mq, nq)\}_{m,n\in\mathbb{Z}}$ and q is the inverse of a very large positive integer.

The discrete partials of a function f on the net are formed by averaging the two difference quotients corresponding to increments ± 1. In a similar vein, it makes sense to define the discrete Laplacian of f at a point (mq, nq) of the grid Λ by

$$(*) \qquad \triangle f(mq,nq) = \left(\sum_{-1\le r,s\le 1} p_{rs} f((m+r)q,(n+s)q)\right) - f(mq,nq),$$

where

$$p_{rs} \begin{cases} 1/4, & (r,s) = (\pm 1, 0) \text{ or } (0, \pm 1), \\ 0, & \text{otherwise.} \end{cases}$$

Thus, in discrete terms, the Dirichlet problem becomes the following: given a function φ on the grid boundary of D, extend φ to a "continuous" function f on $\Lambda \cap D$ satisfying

$$(**) \qquad f(mq,nq) = \sum_{-1\le r,s\le 1} p_{rs} f((m+r)q,(n+s)q).$$

A sample Brownian path ω in this discrete system is a sequence of grid points, each point (mq, nq) being succeeded by one of the four adjacent points $((m \pm 1)q, nq)$, $(mq, (n \pm 1)q)$. The space of all such paths will be denoted by Ω, those starting at the gridpoint (mq, nq) by $\Omega_{(m,n)}$. Given the initial point $\omega(0)$, the entire path is described by a sequence of "jump" symbols drawn from the four-element set $\{(\pm 1, 0), (0, \pm 1)\}$. Giving equal probability $1/4$ to each possible jump, the inverse limit technique is used to construct a probability measure $P_{(m,n)}$ on Ω, supported on $\Omega_{(m,n)}$.

In order to see the probabilistic aspect of the discrete Dirichlet problem, imagine that D is a pond, that $\Lambda \cap D$ labels lily pads and that, on some specific pad (mq, nq), there sits a frog. Consider the following process. The frog leaps to one of the four neighboring lily pads $((m + r)q, (n + s)q)$ with equal probability $p_{rs} = 1/4$. This process is continued until the frog hits one

Figure C.7.1. Random frog

of the pads at the boundary of the pond, say (x, y). At that moment (the first exit time T) the value $\varphi(x, y)$ is recorded, and then assigned to any Brownian path ω agreeing up to this exit time with the path followed by the frog. Let $F(\omega)$ denote this value, defining thereby a function $F : \Omega \to \mathbb{R}$.

The expected value $E_{(m,n)}[F]$ of the function F, relative to the probability measure $P_{(m,n)}$, can be computed by using the elementary facts of conditional probability:

$$\sum \text{Prob}\left\{\text{jump from } (mq, nq) \text{ to an adjacent } (m'q, n'q)\right\} E_{(m',n')}[F].$$

Thus the function $f(mq, nq) = E_{(m,n)}[F]$ is the expected value of the function F that, to a leap sequence starting at (mq, nq), assigns the value of the function φ at the point of the boundary of the pond first visited by the frog. By the above formula, f satisfies the condition $(**)$ for being harmonic on D. Finally, if the frog starts at a boundary point (mq, nq), then the expected value of F is $f(mq, nq) = \varphi(mq, nq)$, and we have found a solution of the discrete Dirichlet problem.

If the behavior of the frog is accepted as a discrete version of the movement of a Brownian particle, then it is reasonable to expect that the solution to the Dirichlet problem on a bounded domain D of the manifold X with boundary data φ will be given by

$$f(x) E_x \left[\varphi(\omega(T_D(\omega)))\right],$$

where T_D is the first exit time from D.

The random frog will now be put to work toward a solution to the Poisson problem, submitting her to the following process. Positioned at time 0 at the point (mq, nq), let her jump at will (at discrete times $t = 0, 1, 2 \ldots$) to one of the neighbouring lily pads with the same probability as before. If at time T she hits a boundary pad, then assign the first exit time $T = T(\omega)$ to the sample Brownian path. While it may or may not be possible to explicitly compute the expectation $E_{(m,n)}[T]$, it turns out that it satisfies an important identity.

As before, if the frog is at (mq, nq) at time t, then at time $t+1$ she is going to be at one of the neighboring lilies $(m'q, n'q) = ((m+s)q, (n+s)q)$ with probability $1/4$. It follows that

$$E_{(m,n)}[T] = \left(\sum_{-1 \leq r,s \leq 1} p_{rs} E_{(m+r,m+s)}[T] \right) + 1.$$

Equivalently, the function $f(mq, nq) = E_{(m,n)}\,[T]$ satisfies the equation

$$\triangle f(mq, nq) = -1$$

at all points of $\Lambda \cap D$. Note also that $T(\omega) = 0$ for paths starting at a boundary point $\omega(0) = (mq, nq)$, so $f(mq, nq) = 0$ and we have solved the Poisson problem.

If it is accepted that the frog mimics a Brownian particle reasonably well, then it should also be accepted that the solution to the Poisson problem for the domain D in the manifold X will be given by the function

$$f(x) = E_x[T_D],$$

where $T_D(\omega)$ is the first exit time of ω from D.

These guesses will be proven in the next section.

C.8. Dynkin's Formula

Let f be a smooth function with compact support on X. Then it is easily verified that

$$E_x\left[f \circ \pi_t\right] - f(x) = E_x\left[\int_0^t \triangle f \circ \pi_s \cdot ds\right],$$

for all $x \in X$ and $t > 0$. This is the simplest version of *Dynkin's formula*, and the general version, as the reader may have guessed, involves replacing the constant stopping time t by an arbitrary stopping time. For full generality, the result will be stated using $\triangle_{\mathrm{i}}$, the infinitesimal generator of the diffusion semigroup, this being an operator that restricts to the Laplacian on the space of smooth, compactly supported functions.

Theorem C.8.1 (Dynkin's formula). *Let f be a function in the domain $\mathcal{D}$ of the infinitesimal generator of the diffusion semigroup. If T is a stopping time such that $P_\bullet\,[T<\infty]=1$, then*

$$E_x\,[f\circ\pi_T]-f(x)=E_x\left[\int_0^T \triangle_{\mathrm{i}} f\circ\pi_t\,dt\right].$$

Proof. Let R_λ be the resolvent operator of the semigroup, as defined in Section B.10. If f is in the domain of the infinitesimal generator of D_t, then by Theorem B.10.14, for each $\lambda>0$ there is a bounded function h such that $f=R_\lambda h$, and such that $\triangle_{\mathrm{i}} f-\lambda f=-h$.

By Fubini's theorem and the fact that h is bounded,

$$\begin{aligned} f(x)=R_\lambda h(x)&=E_x\left[\int_0^\infty e^{-\lambda t} h\circ\pi_t\cdot dt\right]\\ &=E_x\left[\int_0^T e^{-\lambda t}\left(\lambda f-\triangle_{\mathrm{i}} f\right)\circ\pi_t\cdot dt\right]\\ &\quad+E_x\left[\int_T^\infty e^{-\lambda t} h\circ\pi_t\cdot dt\right].\end{aligned}$$

As $\lambda\to0$, the first term converges to

$$-E_x\left[\int_0^T \triangle_{\mathrm{i}} f\circ\pi_t\cdot dt\right].$$

The second term will now be analyzed, starting with the integrand. This is the function on paths ω given by

$$\omega\mapsto\int_{T(\omega)}^\infty e^{-\lambda t}h\circ\pi_t(\omega)\cdot dt.$$

Since $T<\infty$, the change of variables $s=t-T(w)$ is permissible, and the above expression can the be rewritten as

$$e^{-\lambda T(\omega)}\int_0^\infty e^{-\lambda t}h\circ\pi_t\circ\theta_T(\omega)\cdot dt.$$

Recalling that the function $e^{-\lambda T}$ is $\mathfrak{F}_{T+}$-measurable, we see that an application of the strong Markov property C.5.14 yields

$$\begin{aligned} &E_x\left[e^{-\lambda T}\int_0^\infty e^{-\lambda t}h\circ\pi_t\circ\theta_T\cdot dt\right]\\ &\quad=E_x\left[E_x\left[e^{-\lambda T}\int_0^\infty e^{-\lambda t}h\circ\pi_t\circ\theta_T\cdot dt\,\middle|\,\mathfrak{F}_{T+}\right]\right]\\ &\quad=E_x\left[e^{-\lambda T}E_\bullet\left[\int_0^\infty e^{-\lambda T}h\circ\pi_t\cdot dt\right]\circ\pi_T\right]\\ &\quad=E_x\left[e^{-\lambda T}f\circ\pi_T\right],\end{aligned}$$

where the last equality is by the definition of $R_\lambda h$. By the dominated convergence theorem, this converges to $E_x[f \circ \pi_T]$ as $\lambda \to 0$. □

Dynkin's formula makes it possible to express solutions to well-known equations by means of probabilistic formulas. In order do so, we need to extend slightly the range of applicability of such formulas.

Suppose that f is a continuous function on X and that D is a bounded regular domain. Since the discussion that follows is only concerned with the exit time T_D, it may be assumed that f has compact support. If f is smooth and compactly supported on X, Dynkin's formula applies and gives, for $x \in D$,

$$f(x) = E_x[f \circ \pi_{T_D}] - E_x\left[\int_0^{T_D} \triangle_{\mathrm{i}} f \circ \pi_t \cdot dt\right]$$

and, because $\triangle_{\mathrm{i}} f = \triangle f$,

$$f(x) = E_x[f \circ \pi_{T_D}] - E_x\left[\int_0^{T_D} \triangle f \circ \pi_t \cdot dt\right].$$

If f is smooth on D, but not on X, then Dynkin's formula cannot be applied directly, since it may not be true that $f \in \mathfrak{D}$. Assume that $\triangle f$ is bounded on D. Let D_n be an increasing sequence of bounded domains such that $\overline{D}_n \subset D_{n+1}$ and $\bigcup_n D_n = D$. Let φ_n be a smooth function on X that is 1 on $\overline{D}_n$ and 0 on $X \setminus D_{n+1}$. Then $f_n = \varphi_n f$ is a compactly supported smooth function on X and $\triangle f_n = \triangle f$ on D_n. Let T_n denote the first exit time from D_n.

Each f_n is in $\mathfrak{D}$ and, if $x \in D_n$, Dynkin's formula gives

$$\begin{aligned} f(x) = f_n(x) &= E_x[f_n \circ \pi_{T_n}] - E_x\left[\int_0^{T_n} \triangle f_n \circ \pi_t \cdot dt\right] \\ &= E_x[f \circ \pi_{T_n}] - E_x\left[\int_0^{T_n} \triangle f \circ \pi_t \cdot dt\right]. \end{aligned}$$

Due to continuity of paths, $T_n(\omega)$ increases to $T_D(\omega)$ for every path ω starting at x. Thus, by the monotone and bounded convergence theorems and by the continuity of f,

$$\lim_{n\to\infty} E_x[f \circ \pi_{T_n}] = E_x[f \circ \pi_{T_D}],$$

$$\lim_{n\to\infty} E_x\left[\int_0^{T_n} \triangle f \circ \pi_t \cdot dt\right] = E_x\left[\int_0^{T_D} \triangle f \circ \pi_t \cdot dt\right].$$

The following has been proven.

Corollary C.8.2. *Let D be a bounded regular domain in X, and let f be a continuous function on X such that $\triangle f$ is defined, continuous and bounded* *on D. Then*

$$E_x\left[f \circ \pi_{T_D}\right] - f(x)E_x\left[\int_0^{T_D} \triangle f \circ \pi_t \cdot dt\right],$$

for every $x \in D$.

The following exercise is a variation of this argument.

Exercise C.8.3. Let f be a bounded C^2 function on X such that $\triangle f$ is also bounded. Show that

$$E_x\left[f \circ \pi_t\right] = f(x) + E_x\left[\int_0^t \triangle f \pi_s \cdot ds\right].$$

The Dirichlet problem on a bounded regular domain was introduced in Section B.9. An inmediate consequence of the above corollary is a probabilistic formula for the solution to this problem.

Theorem C.8.4. *Let f be a continuous function on ∂D, where D is a bounded regular domain in X. Then $u(x) = E_x\left[f \circ \pi_{T_D}\right]$, $x \in D$, is the continuous extension of f to $\overline{D}$ which is harmonic on D.*

Remark. It was shown in Section B.9 that the solution of the Dirichlet problem can be expressed in terms of a family $\{\eta_x^D\}_{x\in D}$ of probability measures on ∂D. The measure η_x^D is called the harmonic measure of D with respect to x. If $\varphi \in C(\partial D)$, then

$$f(x) = H_D\varphi(x) = \int_{\partial D} f(z)\, d\eta_x^D(z)$$

defines the solution f of the Dirichlet problem. By the above theorem, it is clear that the integral H_D is given by

$$H_D\varphi(x) = E_x\left[\varphi \circ \pi_{T_D}\right],$$

the harmonic measure itself being defined by

$$\eta_x^D(A) = P_x\left[\omega(T_D) \in A; T_D < \infty\right].$$

Since the domain D is bounded, it follows that $P_x\left[T_D < \infty\right] = 1$ (Proposition C.6.2) and η_x^D is a probability measure with support contained in ∂D.

Exercise C.8.5. Show that the probabilistic formula for the solution to the Dirichlet problem leads to a very intuitive proof of the maximum principle.

The probabilistic formula for the solution to the Poisson problem is as follows.

Theorem C.8.6. *Let D be a bounded regular domain and let f be a continuous function on $\overline{D}$. Then the function $h \in C(D)$, vanishing on ∂D and having $\triangle h = -f$ on D, has the probabilistic expression*

$$h(x) E_x\left[\int_0^{T_D} f \circ \pi_t \cdot dt\right].$$

Proof. Since $h \circ \pi_{T_D} = 0$, $\triangle h = -f$ and f is bounded on D, Corollary C.8.2 implies that

$$-h(x) = E_x\left[h \circ \pi_{T_D}\right] - h(x) = E_x\left[-\int_0^{T_D} f \circ \pi_t \cdot dt\right].$$

□

The case in which f is the constant function $f(x) = 1$ for all $x \in \overline{D}$ is of particular interest. This is called the Saint-Venant equation, and Theorem C.8.6 takes the following form.

Corollary C.8.7. *Let D be a bounded domain. The function $x \mapsto E_x\left[T_D\right]$ is smooth on D, continuous on X, equal to 0 on $X \smallsetminus D$, and has $\triangle E_\bullet\left[T_D\right] \equiv -1$ on D.*

The following exercise makes precise the experiment with the bee outlined after Exercise C.6.5

Exercise C.8.8. Let D be a bounded domain in X and let λ be the first eigenvalue of the Laplacian acting on $C_0(D)$. It is known that λ is obtained as the Rayleigh quotient

$$\lambda = \inf_{f \neq 0} \frac{\int_D |\mathrm{grad}\, f|^2}{\int_D |f|^2}.$$

Show that, if $u \in C_0(D)$ is the function with $\triangle u = \mathbf{-1}$ in the corollary above, then $\lambda \geq \sup_{x \in D} u(x)$.

The behavior of the function $E_x\left[T_D\right]$ on a possibly unbounded regular domain D will now be examined.

Proposition C.8.9. *Let D be a regular domain in X and let T_D be the exit time from D. Then the function*

$$f(x) = E_x\left[T_D\right]$$

satisfies $\triangle f \equiv -1$ on D and $f = 0$ on ∂D, or else $f(x) = \infty$ everywhere on D.

Proof. Let $D_1 \subset D_2 \subset \cdots \subset D$ be an increasing sequence of bounded regular domains that exhaust D and let T_n be the exit time from D_n. Then $T_1 \leq T_2 \leq \cdots$ pointwise in $\Omega(X)$, and $\lim_n T_n = T_D$, P_x-almost everywhere in $\Omega(X)$, for each $x \in D$. Since $T_1 \geq 0$ and $E_x[T_1] \geq 0$ for every $x \in D$, the monotone convergence theorem implies that the sequence of continuous functions $E_\bullet[T_n]$ increases pointwise to $E_\bullet[T_D]$ on D.

Suppose that $E_x[T_D]$ is finite at a point $x \in D$. Then there exists a positive integer N such that $x \in D_N$. Let $u_n = E_\bullet[T_n] - E_\bullet[T_N]$. For $n > N$, u_n is harmonic and positive on D_N, and $u_n(y)$ increases to

$$u(y) = E_y[T_D] - E_y[T_N]$$

on D_N. It follows that u is finite in a neighborhood of x in D_N, hence that the set of points of D where $E_\bullet[T_D]$ is finite is open. It is also closed, for if $x_k \to x$ in D, and $u(x_k) < \infty$ for all k, then, for k sufficiently large, a neighborhood of x_k containing x can be found on which the equicontinuity allowed by Harnack's principle can be applied. Therefore u is also finite at x. Since D is connected, either $E_\bullet[T_D]$ is finite everywhere on X, or else it is equal to ∞ on D.

Suppose that $E_\bullet[T_D]$ is finite. It will be shown that it satisfies the equation $\triangle E_\bullet[T_D] = -\mathbf{1}$ on D. If B is a ball in X whose closure is contained in D, it is clear that $T_D = T_B + T_D \circ \theta_{T_B}$. Thus, by Theorem C.5.14 (the extended Markov property) and the remark on page 489,

$$\begin{aligned} E_x[T_D] &= E_x[T_B] + E_x[T_D \circ \theta_{T_B}] \\ & E_x[T_B] + E_x[E_\bullet[T_D] \circ \pi_{T_B}] \\ & E_x[T_B] + H_B(E_\bullet[T_D] \mid \partial B)(x). \end{aligned}$$

The first term of the sum satisfies $\triangle E_x[T_B] = -1$ for all x in B, while the second is a harmonic function on B (note that $E_\bullet[T_D]$ is bounded on $\overline{B}$ if finite on D). Hence $\triangle E_\bullet[T_D] = -\mathbf{1}$ on B. □

Exercise C.8.10. Show directly (without using the Markov property) that, if $E_\bullet[T_D]$ is finite, then $\triangle E_\bullet \equiv -1$ on D.

It is interesting to note that Dynkin's formula makes it possible to represent the infinitesimal generator $\triangle_{\mathrm{i}}$ as a local operator via a stochastic formula. Thus probabilistic formulas can be used directly to solve the Dirichlet and Poisson problems, without relying on the existence and uniqueness results from Appendix B.

Let f be a measurable function on X and let $x \in X$. Define $\triangle_s f(x)$ by the limit (if it exists)

$$\triangle_s f(x) \lim_{U \downarrow x} \frac{E_x[f \circ \pi_{T_U}] - f(x)}{E_x[T_U]}.$$

Here U varies over bounded domains containing x.

Proposition C.8.11. *Let f be a bounded continuous function on X that belongs to the domain $\mathcal{D}$ of the infinitesimal generator $\triangle_{\mathrm{i}}$. Then $\triangle_s f(x)$ exists, for all $x \in X$, and*

$$\triangle_{\mathrm{i}} f(x) = \triangle_s f(x).$$

Proof. By Dynkin's formula,

$$\left| \frac{E_x\left[f \circ \pi_{T_U}\right] - f(x)}{E_x\left[T_U\right]} - \triangle_{\mathrm{i}} f(x) \right|$$
$$\leq \frac{1}{E_x\left[T_U\right]} E_x\left[\int_0^{T_U} \left|\triangle_{\mathrm{i}} f \circ \pi_t - \triangle_{\mathrm{i}} f \circ \pi_0\right| \cdot dt\right].$$

As $\triangle_{\mathrm{i}} f$ is continuous, given $\varepsilon > 0$ there is a neighborhood V of x such that $|\triangle_{\mathrm{i}} f(y) - \triangle_{\mathrm{i}} f(x)| < \varepsilon$, for all $y \in V$. Since $\pi_t(\omega) \in U$ for $t < T_U$, the above quantity is not larger than ε, for all $U \subset V$. □

This result explicitly shows the local nature of the infinitesimal generator $\triangle_{\mathrm{i}}$, something that is not at all clear from its definition.

Exercise C.8.12. Suppose that φ is a continuous function on X (or on the boundary ∂D of the bounded domain D). Show, using the Markov property, that $\triangle_s E_\bullet\left[\varphi \circ \pi_{T_D}\right] = 0$.

Admittedly, this formula says nothing about the boundary values of f, but this is where the regularity of ∂D comes in. Under the ongoing smoothness hypothesis on the boundary of D, we have $P_x\left[T_D = 0\right] = 1$ if $x \in \partial D$. Thus, $\pi_{T_D}(\omega) = x$ for almost all ω; hence the boundary values are the correct ones. It also needs to be shown that f is continuous at boundary points and, again, the regularity of all such points makes this possible. The interested reader will find more on this in Dynkin [**57**].

The Dirichlet problem can be posed also for domains that are not as regular as the ones here. More on this can be found in [**152**], [**31**] and [**57**].

C.9. Local Estimates of Exit Times

The purpose of this section is to justify a technical detail, used in Chapter 2, concerning the behavior of the function $E_\bullet\left[T_D\right]$ on a regular domain D of a complete Riemannian manifold X of bounded geometry (and dimension $\dim X > 1$).

Let a be a lower bound for the sectional curvature of X and let X_a be the corresponding simply connected space form of constant sectional curvature a. Let $\varepsilon > 0$ be a number smaller than the injectivity radius both of X and of X_a. Let $B(x_0, \varepsilon)$ be the geodesic ball of radius ε and center x_0 in X, $B^a(x_0, \varepsilon)$ the geodesic ball of center x_0 and radius ε in X_a. Let $r(x) =$

$d_X(x_0, x)$ denote the radial distance function on $B(x_0, \varepsilon)$, and let r_a be the similarly defined function on $B^a(x_0, \varepsilon)$. The balls $B(x_0, \varepsilon)$ and $B^a(x_0, \varepsilon)$ are identified by means of the exponential map and both are denoted by $B(\varepsilon)$, the identification being unique up to isometry of the tangent spaces. Once the identification is made, it may be assumed that $r_a(x) = r(x)$ for each x in the identified balls.

Let ν denote the volume density function of X and ν_a that of X_a. In polar coordinates, the Laplacian of a function f on $B(x_0, r)$ of the form $f(x) = f(r(x))$ is given by

$$\triangle f = f'' + \frac{\nu'}{\nu} f',$$

where $'$ denotes d/dr.

Let $f = E_\bullet[T_\varepsilon]$ and $f_a = E_\bullet[T^a_\varepsilon]$, where T_ε and T^a_ε are the exit times from $B(x_0, \varepsilon)$ and $B^a(x_0, \varepsilon)$, respectively. The function f (respectively f_a) is continuous on $B(\varepsilon)$, vanishes on $\partial B(\varepsilon)$, and satisfies $\triangle f = -1$ (respectively, $\triangle^a f_a = -1$) on $B(\varepsilon)$.

Because of the high degree of rotational symmetry of a ball about its center in a space of constant curvature, the function f_a is a function of the distance $r_a = r$ on $B(\varepsilon)$. It is also obvious that f_a is a nonincreasing function of r (use the Markov property), that is, $f'_a \geq 0$.

By Theorem B.2.2, the volume densities of X and X_a in polar coordinates satisfy the inequality

$$\frac{\nu'_a}{\nu_a} \geq \frac{\nu'}{\nu}.$$

Therefore, because

$$\begin{aligned}\triangle f_a &= f''_a + \frac{\nu'}{\nu} f'_a \\ &= \triangle^a f_a + \left\{ \frac{\nu'}{\nu} - \frac{\nu'_a}{\nu_a} \right\} f'_a\end{aligned}$$

and since $\triangle^a f_a = \triangle f \equiv 1$, it follows that

$$\triangle(f_a - f) = \left\{ \frac{\nu'}{\nu} - \frac{\nu'_a}{\nu_a} \right\} f'_a \geq 0.$$

Since $f_a - f = 0$ on $\partial B(\varepsilon)$, the maximum principle implies that $f_a - f \leq 0$, thereby establishing the following result.

Proposition C.9.1. *Let a be a lower bound for the sectional curvature of X and let $\varepsilon > 0$ be smaller than the injectivity radius. If T_ε and T^a_ε are the exit times for the ball $B(\varepsilon)$ in the manifold X and in the space of constant curvature a, respectively, then*

$$E_x[T_\varepsilon] \geq E_x[T^a_\varepsilon],$$

for all $x \in B(\varepsilon)$.

Exercise C.9.2. By the same argument, show that, if b is an upper bound for the sectional curvature of X, then $E_x [T_\varepsilon] \leq E_x \left[T_\varepsilon^b\right]$ for $x \in B(\varepsilon)$.

Remark. A. Debiard, B. Gaveau and E. Mazet [**45**] showed (using Ito's stochastic calculus, which is very advanced for the present level) that

$$P_x^a [T_\varepsilon^a > t] \leq P_x [T_\varepsilon > t]$$

for $x \in B(\varepsilon)$. Proposition C.9.1 also follows from this inequality.

Exercise C.9.3. Let X be a manifold of constant curvature κ. Compute $E_\bullet [T_\varepsilon]$, where ε is smaller than the injectivity radius.

The first author is grateful to J. A. Álvarez López for a discussion on exit times that led to the following exercises.

Exercise C.9.4. Let X be the Euclidean plane and let D_α be a domain bounded by two rays that make an angle α. Compute $E_\bullet [T_{D_\alpha}]$.

Exercise C.9.5. Let D be a horoball in the two-dimensional hyperbolic plane of constant curvature -1. Compute $E_\bullet [T_D]$.

Another estimate for local functions will now be proved. It is also obtained in [**45**] by different methods. The notation introduced at the begining of this section remains in force. Let $\delta < \varepsilon$. Let $u^\varepsilon(x,t)$ be the solution to the heat equation on $B(\varepsilon)$, with initial condition $\chi_\delta = \chi_{B(x_0,\delta)}$. Let $u^a(x,t)$ be the solution in the space of constant curvature a.

Let $q_\varepsilon(x,y;t)$ (respectively, $q_\varepsilon^a(x,y;t)$) denote the heat kernel of $B(\varepsilon)$ in X (respectively, of $B(\varepsilon)$ in X_a). The function u has the expression

$$u(x,t) = E_x \left[\chi_\varepsilon; t < T_{\iota/2}\right] \int_{B(\varepsilon)} \chi_\delta(y) q_\varepsilon(x,y;t) \cdot \mathrm{vol}(y),$$

and there is a corresponding similar expression for u^a.

By the same reasoning as above, $u^a(x,t)$ is a function only of the distance $r(x) = d(x_0,x)$, because the initial condition χ_δ has that property. A similar reasoning shows that $u^a(x,t)$ is a decreasing function of r, and similar calculations show that

$$\triangle \left(u(x,t) - u^a(x,t)\right) \geq \frac{d}{dt} \left(u(x,t) - u^a(x,t)\right).$$

Since the initial conditions are the same, $u(x,0) = u^a(x,0)$. Therefore, the maximum principle for parabolic equations (Lemma B.6.5) implies that $u(x,t) \geq u^a(x,t)$, for $t \geq 0$.

Because the number a is a lower bound for the sectional curvatures of X, Lemma B.2.4 gives $V(x_0, \varepsilon) \leq V^a(\varepsilon)$. Therefore,

$$\frac{1}{V^a(\varepsilon)} \int_{B(\varepsilon)} q^a(x, y; t) \cdot \mathrm{vol}_a(y) \leq \frac{1}{V(x_0, \varepsilon)} \int_{B(\varepsilon)} q(x, y; t) \cdot \mathrm{vol}(y)$$

and, letting $\varepsilon \to 0$, we see that the densities q and q^a are related by the inequality

$$q^a(x_0, y; t) \leq q(x_0, y; t).$$

A similar argument applies to the case in which b is an upper bound for the sectional curvatures, and thus the following result holds.

Proposition C.9.6. *Let X be a Riemannian manifold. Let $x \in X$ have injectivity radius $> \varepsilon$, and let the sectional curvatures of X in $B(x, \varepsilon)$ be in the interval $[a, b]$. If $q_\varepsilon(x, y; t)$ is the heat kernel of the ball $B(x, \varepsilon)$ and $q_\varepsilon^\kappa(x, y; t)$ is that of the ball of radius ε in the space of constant curvature κ, then*

$$q_\varepsilon^a(x, y; t) \leq q_\varepsilon(x, y; t) \leq q_\varepsilon^b(x, y; t).$$

Using this result, it is possible to give a proof of the regularity of boundary points of a regular domain D in a manifold of bunded geometry. The details are left to the reader.

Finally, the behavior of $E_\bullet\,[T_D]$ on a domain D with compact boundary is examined.

Proposition C.9.7. *Let D be an unbounded regular domain in X with compact boundary. If the function $f(x) = E_x\,[T_D]$ is finite, then $f(x) \to \infty$ as $x \to \infty$ in D.*

Proof. Since ∂D is compact, $x \to \infty$ in D if and only if $d_X(x, \partial D) \to \infty$. Fix a number $\varepsilon > 0$ smaller than the injectivity radius of X and let

$$D = D_0 \supset D_1 \supset D_2 \supset \cdots \supset D_k \supset \cdots$$

be a filtration by domains, each with smooth, compact boundary, such that

(1) $d_X(x, \partial D_{k-1}) \geq \varepsilon$, for all $x \in \overline{D}_k$ and all $k \geq 1$;

(2) $\overline{D} \setminus D_k$ is compact for all $k \geq 1$;

(3) $\bigcap_{k \geq 0} D_k = \emptyset$.

It will be enough to show the existence of a constant $C > 0$ such that $f(x) \geq kC$, for all $x \in D_k$.

Given $x \in X$, the maximum principle implies that there exists a constant $C > 0$ such that $E_x(T_{x,\varepsilon}) \geq C$, where $T_{x,\varepsilon}$ is the exit time of $B(x, \varepsilon)$. *A priori*, this number depends on x. In the space X_a, however, the homogeneity implies that C is independent of x (alternatively, use the result of

Exercise C.9.3) and Proposition C.9.1 implies that this value of C works for all $x \in X$. By the choice of the filtration, it follows that

$$E_x\left[T_{D_k}\right] \geq E_x\left[T_{x,\varepsilon}\right] \geq C,$$

for all $x \in \overline{D}_{k+1}$.

It will now be shown, by induction on k, that $f(x) \geq kC$, for all $x \in \overline{D}_k$ and all $k \geq 1$. The case $k = 1$ is immediate by the previous paragraph. Thus assume, inductively, that the assertion holds for a certain value of k. Since $T_D = T_{D_k} + T_D \circ \theta_{T_{D_k}}$, an appeal to the previous paragraph and to Theorem C.5.14 shows that, for $x \in \overline{D}_{k+1}$,

$$\begin{aligned} E_x\left[T_D\right] E_x\left[T_{D_k}\right] + E_x\left[T_D \circ \theta_{T_{D_k}}\right] \\ \geq C + E_x\left[T_D \circ \theta_{T_{D_k}}\right] \\ = C + E_x\left[E_\bullet\left[T_D\right] \circ \pi_{T_{D_k}}\right]. \end{aligned}$$

By the induction hypothesis, the function $E_\bullet\left[T_D\right] \circ \pi_{T_{D_k}}$, restricted to $\Omega(X)_x$, is bounded below by kC and the assertion follows. □

Appendix D

Planar Foliations

Throughout this appendix, M is a simply connected $(n+1)$-manifold without boundary and $\mathcal{F}$ is a smooth foliation by n-planes (called simply a "planar foliation"). We must assume that the leaves are diffeomorphic, not merely homeomorphic, to the standard $\mathbb{R}^n$. Of course, this is restrictive only in the case $n = 4$. Our goal is to prove the theorem of F. Palmeira (Theorem 9.1.10) that M is diffeomorphic to $\mathbb{R}^{n+1}$. In fact, this is a corollary of a stronger result of Palmeira, namely that a simply connected $(n+1)$-manifold, foliated by n-planes, is determined up to conjugacy by the leaf space. Unlike the corollary, this stronger result is only true for $n \geq 2$.

It is remarked in [**143**, p. 110] that the proof can be carried out for foliations integral to C^0 plane fields, but this would seem to involve some rather delicate and distracting considerations. It is not hard to weaken the smoothness condition on $\mathcal{F}$ to class C^2, but certain isotopies and embeddings will then be only of class C^1. Whether noted explicitly or not, all isotopies, imbeddings, conjugacies, *etc.* will be smooth of class C^∞.

Note that simple connectivity of M implies that $\mathcal{F}$ is both transversely orientable and tangentially orientable.

D.1. The Space of Leaves

The quotient space $M/\mathcal{F}$ is called the space of leaves. For general foliations, this space is so nasty as to be virtually useless, but in our case it will be a (generally non-Hausdorff) 1-manifold.

Proposition D.1.1. *No transverse curve to $\mathcal{F}$ can meet the same leaf twice. Consequently, the leaves of $\mathcal{F}$ are closed in M.*

Proof. If such a transverse curve is not closed, a subarc exists that begins and ends on the same leaf. Since $\mathcal{F}$ is transversely orientable, the "waterfall" construction [I, Lemma 3.3.7] yields a closed transversal. This transversal is nullhomotopic since M is simply connected, so some leaf of $\mathcal{F}$ has nontrivial holonomy [I, Proposition 7.3.2]. (We emphasize that M was not assumed to be compact in the proof of that proposition.) Since the leaves are simply connected, this is a contradiction. If a leaf L were not closed in M, it would accumulate at some point x not on L. Thus, the leaf would have more than one plaque (indeed, infinitely many) in a foliated chart about x, and a transverse arc could be constructed from one of these plaques to another. □

Corollary D.1.2. *The leaf space $M/\mathcal{F}$ is a simply connected, second countable, differentiable* 1*-manifold, oriented by a choice of transverse orientation of $\mathcal{F}$.*

Proof. The quotient map $\pi : M \to M/\mathcal{F}$ is continuous; hence $M/\mathcal{F}$ is connected. If $U \subseteq M$ is open, then $\pi^{-1}(\pi(U))$ is the $\mathcal{F}$-saturation of U, hence open. By the definition of the quotient topology, $\pi(U)$ is open in $M/\mathcal{F}$; hence π is an open map and, in fact, every open subset of $M/\mathcal{F}$ is the image of an open subset of M. Thus, π carries a countable base for the topology of M to a countable family of open subsets of $M/\mathcal{F}$ that is easily checked to be a base of the quotient topology. Let $L \in M/\mathcal{F}$ and let $J \subset M$ be an open transverse arc to $\mathcal{F}$ passing through the leaf L. Observe that the $\mathcal{F}$-saturation of any relatively open subset of J is open in M; hence $\pi|J$ is an open map. By Proposition D.1.1, $\pi|J$ is also one-to-one, hence is an embedding of J as an open neighborhood of L in $M/\mathcal{F}$. This is a coordinate chart, canonically oriented by the transverse orientation of $\mathcal{F}$. The local coordinate changes in $M/\mathcal{F}$ will be smooth and orientation-preserving. It remains for us to prove simple connectivity. Let $\sigma : [0,1] \to M/\mathcal{F}$ be a loop. By compactness of $[0,1]$, find a partition $0 = t_0 < t_1 < \cdots < t_r = 1$ such that each $\sigma_i = \sigma|[t_{i-1}, t_i]$ takes its image in a Euclidean neighborhood in $M/\mathcal{F}$ of the form $\pi(J_i)$, where J_i is a transverse arc. Let $L_i = \pi^{-1}(\sigma(t_i))$, $0 \le i \le r$, remarking that $L_0 = L_r$. Thus, σ_i has a unique lift $\widetilde{\sigma}_i$ to a continuous path in J_i from the leaf L_{i-1} to the leaf L_i, $1 \le i \le r$. Piece together a continuous loop

$$\widetilde{\sigma} = \widetilde{\sigma}_1 + \tau_1 + \cdots + \widetilde{\sigma}_r + \tau_r,$$

where τ_i is a suitable path in L_i, $1 \le i \le r$. Then $\pi \circ \widetilde{\sigma}$ is homotopic to σ. Since M is simply connected, this implies that σ is nullhomotopic, hence that $M/\mathcal{F}$ is simply connected. □

The following exercises are straightforward.

Exercise D.1.3. Consider simply connected manifolds M with nonempty boundary that are foliated by hyperplanes. Assume that $\partial M = \partial_\tau M$ and prove that that $M/\mathcal{F}$ is a simply connected 1-manifold with boundary.

Exercise D.1.4. The theory of covering spaces and fundamental groups works quite well for non-Hausdorff manifolds (*cf.* [**125**, Chapter 5]). Use this to prove that a connected 1-manifold V (with or without boundary) is simply connected if and only if every nonboundary point x separates the manifold. In this case, show that $V \smallsetminus \{x\}$ has exactly two components and, if U is a connected Euclidean neighborhood of x, the two components of $U \smallsetminus \{x\}$ lie in distinct components of $V \smallsetminus \{x\}$.

Example D.1.5. By the Poincaré-Bendixson theorem, foliations $\mathcal{F}$ of $\mathbb{R}^2$ have no compact leaves. Thus, $\mathbb{R}^2/\mathcal{F}$ is an orientable, simply connected 1-manifold. We will fix a leafwise orientation on $\mathcal{F}$. Considering the two foliations in Figure D.1.1, we see that topologically nonconjugate foliations of $\mathbb{R}^2$ can have homeomorphic leaf spaces. (The oriented transversals in the figure determine an orientation-preserving homeomorphism of leaf spaces.) However, for foliations of $\mathbb{R}^2$, there is a further structure on the leaf space, an order relation on certain sets of *branch points* (see the definition below), and the foliations are classified by $\mathbb{R}^2/\mathcal{F}$ equipped with this order.

Definition D.1.6. A point x in a 1-manifold V is called a branch point if there exists a distinct point y in V such that every neighborhood of y meets every neighborhood of x. If V is oriented, a pair of (possibly identical) branch points are said to coincide on the right (respectively, on the left) if the neighborhoods have nonempty intersection bordering x and y on the right (respectively, on the left).

If x and y are branch points that coincide on the right, write $x \sim_+ y$ and observe that this is an equivalence relation on the set of branch points. Similarly, left coincidence $\sim_-$ is an equivalence relation. Simple connectivity (*cf.* Exercise D.1.4) forbids distinct branch points from coinciding both on the right and the left. Notice that, if x and y are distinct branch points in $\mathbb{R}^2/\mathcal{F}$ such that either $x \sim_+ y$ or $x \sim_- y$, then there is a path

$$\sigma : [0,1] \to \mathbb{R}^2/\mathcal{F}$$

from x to y that meets no other branch points in the same equivalence class. The leafwise orientation of $\mathcal{F}$ defines a continuous map $\rho : \mathbb{R}^2 \to S^1$, and the curve $\rho \circ \widetilde{\sigma}$ (where $\widetilde{\sigma}$ is a lift of σ) traverses S^1 in either a clockwise or a counterclockwise manner. In fact, this property is independent of the choices of σ and $\widetilde{\sigma}$ and defines an order on each $\sim_+$ equivalence class and on each $\sim_-$ equivalence class. A brief study of Figure D.1.1 reveals that, while there is an orientation-preserving homeomorphism between the leaf

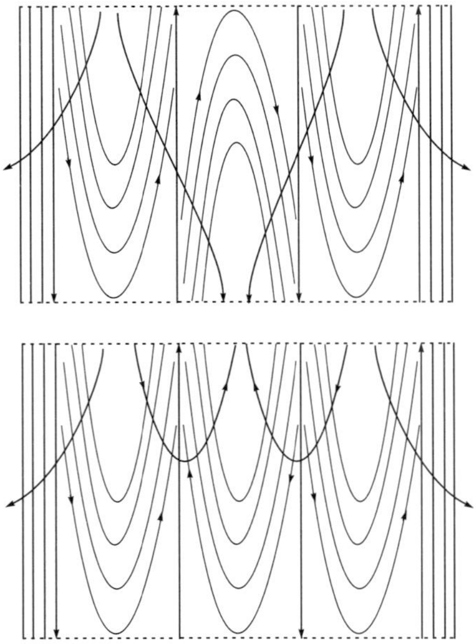

Figure D.1.1. Nonconjugate foliations of $\mathbb{R}^2$ with homeomorphic leaf spaces

spaces, this homeomorphism reverses the order in the $\sim_+$ equivalence class while preserving the order in each $\sim_-$ equivalence class. An orientation-preserving homeomorphism that preserves the order in all these equivalence classes will be called an order-preserving homeomorphism. The classification of all foliations of $\mathbb{R}^2$, achieved in the early 1940s by W. Kaplan [**109, 110**], amounts to the following theorem (*cf.* [**87**, p. 125]).

Theorem D.1.7. *The conjugacy classes of transversely oriented, smooth foliations of* $\mathbb{R}^2$ *correspond bijectively to the the order-preserving diffeomorphism classes of oriented, simply connected* 1*-manifolds.*

The only part of this result that we will need is the following.

Corollary D.1.8. *Every simply connected, differentiable* 1*-manifold is the leaf space of some smooth foliation of the plane.*

Exercise D.1.9. Describe a foliation $\mathcal{F}$ of $\mathbb{R}^2$ such that the set of branch points is dense in $\mathbb{R}^2/\mathcal{F}$.

Exercise D.1.10. Let V be a simply connected 1-manifold, possibly with boundary, and let $U \subset V$ be an open subset such that $U \cong \mathbb{R}$. Show that, for each point $x \in \partial U$ (the set-theoretic boundary of U in V), either there is a unique point $z \in U$ such that x and z are not separated in V, or else

$U \cup \{x\}$ is homeomorphic to a half-open interval $(-\infty, \infty]$ or $[-\infty, \infty)$. In the case of a Reeb strip (*cf.* Figure D.1.1), the leaf space has interior $U \cong \mathbb{R}$ with ∂U a pair of points x_1 and x_2. Decide which of the above possibilities holds for x_i, $i = 1, 2$.

Exercise D.1.11. Let V be a simply connected 1-manifold, possibly with boundary, and let $U \subset V$ be an open, connected, Hausdorff submanifold. Prove that ∂U is at most countably infinite.

It is a remarkable fact that, if $\dim M > 2$, then the diffeomorphism class of the leaf space $M/\mathcal{F}$ of a planar foliation does determine the foliation up to topological conjugacy. Here is the formal statement.

Theorem D.1.12 (Palmeira). *Let M and M' be simply connected $(n+1)$-manifolds, $n \geq 2$, with respective planar foliations $\mathcal{F}$ and $\mathcal{F}'$. Then there is a diffeomorphism $h : M/\mathcal{F} \to M'/\mathcal{F}'$ if and only if there is such a diffeomorphism $H : M \to M'$ carrying $\mathcal{F}$ to $\mathcal{F}'$ and inducing h.*

The "if" part of this theorem is easy, but the "only if" part is deep.

Corollary D.1.13. *If M is a simply connected $(n+1)$-manifold, $n \geq 2$, and $\mathcal{F}$ is a planar foliation of M, then there is a foliation $\mathcal{F}^*$ of $\mathbb{R}^2$ such that the foliated manifolds $(M, \mathcal{F})$ and $(\mathbb{R}^{n+1}, \mathbb{R}^{n-1} \times \mathcal{F}^*)$ are smoothly conjugate.*

Here, of course, $\mathbb{R}^{n-1} \times \mathcal{F}^*$ is the foliation with leaves $\mathbb{R}^{n-1} \times L$, as L varies over the set of leaves of $\mathcal{F}^*$. Since the leaf space of $\mathcal{F}^*$ and that of $\mathbb{R}^{n-1} \times \mathcal{F}^*$ are canonically diffeomorphic, this corollary is an immediate consequence of Theorem D.1.12 and Corollary D.1.8. In particular, this will establish Theorem 9.1.10

Remark. The above corollary has been extended to planar laminations of $\mathbb{R}^3$ by D. Gabai and W. Kazez [**76**].

Exercise D.1.14. If $\mathcal{F}$ and $\mathcal{F}'$ are the foliations shown in Figure D.1.1, can you describe a diffeomorphism of $\mathbb{R}^3$ to itself that carries $\mathbb{R} \times \mathcal{F}$ to $\mathbb{R} \times \mathcal{F}'$? A qualitative description, without formulas, suffices.

The rest of this appendix will be devoted to proving Theorem D.1.12. The casual reader who prefers to skim the proof may not notice where the hypothesis that $n \geq 2$ is used. It is needed exactly for Corollary D.2.6. The critical fact is that connected manifolds of dimension $n \geq 2$ cannot be disconnected by removing a finite family of disjoint, closed n-balls from the interior.

D.2. Basic Isotopies

We will need a technical result about extending embeddings, treated in [**143**, Appendix] for the C^∞ case. This is a parametrized version of an extension theorem of R. Palais [**142**]. The theorem of Palais is an easy consequence of the fact that orientation-preserving embeddings of D^n into an open, connected, oriented n-manifold are unique up to compactly supported ambient isotopy. A very simple proof of this will be found in [**99**, Chapter 8], and this proof extends easily to give Palmeira's parametrized extension theorem.

Definition D.2.1. Let P and N be oriented n-manifolds, not necessarily compact and possibly with boundary, and let $J \subseteq \mathbb{R}$ be a possibly degenerate interval. A smooth map

$$f : N \times J \to P \times J$$

is level-preserving if $f(N \times \{t\}) \subseteq P \times \{t\}$, $\forall t \in J$. This map is a level-preserving embedding if it is required that $f|N \times \{t\}$ be a C^∞ embedding, $\forall t \in J$

The set of all maps as in this definition will be denoted by

$$C^J(N \times J, P \times J).$$

The subspace of level-preserving C^∞ embeddings will be denoted by

$$\mathrm{Emb}^J(N \times J, P \times J),$$

and the (possibly empty) subset of surjective embeddings will be denoted by

$$\mathrm{Diff}^J(N \times J, P \times J).$$

As is standard, a subscript "+" will refer to the subsets of orientation-preserving elements.

For isotopies f_s in $\mathrm{Emb}^J(N \times J, P \times J)$, the *track* of the isotopy will be an element

$$F \in \mathrm{Emb}^J(([0,1] \times N) \times J, ([0,1] \times P) \times J)$$

of the form

$$F(s,x,t) = (s, f_s(x,t), t) = (s, f_{s,t}(x), t).$$

If $N = P$ and $f_0 = \mathrm{id}_{P \times J}$, then the isotopy is called a (level-preserving) *diffeotopy* of $P \times J$. Isotopies of the form $f_s = \varphi_s \circ f$, where φ_s is a diffeotopy of $P \times J$ and $f \in \mathrm{Emb}^J(N \times J, P \times J)$, will be called *ambient isotopies* of f.

The following is standard in case J degenerates to a single point. See, for instance, [**99**, Chapter 8, §1]. For the general case, one modifies the proof, which uses the standard theorem for existence and uniqueness of solutions of ordinary differential equations, by appealing to the version of that theorem with parameters. In our case, J is the parameter space.

Proposition D.2.2. *If $N \subset \operatorname{int} P$ is a smooth, compact submanifold (possibly with boundary) and*

$$f \in \operatorname{Emb}_+^J(N \times J, (\operatorname{int} P) \times J)$$

is isotopic in this space to the inclusion, then there is an ambient, level-preserving isotopy of f to the inclusion map. Furthermore, this isotopy is supported outside of a neighborhood of $\partial P \times J$. In particular, f extends to a diffeomorphism

$$\widetilde{f} \in \operatorname{Diff}_+^J(P \times J, P \times J)$$

that is the identity in a small enough neighborhood of $\partial P \times J$ and is isotopic to the identity through diffeomorphisms of the same type.

Here, one uses the trivial fact that the inclusion map extends as the identity, using the ambient isotopy to conclude that f also extends as desired, *cf.* [**99**, p. 180, Theorem 1.5].

Observe that the diffeomorphisms $\widetilde{f}$ in Proposition D.2.2 form a subgroup $G \subset \operatorname{Diff}_+^J(P \times J, P \times J)$. In case $P = D^n$ is the unit n-disk, we will denote this subgroup by G_n.

In what follows, it will be convenient to fix a smooth identification $\operatorname{int} D^n = \mathbb{R}^n$, where D^n is the closed unit n-disk. We can set up this diffeomorphism radially so that round n-disks centered at the origin in $\operatorname{int} D^n$ are also such, with different radii, in $\mathbb{R}^n$. Let $B^n \subset \mathbb{R}^n$ denote any compact round disk centered at the origin.

Proposition D.2.3. *If $f \in \operatorname{Emb}_+^J(B^n \times J, \mathbb{R}^n \times J)$, then there is $\varphi \in G_n$ that extends f.*

In the case that J degenerates to a single point, this is the extension theorem of Palais. This proposition will be an immediate consequence of Proposition D.2.2 and the following.

Lemma D.2.4. *The map f in Proposition* D.2.3 *is isotopic to the inclusion map through elements of* $\operatorname{Emb}_+^J(B^n \times J, \mathbb{R}^n \times J)$.

Proof. We carry out this isotopy in three stages. First, we perform an isotopy of f to an embedding that is the identity on $\{0\} \times J$. We then perform a linearizing isotopy to an embedding $(x, t) \mapsto (A(t)x, t)$, where $A : J \to \operatorname{Gl}_+(n)$ is of class C^∞. Finally, a suitable C^∞ homotopy of this path to the constant path $I(t) \equiv I \in \operatorname{Gl}_+(n)$ completes the isotopy.

The map $f|\{0\} \times J$ is isotopic to the inclusion map. Just use the translation isotopy $(0, t) \mapsto (f_t(0) - sf_t(0), t)$, $0 \leq s \leq 1$. By Proposition D.2.2, this extends to an ambient isotopy h_s of the desired type on $D^n \times J$, and the composition $h_s \circ f$ gives the first stage of our isotopy.

We now assume that $f(0,t) = (0,t)$, for all $t \in J$. The linearizing isotopy is defined by

$$f_{s,t}(x) = \begin{cases} f_t(sx)/s, & 0 < s \leq 1, \\ Df_t(0) \cdot x, & s = 0, \end{cases}$$

where $Df_t(0)$ is the Jacobian matrix of f_t at $x = 0$ (*cf.* [**99**, p. 186]). Thus, the C^∞ path $A : J \to \mathrm{Gl}_+(n)$ resulting from the stage two isotopy is

$$A(t) = Df_t(0).$$

This path lies in $\mathrm{Gl}_+(n)$ because f is orientation-preserving. It will be convenient to replace $A(t)$ with $A(0)^{-1}A(t)$. Since $\mathrm{Gl}_+(n)$ is path connected, this is isotopic to $A(t)$, and so we are allowed to assume that $A(0) = I$.

We now assume that

$$f(x,t) = (A(t) \cdot x, t),$$

where $A : J \to \mathrm{Gl}_+(n)$ and $A(0) = I$. The isotopy

$$f_s(x,t) = (A((1-s)t) \cdot x, t)$$

connects $f_0 = f$ to $f_1 = \mathrm{id}\,|(B^n \times J)$. □

We now generalize Proposition D.2.3, replacing $\mathbb{R}^n$ with $\operatorname{int} P$, where P is a connected, oriented n-manifold. Fix a smoothly embedded n-disk $B^n \subset \operatorname{int} P$, where $n = \dim P$.

Proposition D.2.5. *If P is connected and*

$$f \in \mathrm{Emb}_+^J(B^n \times J, (\operatorname{int} P) \times J),$$

then there is a $\varphi \in G$ that extends f.

Proof. Slightly enlarge B^n to an embedded disk D^n so that $B^n \subset \operatorname{int} D^n$. We coordinatize these disks so that they are concentric. For notational simplicity, let $W = (\operatorname{int} D^n) \times J$ and $V = (\operatorname{int} P) \times J$.

Let $x_0 \in B^n$ be the center of the embedded disk and set $\sigma(t) = (x_0, t)$, as t ranges over J. Let $\tau = f \circ \sigma$. Using the connectivity of P, we will show that there is a level-preserving, isotopy of τ to σ. Indeed, write

$$\tau(t) = f(x_0, t) = (g(t), t),$$

where $g : J \to \operatorname{int} P$ is smooth. Without loss of generality, assume that $0 \in J$ and set $\tau_s(t) = (g(st), t)$, $0 \leq s \leq 1$. This defines an isotopy of τ to the inclusion $\iota : \{g(0)\} \times J \hookrightarrow (\operatorname{int} P) \times J$. The fact that P is connected implies that there is a smooth path $\eta(s)$ in $\operatorname{int} P$ from x_0 to $g(0)$. Then $\iota_s(t) = (\eta(s), t)$ completes the isotopy of τ to σ. By Proposition D.2.2, we realize this as an ambient isotopy.

We can now assume that f is the identity on $\{x_0\} \times J$. Then there is a tubular neighborhood $U \subset (\operatorname{int} B^n) \times J$ of $\{x_0\} \times J$ such that $f(U) \subset W$. A radial isotopy deforms $B^n \times J$ into U, and composition of this isotopy with f gives a level-preserving isotopy of f to h such that

$$h(B^n \times J) \subset U \subset (\operatorname{int} D^n) \times J.$$

We are now reduced to the situation of Lemma D.2.4, obtaining an isotopy of h to the inclusion and concluding that the desired extension exists by Proposition D.2.2. □

Corollary D.2.6. *Let P be a connected, oriented n-manifold, $n \geq 2$, and let $B_1, \dots, B_r$ be a disjoint family of closed, oriented n-balls. Then level-preserving and orientation-preserving smooth embeddings*

$$f : \bigsqcup_{i=1}^{r} B_i \times J \hookrightarrow (\operatorname{int} P) \times J$$

are unique up to composition with elements of G.

Proof. After choosing an embedding of the family of disks into $\operatorname{int} P$, we view

$$\bigsqcup_{i=1}^{r} B_i \times J \subset (\operatorname{int} P) \times J.$$

It will be enough to show that f is ambiently isotopic through level-preserving embeddings to the inclusion map. We proceed by induction on r, remarking that the case $r = 1$ is just Proposition D.2.5. If $r > 1$, the inductive hypothesis allows us to assume that the restriction of f to $B_i \times J$ is the inclusion, $1 \leq i \leq r - 1$. Then we can replace P with

$$P' = P \smallsetminus \bigsqcup_{i=1}^{r-1} \operatorname{int} B_i.$$

Since $n \geq 2$, this is connected, and an application of Proposition D.2.5 then allows us to perform a level-preserving, ambient isotopy of $f|B_r \times J$ to the inclusion in $P' \times J$. □

The following is basically another way of looking at Corollary D.2.6.

Corollary D.2.7. *Let P and f be as in Corollary* D.2.6 *and let*

$$\varphi, \varphi' : D^n \times J \to (\operatorname{int} P) \times J$$

be level-preserving and orientation-preserving embeddings with images disjoint from $\operatorname{im} f$ *(but not necessarily disjoint from each other). Then there is $\Phi \in G$ which is the identity on* $\operatorname{im} f$*, is isotopic to the identity through such diffeomorphisms, and is such that $\Phi \circ \varphi' = \varphi$.*

Proof. By Corollary D.2.6, we can recoordinatize $P \times J$ in a level-preserving and orientation-preserving way so that f becomes the inclusion map

$$\bigsqcup_{i=1}^{r} B_i \times J \hookrightarrow (\operatorname{int} P) \times J$$

relative to an embedding of $B_1, \dots, B_r$ as disjoint n-balls in $\operatorname{int} P$. Now replace P with the complement P' of the union of the interiors of these balls, and let G' denote the group of level-preserving diffeomorphisms of $P' \times J$ that are the identity in a small enough neighborhood of $\partial P' \times J$ and are isotopic to the identity through diffeomorphisms of the same type. We can view φ and φ' as embeddings into $(\operatorname{int} P') \times J$ and apply Corollary D.2.6 to find $\Phi' \in G'$ such that $\Phi' \circ \varphi' = \varphi$. But Φ' extends via the identity over the image of f to give the required $\Phi \in G$. □

D.3. The Hausdorff Case

The only simply connected Hausdorff manifold is the real line. The following is an important special case of Corollary D.1.8.

Proposition D.3.1. *If $\mathcal{F}$ is a planar foliation of a simply connected manifold M, $\dim M = n+1$, then $\mathcal{F}$ is smoothly conjugate to the foliation of $\mathbb{R}^{n+1}$ by parallel hyperplanes if and only if $M/\mathcal{F} \cong \mathbb{R}$.*

The "only if" part is trivial. We also note that this result is true even if $n = 1$.

If X is a smooth, nowhere vanishing vector field on M, we denote by $\mathcal{L}_X$ the 1-dimensional foliation tangent to X.

Lemma D.3.2. *Let X be a C^∞ vector field on M, everywhere transverse to the planar foliation $\mathcal{F}$ and such that every leaf of $\mathcal{L}_X$ meets every leaf of $\mathcal{F}$. Then $\mathcal{F}$ is smoothly conjugate to the foliation of $\mathbb{R}^n$ by parallel hyperplanes.*

Proof. Indeed, fix a leaf ℓ of $\mathcal{L}_X$ and a leaf L of $\mathcal{F}$. Since ℓ meets every leaf of $\mathcal{F}$, but cannot meet the same leaf twice (Proposition D.1.1), there is a well-defined submersion

$$p_{\mathcal{F}} : M \to \ell,$$

defined by projection along the leaves of $\mathcal{F}$. Since every leaf of $\mathcal{L}_X$ meets L exactly once, there is a C^∞ submersion

$$p_X : M \to L,$$

defined by projection along the leaves of $\mathcal{L}_X$. After identifying ℓ with $\mathbb{R}$ and L with $\mathbb{R}^{n-1}$, we see that $p_X \times p_{\mathcal{F}}$ is the desired conjugacy. □

Thus, in order to prove the "if" part of Proposition D.3.1, we only need to show that, if $M/\mathcal{F} \cong \mathbb{R}$, then there is a vector field as in Lemma D.3.2. This is not obvious. The proof will proceed by choosing an arbitrary vector field X, everywhere transverse to $\mathcal{F}$, and then performing a possibly infinite sequence of local modifications of this field, producing a sequence $\{X_k\}_{k=1}^{\infty}$ of fields converging smoothly to a field X_∞ with the desired properties.

In analogy with Definition D.2.1, we will be considering smooth embeddings

$$\varphi : D^{n-1} \times [b^-, b^+] \to M$$

with the following properties:

(1) φ carries $D^{n-1} \times \{t\}$ into a leaf of $\mathcal{F}$, $b^- \leq t \leq b^+$.

(2) The curves $\varphi : \{x\} \times [b^-, b^+] \to M$ are transverse to $\mathcal{F}$, $\forall\, x \in D^{n-1}$.

Definition D.3.3. The embedded solid cylinder as above will be called a *normal plug.* The components of $\varphi(D^{n-1} \times \{b^\pm\})$ are called the *bases* of the plug and $\varphi(\partial D^{n-1} \times [b^-, b^+])$ is called the *wall.*

Thus, a normal plug is just a (closed) foliated chart for $\mathcal{F}$. In fact, it is bifoliated, but the transverse foliation is only defined in the plug. In order to construct a normal plug containing a point y of M in its interior, choose an $\mathcal{F}$-transverse, C^∞ vector field in a small neighborhood of y and smoothly reparametrize the local flow lines so that all meet the same leaf at the same time.

Lemma D.3.4. *If $K \subset M$ is compact and connected, there is a normal plug containing K in its interior.*

The idea will be to assemble a finite family of normal plugs with interiors covering K and use these to construct a single normal plug engulfing K. The assumption that $M/\mathcal{F} \cong \mathbb{R}$ will be essential for this. Before going into the details, we use this result for the following.

Proof of Proposition D.3.1. Let $\{K_j\}_{j=1}^{\infty}$ be a sequence of compact, connected subsets of M such that $K_j \subset \operatorname{int} K_{j+1}$, $j \geq 1$, and such that M is the increasing union of these sets. Inductively applying Lemma D.3.4, we create a sequence $\{C_i\}_{i=1}^{\infty}$ of normal plugs by the following process:

(a) $K_1 \subset \operatorname{int} C_1$;

(b) if C_k has been selected, $1 \leq k \leq i$, let $j \geq i+1$ be so large that $C_k \subseteq K_j$, $1 \leq k \leq i$, and choose C_{i+1} so that $K_j \subset \operatorname{int} C_{i+1}$.

Thus, these plugs are nested, each in the interior of the next, and exhaust the manifold.

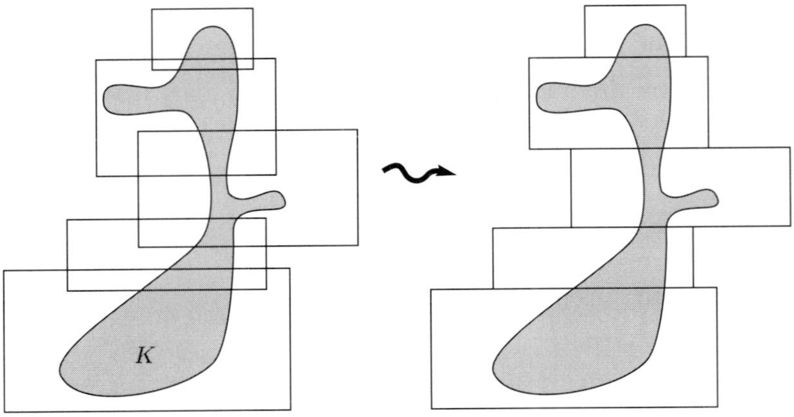

Figure D.3.1. Converting the plug cover to a stack of plugs

Beginning with a smooth, $\mathcal{F}$-transverse vector field X, orient $\mathcal{F}$ transversely by X and create the sequence $\{X_k\}_{k=0}^{\infty}$ inductively as follows. Set $X_0 = X$. If X_k has been defined, let U_{k+1} be a neighborhood of the wall Σ_{k+1} of C_{k+1}, chosen so that $C_k \cap U_{k+1} = \emptyset$. Let v_{k+1} be a smooth, positively directed, $\mathcal{F}$-transverse vector field defined only on U_{k+1} and tangent to Σ_{k+1}. Finally, assemble v_{k+1} and X_k into a single transverse vector field X_{k+1} via a smooth partition of unity subordinate to $\{U_{k+1}, M \smallsetminus \Sigma_{k+1}\}$. Notice that, for $j > k$, X_j agrees with X_k in a neighborhood of C_k. Hence the pointwise limit

$$X_\infty = \lim_{k\to\infty} X_k$$

exists and agrees with X_k in a neighborhood of C_k, for all $k \geq 1$. It is also evident that any orbit of X_k passing through a point of C_k can only enter C_k through the bottom base and can only escape C_k by exiting the plug through the top base. Thus, X_∞ also has this property, for each $k \geq 1$. Let ℓ be a leaf of $\mathcal{L}_{X_\infty}$ and let L be a leaf of $\mathcal{F}$. Choose k large enough so that both ℓ and L meet C_k. Since ℓ can only enter C_k through the bottom and can only exit C_k by passing through the top, it meets every leaf that meets C_k. In particular, $\ell \cap L \neq \emptyset$. As remarked earlier, Proposition D.3.1 follows. □

Proof of Lemma D.3.4. Let $K \subset M$ be compact and connected. Given a point $x \in K$, find a smoothly embedded $(n-1)$-disk D_x in the leaf through x, chosen so that $K \cap D_x \subset \operatorname{int} D_x$. Using the local flow of any smooth transverse vector field, as already indicated, we can construct a normal plug containing x in its interior and such that K does not meet the wall of the

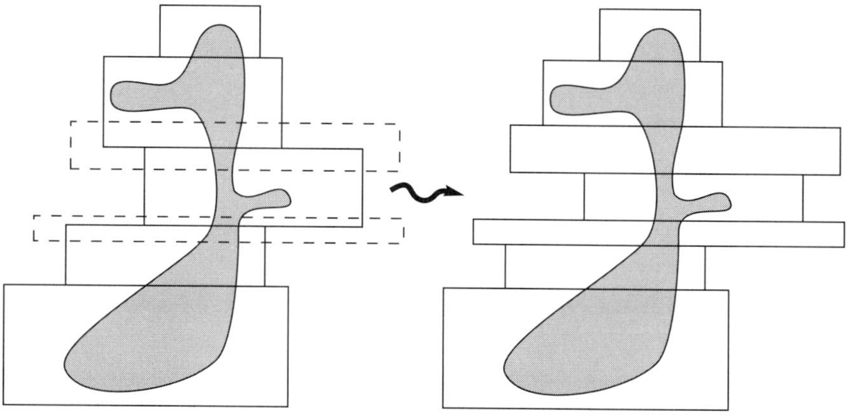

Figure D.3.2. Producing an unstaggered stack of plugs

plug. By compactness, we cover K with the interiors of finitely many such plugs $C_1, \dots, C_r$.

The projection $p : M \to M/\mathcal{F} = \mathbb{R}$ carries K onto a compact interval J and it carries each C_k onto a compact interval I_k. The interiors of the intervals I_k form an open cover of J, Hence one can shrink each I_k in such a way that any two intersect in at most one common endpoint and so that the resulting compact intervals cover J. Shrinking I_k can be achieved by shrinking C_k in the transverse direction, this operation on the plugs being depicted schematically in Figure D.3.1. The result is a "stack" of plugs which is generally "staggered". That is, two interfacing plugs may have bases in a common leaf L, neither of which is contained in the interior of the other. To remedy this, choose a disk in L that engulfs both of these bases in its interior and thicken this slightly to a plug. Now some further transverse shrinkings, as illustrated in Figure D.3.2, produce an unstaggered stack of plugs.

We now show how to "cobble" together these plugs to a single one. Consider two adjacent plugs, C_k and C_{k+1}, in the stack, assuming for definiteness that the top base B_k^+ of C_k is contained in the interior of the bottom base B_{k+1}^- of C_{k+1}. Thicken C_{k+1} slightly to a normal plug C by lowering B_{k+1}^- as indicated in Figure D.3.3. Fix $B^{n-1} \subset \operatorname{int} D^{n-1}$ as in Proposition D.2.3 and let the embeddings of these respective plugs be of the form

$$g_k : B^{n-1} \times J_k \to C_k \subset M,$$
$$g_{k+1} : D^{n-1} \times J_{k+1} \to C_{k+1} \subset M,$$
$$g : D^{n-1} \times J \to C \subset M,$$

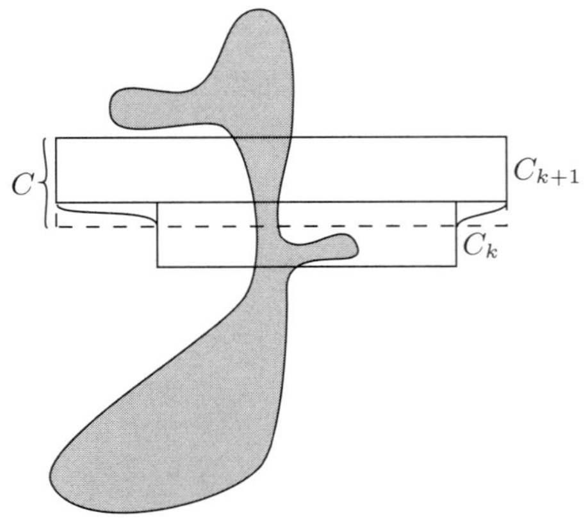

Figure D.3.3. Cobbling together adjacent plugs

where $J_{k+1} \subset J$ and the upper endpoints of these intervals coincide. The composition $g^{-1} \circ g_k$ can be interpreted in the overlap as a level-preserving embedding $f : B^{n-1} \times I \to \operatorname{int} D^{n-1} \times I$, where I is a closed interval and we have taken the liberty of suitably rechoosing the transverse parameter. One thinks of g^{-1} (restricted to the overlap) as a coordinate system, relative to which f is the embedding of $C \cap C_k$. If φ is the extension given by Proposition D.2.3, then φ^{-1} recoordinatizes the picture so that the embedding can be viewed as the inclusion $B^{n-1} \times I \subset D^{n-1} \times I$. It is now trivial to cobble together the plugs as indicated in Figure D.3.3. Finite repetition of this procedure produces the desired normal plug containing K in its interior. □

D.4. Decomposing the Foliation

We will decompose the foliated manifolds $(M, \mathfrak{F})$ and $(M', \mathfrak{F}')$ along a countable, discrete family of leaves so that the components are all as in the previous section. If there is a diffeomorphism $h : M/\mathfrak{F} \to M'/\mathfrak{F}'$, we can choose these decompositions to be respected by h. By Proposition D.3.1, h will lift to a conjugacy of foliations on each component. In the following section, these conjugacies will then be extended compatibly across the decomposing leaves. The reader who has successfully carried out Exercise D.1.14 should already have some idea as to how this extension will be constructed.

Definition D.4.1. Let V be a connected, 1-dimensional manifold. A collection $\{V_i\}_{i=1}^{N}$, $1 \le N \le \infty$, of disjoint open subsets is a *normal decomposition* of V if

(1) $V_i \cong \mathbb{R}$, $1 \le i < N + 1$;

(2) $\{\overline{V}_i\}_{i=1}^N$ covers V;

(3) for all $i \neq j$, $\overline{V}_i \cap \overline{V}_j$ is either empty or a singleton called a *cut point*; and

(4) the set of cut points is exactly the complement of $\bigcup_{i=1}^N V_i$ and is a closed, discrete subset of V.

Proposition D.4.2. *If V is a simply connected 1-manifold, then V admits a normal decomposition.*

Evidently, if $V = M/\mathcal{F} = M'/\mathcal{F}'$, this will lift to decompositions of $(M, \mathcal{F})$ and $(M', \mathcal{F}')$ of the required type. We will give the proof of Proposition D.4.2 after some preliminary lemmas. We fix the hypotheses of the proposition.

Lemma D.4.3. *Each open, connected, Hausdorff submanifold $U \subset V$ is diffeomorphic to $\mathbb{R}$.*

Proof. By the standard classification of Hausdorff 1-manifolds, the only alternative is that $U \cong S^1$. Assume this and let $x \in U$. By Exercise D.1.4, $V \smallsetminus \{x\}$ is not connected and so we can pick points $y \in U \smallsetminus \{x\}$ and z in a component W of $V \smallsetminus x$ not meeting $U \smallsetminus \{x\}$. Since V is locally path-connected, it follows that V is path-connected and W is a path-connected component of $V \smallsetminus \{x\}$. Let σ be an immersed path in V joining z to y. Thus, σ must pass through $x = \sigma(t_0)$. By the continuity of σ, there is a neighborhood $(t_0 - \varepsilon, t_0 + \varepsilon) \subset \operatorname{dom} \sigma$ carried diffeomorphically by σ to an open arc in U containing x. It follows easily that there is a subarc of σ, starting at z, ending at a point of U and not meeting x. This contradiction proves that $U \cong \mathbb{R}$. □

Lemma D.4.4. *Let U_0 and U_1 be disjoint, open, connected Hausdorff submanifolds of V. Then $\overline{U}_0 \cap \overline{U}_1$ is either empty or a singleton.*

Proof. If x and y are distinct points of this intersection, let V_0 and V_1 be the two components of $V \smallsetminus x$ given by Exercise D.1.4. Fix an orientation of V such that V_0 borders x on the left, V_1 on the right. By Exercise D.1.10, there are open, nonempty subintervals $J_i \subseteq U_i$, $i = 1, 2$, each bordering x on one side. If both border x on the left (respectively, on the right), then $U_0 \cap U_1 \neq \emptyset$, contrary to hypothesis. Thus, we can choose the indexing so that $U_i \subseteq V_i$, $i = 0, 1$. But Exercise D.1.10 implies that $U_0 \cup \{y\} \cup U_1$ is connected. Since this set lies in $V_0 \cup V_1$, we have a contradiction. □

Lemma D.4.5. *Let U_0 and U_1 be open, connected Hausdorff submanifolds of V. Then $U_1 \cap \partial U_0$ has cardinality c at most 2. In any event, $U_1 \cap U_0$ is either empty or an open interval and, in the latter case, $U_1 \cap \overline{U}_0$ is the corresponding relatively closed subinterval of U_1.*

Proof. Suppose that x, y and z are distinct points of $U_1 \cap \partial U_0$. Since $U_1 \cong \mathbb{R}$ (Lemma D.4.3), we can assume that z separates x and y in U_1. Since $z \in \partial U_0$, it does not separate U_0. On the other hand, Exercise D.1.4 requires that z separate V into two components V_0 and V_1 such that $x \in V_0$ and $y \in V_1$. But x can be connected to y by a path, the interior of which lies in U_0 and hence misses z, a contradiction. For the final assertions, the reader can consider separately the cases that $c = 0, 1,$ or 2. In each case, appeal to Exercise D.1.10 (and, if $c = 2$, simple connectivity) to check the claim. □

Lemma D.4.6. *Each point $x \in V$ has an open, connected, Hausdorff neighborhood U, maximal with respect to these properties. Each point $y \in \partial U$ is a branch point of V, and there is a unique point $y' \in U$ such that every neighborhood of y in V meets every neighborhood of y'. Finally, there is an at most countably infinite cover $\{U_i\}_{i=1}^Q$ of V by such open submanifolds.*

Proof. Since V is a 1-manifold, every point $x \in V$ has an open, connected Hausdorff neighborhood. We extend this to a maximal such neighborhood via Zorn's lemma. Indeed, the family $\mathcal{W}$ of such neighborhoods of x is partially ordered by inclusion. If $\{W_\alpha\}_{\alpha \in \mathfrak{A}}$ is a linearly ordered subfamily, then evidently the union

$$W = \bigcup_{\alpha \in \mathfrak{A}} W_\alpha$$

is an element of $\mathcal{W}$. Thus, this family is inductive and must contain a maximal element U. Let $y \in \partial U$. By the maximality of U, $U \cup \{y\}$ cannot be a half-open interval; hence Exercise D.1.10 gives the second assertion. Since $x \in V$ is arbitrary, there is a cover of V by such open submanifolds and second countability allows us to pass to a countable subcover. □

Proof of Proposition D.4.2. Let $\{U_i\}_{i=1}^Q$ be an enumeration of the cover given by Lemma D.4.6, where $1 \leq Q \leq \infty$. Set

$$W_k = U_k \smallsetminus \bigcup_{i=1}^{k-1} \overline{U}_i, \quad 1 \leq k < Q + 1,$$

obtaining a countable, disjoint family of open subsets of V. Since $\{\overline{U}_k\}_{k=1}^Q$ covers V, it follows easily that $\{\overline{W}_k\}_{k=1}^Q$ also covers V. From Lemma D.4.5 and our construction, it follows that each W_k falls into a finite family $W_{k,1}, W_{k,2}, \ldots, W_{k,r_k}$ of components, each of which is an open subset of the open Hausdorff submanifold U_k. As k ranges over all admissible values, we use the lexicographic order on these components, obtaining a sequence $\{V_i\}_{i=1}^N$ of disjoint, open, connected Hausdorff manifolds such that

$$\bigcup_{i=1}^N \overline{V}_i = \bigcup_{k=1}^Q \overline{W}_k = V.$$

By Lemma D.4.3, $V_i \cong \mathbb{R}$, $\forall\, i$, and we have verified properties (1) and (2) of Definition D.4.1. Property (3) is given by Lemma D.4.4 and it remains for us to check property (4). Since both $\{U_k\}_{k=1}^{Q}$ and $\{\overline{V}_i\}_{i=1}^{N}$ cover V, we see that $\{U_k \cap \overline{V}_i\}_{i,k=1}^{Q,N}$ also covers V. But $U_k \cap \overline{V}_i$ is a relatively closed subinterval of U_k. Let K be the set of endpoints of these relatively closed subintervals. Since the V_i's are open and disjoint, no point of K can belong to $\bigcup_{i=1}^{N} V_i$, and it follows that K is exactly the complement of this set, hence is closed in V. For each $k \geq 1$, $U_k \cap V_i = \emptyset$ if i is sufficiently large. Thus, we see that $K \cap U_k$ is finite, for each k, and so K is discrete. Also, no cut point can belong to any V_i, and so the set of cut points is a subset $C \subseteq K$. Conversely, if $x \in K$, let U be a connected Euclidean neighborhood of x. Since K is discrete, we can assume that $K \cap U = \{x\}$. Thus, there are distinct indices i, j such that $V_i \cap U$ borders x on the left and $V_j \cap U$ borders x on the right. By Lemma D.4.4, $x \in C$, and it follows that $C = K$. □

Corollary D.4.7. *Let $(M, \mathcal{F})$ be a simply connected, planar-foliated manifold without boundary. Then there is a countable family $\{M_i\}_{i=1}^{N}$ of $\mathcal{F}$-saturated submanifolds with boundary such that*

(1) $(\operatorname{int} M_i)/\mathcal{F} \cong \mathbb{R}$, $1 \leq i < N+1$;

(2) $\{M_i\}_{i=1}^{N}$ *covers* M;

(3) *if* $i \neq j$, $M_i \cap M_j = L_{ij}$ *is either empty or a single common boundary leaf called a cut leaf; and*

(4) *the family of cut leaves is discrete in* M*, and the union* K *of these leaves is exactly the complement of the open set* $\bigcup_{i=1}^{N} \operatorname{int} M_i$.

Indeed, if $p : M \to M/\mathcal{F}$ is the natural projection and if $\{V_i\}_{i=1}^{N}$ is a normal decomposition of $M/\mathcal{F}$, we take

$$M_i = p^{-1}(\overline{V}_i), \quad 1 \leq i < N+1.$$

The four properties correspond exactly to the properties of a normal decomposition. Since K is closed, discreteness of the family of cut leaves implies that every sequence $\{x_i\}_{i=1}^{\infty}$ in K, having at most one point in each cut leaf, diverges in M.

D.5. Construction of the Diffeomorphism

We assume that $(M, \mathcal{F})$ and $(M', \mathcal{F}')$ are simply connected, planar-foliated $(n+1)$-manifolds. We are given a diffeomorphism $h : M/\mathcal{F} \to M'/\mathcal{F}'$ and must show how to lift h to a C^∞ conjugacy

$$H : (M, \mathcal{F}) \to (M', \mathcal{F}').$$

We will fix tangential and transverse orientations of these foliations and require that H respect these orientations. The transverse orientations induce orientations on the leaf spaces preserved by h.

Fix a normal decomposition of $M/\mathfrak{F}$ and use this to define a decomposition $\{M_i\}_{i=1}^N$ of M as in Corollary D.4.7. Since M is connected, we can number the elements of this decomposition so that $M_i \cap M_{i+1}$ is a cut leaf, $1 \le i < N+1$. Via the diffeomorphism h, we obtain a corresponding normal decomposition of $M_i'/\mathfrak{F}'$ and the associated decomposition $\{M_i'\}_{i=1}^N$ of M'. Set

$$\begin{aligned} N_0 &= \emptyset, \\ N_k &= M_1 \cup M_2 \cup \cdots \cup M_k, \quad 1 \le k < N+1, \end{aligned}$$

and define N_k' analogously, $0 \le k < N+1$. These submanifolds are saturated, and we can assume inductively that $h_k = h|(N_k/\mathfrak{F})$ has been lifted to a C^∞ conjugacy

$$H_k : (N_k, \mathfrak{F}|N_k) \to (N_k', \mathfrak{F}'|N_k'),$$

the case $k = 0$ being vacuously true. The rest of this section is devoted to constructing a C^∞ extension of H_k to a lift H_{k+1} of h_{k+1}. There is generally a smoothness problem at the cut leaves interfacing N_k and M_{k+1}. This will require a mild modification of H_k near these cut leaves in order to guarantee that H_{k+1} is a diffeomorphism. These successive extensions then fit together to give the required lift H of h. For this, one needs part (4) of Corollary D.4.7.

Generally, H_k may be defined on some, but not all, of the components of ∂M_{k+1}. Since h defines a one-to-one correspondence $L \leftrightarrow L'$ between the set of cut leaves L in M and the set of cut leaves L' in M', we can choose orientation-preserving diffeomorphisms between the corresponding boundary leaves of M_{k+1} and M_{k+1}' on which H_k is not already defined. In this way, $H_{k+1}|\partial M_{k+1}$ has been defined, and we must complete this extension to all of M_{k+1}, hence to N_{k+1}.

Let $\{L_i\}_{i=1}^Q$ be the family of components of ∂M_{k+1}, the component of $\partial M_{k+1}'$ corresponding to L_i being denoted, as above, by L_i'.

Definition D.5.1. The boundary component L_i is a left boundary leaf if M_{k+1} lies to the right of L_i, relative to the transverse orientation of $\mathfrak{F}$. Otherwise, L_i is a right boundary leaf.

Remark that L_i' is a left boundary leaf if and only if L_i is a left boundary leaf.

Fixing a complete Riemannian metric on M, define choices of disjoint, normal neighborhoods $V_i = V(L_i)$ in M of these components of ∂M_{k+1}, the normal fibers being length-minimizing geodesic arcs oriented compatibly with the transverse orientation of $\mathcal{F}$. We can assume that these fibers are uniformly bounded in length, say, by 2, all being parametrized by directed arclength from L_i. We can assume that some fiber J_i is so parametrized as $(-1,1)$. Similar constructions are carried out for L_i' in M'.

Lemma D.5.2. *If $x_i \in V_i$, $1 \le i < Q+1$, then the sequence $\{x_i\}_{i=1}^Q$ does not contain a subsequence that converges in M. The analogous assertion holds in M'.*

Proof. Indeed, one chooses $y_i \in L_i$ at Riemannian distance at most 1 from x_i, $1 \le i < Q+1$. If the x_i's admit a subsequence converging to $x \in M$, the corresponding y_i's ultimately lie in the ball about x of radius 2. This ball is compact since the metric is complete; hence $\{y_i\}_{i=1}^Q$ admits a convergent subsequence, contradicting (4) in Corollary D.4.7. □

Because of part (4) of Corollary D.4.7, we can assume not only that these normal neighborhoods are disjoint, but that the only cut leaf that meets V_i is L_i. This neighborhood is generally not $\mathcal{F}$-saturated. There is a fiber J_i of V_i parametrized as $(-1,1)$ so that

$$M_{k+1} \cap J_i = \begin{cases} [0,1), & \text{if } L_i \text{ is a left boundary leaf,} \\ (-1,0], & \text{if } L_i \text{ is a right boundary leaf.} \end{cases}$$

Let U_i be the $\mathcal{F}$-saturation of J_i and appeal to Proposition D.3.1 to define a C^∞ diffeomorphism

$$\varphi_i : U_i \to L_i \times J_i$$

carrying $\mathcal{F}|U_i$ to the product foliation. The right and left halves of U_i (containing the leaf L_i) will be denoted by $U_i^\pm$, respectively. Relative to φ_i, we can identify U_i^+ as $L_i \times [0,1)$ and U_i^- as $L_i \times (-1,0]$.

In order to keep indices from getting out of hand, we fix attention on $L = L_i$, denoting $L_i', U_i, J_i, V_i, \varphi_i$, *etc.*, by L', U, J, V, φ, *etc.*, respectively. For definiteness, we will assume that $U^+ \subset M_{k+1}$. That is, L is a left boundary leaf. In the opposite case, obvious modifications are made in all that follows.

By projection to the quotient, we identify J as an open subarc of $M/\mathcal{F}$, carry it by h to an open subarc of $M'/\mathcal{F}'$ and then lift this to an $\mathcal{F}'$-transverse arc J' in M', also parametrized (via h) as $(-1,1)$, meeting L' in $\{0\}$ and meeting no other cut leaf. Let U' be the $\mathcal{F}'$-saturation of J' and fix the diffeomorphism

$$\varphi' : U' \to L' \times J'.$$

As for the neighborhood U, we define the positive and negative sides $U'^\pm$ of U'. Via these coordinatizations of U and U', we extend H_{k+1} (already defined on L) to a C^∞ diffeomorphism of U^+ onto U'^+ by defining

$$\begin{aligned} G &: L \times [0,1) \to L' \times [0,1), \\ G(x,t) &= (H_{k+1}(x), t), \\ H_{k+1} &= (\varphi')^{-1} \circ G \circ \varphi. \end{aligned}$$

In case H_k was already defined on U^-, there may be a "shear" effect along L that prevents smoothness when we try to extend $H_k|U^-$ by H_{k+1} on U^+.

Lemma D.5.3. *If H_k is defined on U^-, there is a normal one-sided neighborhood W^- of L in U^-, disjoint from suitable normal neighborhoods of the other boundary leaves of N_k, together with a modification of H_k to a C^∞ conjugacy*

$$\widetilde{H}_k : (N_k, \mathcal{F}|N_k) \to (N'_k, \mathcal{F}'|N'_k)$$

that agrees with H_k outside of W^- and is smoothly extended by H_{k+1}.

Proof. The formula for $H_k|U^-$, relative to the coordinatizations φ and φ', will be of the form

$$(x,t) \in L \times (-1,0] \mapsto (f(x,t),t) \in L' \times (-1,0],$$

where f is of class C^∞. We select a C^∞ function

$$\lambda : (-1,0] \to (-1,0]$$

that vanishes identically near 0 and is the identity near -1, and we then replace the above formula for $H_k|U^-$ with

$$(x,t) \mapsto (f(x,\lambda(t)),t).$$

While this is smoothly extended by H_{k+1}, the saturated neighborhood U^- may meet every neighborhood of other boundary leaves of N_k, so we must suitably localize this definition to a nonsaturated, one-sided normal neighborhood $W \subset U^-$ that misses suitable normal neighborhoods of the other components of ∂N_k. Select such a neighborhood W and another such normal neighborhood O such that $\overline{O} \subset W$. If $\{\mu_0, \mu_1\}$ is a smooth partition of unity subordinate to the open cover $\{\varphi(W), \varphi(U^- \smallsetminus \overline{O})\}$ of $L \times (-1,0]$, we define

$$\lambda(x,t) = \mu_0(x,t)\lambda(t) + \mu_1(x,t)t,$$

noting that $\lambda(x,t) = \lambda(t)$ on $\varphi(O)$ and $\lambda(x,t) = t$ outside of $\varphi(W)$. The formula for the modified diffeomorphism $\widetilde{H}_k$ relative to the coordinatizations φ and φ' will be

$$(x,t) \mapsto (f(x,\lambda(x,t)),t).$$

□

The saturated neighborhood U^+ is not generally disjoint from arbitrary neighborhoods of other components of ∂M_{k+1}, so we restrict the extension H_{k+1} to the one-sided normal neighborhood $V^+ = V \cap U^+$ of L. Allowing L to range over all of the components $\{L_i\}_{i=1}^{Q}$ of ∂M_{k+1}, we obtain an extension H_{k+1} of H_k to a neighborhood of ∂M_{k+1}. Our goal, then, is to modify these local extensions outside suitable neighborhoods of the boundary leaves so that they extend to a conjugation, again denoted by H_{k+1}, of $\mathcal{F}|M_{k+1}$ to $\mathcal{F}'|M'_{k+1}$ that lifts h over $M_{k+1}/\mathcal{F}$.

The neighborhood of the boundary leaf that we will construct will be called a *cylindrical collar*. We will need cylindrical collars of interior leaves also, so we define the notion in general.

Definition D.5.4. If L is a leaf of $\mathcal{F}|M_{k+1}$ and V^+ is a one-sided normal neighborhood in M_{k+1} of L on the right, a right cylindrical collar of L will be a set $B \subset V^+ \smallsetminus L$, the image of an embedding $D^n \times (0, b] \hookrightarrow V^+$ that carries each factor $D^n \times \{w\}$ into a leaf L_w of $\mathcal{F}$ and is such that each $x \in L$ has a neighborhood W_x in V^+ with $W_x \smallsetminus L \subset B$. Left cylindrical collars are defined analogously.

Left boundary leaves will have only right cylindrical collars, and *vice versa.*

Lemma D.5.5. *Every leaf L of $\mathcal{F}|M_{k+1}$ admits cylindrical collars. If L is an interior leaf, it admits such collars on both sides. If L is a boundary leaf and L' the corresponding boundary leaf in M'_{k+1}, there are respective cylindrical collars B and B' such that $H_{k+1}(B) = B'$.*

Proof. For definiteness, we assume that L has a right side V^+ in M_{k+1} and we work on that side. Even if L is not a boundary leaf, the saturation U^+ of V^+ admits a foliated diffeomorphism $\varphi : U^+ \to L \times [0, 1)$.

Since L is diffeomorphic to the standard coordinate space $\mathbb{R}^n$, we exhaust L by a sequence $\{D_j\}_{j=0}^{\infty}$ of subsets diffeomorphic to D^n, chosen so that $D_j \subset \operatorname{int} D_{j+1}$. In case L is a boundary leaf, notice that H_{k+1} carries this to a similar nest of n-disks in L', also exhausting that leaf. Now choose $b_j \downarrow 0$ so that $1 > b_0 = b$ and, in the coordinatization φ of $U^+ \cong L \times [0, 1)$, the cylinder $D_j \times [0, b_j]$ is contained in V^+. The infinite union

$$D_0 \times [b_1, b_0] \cup D_1 \times [b_2, b_1] \cup \cdots \cup D_m \times [b_{m+1}, b_m] \cup \cdots$$

is an infinite, nonstaggered stack of normal plugs (see Figure D.5.1) that can be smoothly cobbled together, as illustrated in Figure D.5.2, using the construction in the proof of Lemma D.3.4. This produces a half-open normal plug B. Since every point $x \in L$ lies in the interior of D_j, for all large values of j, it is clear that B is a cylindrical collar. In case L is a boundary leaf, define the cylindrical collar B' of L' to be $H_{k+1}(B)$. $\square$

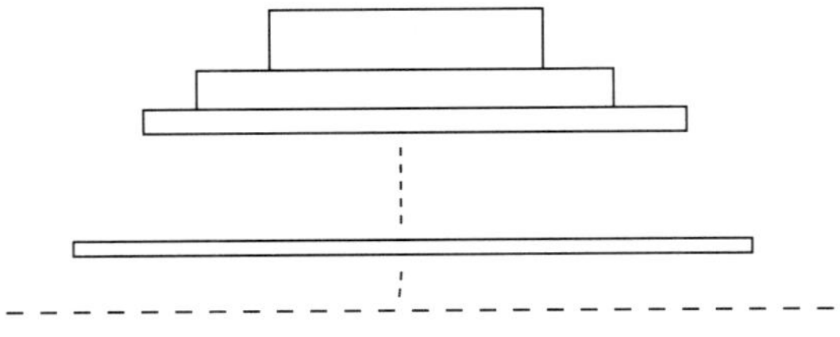

Figure D.5.1. An infinite, nonstaggered stack of normal plugs

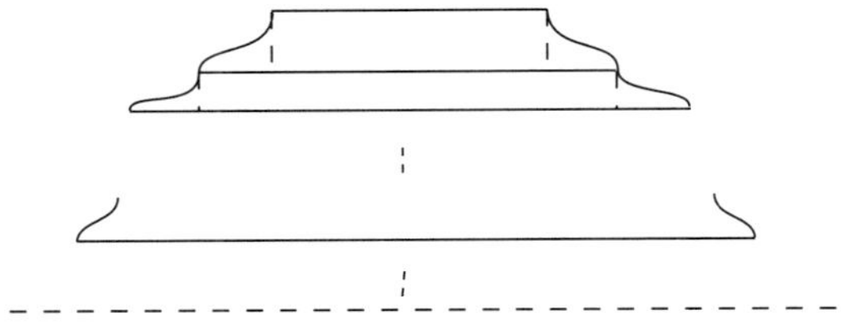

Figure D.5.2. Cobbling together the plugs to form the cylindrical collar B of L

Remark. Note that the "cobbling" process prevents the one-dimensional fibers on and near the boundary of B from agreeing with the normal fibers of $V^+ \setminus L$, but that near any point of L these fibers may be assumed to agree. This, however, is not particularly consequential. The fibers of the cylindrical collar will replace the normal fibers that defined V^+, this original normal neighborhood no longer being of use.

Definition D.5.6. If a cylindrical collar is coordinatized as $D^n \times (0, b]$, the subcollar $B_c \cong D^n \times (0, c]$, $0 < c < b$, is said to be a truncation of B. Similarly, if $B \cong D^n \times [b, 0)$, truncations are of the form $B_c \cong D^n \times [c, 0)$, $b < c < 0$.

For the boundary leaves, it will be convenient to choose another cylindrical collar $\widetilde{B}$ of L so that $\widetilde{B} \subset \operatorname{int} B$ by a level-preserving inclusion. Again, we set $\widetilde{B}' = H_{k+1}(\widetilde{B})$.

The plan, then, is to modify the extension H_{k+1} in each B_i, leaving it unchanged in a truncation of $\widetilde{B}_i$, so that it can be completed to an extension, again denoted by H_{k+1}, over all of M_{k+1}.

By Proposition D.3.1, we construct diffeomorphisms

$$\psi : \operatorname{int} M_{k+1} \to \mathbb{R}^{n+1},$$
$$\psi' : \operatorname{int} M'_{k+1} \to \mathbb{R}^{n+1},$$

carrying the respective foliations $\mathcal{F}|\operatorname{int} M_{k+1}$ and $\mathcal{F}'|\operatorname{int} M'_{k+1}$ onto the foliation by parallel hyperplanes $\mathbb{R}^n \times \{z\}$. This can be done so that

$$H = (\psi')^{-1} \circ \psi : \operatorname{int} M_{k+1} \to \operatorname{int} M'_{k+1}$$

lifts $h|(\operatorname{int} M_{k+1})/\mathcal{F}$. Indeed, Proposition D.3.1 allows us to choose a lift H of $h|(\operatorname{int} M_{k+1})/\mathcal{F}$, and we then set

$$\psi' = \psi \circ H^{-1}.$$

Remark. A major problem with our program will be that the above choice of H may not carry each B_i onto the corresponding B'_i. An example can be constructed by referring to Figure D.1.1. After crossing this figure with a copy of $\mathbb{R}$, one can produce a natural conjugation H of the interior of the center "Reeb sandwich" of the first foliation to that of the second that lifts the homeomorphism h between the leaf spaces but interchanges the cylindrical collars of the two boundary leaves. Merely perform a rotation through π radians about a suitable axis perpendicular to the plane of the figure. In this case another choice of the lift of h is possible that preserves these cylindrical collars, as the reader who solved Exercise D.1.14 may have noted already. If we can show that this is generally the case, the following lemma will complete the proof of Theorem D.1.12.

Lemma D.5.7. *If the diffeomorphisms ψ and ψ' can be chosen so that H carries each B_i onto the corresponding B'_i, then the diffeomorphisms H_{k+1}, restricted to truncations of the collars $\widetilde{B}_i$, $1 \le i < Q+1$, extend compatibly over* $\operatorname{int} M_{k+1}$ *so as to define a smooth conjugation of $\mathcal{F}|M_{k+1}$ to $\mathcal{F}'|M'_{k+1}$.*

Proof. Fix a boundary component $L = L_i$, dropping the index i from the notation for this leaf, its corresponding (via h) leaf L' and the corresponding cylindrical neighborhoods. We coordinatize B as $D^n \times (0, b+\eta]$ and view $\widetilde{B} \subset B$ as a level-preserving embedding

$$B^n \times (0,b] \xrightarrow{f} (\operatorname{int} D^n) \times (0,b] \hookrightarrow (\operatorname{int} D^n) \times (0, b+\eta],$$

where, of course, B^n is another closed n-ball. By temporarily ignoring the interval $(b, b+\eta]$ and applying Proposition D.2.3, we recoordinatize $D^n \times (0,b]$ so as to view f as the inclusion

$$B = B^n \times (0,b] \hookrightarrow D^n \times (0,b],$$

with $B^n \subset \operatorname{int} D^n$ and concentric with D^n.

Since, by hypothesis, H carries B onto B', we see that

$$\theta = H^{-1} \circ H_{k+1} : B \to B$$

is a level-preserving diffeomorphism. In particular, θ restricts to a level-preserving embedding

$$\widetilde{\theta} : B^n \times (0, b] \to (\operatorname{int} D^n) \times (0, b],$$

and Proposition D.2.3 allows us to extend $\widetilde{\theta}$ to a level-preserving diffeomorphism

$$\overline{\theta} : D^n \times (0, b] \to D^n \times (0, b],$$

which is the identity near $\partial D^n \times (0, b]$ and is isotopic to the identity through such diffeomorphisms. This isotopy allows us to extend $\overline{\theta}$ across $D^n \times [b, b+\eta]$ (using the second factor as the isotopy parameter) so as to be the identity near

$$D^n \times \{b + \eta\} \cup \partial D^n \times (0, b + \eta].$$

The minor smoothness problem at the level $t = b$ is handled by a standard trick already employed in the proof of Lemma D.5.3. This will require that $\widetilde{B}$ be replaced with a truncation $B^n \times (0, b - \varepsilon]$ for an arbitrarily small $\varepsilon > 0$. It should be clear that the diffeomorphism $H \circ \overline{\theta}$ extends the restriction of H_{k+1} to this truncation and agrees with H outside of a closed neighborhood of $\widetilde{B}$ in $\operatorname{int} B$.

Proceeding in this way at each boundary leaf $L = L_i$ completes the proof of the lemma. □

We must modify the diffeomorphism

$$\psi' : \operatorname{int} M'_{k+1} \to \mathbb{R}^{n+1}$$

so that $H = (\psi')^{-1} \circ \psi$ satisfies the hypotheses of Lemma D.5.7.

Recall that ψ conjugates $\mathcal{F}|\operatorname{int} M_{k+1}$ to the foliation of $\mathbb{R}^{n+1} = \mathbb{R}^n \times \mathbb{R}$ having leaves $\mathbb{R}^n \times \{z\}$, $z \in \mathbb{R}$. Thus, for each real number z, denote by L_z the leaf of $\mathcal{F}$ in $\operatorname{int} M_{k+1}$ such that $\psi(L_z) = \mathbb{R}^n \times \{z\}$. Note that, if L'_z is the similarly defined leaf of $\mathcal{F}'$ in $\operatorname{int} M'_{k+1}$, then $\psi'(L'_z) = \mathbb{R}^n \times \{z\}$ and $H(L_z) = L'_z$.

Definition D.5.8. The leaves L_z and L'_z, as well as $\mathbb{R}^n \times \{z\}$, will be said to be at level z.

Thus, the maps ψ, ψ' and H are level-preserving in this new sense. Our modification of ψ', hence of H, will also be level-preserving.

The transverse foliation of $\mathbb{R}^{n+1}$ by leaves $\{v\} \times \mathbb{R}$, $v \in \mathbb{R}^n$, pulls back via ψ^{-1} to an $\mathcal{F}$-transverse, one-dimensional foliation $\mathcal{L}$ of $\operatorname{int} M_{k+1}$, each leaf of which meets each leaf of $\mathcal{F}|\operatorname{int} M_{k+1}$ exactly once. Note that these leaves are oriented by the transverse orientation of $\mathcal{F}$.

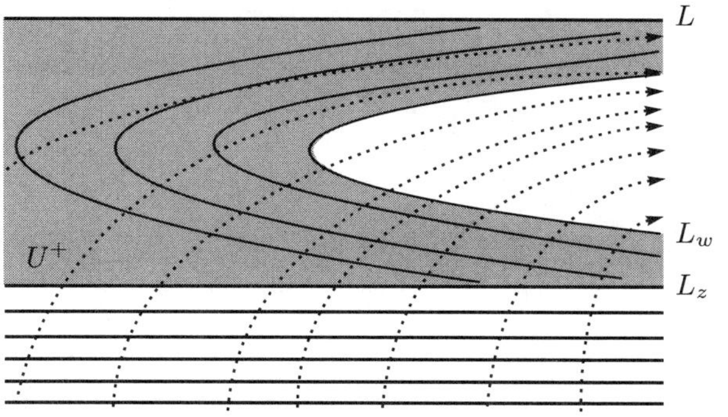

Figure D.5.3. The leaves of $\mathcal{L}$ entering U^+ through L_z and skimming by L

Let L be a boundary leaf of $\mathcal{F}|M_{k+1}$. By Exercise D.1.10, either there is a unique leaf L_z of $\mathcal{F}|\operatorname{int} M_{k+1}$ such that L and L_z are not separated in the leaf space $M_{k+1}/\mathcal{F}$, or else L completes $\operatorname{int} M_{k+1}/\mathcal{F} = (-\infty, \infty)$ to a half-open interval $(-\infty, \infty]$ or $[-\infty, \infty)$.

Definition D.5.9. In the above situations, the leaf $L \subset \partial M_{k+1}$ will be said to be at level z, ∞ or $-\infty$, respectively.

Observe that only boundary leaves can have levels $\pm\infty$.

Example D.5.10. In Figure D.5.3, a 2-dimensional example (easily made 3-dimensional by crossing with $\mathbb{R}$) is given. Here, L is a left boundary leaf at finite level z and U_+ is the shaded region. The dotted curves are the directed leaves of $\mathcal{L}$, which enter U^+ through the interior leaf L_z, skim by the boundary leaf L and exit U^+ through a leaf L_w that is separated from L in the leaf space. In case L is a right boundary leaf, $U \cap M_{k+1} = U^-$ and the arrows on the leaves of $\mathcal{L}$ are reversed. They enter U^- through L_w and exit through L_z. In Figure D.5.4, there are two left boundary leaves L_1 and L_2, with $U_1^+ = U_2^+ = U^+$ the shaded region. We draw $\mathcal{L}$ so that its leaves all exit U^+ through a leaf L_w that is separated in the leaf space from both boundary leaves. We can think that the leaves of $\mathcal{L}$ enter the shaded region at $-\infty$. In this case, each boundary leaf L_1 and L_2 lies at level $-\infty$. Reversing the transverse orientation of $\mathcal{F}$ reverses the orientation of $\mathcal{L}$ so that its leaves enter the shaded region at "time" w, exit at ∞, and the boundary leaves are at level ∞.

Exercise D.5.11. Prove that the situations illustrated in the above example are quite general. That is, if L is a left boundary leaf of M_{k+1}, then the level $z \in [-\infty, \infty)$ of L corresponds to the "entrance time" of the leaves of $\mathcal{L}$ into

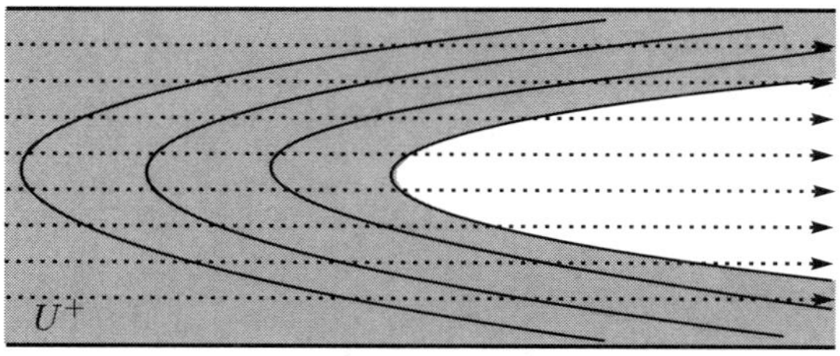

Figure D.5.4. Both boundary leaves are at level $-\infty$

U^+, and if L is a right boundary leaf, the level $z \in (-\infty, \infty]$ is the "exit time" from U^-. Prove also that, if L has level z, then the corresponding leaf $L' \subset \partial M'_{k+1}$ also has level z.

Remark. The cylindrical collar B that we have built for a left boundary leaf L at finite level z lies entirely in V^+, hence in U^+. By the above exercise, $L_z \cap U^+ = \emptyset$, so this cylindrical collar does not meet L_z. A similar remark holds for the cylindrical collars of right boundary leaves at finite level.

Fix a finite value of z that occurs as the level of some boundary leaf and let $\{L_{i_q}\}_{q=1}^{Q_z}$ be the subsequence of left (respectively, right) boundary leaves that lie at this level, $1 \le Q_z \le \infty$. There is nothing to prevent this sequence from being infinite.

Lemma D.5.12. *There is a right (respectively, left) cylindrical collar B_z for L_z that meets no cylindrical collar B_{i_q} and such that $H(B_z)$ meets no cylindrical collar B'_{i_q}, $1 \le q < Q_z + 1$.*

Proof. For definiteness, consider the sequence of left boundary leaves at level z. Let $\{D_r\}_{r=0}^{\infty}$ be the exhaustion of L as in the proof of Lemma D.5.5. As in that proof, consider normal neighborhoods $D_r \times [0, b_r]$, where

$$D_r \times \{0\} = D_r \subset L.$$

For convenience, the normal fibers can be taken as subarcs of leaves of $\mathfrak{L}$. We claim that the sequence $b_r \downarrow 0$ can be chosen so that, for each $r \ge 0$, $D_r \times [0, b_r]$ does not meet B_{i_q}, $1 \le q < Q_z + 1$. Indeed, we fix a value of r for which no such $b_r > 0$ exists and deduce a contradiction. Let $r_j \downarrow 0$ and choose $x_j \in B_{i_{q_j}}$ that also lies in $D_r \times [0, r_j]$. Since D_r is compact, we pass to a subsequence, if necessary, so that $x_j \to x \in D_r \subset L_z$. In case infinitely many terms of this sequence lie in one B_{i_q}, the fact that this collar is relatively closed in $M_{k+1} \smallsetminus L_{i_q}$ implies that $x \in B_{i_q}$, contradicting the fact that this collar cannot meet L_z. Thus, we can extract a convergent subsequence with each point in a distinct collar B_{i_q} and this contradicts Lemma D.5.2. The (right) cylindrical collar B_z constructed from this data is disjoint from every B_{i_q}. Since B_z is a cylindrical collar for L_z, $H(B_z)$ is such a collar for L'_z. Applying the same argument as above, one can build a smaller cylindrical collar $B'_z \subset H(B_z)$ that misses every B'_{i_q}. Replace B_z with $H^{-1}(B'_z)$ and observe that all assertions follow. □

We are nearly ready to modify $\psi' : \operatorname{int} M'_{k+1} \to \mathbb{R}^{n+1}$ so that

$$H = (\psi')^{-1} \circ \psi$$

satisfies the hypotheses of Lemma D.5.7. The modification will be carried out inductively on the sequence $\{B'_i\}_{i=1}^{Q}$. Lemma D.5.13 will begin the induction and the inductive step will be the proof of Lemma D.5.14.

It will be convenient to recoordinatize the transverse parameter of cylindrical collars by level. Thus, a right cylindrical collar B of a leaf L will be viewed as the image of an embedding

$$f : D^n \times (z, b] \hookrightarrow \operatorname{int} M_{k+1},$$

where $f(D^n \times \{u\}) \subset L_u$, $z < u \leq b$. Similarly, a left cylindrical collar will be the image of

$$f : D^n \times [b, z) \hookrightarrow \operatorname{int} M_{k+1},$$

where the transverse parameter corresponds to the levels of leaves. Analogous representations of cylindrical collars will be used in M'_{k+1}.

In the following discussion, every truncation of B_i and B'_i will be accompanied by a suitable truncation of $\widetilde{B}_i$ and $\widetilde{B}'_i$ so as to preserve the inclusions.

Lemma D.5.13. *There is a level-preserving diffeomorphism*

$$\Phi_1 : \mathbb{R}^n \times \mathbb{R} \to \mathbb{R}^n \times \mathbb{R},$$

together with truncations $\widehat{B}'_1$ and $\widehat{B}_1$ of B'_1 and B_1, respectively, so that

$$\Phi_1 \circ \psi'(\widehat{B}'_1) = \psi(\widehat{B}_1).$$

Proof. We consider two cases, according to whether the level z of L'_1 and L_1 is infinite or finite.

In the infinite case, assume for definiteness that the level is ∞. Let the embeddings of the respective cylindrical collars B_1 and B'_1 be

$$f_1 : D^n \times [b_1, \infty) \hookrightarrow \operatorname{int} M_{k+1},$$
$$f'_1 : D^n \times [b_1, \infty) \hookrightarrow \operatorname{int} M'_{k+1}.$$

Then $\varphi'_1 = \psi' \circ f'_1$ and $\varphi_1 = \psi \circ f_1$ can be viewed as level-preserving maps into $\mathbb{R}^n \times [b_1, \infty)$. Fix the truncations $\widehat{B}'_1$ and $\widehat{B}_1$ of these collars, parametrized as $D^n \times [b_1 + \varepsilon, \infty)$. We will define the desired diffeomorphism Φ_1 on $\mathbb{R}^n \times [b_1, \infty)$

so as to be the identity on $\mathbb{R}^n \times [b_1, b_1 + \delta]$, where $0 < \delta < \varepsilon$. Then Φ_1 will extend by the identity over $\mathbb{R}^n \times (-\infty, b_1]$. Appeal to Corollary D.2.6, with P a closed n-disk, $\mathbb{R}^n = \operatorname{int} P$, $J = [b_1 + \varepsilon, \infty)$, and define $\Phi_1 \in G$ as desired. Since Φ_1 is isotopic to the identity in G, we can use the parameter interval $[b_1, b_1 + \varepsilon]$ as the isotopy parameter to extend Φ_1 over $\mathbb{R}^n \times [b_1, b_1 + \varepsilon]$ so as to be equal to the identity on $\mathbb{R}^n \times [b_1, b_1 + \delta]$.

In the case that L_1 and L_1' are at finite level z, assume, for definiteness, that they are right boundary leaves. One would like to mimic the above argument, defining $\Phi_1 \in G$ as desired on $\mathbb{R}^n \times [b_1, z)$ so that it extends by the identity over the rest of $\mathbb{R}^n \times \mathbb{R}$. The extension past level b_1 will be achieved exactly as above. The problem posed by the finiteness of z will be solved via the collar $\psi(B_z) = \psi'(B_z')$. Let B_z and B_z' be realized as embeddings

$$g_1 : D^n \times [c, z) \hookrightarrow \operatorname{int} M_{k+1},$$
$$g_1' : D^n \times [c, z) \hookrightarrow \operatorname{int} M_{k+1}',$$

chosen so that

$$\theta_1 = \psi \circ g_1 = \psi' \circ g_1' = \theta_1'.$$

The collars B_1 and B_1' are the respective images of embeddings

$$f_1 : D^n \times [b_1, z) \hookrightarrow \operatorname{int} M_{k+1},$$
$$f_1' : D^n \times [b_1, z) \hookrightarrow \operatorname{int} M_{k+1}'.$$

After suitably truncating these collars, we can assume that $b_1 > c$. Again, we set

$$\varphi_1 = \psi \circ f_1,$$
$$\varphi_1' = \psi' \circ f_1'.$$

Thus, since $\psi'(B_1')$ and $\psi'(B_z')$ are disjoint, as are $\psi(B)$ and $\psi(B_z)$, we can apply Corollary D.2.7, letting $\theta_1 = \theta_1'$ play the role of f, to find a level-preserving diffeomorphism

$$\Phi_1 : \mathbb{R}^n \times [b_1, z) \to \mathbb{R}^n \times [b_1, z)$$

such that

$$\Phi_1 \circ \varphi_1' = \varphi_1,$$

Φ_1 being the identity on the image

$$C_z = \psi'(B_z') = \psi(B_z).$$

By the definition of a cylindrical collar, it is clear that every point of $\mathbb{R}^n \times \{z\}$ has a neighborhood meeting $\mathbb{R}^n \times (-\infty, z)$ only in the set C_z, so Φ_1 can be extended smoothly by the identity over $\mathbb{R}^n \times [z, \infty)$. The smooth extension over $\mathbb{R}^n \times [c, b_1]$, thence over the rest of $\mathbb{R}^n \times \mathbb{R}$, is managed as above, appealing to Corollary D.2.7 rather than to Corollary D.2.6. □

We replace ψ' with $\psi'_1 = \Phi_1 \circ \psi$. The plan is to define a sequence

$$\psi'_i = \Phi_i \circ \Phi_{i-1} \circ \cdots \circ \Phi_1 \circ \psi'$$

so that

$$\psi'_i(\widehat{B}'_j) = \psi(\widehat{B}_j), \quad 1 \le j \le i,$$

for suitable truncations of these cylindrical collars, and such that ψ'_i agrees with ψ'_j on $\widehat{B}'_j$, $1 \le j \le i$. In case $Q < \infty$, this will be a finite sequence and ψ'_Q will be the desired modification of ψ'. If $Q = \infty$, we will need to guarantee that every point of $\mathbb{R}^{n+1}$ has a neighborhood in which all but finitely many of the level-preserving diffeomorphisms Φ_j are equal to the identity. It will then follow that the pointwise limit

$$\psi'_\infty = \lim_{i\to\infty} \psi'_i$$

exists and is a diffeomorphism of $\operatorname{int} M'_{k+1}$ onto $\mathbb{R}^{n+1}$, which, together with

$$H_\infty = (\psi'_\infty)^{-1} \circ \psi : \operatorname{int} M_{k+1} \to \operatorname{int} M'_{k+1},$$

satisfies the hypotheses of Lemma D.5.7 on ψ' and H. This will complete the proof of Theorem D.1.12.

We consider the case $Q = \infty$, the finite case being easier since the induction will be finite. Fix an exhaustion of $\mathbb{R}^{n+1}$ by an increasing sequence $\{K_i\}_{i=1}^\infty$ of compact sets, requiring that $K_i \subset \operatorname{int} K_{i+1}$ for all $i \ge 1$. We will require that $\Phi_i | K_i$ be the identity and, since Φ_1 was already defined without reference to K_1, it will be convenient to take $K_1 = \emptyset$. This will guarantee that all but finitely many Φ_i are supported outside suitable neighborhoods of arbitrary points of $\mathbb{R}^{n+1}$. Of course, no such system will be needed when $Q < \infty$.

Lemma D.5.14. *There exists a sequence $\{\Phi_i\}_{i=1}^Q$ of level-preserving diffeomorphisms of $\mathbb{R}^n \times \mathbb{R}$ satisfying all of the above requirements.*

Proof. The construction will be inductive, the case $i = 1$ being given by Lemma D.5.13. Assume, then, that $\Phi_1, \ldots, \Phi_m$ have been defined with all appropriate properties for some $m \ge 1$. For the inductive step, we again distinguish the cases in which the level z of L_{m+1} and L'_{m+1} is infinite or finite.

For the infinite case, assume for the sake of definiteness that $z = \infty$. By the inductive hypothesis,

$$\psi'_m(\widehat{B}'_i) = \psi(\widehat{B}_i), \quad 1 \le i \le m.$$

For notational convenience, temporarily renumber so that $L_1, \ldots, L_r$ are the leaves having levels $< \infty$ (r may be 0). We can suppose that the truncations

$\widehat{B}_{m+1}$ and $\widehat{B}'_{m+1}$ are the images of level-preserving embeddings

$$f_{m+1} : D^n \times [b_{m+1}, \infty) \to \operatorname{int} M_{k+1},$$
$$f'_{m+1} : D^n \times [b_{m+1}, \infty) \to \operatorname{int} M'_{k+1},$$

where b_{m+1} is chosen so large that neither K_{m+1} nor $\psi(\widehat{B}_i)$ meets the submanifold $\mathbb{R}^n \times [b_{m+1}, \infty)$, $1 \le i \le r$, while $\psi(\widehat{B}_i)$ meets every level in this set, $r + 1 \le i \le m$. An easy application of Corollary D.2.7 now provides the desired Φ_{m+1}, defined on the submanifold $\mathbb{R}^n \times [b_{m+1} + \varepsilon, \infty)$, for some $\varepsilon > 0$, and our usual trick allows us to extend this over all of $\mathbb{R}^n \times \mathbb{R}$ so as to be the identity on $\mathbb{R}^n \times (-\infty, b_{m+1} + \varepsilon/2]$.

In the finite case, assume, for definiteness, that L_{m+1} and L'_{m+1} are right boundary leaves, so the truncated collars $\widehat{B}_{m+1}$ and $\widehat{B}'_{m+1}$ are the images of level-preserving embeddings

$$f_{m+1} : D^n \times [b_{m+1}, z) \to \operatorname{int} M_{k+1},$$
$$f'_{m+1} : D^n \times [b_{m+1}, z) \to \operatorname{int} M'_{k+1},$$

and we must choose b_{m+1} suitably. This time, choose the numbering of the leaves so that L_i is not a right boundary leaf at level z *and* $\psi(\widehat{B}_i)$ does not meet the level $\mathbb{R}^n \times \{z\}$, exactly for $1 \le i \le r$. Again it is possible that $r = 0$. Thus, we can choose b_{m+1} so that these cylinders do not meet $\mathbb{R}^n \times [b_{m+1}, z)$. (We emphasize that cylindrical collars of left boundary leaves at level z are among these first r collars.) Again temporarily renumbering, we can assume that L_i is a right boundary leaf at level z, $r + 1 \le i \le q$ (where $q = r$ is allowed). Choosing b_{m+1} closer to z, we can assume that $\psi(B_i) = \psi'_m(B'_i)$ meets each $\mathbb{R}^n \times \{u\}$, $b_{m+1} \le u < z$. Finally, again choosing b_{m+1} closer to z, if necessary, we can assume that the intersections $\psi(B_i) \cap \mathbb{R}^n \times [b_{m+1}, z)$ are engulfed by the collar $\psi(B_z)$, $q+1 \le i \le m$, as is the (possibly empty) set $K_{m+1} \cap \mathbb{R}^n \times [b_{m+1}, z)$. Thus, the disjoint collars $\psi(B_z)$ and $\psi(\widehat{B}_i) = \psi'_m(\widehat{B}'_i)$, $r + 1 \le i \le q$, intersected with $\mathbb{R}^n \times [b_{m+1}, z)$, can play the role of the image of f in Corollary D.2.7 and the required Φ_{m+1} can be defined on $\mathbb{R}^n \times [b_{m+1}, z)$. The usual tricks then allow us to modify this diffeomorphism in $\mathbb{R}^n \times [b_{m+1}, b_{m+1} + \varepsilon)$ and extend via the identity over all of $\mathbb{R}^{n+1}$. □

Bibliography

1. J. W. Alexander, *The combinatorial theory of complexes*, Ann. of Math. **31** (1930), 292–320.
2. ———, *A lemma on systems of knotted curves*, Proc. Nat. Acad. Sci. **9** (1923), 93–95.
3. J. A. Álvarez López and A. Candel, *Generic geometry of leaves*, Preprint, 1999. (Revised version in preparation.)
4. ———, *Equicontinuous foliated spaces*, Preprint, 2001.
5. ———, *Topological description of Riemannian foliations with dense leaves*, Preprint, 2002.
6. T. Aubin, *Nonlinear Analysis on Manifolds. Monge–Ampère Equations*, Grundlehren der Mathematische Wissenschaften, Vol. 252, Springer-Verlag, New York, NY, 1982.
7. M. Berger, P. Gauduchon, and E. Mazet, *Le Spectre d'une Variété Riemannienne*, Lecture Notes in Mathematics, vol. 194, Springer-Verlag, Berlin, 1971.
8. S. A. Bleiler and A. J. Casson, *Automorphisms of Surfaces after Nielsen and Thurston*, Cambridge Univ. Press, Cambridge, 1988.
9. R. Bott, *Lectures on characteristic classes and foliations*, Lecture Notes in Math., 279, Springer-Verlag, New York, NY, 1972, pp. 1–94.
10. ———, *On some formulas for the characteristic classes of group–actions*, Lecture Notes in Math., 652, Springer-Verlag, New York, NY, 1978, pp. 25–56.
11. R. Bott and L. Tu, *Differential Forms in Algebraic Topology*, Springer–Verlag, New York, NY, 1982.
12. N. Bourbaki, *Éléments de mathématique. Fasc. XIII. Livre VI: Intégration. Chapitres 1, 2, 3 et 4: Inégalités de convexité, Espaces de Riesz, Mesures sur les espaces localement compacts, Prolongement d'une mesure, Espaces L^p*, Deuxième édition, revue et augmentée. Actualités Scientifiques et Industrielles, No. 1175, Hermann, Paris, 1965.
13. R. Bowen, *Weak mixing and unique ergodicity of homogeneous spaces*, Israel J. of Math. **23** (1979), 337–342.
14. L. G. Brown, P. Green, and M. A. Rieffel, *Stable isomorphism and strong Morita equivalence of C^*-algebras*, Pacific J. Math. **71** (1977), no. 2, 349–363.

15. B. Burde and H. Zieschang, *Knots,* Studies in Mathematics **5**, de Gruyter, New York, NY, 1985.

16. A. Candel, *Uniformization of surface laminations*, Ann. Sci. École Norm. Sup. **26** (1993), 489–516.

17. ———, *C^*-algebras of proper foliations*, Proc. Amer. Math. Soc. **124** (1996), no. 3, 899–905.

18. ———, *The harmonic measures of Lucy Garnett*, To appear in Adv. Math.

19. J. Cantwell and L. Conlon, *Growth of leaves*, Comment. Math. Helv. **53** (1978), 93–111.

20. ———, *The Dynamics of Open Foliated Manifolds and a Vanishing Theorem for the Godbillon–Vey Class*, Advances in Math. **53** (1984), 1–27.

21. ———, *Every surface is a leaf*, Topology **26** (1987), 265–285.

22. ———, *Foliations and subshifts*, Tôhoku Math J. **40** (1988), 165–187.

23. ———, *Leaves of Markov local minimal sets in foliations of codimension one*, Publicacions Matemàtiques **33** (1989), 461–484.

24. ———, *Generic leaves*, Comment. Math. Helv. **73** (1998), 306–336.

25. ———, *Isotopies of foliated 3–manifolds without holonomy*, Adv. in Math. **144** (1999), 13–49.

26. ———, *Endsets of exceptional leaves; a theorem of G. Duminy*, Proceedings of the Euroworkshop on Foliations, Geometry and Dynamics, World Scientific, 2002, pp. 225–261.

27. I. Chavel, *Eigenvalues in Riemannian Geometry*, Academic Press, Orlando, FL, 1984, Including a chapter by Burton Randol, With an appendix by Jozef Dodziuk.

28. ———, *Isoperimetric Inequalities. Differential Geometric and Analytic Perspectives*, Cambridge University Press, Cambridge, 2001.

29. S.Y. Cheng, P. Li, and S.-T. Yau, *On the upper estimate of the heat hernel of a complete Riemannian manifold*, Amer. J. Math. **103** (1981), no. 5, 1021–1063.

30. G. Choquet, *Lectures on Analysis. Vol. II: Representation Theory*, W. A. Benjamin, Inc., New York-Amsterdam, 1969.

31. K. L. Chung, *Lectures from Markov Processes to Brownian Motion*, Grundlehren der Mathematische Wissenschaften, Vol. 249, Springer-Verlag, New York, NY, 1982.

32. L. Conlon, *Differentiable Manifolds; Second Edition*, Birkhäuser, Boston, Massachusetts, 2001.

33. A. Connes, *The von Neumann algebra of a foliation*, Mathematical problems in theoretical physics (Proc. Internat. Conf., Univ. Rome, Rome, 1977), Lecture Notes in Physics, no. 80, Springer, Berlin, 1978, pp. 145–151.

34. ———, *Sur la théorie non commutative de l'intégration*, Algèbres d'opérateurs (Sém., Les Plans-sur-Bex, 1978), Springer–Verlag, Berlin, 1979, pp. 19–143.

35. ———, *Feuilletages et algèbres d'opérateurs*, Lecture Notes in Math., no. 842, Springer-Verlag, Berlin, 1981, pp. 139–155.

36. ———, *A survey of foliations and operator algebras*, Proc. Symp. Pure Math., vol. **38**, Amer. Math. Soc., 1982, pp. 521–628.

37. ———, *Cyclic cohomology and the transverse fundamental class of a foliation*, Geometric methods in operator algebras (Kyoto, 1983), Longman Sci. Tech., Harlow, 1986, pp. 52–144.

38. ———, *Noncommutative Geometry*, Academic Press Inc., San Diego, CA, 1994.

39. M. Crainic and I. Moerdijk, *A remark on sheaf theory for non-Hausdorff manifolds*, Tech. Report 1119, Utrecht University, 1999.
40. ———, *A homology theory for étale groupoids*, J. Reine Angew. Math. **521** (2000), 25–46.
41. R. Crowell, *Genus of alternating link types*, Ann. of Math. **69** (1959), 258–275.
42. F. Alcalde Cuesta, G. Hector, and P. Schweitzer, *Sur l'existence de feuilles compacts en codimension* 1, in preparation.
43. K. R. Davidson, *C^*-Algebras by Example*, Fields Institute Monographs, vol. 6, American Mathematical Society, Providence, R.I., 1996.
44. G. de Rham, *Variétés Différentiables*, Hermann, Paris, 1960.
45. A. Debiard, B. Gaveau, and E. Mazet, *Théorèmes de comparaison en géométrie riemannienne*, Publ. Res. Inst. Math. Sci. **12** (1976/77), no. 2, 391–425.
46. M. Dehn, *Die Gruppe der Abbildungsklassen*, Acta Math. **69** (1938), 135–206, Translated and reprinted in "Papers on Group Theory and Topology by Max Dehn", Springer–Verlag, 1987, pp. 256–362.
47. J. Dixmier, *C^*-Algebras*, North-Holland Publishing Co., Amsterdam, 1977, Translated from the French by Francis Jellett, North-Holland Mathematical Library, Vol. 15.
48. ———, *Von Neumann Algebras*, North-Holland Publishing Co., Amsterdam, 1981, With a preface by E. C. Lance, Translated from the second French edition by F. Jellett.
49. J. Dixmier and A. Douady, *Champs continus d'espaces hilbertiens et de C^*-algèbres*, Bull. Soc. Math. France **91** (1963), 227–284.
50. J. Dodziuk, *Maximum principle for parabolic equations and the heat flow on open manifolds*, Indiana Univ. Math. J. **32** (1983), 703–716.
51. R. Dudley, *Real Analysis and Probability*, Wadsworth & Brooks/Cole Advanced Books & Software, Pacific Grove, CA, 1989.
52. G. Duminy, *L'invariant de Godbillon–Vey d'un feuilletage se localise dans les feuilles ressort*, preprint (1982).
53. N. Dunford and J. T. Schwartz, *Linear Operators, Part I: General Theory*, John Wiley & Sons, Inc., New York, NY, 1988, With the assistance of William G. Bade and Robert G. Bartle, Reprint of the 1958 original, A Wiley-Interscience Publication.
54. A. H. Durfee, *Foliations of odd dimensional spheres*, Ann. of Math. **97** (1973), 407–411.
55. A. H. Durfee and H. B. Lawson, *Fibered knots and foliations of highly connected manifolds*, Inv. Math. **17** (1972), 203–215.
56. E. Dynkin, *Theory of Markov Processes*, Prentice–Hall, Englewood Cliffs, NJ, 1961, Translated from the Russian by D. E. Brown; edited by T. Kövóry.
57. ———, *Markov Processes. Vols. I, II*, Translated with the authorization and assistance of the author by J. Fabius, V. Greenberg, A. Maitra, G. Majone. Die Grundlehren der Mathematischen Wissenschaften, Bände 121, Vol. 122, Academic Press Inc., Publishers, New York, NY, 1965.
58. E. G. Effros and F. Hahn, *Locally compact transformation groups and C^*-algebras*, Memoirs of the American Mathematical Society, no. 75, American Mathematical Society, Providence, RI, 1967.
59. C. Ehresmann, *Structures feuilletées*, Proc. 5th Canadian Math. Congress, Univ. of Toronto Press, 1963, pp. 109–172.

60. D. B. A. Epstein, *The simplicity of certain groups of homeomorphisms*, Compositio Math. **22** (1970), 165–173.

61. ———, *Periodic flows on 3–manifolds*, Ann. of Math. **95** (1972), 66–82.

62. T. Fack, *Quelques remarques sur le spectre des C^*-algèbres de feuilletages*, Bull. Soc. Math. Belg. Sér. B **36** (1984), no. 1, 113–129.

63. T. Fack and G. Skandalis, *Sur les représentations et idéaux de la C^*-algèbre d'un feuilletage*, J. Operator Theory **8** (1982), 95–129.

64. T. Fack and X. Wang. *The C^*-algebras of Reeb foliations are not AF-embeddable*, Proc. Amer. Math. Soc. **108** (1990), 941–946.

65. P. A. Fillmore, *A User's Guide to Operator Algebras*, John Wiley & Sons, Inc., New York, NY, 1996, A Wiley-Interscience Publication.

66. G. B. Folland, *Introduction to Partial Differential Equations*, Princeton University Press, Princeton, NJ, 1976, Preliminary informal notes of university courses and seminars in mathematics, Mathematical Notes.

67. ———, *Real Analysis. Modern Techniques and their Applications*, second ed., John Wiley & Sons Inc., New York, NY, 1999, A Wiley-Interscience Publication.

68. H. Furstenberg, *Recurrence in Ergodic Theory and Combinatorial Number Theory*, Princeton University Press, Princeton, NJ, 1981.

69. S. A. Gaal, *Linear Analysis and Representation Theory*, Die Grundlehren der Mathematischen Wissenschaften, Band 198, Springer-Verlag, New York, NY, 1973.

70. D. Gabai, *Foliations and the topology of 3–manifolds*, J. Diff. Geom. **18** (1983), 445–503.

71. ———, *Foliations and genera of links*, Topology **23** (1984), 381–394.

72. ———, *Genera of the arborescent links*, Mem. Amer. Math. Soc. **59** (1986), 1–98.

73. ———, *Foliations and the topology of 3–manifolds* II, J. Diff. Geom. **26** (1987), 461–478.

74. ———, *Foliations and the topology of 3–manifolds* III, J. Diff. Geom. **26** (1987), 479–536.

75. ———, *Combinatorial volume preserving flows and taut foliations*, Comment. Math. Helv. **75** (2000), 109–124.

76. D. Gabai and W. Kazez, *Homotopy, isotopy and genuine laminations of 3-manifolds*, Geometric Topology (William Kazez, ed.) **1** (1997), 123–138.

77. L. Garnett, *Foliations, the ergodic theorem and Brownian motion*, J. Funct. Anal. **51** (1983), no. 3, 285–311.

78. E. Ghys, *Une variété qui n'est pas une feuille*, Topology **24** (1985), 67–73.

79. ———, *Classe d'Euler et minimal exceptionnel*, Topology **26** (1987), 57–73.

80. ———, *Topologie des feuilles génériques*, Ann. of Math. **141** (1995), 387–422.

81. E. Ghys and V. Sergiescu, *Sur un groupe remarquable de difféomorphismes du cercle*, Comment. Math. Helv. **62** (1987), 185–239.

82. J. Glimm, *Families of induced representations*, Pacific J. Math. **12** (1962), 885–911.

83. E. C. Gootman and J. Rosenberg, *The structure of crossed product C^*-algebras: a proof of the generalized Effros-Hahn conjecture*, Invent. Math. **52** (1979), no. 3, 283–298.

84. M. Gromov, *Asymptotic invariants of infinite groups*, Geometric group theory, Vol. 2 (Sussex, 1991) (Graham A. Niblo and Martin A. Roller, eds.), London Math. Soc. Lecture Note Ser., no. 182, Cambridge Univ. Press, Cambridge, 1993, pp. 1–295.

85. A. Haefliger, *Feuilletages sur les variétés ouvertes*, Topology **9** (1970), 183–194.

86. ———, *Homotopy and integrability*, Lecture Notes in Math., 197, Springer–Verlag, 1971.

87. A. Haefliger and G. Reeb, *Variétes (non separées) à une dimension et structures feuilletées du plan*, Enseignement Math. (2) **3** (1957), 107–125.

88. P. Hahn, *The regular representations of measure groupoids*, Trans. Amer. Math. Soc. **242** (1978), 35–72.

89. J. Hass, *Minimal surfaces in foliated manifolds*, Comment. Math. Helv. **61** (1986), 1–32.

90. J. Hass and P. Scott, *The existence of least area surfaces in 3–manifolds*, Trans. Amer. Math. Soc. **310** (1988), 87–114.

91. G. Hector and U. Hirsch, *Introduction to the Geometry of Foliations, Part B*, Vieweg and Sohn, Braunschweig, 1983.

92. J. Heitsch and S. Hurder, *Secondary classes, Weil operators and the geometry of foliations*, J. Diff. Geom. **20** (1984), 291–309.

93. J. Hemple, 3*–Manifolds*, Princeton Univ. Press and Univ. of Tokyo Press, Princeton, NJ, and Tokyo, 1976.

94. M. Herman, *Sur la conjugaison différentiable des difféomorphismes du cercle à des rotations*, Publ. Math. I.H.E.S. **49** (1979), 5–233.

95. M. Herman and F. Sergeraert, *Sur un théorème d'Arnold et Kolmogorov*, C. R. Acad. Sci. Paris **273** (1971), A409–A411.

96. E. Hille and R. S. Phillips, *Functional Analysis and Semi-groups*, American Mathematical Society Colloquium Publications, vol. XXXI, American Mathematical Society, Providence, RI, 1957.

97. M. Hilsum and G. Skandalis, *Stabilité des C^*-algèbres de feuilletages*, Ann. Inst. Fourier (Grenoble) **33** (1983), no. 3, 201–208.

98. M. W. Hirsch, *A stable analytic foliation with only exceptional minimal sets*, Dynamical Systems, Warwick, 1974, Lecture Notes in Math., 468, Springer–Verlag, 1975, pp. 9–10.

99. ———, *Differential Topology*, Springer-Verlag, New York, NY, 1976.

100. H. Hopf, *Enden offener Räume und unendliche diskontinuerliche Gruppen*, Comm. Math. Helv. **16** (1944), 81–100.

101. S. Hurder, *Vanishing of secondary classes for compact foliations*, J. London Math. Soc. **28** (1983), 175–183.

102. ———, *The Godbillon measure of amenable foliations*, J. Diff. Geom. **23** (1986), 347–365.

103. S. Hurder and A. Katok, *Ergodic theory and Weil measures for foliations*, Ann. of Math. **126** (1987), 221–275.

104. S. Hurder and R. Langevin, *Dynamics and the Godbillon-Vey class of C^1 foliations*, preprint, 2000.

105. D. Husemoller, *Fibre Bundles*, McGraw-Hill, New York, NY, 1966.

106. K. Itô and Jr. H. P. McKean, *Diffusion Processes and their Sample Paths*, Springer–Verlag, Berlin, 1974, Second printing, corrected, Die Grundlehren der Mathematischen Wissenschaften, Band 125.

107. W. Jaco, *Lectures on Three–Manifold Topology*, Regional Conference Series in Mathematics, vol. 43, American Mathematical Society, Providence, RI, 1980.

108. M. Kac, *On the notion of recurrence in discrete stochastic processes*, Bulletin of the Amer. Math. Soc. **53** (1947), 1002–1010.

109. W. Kaplan, *Regular curve families filling the plane I*, Duke Math. J. **7** (1940), 154–185.

110. ______, *Regular curve families filling the plane II*, Duke Math. J. **8** (1941), 11–46.

111. A. Katok and B. Hasselblatt, *Introduction to the Modern Theory of Dynamical Systems*, Cambridge University Press, Cambridge, England, 1995.

112. M. Kellum, *Uniformly quasi–isometric foliations*, Ergodic Theory Dynam. Systems **13** (1993), 101–122.

113. J. R. Kinney, *Continuity properties of sample functions of Markov processes*, Trans. Amer. Math. Soc. **74** (1953), 280–302.

114. A. A. Kirillov, *Elements of the Theory of Representations*, Springer-Verlag, Berlin, 1976, Translated from the Russian by Edwin Hewitt, Grundlehren der Mathematischen Wissenschaften, Band 220.

115. W. Krieger, *On ergodic flows and the isomorphism of factors*, Math. Ann. **223** (1976), no. 1, 19–70.

116. N. Kryloff and N. Bogoliuboff, *La théorie générale de la mesure dans son application à l'étude des systèmes dynamiques de la mécanique non linéaire*, Annals of Math. **38** (1937), no. 2, 65–113.

117. F. Laudenbach and S. Blank, *Isotopie de formes fermées en dimension trois*, Inv. Math. **54** (1979), 103–177.

118. H. B. Lawson, *Codimension one foliations of spheres*, Ann. of Math. **94** (1971), 494–503.

119. G. Levitt, *Feuilletages des variétés de dimension* 3 *qui sont des fibrés en cercles*, Comment. Math. Helv. **53** (1978), 572–594.

120. P. Li and S. T. Yau, *On the parabolic heat kernel of the Schrödinger operator*, Acta Math. **156** (1986), 153–201.

121. W. B. R. Lickorish, *A representation of orientable combinatorial* 3*–manifolds*, Ann. of Math. **76** (1962), 531–538.

122. ______, *Homeomorphisms of nonorientable two–manifolds*, Proc. Camb. Phil. Soc. **59** (1963), 307–317.

123. ______, *A foliation for* 3*–manifolds*, Ann. of Math. **82** (1965), 414–420.

124. G. W. Mackey, *Ergodic theory and virtual groups*, Math. Ann. **166** (1966), 187–207.

125. W. S. Massey, *Algebraic Topology: An Introduction*, Harcourt, Brace & World, New York, NY, 1967.

126. J. Mather, *Commutators of diffeomorphisms*, Comment. Math. Helv. **49** (1974), 512–528.

127. ______, *Commutators of diffeomorphisms. II*, Comment. Math. Helv. **50** (1975), 33–40.

128. K. Millett, *Generic properties of proper foliations*, Fund. Math. **128** (1987), 131-138.

129. J. Milnor, *On the existence of a connection with curvature zero*, Comment. Math. Helv. **32** (1957), 215–223.

130. J. W. Milnor and J. D. Stasheff, *Characteristic Classes*, Princeton University Press, Princeton, NJ, 1974.

131. S. Minakshisundaram, *Eigenfunctions on Riemannian manifolds*, J. Indian Math. Soc. (N.S.) **17** (1953), 159–165.

132. P. Molino, *Riemannian Foliations*, Birkhaüser, Boston, MA, 1988, Translated from the French by Grant Cairns, With appendices by Cairns, Y. Carrière, É. Ghys, E. Salem and V. Sergiescu.

133. C. C. Moore, *Ergodic theory and von Neumann algebras*, Operator algebras and applications, Part 2 (Kingston, Ont., 1980) (Richard V. Kadison, ed.), Proceedings Symposia in Pure Mathematics, vol. 38, Amer. Math. Soc., Providence, RI, 1982, pp. 179–226.

134. C. C. Moore and C. Schochet, *Global Analysis on Foliated Spaces*, Mathematical Sciences Research Institute, vol. 9, Springer-Verlag, New York, NY, 1988.

135. R. Mosher and M. Tangora, *Cohomology Operations and Applications in Homotopy Theory*, Harper and Row, New York, NY, 1968.

136. R. Moussu and F. Pelletier, *Sur le théorème de Poincaré–Bendixson*, Ann. Inst. Fourier, Grenoble **24** (1974), 131–148.

137. J. Munkres, *Elementary Differential Topology*, Princeton University Press, Princeton, NJ, 1966.

138. K. Murasugi, *On the genus of the alternating knot, I, II*, J. Math. Soc. Japan **10** (1958), 94–105, 235–248.

139. M. A. Naĭmark, *Normed Rings*, English ed., Wolters-Noordhoff Publishing, Groningen, 1970, Translated from the first Russian edition by Leo F. Boron.

140. T. Natsume, *The C^*-algebras of codimension one foliations without holonomy*, Math. Scand. **56** (1985), no. 1, 96–104.

141. S. P. Novikov, *Topology of foliations*, Trans. Moscow Math. Soc. **14** (1965), 268–305.

142. R. Palais, *Extending diffeomorphisms*, Proc. Amer. Math. Soc. **11** (1960), 274–277.

143. F. Palmeira, *Open manifolds foliated by planes*, Ann. of Math. **107** (1978), 109–131.

144. J. Pasternack, *Topological obstructions to integrability and Riemannian geometry of foliations*, Ph.D. thesis, Princeton University, 1970.

145. G. K. Pedersen, *C^*-algebras and their Automorphism Groups*, Academic Press Inc. [Harcourt Brace Jovanovich Publishers], London, 1979.

146. ———, *Analysis Now*, Springer-Verlag, New York, NY, 1989.

147. K. Petersen, *Ergodic Theory*, Cambridge Studies in Advanced Mathematics, vol. 2, Cambridge University Press, Cambridge, 1983.

148. P. Peterson, *Riemannian Geometry*, Graduate Texts in Mathematics, vol. 171, Springer–Verlag, New York, NY, 1998.

149. R. R. Phelps, *Lectures on Choquet's Theorem*, second ed., Lecture Notes in Mathematics, no. 1757, Springer-Verlag, Berlin, 2001.

150. D. Pixton, *Nonsmoothable, nonstable group actions*, Trans. Amer. Math. Soc. **229** (1977), 259–268.

151. V. Poénaru, *Homotopy theory and differentiable singularities*, Lecture Notes in Math., 197, Springer-Verlag, New York, NY, 1971, pp. 106–132.

152. S. C. Port and C. J. Stone, *Brownian Motion and Classical Potential Theory*, Probability and Mathematical Statistics, Academic Press [Harcourt Brace Jovanovich Publishers], New York, NY, 1978.

153. M. H. Protter and H. F. Weinberger, *Maximum Principles in Differential Equations*, Prentice-Hall, Inc., Englewood Cliffs, NJ, 1967.

154. N. V. Quê and R. Roussarie, *Sur l'isotopie des formes fermées en dimension* 3, Inv. Math. **64** (1981), 69–87.

155. J. Renault, *A Groupoid Approach to C^*-Algebras*, Lecture Notes in Mathematics, no. 793, Springer-Verlag, Berlin, 1980.

156. M. A. Rieffel, *Morita equivalence for C^*-algebras and W^*-algebras*, J. Pure Appl. Algebra **5** (1974), 51–96.

157. J. Roe, *Coarse cohomology and index theory on complete Riemannian manifolds*, Mem. Amer. Math. Soc. **104** (1993), no. 497, x+90.

158. D. Rolfsen, *Knots and Links*, Publish or Perish, Inc., Berkeley, CA, 1976.

159. H. Rosenberg, *Foliations by planes*, Topology **7** (1968), 131–138.

160. R. Roussarie, *Plongements dans les variétés feuilletées et classification de feuilletages sans holonomie*, I.H.E.S. Publ. Math. **43** (1973), 101–142.

161. W. Rudin, *Real and Complex Analysis*, 2nd ed., McGraw–Hill, New York, NY, 1974.

162. J.-L. Sauvageot, *Idéaux primitifs de certains produits croisés*, Math. Ann. **231** (1977/78), no. 1, 61–76.

163. P. Schweitzer, *Some problems in foliation theory and related areas*, Lecture Notes in Math., 652, Springer–Verlag, 1978, pp. 240–252.

164. L. C. Siebenmann, *Deformations of homeomorphisms on stratified sets*, Comment. Math. Helv. **47** (1972), 123–163.

165. V. V. Solodov, *Components of topological foliations*, Mat. Sb. (N.S.) **119** (1982), 340–354, Russian; English transl., Math. USSR Sb. **47** (1984), 329-343.

166. H. Takai, *C^*-algebras of Anosov foliations*, Operator Algebras and their Connections with Topology and Ergodic Theory (Buşteni, 1983), Lecture Notes in Math., 1132, Springer–Verlag, 1985, pp. 509–516.

167. M. Takamura T. Inaba, T. Nishimori and N. Tsuchiya, *Open manifolds which are non–realizable as leaves*, Kodai Math. J. **8** (1985), 112–119.

168. M. Takesaki, *Theory of Operator Algebras. I*, Springer-Verlag, New York, NY, 1979.

169. ______, *Structure of factors and automorphism groups*, CBMS Regional Conference Series in Mathematics, vol. 51, Published for the Conference Board of the Mathematical Sciences, Washington, D.C., by the Amer. Math. Soc., Providence, RI, 1983.

170. I. Tamura, *Every odd dimensional homotopy sphere has a foliation of codimension one*, Comment. Math. Helv. **47** (1972), 164–170.

171. ______, *Spinnable structures on differentiable manifolds*, Proc. Japan. Acad. **48** (1972), 293–296.

172. W. Thurston, *The theory of foliations in codimension greater than one*, Comment. Math. Helv. **49** (1974), 214–231.

173. ______, *A local construction of foliations for three–manifolds*, Proc. Sympos. Pure Math., vol. XXVII, Amer. Math. Soc., 1975, pp. 315–319.

174. ______, *Existence of codimension–one foliations*, Ann. of Math. **104** (1976), 249–268.

175. ______, *A norm for the homology of three–manifolds*, Mem. Amer. Math. Soc. **59** (1986), No. 339, pp. 99–130.

176. A. M. Torpe, *K-theory for the leaf space of foliations by Reeb components*, J. Funct. Anal. **61** (1985), no. 1, 15–71.

177. T. Tsuboi, *On the foliated products of class C^1*, Ann. of Math. **130** (1989), 227–271.

178. B. L. van der Waerden, *Modern Algebra*, Frederick Unger, New York, NY, 1970.

179. J. W. Vick, *Homology Theory*, Springer–Verlag, New York, NY, 1982.

180. A. H. Wallace, *Modifications and cobounding manifolds*, Canad. J. Math. **12** (1960), 503–528.

181. X. Wang, *On the C^*-Algebras of Foliations in the Plane*, Lecture Notes in Mathematics, no. 1257, Springer-Verlag, Berlin, 1987.

182. A. Weil, *Les familles de courbes sur le tore*, Collected Papers, vol. I, Springer-Verlag, 1979, pp. 113–115.

183. R. L. Wheeden and A. Zygmund, *Measure and Integral*, Pure and Applied Mathematics, vol. 43, Marcel Dekker Inc., New York, NY, 1977.

184. D. P. Williams, *The topology on the primitive ideal space of transformation group C^*-algebras and C.C.R. transformation group C^*-algebras*, Trans. Amer. Math. Soc. **266** (1981), no. 2, 335–359.

185. E. Winkelnkemper, *Manifolds as open books*, Bull. Amer. Math. Soc. **79** (1973), 45–51.

186. ———, *The graph of a foliation*, Ann. Global Anal. Geom. **1** (1983), no. 3, 51–75.

187. ———, *The number of ends of the universal leaf of a Riemannian foliation*, Differential geometry (College Park, MD, 1981/1982), Birkhäuser, Boston, MA, 1983, pp. 247–254.

188. J. Wood, *Foliations on 3–manifolds*, Ann. of Math. **89** (1969), 336–358.

189. ———, *Bundles with totally disconnected structure group*, Comment. Math. Helv. **46** (1971), 257–273.

190. F. Wright, *Mean least recurrence time*, J. London Math. Soc. **36** (1961), 382–384.

191. K. Yosida, *Functional Analysis*, Sixth ed., Springer-Verlag, New York, NY, 1980, Die Grundlehren der Mathematischen Wissenschaften, Band 123.

192. G. Zeller-Meier, *Produits croisés d'une C^*-algèbre par un groupe d'automorphismes*, J. Math. Pures Appl. (9) **47** (1968), 101–239.

Index

Titles in This Series

60 **Alberto Candel and Lawrence Conlon,** Foliations II, 2003
59 **Steven H. Weintraub,** Representation theory of finite groups: algebra and arithmetic, 2003
58 **Cédric Villani,** Topics in optimal transportation, 2003
57 **Robert Plato,** Concise numerical mathematics, 2003
56 **E. B. Vinberg,** A course in algebra, 2003
55 **C. Herbert Clemens,** A scrapbook of complex curve theory, second edition, 2003
54 **Alexander Barvinok,** A course in convexity, 2002
53 **Henryk Iwaniec,** Spectral methods of automorphic forms, 2002
52 **Ilka Agricola and Thomas Friedrich,** Global analysis: Differential forms in analysis, geometry and physics, 2002
51 **Y. A. Abramovich and C. D. Aliprantis,** Problems in operator theory, 2002
50 **Y. A. Abramovich and C. D. Aliprantis,** An invitation to operator theory, 2002
49 **John R. Harper,** Secondary cohomology operations, 2002
48 **Y. Eliashberg and N. Mishachev,** Introduction to the h-principle, 2002
47 **A. Yu. Kitaev, A. H. Shen, and M. N. Vyalyi,** Classical and quantum computation, 2002
46 **Joseph L. Taylor,** Several complex variables with connections to algebraic geometry and Lie groups, 2002
45 **Inder K. Rana,** An introduction to measure and integration, second edition, 2002
44 **Jim Agler and John E. McCarthy,** Pick interpolation and Hilbert function spaces, 2002
43 **N. V. Krylov,** Introduction to the theory of random processes, 2002
42 **Jin Hong and Seok-Jin Kang,** Introduction to quantum groups and crystal bases, 2002
41 **Georgi V. Smirnov,** Introduction to the theory of differential inclusions, 2002
40 **Robert E. Greene and Steven G. Krantz,** Function theory of one complex variable, 2002
39 **Larry C. Grove,** Classical groups and geometric algebra, 2002
38 **Elton P. Hsu,** Stochastic analysis on manifolds, 2002
37 **Hershel M. Farkas and Irwin Kra,** Theta constants, Riemann surfaces and the modular group, 2001
36 **Martin Schechter,** Principles of functional analysis, second edition, 2002
35 **James F. Davis and Paul Kirk,** Lecture notes in algebraic topology, 2001
34 **Sigurdur Helgason,** Differential geometry, Lie groups, and symmetric spaces, 2001
33 **Dmitri Burago, Yuri Burago, and Sergei Ivanov,** A course in metric geometry, 2001
32 **Robert G. Bartle,** A modern theory of integration, 2001
31 **Ralf Korn and Elke Korn,** Option pricing and portfolio optimization: Modern methods of financial mathematics, 2001
30 **J. C. McConnell and J. C. Robson,** Noncommutative Noetherian rings, 2001
29 **Javier Duoandikoetxea,** Fourier analysis, 2001
28 **Liviu I. Nicolaescu,** Notes on Seiberg-Witten theory, 2000
27 **Thierry Aubin,** A course in differential geometry, 2001
26 **Rolf Berndt,** An introduction to symplectic geometry, 2001
25 **Thomas Friedrich,** Dirac operators in Riemannian geometry, 2000
24 **Helmut Koch,** Number theory: Algebraic numbers and functions, 2000

For a complete list of titles in this series, visit the
AMS Bookstore at **www.ams.org/bookstore/**.